The IMA Volumes in Mathematics and Its Applications

Volume 23

Series Editors
Avner Friedman   Willard Miller, Jr.

Institute for Mathematics and its Applications
IMA

The **Institute for Mathematics and its Applications** was established by a grant from the National Science Foundation to the University of Minnesota in 1982. The IMA seeks to encourage the development and study of fresh mathematical concepts and questions of concern to the other sciences by bringing together mathematicians and scientists from diverse fields in an atmosphere that will stimulate discussion and collaboration.

The IMA Volumes are intended to involve the broader scientific community in this process.

Avner Friedman, Director
Willard Miller, Jr., Associate Director

* * * * * * * * * *

IMA PROGRAMS

1982-1983	**Statistical and Continuum Approaches to Phase Transition**
1983-1984	**Mathematical Models for the Economics of Decentralized Resource Allocation**
1984-1985	**Continuum Physics and Partial Differential Equations**
1985-1986	**Stochastic Differential Equations and Their Applications**
1986-1987	**Scientific Computation**
1987-1988	**Applied Combinatorics**
1988-1989	**Nonlinear Waves**
1989-1990	**Dynamical Systems and Their Applications**
1990-1991	**Phase Transitions and Free Boundaries**

* * * * * * * * * *

SPRINGER LECTURE NOTES FROM THE IMA:

The Mathematics and Physics of Disordered Media
Editors: Barry Hughes and Barry Ninham
(Lecture Notes in Math., Volume 1035, 1983)

Orienting Polymers
Editor: J.L. Ericksen
(Lecture Notes in Math., Volume 1063, 1984)

New Perspectives in Thermodynamics
Editor: James Serrin
(Springer-Verlag, 1986)

Models of Economic Dynamics
Editor: Hugo Sonnenschein
(Lecture Notes in Econ., Volume 264, 1986)

L. Auslander F.A. Grünbaum J.W. Helton
T. Kailath P. Khargonekar S. Mitter
Editors

Signal Processing

Part II: Control Theory and Applications

Edited by
F.A. Grünbaum J.W. Helton P. Khargonekar

With 94 Illustrations

Springer-Verlag
New York Berlin Heidelberg
London Paris Tokyo Hong Kong

F. Alberto Grünbaum
Department of Mathematics
University of California—Berkeley
Berkeley, CA 94720, USA

J. William Helton
Department of Mathematics
University of California—San Diego
La Jolla, CA 92093, USA

Pramod P. Khargonekar
Department of Electrical Engineering
University of Minnesota
Minneapolis, MN 55455, USA

Series Editors

Avner Friedman
Willard Miller, Jr.
Institute for Mathematics and Its Applications
University of Minnesota
Minneapolis, MN 55455, USA

Mathematics Subject Classification: 43xx, 82xx, 93Bxx, 93Cxx 94Axx

Library of Congress Cataloging-in-Publication Data
(Revised for volume 2)
Signal processing.
(The IMA volumes in mathematics and its applications; v. 22–23.)
"Based on lectures delivered during a six week program held at the IMA from June 27 to August 5, 1988"—Foreword.
Contents: pt. 1. Signal processing theory—v. 2. Control theory and applications.
1. Signal processing—Mathematics. I. Auslander, Louis. II. University of Minnesota. Institute for Mathematics and Its Applications. III. Series.
TK5102.5.S535 1990 621.382'2 89-26216

Camera-ready copy prepared by the IMA using TEX.
Printed and bound by Edwards Brothers, Inc., Ann Arbor, Michigan.
Printed in the United States of America.

9 8 7 6 5 4 3 2 1 Printed on acid-free paper.

ISBN 0-387-97230-7 Springer-Verlag New York Berlin Heidelberg
ISBN 3-540-97230-7 Springer-Verlag Berlin Heidelberg New York

The IMA Volumes
in Mathematics and its Applications

Current Volumes:

Volume 1: Homogenization and Effective Moduli of Materials and Media
Editors: Jerry Ericksen, David Kinderlehrer, Robert Kohn, J.-L. Lions

Volume 2: Oscillation Theory, Computation, and Methods of Compensated Compactness
Editors: Constantine Dafermos, Jerry Ericksen,
David Kinderlehrer, Marshall Slemrod

Volume 3: Metastability and Incompletely Posed Problems
Editors: Stuart Antman, Jerry Ericksen, David Kinderlehrer, Ingo Muller

Volume 4: Dynamical Problems in Continuum Physics
Editors: Jerry Bona, Constantine Dafermos, Jerry Ericksen, David Kinderlehrer

Volume 5: Theory and Applications of Liquid Crystals
Editors: Jerry Ericksen and David Kinderlehrer

Volume 6: Amorphous Polymers and Non-Newtonian Fluids
Editors: Constantine Dafermos, Jerry Ericksen, David Kinderlehrer

Volume 7: Random Media
Editor: George Papanicolaou

Volume 8: Percolation Theory and Ergodic Theory of Infinite Particle Systems
Editor: Harry Kesten

Volume 9: Hydrodynamic Behavior and Interacting Particle Systems
Editor: George Papanicolaou

Volume 10: Stochastic Differential Systems, Stochastic Control Theory and Applications
Editors: Wendell Fleming and Pierre-Louis Lions

Volume 11: Numerical Simulation in Oil Recovery
Editor: Mary Fanett Wheeler

Volume 12: Computational Fluid Dynamics and Reacting Gas Flows
Editors: Bjorn Engquist, M. Luskin, Andrew Majda

Volume 13: Numerical Algorithms for Parallel Computer Architectures
Editor: Martin H. Schultz

Volume 14: Mathematical Aspects of Scientific Software
Editor: J.R. Rice

Volume 15: Mathematical Frontiers in Computational Chemical Physics
Editor: D. Truhlar

Volume 16: Mathematics in Industrial Problems
by Avner Friedman

Volume 17: Applications of Combinatorics and Graph Theory to the Biological and Social Sciences
Editor: Fred Roberts

Volume 18: q-Series and Partitions
Editor: Dennis Stanton

Volume 19: Invariant Theory and Tableaux
Editor: Dennis Stanton

Volume 20: Coding Theory and Design Theory Part I: Coding Theory
Editor: Dijen Ray-Chaudhuri

Volume 21: Coding Theory and Design Theory Part II: Design Theory
Editor: Dijen Ray-Chaudhuri

Volume 22: Signal Processing: Part I
Editors: L. Auslander, F.A. Grünbaum, W. Helton, T. Kailath, P. Khargonekar and S. Mitter

Volume 23: Signal Processing: Part II
Editors: L. Auslander, F.A. Grünbaum, W. Helton, T. Kailath, P. Khargonekar and S. Mitter

Volume 24: Mathematics in Industrial Problems, Part 2
by Avner Friedman

Forthcoming Volumes:

1988-1989: *Nonlinear Waves*

Solitons in Physics and Mathematics

Two Phase Waves in Fluidized Beds, Sedimenation, and Granular Flows

Nonlinear Evolution Equations that Change Type
Computer Aided Proofs in Analysis
Multidimensional Hyperbolic Problems and Computations (2 Volumes)
Microlocal Analysis and Nonlinear Waves

Summer Program 1989: *Robustness, Diagnostics, Computing and Graphics in Statistics*
Robustness, Diagnostics in Statistics (2 Volumes)
Computing and Graphics in Statistics

1989-1990: *Dynamical Systems and Their Applications*
Introduction to Dynamical Systems

FOREWORD

The two volumes of Signal Processing are based on lectures delivered during a six week program held at the IMA from June 27 to August 5, 1988. The first two weeks of the program dealt with general areas and methods of Signal Processing. The problem areas included imaging and analysis of recognition, x-ray crystallography, radar and sonar, signal analysis and 1-D signal processing, speech, vision, and VLSI implementation. The methods discussed included harmonic analysis and wavelets, operator theory, algorithm complexity, filtering and estimation, and inverse scattering. The topics of weeks three and four were digital filter, VLSI implementation, and integrable circuit modelling. In week five the concentration was on robust and nonlinear control with aerospace applications, and in week six the emphasis was on problems in radar, sonar and medical imaging.

Because of the large overlap between the various one-week and two-week segments of the program, we found it more convenient to divide the material somewhat differently. Part I deals with general signal process theory and Part II deals with (i) application of signal processing, (ii) control theory related themes.

We are grateful to the scientific organizers: Tom Kailath (Chairman), Louis Auslander, F. Alberto Grunbaum, J. William Helton, Pramod P. Khargonekar and Sanjoy K. Mitter.

We are also grateful for the generous support given to the IMA program by the Office of Naval Research, the Air Force Office of Scientific Research, the Army Research Office and the National Security Agency.

Signal Processing is undergoing tremendous developments; it is our hope that these two volumes will serve as a source of information and stimulation to mathematical scientists who wish to get acquainted with this field.

Avner Friedman
Willard Miller, Jr.

CONTENTS

SIGNAL PROCESSING: PART II

I – CONTROL THEORY

II – APPLICATIONS OF SIGNAL PROCESSING

CONTENTS

I: CONTROL THEORY

SENSITIVITY MINIMIZATION AND BITANGENTIAL NEVANLINNA–PICK INTERPOLATION IN CONTOUR INTEGRAL FORM*

JOSEPH A. BALL† ISRAEL GOHBERG‡ LEIBA RODMAN‡†

Abstract. We show how the sensitivities $S = (I + PK)^{-1}$ associated with a stabilizing compensator K for a regular square plant P can be characterized as all solutions of an interpolation problem expressible directly from a realization of the plant P. This leads to a solution of the weighted sensitivity minimization problem (1-block problem) directly in terms of a realization of the plant and solutions of a couple of Lyapunov equations, without recourse to the Youla parametrization of stabilizing compensators. For the special case of the 1-block problem, this approach gives an elementary alternative to the recent results of Doyle, Francis, Glover and Khargonekar [DFGK] for the more general 4-block problem.

CONTENTS

1. The sensitivity minimization problem. A special case of the standard problem in H^∞ control theory (see [Fr]) concerns the setup depicted in Figure 1. Here v_1, v_2, u and y are vector-valued signals. The signal u is the control input, y the measured output, v_1 and v_2 disturbances. The transfer matrix functions P and K we take to be rational matrix functions.

*These results were presented by the first author at the Signal Processing Workshop, Institute for Mathematics and its Applications, University of Minnesota, July, 1988.

†Department of Mathematics, Virginia Tech, Blacksburg, VA 24061. Partially supported by NSF grant.

‡School of Mathematical Sciences, Tel-Aviv University, Tel-Aviv, 69978, ISRAEL. Partially supported by the Air Force Office for Scientific Research grant AFOSR–87–0287.

‡†Department of Mathematics, College of William and Mary, Williamsburg, VA 23185. Partially supported by NSF grant.

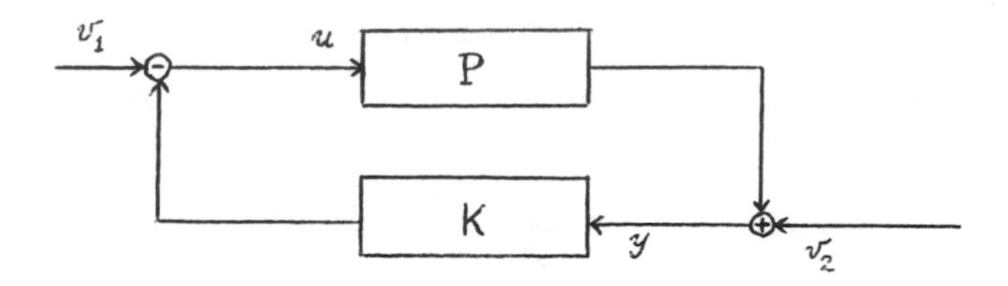

FIGURE 1

The system diagram stands for the system of equations

$$v_1 = u + Ky$$
$$v_2 = -Pu + y$$

or in matrix form

$$\begin{bmatrix} v_1 \\ v_2 \end{bmatrix} = \begin{bmatrix} I & K \\ -P & I \end{bmatrix} \begin{bmatrix} u \\ y \end{bmatrix} \tag{1.1}$$

The system is well posed if for any choice of signals v_1, v_2 there exist signals u, y so that the equations (1.1) are satisfied. Stability of vector-valued signal w in our frequency domain setup we take to mean that it be in vector-valued H^2 of the right half plane, i.e. w should be analytic as in the right half plane and

$$\|w\|_2^2 = \sup_{\xi>0} \int_{-\infty} \|w[\xi + jw)\|^2 dw < \infty$$

The system in Figure 1 is said to be internally stable if u and y are in H^2 whenever v_1 and v_2 are in H^2; equivalently the system is internally stable if and only if the matrix function $\begin{bmatrix} I & K \\ -P & I \end{bmatrix}$ is invertible with inverse having no poles in the closed right half plane (including infinity). By a Schur complement argument, one can see that $\begin{bmatrix} I & K \\ -P & I \end{bmatrix}$ is invertible if and only if $(I + PK)^{-1}$ is invertible, and then the inverse can be computed explicitly as

$$\begin{bmatrix} I & K \\ -P & I \end{bmatrix}^{-1} = \begin{bmatrix} (I + KP)^{-1} & -K(I + PK)^{-1} \\ (I + PK)^{-1}P & (I + PK)^{-1} \end{bmatrix}$$

Thus *internal stability means*: the four rational matrix functions $(I+KP)^{-1}, -K(I+PK)^{-1}, (I + PK)^{-1}P$ and $(I + PK)^{-1}$ should be proper (i.e. analytic at infinity) and have all poles in the open left half plane. The particular transfer function $(I+PK)^{-1}$ is the transfer map from v_2 to y. The *sensitivity minimization problem*, introduced by Zames (see [Fr]), is to minimize the effect of the disturbance v_2 on the measured output in a worst-case sense, i.e.

$$\min_{K \text{ stabilizes } P} \sup_{\substack{v_2 \in H^2 \\ v_1 = 0}} \frac{\|y\|_2}{\|v\|_2}$$

The problem can also be given other engineering interpretations, for example, the tracking problem. Since the H^∞-norm is the induced operator norm on H^2, this becomes

$$\min_{K \text{ stabilizes } P} \|(I+PK)^{-1}\|_\infty$$

where in general

$$\|W\|_\infty = \sup_{-\infty < w < \infty} \|W(jw)\|$$

and the norm on the right is the operator or spectral norm. In engineering practice one chooses weight functions W_1 and W_2 (usually taken to be rational with no poles or zeros in the closed right half plane) and seeks to minimize the weighted sensitivity norm

$$\min_{K \text{ stabilizes } P} \|W_1(I+PK)^{-1}W_2\|_\infty.$$

The usual approach to this problem (see [Fr]) is to use the Youla linear fractional parametrization of stabilizing compensators to convert the problem to a more affine form equivalent to a matrix Nevanlinna-Pick interpolation problem the so-called "model-matching" problem. This involves computation of a double coprime factorization of the plant, and solving Bezout equations. If one then converts to the Nehari problem, one must also compute some inner-outer factorizations. There has been some work [BR2, K, R] showing how state space formulas for the solution can be obtained directly from the data of the model matching problem. Here we show how the data for the Nevanlinna-Pick interpolation problem can be found directly from a state space realization of the plant P. For the scalar case this is well-known [YBL]; the precise form of the resulting bitangential Nevanlinna-Pick interpolation problem (directly from pole and zero data of the plant) was pointed out in [BL]. We show how these matricial interpolation conditions can be derived directly, without passing through the Youla parametrization. Moreover, we show how usually clumsy high order interpolation conditions can be expressed in a streamlined contour integral form introduced by Nudelman [N1, N2], and how all interpolants of norm less than some tolerance can be parametrized via a linear fractional map given explicitly in terms of a state space realization. Our approach to interpolation is based on a synthesis of ideas from many sources, in particular [BH], [BR1], [BGK] and [GKLR], and will be presented in a self-contained systematic way in [BGR4]. Preliminary accounts of various aspects include [BGR1], [BGR2].

The problem of weighted sensitivity norm minimization can be handled in the same way. Our analysis can be viewed as doing for the easier sensitivity minimization (or "1-block") problem what the recent highly significant work of [DFGK] does for the more general "4 block" problem; our approach, while less general, we believe to have some appeal due to its direct and elementary character, and illustrates a general technique that can be applied to other interpolation problems as well.

2. Interpolation characterization of stable sensitivities: the simple case. The problem we consider in this section is that of characterizing all sensitivity functions $S = (I + PK)^{-1}$ associated with stabilizing compensators K. More precisely, the mathematical problem to be solved in this section is: *given the rational matrix function P,* (with no poles and zeros on the imaginary axis including ∞) *describe all rational matrix functions S of the form*

$$S = (I + PK)^{-1}$$

where K is any rational matrix function such that the four functions

$$S = (I + PK)^{-1}, K(I + PK)^{-1}, (I + PK)^{-1}P, (I + KP)^{-1}$$

are all stable. (i.e. have all poles in the open left half plane and no pole at ∞). Our techniques in general apply to the case where the plant P is a rational square (say $n \times n$) matrix function with $\det P$ not vanishing identically, and such that both P and P^{-1} have no poles on the $j\omega$ axis (including infinity).

To keep the preliminary discussion elementary, we assume that all poles and zeros of P in the right half plane are simple; by this we mean that in the Smith-McMillan form for $P(z)$ at a point z_0 in the right-half plane

$$P(z) = E(z) \cdot \operatorname{diag}\left((z - z_0)^{\mu_1}, \ldots, (z - z_0)^{\mu_n}\right) F(z)$$

(where E and F are analytic and invertible at z_0), at most one index μ_j is positive in which case it is 1, and at most one index μ_k is negative in which case it is -1. Thus all but at most two indices are nonzero at each point z_0 in the right half plane. Let $z_1, \ldots, z_p$ be the zeros of $P(z)$ in the open right half plane Π^+ (i.e. points z_0 where some index is 1) and let $w_1, \ldots, w_q$ be the poles of $P(z)$ in Π^+ (i.e. points z_0 where some index is -1); note that we allow some points ξ_{ij} to be both a zero and a pole ($z_i = \xi_{ij} = w_j$).

Intuitively, a consequence of our assumptions is that the zero and poles of P occur only in certain directions. This we make precise in the following way. If z_i is a zero but not a pole of $P(z)$, we say that the row vector $x_i \neq 0$ is a left null vector for $P(z)$ at z_i if

$$x_i P(z_i) = 0.$$

If z_i is also a pole, then in general we need a *left null function* $x_i(z)$, i.e. a row vector function $x_i(z)$ analytic at z_0 for which

i) $x_i(z_0) \neq 0$

ii) $x_i(z)P(z)$ is analytic at z_0

and

iii) $\quad x_i(z)P(z)\big|_{z=z_0} \quad = 0.$

If $x_i(z)$ has Taylor expansion

$$x_i(z) = \sum_{j=0}^{\infty} x_{ij}(z - z_i)^j$$

at z_i, we refer to $x_i = x_{i0}$ as a *left null vector* for $P(z)$ at z_0 and the pair (x_{i0}, x_{i1}) as an *augmented null* chain for $P(z)$ at z_0. The directions of poles of $P(z)$ is measured by so-called "pole vectors," A column vector u_j is said to be a *pole vector* for $P(x)$ at the pole w_j if there exist a column vector function $\phi(z)$ analytic at w_j such that $P(z)\phi(z)$ has the form

$$P(z)\phi(z) = (z - w_j)^{-1}u_j + [\text{analytic at } w_j].$$

If the point $\xi_{ij} = z_i = w_j$ is both a zero and a pole for $P(z)$, the only piece of information regarding the vector x_{i1} in an augmented null chain for $P(z)$ at ξ_{ij} which we shall need is the scalar quantity

$$\rho_{ij} = x_{i1}u_j.$$

Note that ρ_{ij} carries information as to how the pole and zero at ξ_{ij} interact with each other; for this reason we call ρ_{ij} the *null-pole coupling number* for $P(z)$ at ξ_{ij}. The following result shows how the null and pole vector together with the associated coupling number can be used to solve a factorization problem.

PROPOSITION 2.1. *Suppose the rational matrix function $P(z)$ has a simple zero and a simple pole at the point $\xi_{ij} = z_i = w_j$ with left null vector x_i, right pole vector u_j and associated coupling number ρ_{ij}. Then a rational vector function $f(z)$ factors as*

$$f(z) = P(z)g(z)$$

where g is a rational vector function which is analytic at ξ_{ij} if and only if there is a scalar constant α for which

$$f(z) = \alpha(z - \xi_{ij})^{-1}u_j + h(z)$$

where h is analytic at ξ_{ij} and satisfies

$$x_i h(\xi_{ij}) = \rho_{ij}\alpha.$$

Proof. See [BR1].

We can now state our characterization of sensitivities $S = [I + PK]^{-1}$ formed from stabilizing compensators for the case of a plant P having only simple zeros and poles in the open right half plane. We denote by $\mathbb{S}$ the set of all *stable* rational matrix functions, i.e. those that are analytic in the closed right half-plane and at infinity.

THEOREM 2.2. *Let $P(z)$ be a regular rational $n \times n$ matrix function with no poles and no zeros on the imaginary axis (including infinity) and having only simple zeros $z_1, \ldots, z_p$ and simple poles $w_1, \ldots w_q$, in the open right half plane. Let x_i be a left null vector for $P(z)$ at z_{ij}, u_j a right pole vector for $P(z)$ at w_j and ρ_{ij} the associated null-pole coupling number for $P(z)$ at ξ_{ij} if $\xi_{ij} = z_i = w_j$, for $1 \leq i \leq p, \quad 1 \leq j \leq q$. Then the regular rational matrix function $S(z)$ has the form*

$$S(z) = \left(I + P(z)K(z)\right)^{-1}$$

for a stabilizing compensator $K(z)$ if and only if

(i) $S \in \mathcal{S}$

(ii) $x_i S(z_i) = x_i$ for $i \leq i \leq p$

(iii) $S(w_j)u_j = 0$ for $1 \leq j \leq q$

and

(iv) $x_i S'(\xi_{ij})u_j = -\rho_{ij}$ *whenever*

$$\xi_{ij} = z_i = w_j$$

Proof. We have seen that K is a stabilizing compensator if and only if the four functions $(I+PK)^{-1} = S, (I+PK)^{-1}P = SP$, $-K(I+PK)^{-1} = P^{-1}(S-I)$ and $(I + KP)^{-1} = P^{-1}SP$ are all stable. Conversely if S is a stable regular rational matrix function such that the three auxiliary matrix functions SP, $P^{-1}(S-I)$ and $P^{-1}SP$ are all stable, then $K = P^{-1}(S^{-1} - I)$ is a stabilizing compensator and $S = (I + PK)^{-1}$. Thus the problem is to characterize the regular stable functions S for which the three auxiliary functions SP, $P^{-1}(S-I)$ and $P^{-1}S$ are all stable.

We first analyze SP. Since S is stable, the only possible poles of SP in Π^+ are at the poles $w_1, \ldots, w_q$ of P in Π^+. The poles of P in Π^+ occur at $w_1, \ldots, w_q$ and by assumption are all simple with associated pole vector u_j at w_j. This means that $P(z)$ has Laurent expansion at w_j.

$$P(z) = (z - w_j)^{-1}X_j + [\text{analytic at}\, w_j]$$

where

$$Im\ X_j = \text{span}\ \{u_j\}.$$

From this one easily sees that $S \in \mathcal{S}$ satisfying the interpolation condition (iii) is equivalent to $S \cdot P$ being stable.

Next consider $P^{-1}(S - I)$. By assumption the zeros of P in Π^+ occur at $z_1, \ldots, z_p$ and are all simple with null vector x_i at the zero z_i. As $S - I$ is stable, the only possible poles of $P^{-1}(S-I)$ in Π^+ are at the poles of P^{-1} in Π^+; namely, at the zeros $z_1, \ldots, z_p$ of P in Π^+. If $x_i(z) = x_i + (z - z_i)x_{i1}$ is a left null function associated with left null vector x_i for $P(z)$ at z_i, then

$$x_i(z)P(z) = (z - z_i)f(z)$$

where f is a row vector function analytic at z_i. Multiplying both sides by $P(z)^{-1}(z-z_i)^{-1}$ gives

$$x_i(z)(z-z_i)^{-1} = f(z)\; P(z)^{-1}.$$

This shows that the row vector x_i is a *left pole vector* for $P(z)^{-1}$ at z_i. As the poles z_i of $P(z)^{-1}$ is simple (since the zero z_i of $P(z)$ is simple), we conclude that $P(z)^{-1}$ has Laurent expansion at z_i

$$P(z)^{-1} = (z-z_i)^{-1}Y_i + [\text{analytic at } z_i]$$

where the left image $l\text{-}Im\; Y_i = \{y\; Y_i :\; y \text{ is an } 1\times n \text{ row vector}\}$ is given by

$$l\text{-}Im\; Y_i = \text{span}\;\{x_i\}.$$

From this we easily see that $S \in \mathcal{S}$ satisfying the interpolation condition (ii) is equivalent to $P^{-1}(S-I)$ being stable.

Finally we consider $P^{-1}SP = P^{-1}(S-I)P+I$. For the analysis of this condition we assume that $S \in \mathcal{S}$ and the interpolation conditions (ii) and (iii) hold. Clearly $P^{-1}SP \in \mathcal{S}$ is equivalent to $P^{-1}(S-I)P \in \mathcal{S}$. If $S \in \mathcal{S}$ then the only possible poles of $P^{-1}(S-I)P$ in Π^+ are at the poles of P^{-1} and P in Π^+ (i.e., $z_1,\ldots,z_p,w_1,\ldots,w_q$). If the zero z_i of P is not also pole of P, then P is analytic at z_i and we saw above that $P^{-1}(S-I)$ is analytic exactly when the interpolation condition (ii) holds; hence in this case the product $P^{-1}(S-I)\cdot P$ is analytic at z_i. On the other hand, if the pole w_j of P is not also a zero, then P^{-1} is analytic at w_j and we saw above that SP is analytic at w_j exactly when the interpolation condition (iii) holds; thus in this case $P^{-1}(S-I)P = P^{-1}\cdot SP - I$ is analytic at w_j.

Finally if $\xi_{ij} = z_i = w_j$ is both a zero and a pole, we use the interpolation condition (iv) and Proposition 2.1. Note that $P^{-1}(S-I)P$ is analytic at ξ_{ij} if and only if

$$(S-I)P\phi = P\cdot[\text{analytic at } \xi_{ij}] \tag{2.1}$$

for every rational column vector function ϕ analytic at ξ_{ij}. By Proposition 2.1 we know that $P\phi$ has the form

$$P\phi = \alpha(z-\xi_{ij})^{-1}u_j + h(z) \tag{2.2}$$

where h is analytic at ξ_{ij} with

$$x_i\; h(\xi_{ij}) = \rho_{ij}\alpha. \tag{2.3}$$

Then $(S-I)P\phi$ assumes the form

$$\begin{aligned}(S-I)P\phi = &-\alpha(z-\xi_{ij})^{-1}u_j\\ &+[\alpha(z-\xi_{ij})^{-1}S(z)u_j\\ &+S(z)h(z)-h(z)]\end{aligned}$$

From the interpolation condition (iii), $(z-\xi_{ij})^{-1}S(z)u_j$ is analytic at ξ_{ij} and hence

(2.5) $$(z-\xi_{ij})^{-1}\ S(z)u_j\big|_{z=\xi_{ij}} = S'(\xi_{ij})u_j$$

Now by (2.5) and Proposition 2.1 we see that $(S-I)P\phi = P\cdot[\text{analytic at } \xi_{ij}]$ if and only if

(2.6) $$x_i[\alpha S'(\xi_{ij})u_j + S(\xi_{ij})h(\xi_{ij}) - h(\xi_{ij})] = -\rho_{ij}\alpha.$$

By the interpolation condition (ii) $x_i\ S(\xi_{ij})h(\xi_{ij}) = x_i h(\xi_{ij})$ while $x_i h(\xi_{ij}) = \rho_{ij}\alpha$ by (2.3). Thus (2.6) collapses to

$$\alpha\ x_i\ S'(\xi_{ij})u_j = -\alpha\rho_{ij}.$$

This holding true for all scalars α is just the interpolation condition (iv). We conclude that, given that (i), (ii) and (iii) hold, (iv) is equivalent to $P^{-1}SP \in \mathcal{S}$ as needed. □

REMARK. For the case where the zeros $z_1, \ldots, z_p$ of the plant $P(z)$ in Π^+ are disjoint from the poles $w_1, \ldots, w_q$, it follows from the equivalences in the proof of Theorem 2.2 that the fourth function $(I+KP)^{-1}$ is stable once the other three $(I+PK)^{-1}$, $(I+PK)^{-1}P$ and $-K(I+PK)^{-1}$ are all stable. In particular this applies in the scalar case; of course one also sees this in the scalar case immediately from the identity $(I+KP)^{-1} = (I+PK)^{-1}$. For the general matrix case where zeros and poles of P may intersect, we see that the stability of $(I+KP)^{-1}$ does not necessarily follows from the stability of $(I+PK)^{-1}$, $(I+PK)^{-1}P$ and $K(I+KP)^{-1}$.

3. Null-pole structure. This section is auxiliary character. We recall here some results on the null–pole structure for rational matrix functions (proved mainly in [BGR3, BGR5]) that will be used later on. No attempt is made to present these results in the general framework; rather, they are stated in the form which is convenient here.

Let $P(z)$ be $n\times n$ rational matrix function without poles on the imaginary axis and at infinity. We start with the notion of the null-pole triple of P with respect to the open right half-plane Π^+. Let

$$P(z) = D + C(zI-A)^{-1}B$$

be a minimal realization for $P(z)$. Because of our assumptions on $P(z)$ the matrix D is invertible and the matrices A and $A^x = A - BD^{-1}C$ have no pure imaginary eigenvalues. Let M (resp. M^x) be the spectral invariant subspace for A corresponding to the eigenvalues of A (resp. of A^x) in Π^+ and let Q^x be the Riesz projection on M^x. The triple

(3.1) $$(C|M,\ A|M; A^x|M^x, Q^xBD^{-1}; Q^x|M)$$

is called a *null-pole triple* of P with respect to Π^+. The pair of matrices $(C|M,\ A|M)$ is called *right pole pair* of P (with respect to Π^+), the pair $(A^x|M^x, Q^xBD^{-1})$ is called *left null pair* (with respect to Π^+), and $Q^x|M$ is called the coupling matrix. These notions were developed in [BGK, BGR2, BR2, GKLR, GK], see also [BGR1, BGR3]. Intuitively, the pair $(C|M, A|M)$ describes the pole structure in Π^+ of P from the right, the pair $(A^x|M^x,\ Q^x BD^{-1})$ describes the null structure of P in Π^+ from the left, and $Q^x|M$ describe the interaction between these structures. These intuitive notions can be made precise in terms of null and pole functions, and a thorough exposition of these and related notions will be found in [BGR4].

Here we will be content with exposition of some very basic features of null-pole triples.

It is convenient to consider, together with (3.1), all triples *similar* to it. Here the similarity of two triples $(C', A'_\pi; A'_\zeta, B'; \Gamma')$ and $(C'', A''_\pi; A''_\zeta, B''; \Gamma'')$ means that for some invertible matrices E and F the equalities

$$C'' = C'E;\ \ A''_\pi = E^{-1}A'_\pi E; A''_\zeta = F^{-1}A'_\zeta F;\ \ B'' = F^{-1}B_\zeta; \Gamma'' = E^{-1}\Gamma' F$$

hold. Any triple similar to (3.1) will be also called a null-pole triple of $P(z)$ with respect to Π^+.

Before we proceed, let us write down the null-pole triple for a simple case treated in the previous section.

Example 3.1. Assume that all poles and zeros of $P(z)$ in Π^+ are simple. Let $z_1, \ldots, z_p$ be the distinct zeros of $P(z)$ in Π^+ with corresponding left null vectors $x_1, \ldots, x_p$, and let $w_1, \ldots, w_q$ be the distinct poles of $P(z)$ in Π^+ with corresponding pole vectors $u_1, \ldots, u_q$(we allow the case when the sets $\{z_1, \ldots, z_p\}$ and $\{w_1, \ldots, w_q\}$ are not disjoint). If $\xi_{ij} = z_i = w_j$, let ρ_{ij} be the null-pole coupling number of $P(z)$ at ξ_{ij}. Then to produce a null-pole triple of $P(z)$ with respect to Π^+ we put

$$A_\zeta = \begin{bmatrix} z_1 & & & 0 \\ & z_2 & & \\ & & \ddots & \\ 0 & & & z_p \end{bmatrix}; \qquad B = \begin{bmatrix} x_1 \\ x_2 \\ \vdots \\ x_p \end{bmatrix}$$

$$A_\pi = \begin{bmatrix} w_1 & & & 0 \\ & w_2 & & \\ & & \ddots & \\ 0 & & & w_q \end{bmatrix}; \qquad C = [u_1 u_2 \ldots u_q].$$

and

$$\Gamma = [\Gamma_{ij}]_{1 \le i \le p; 1 \le j \le q}\ ,$$

where

$$\Gamma_{ij} = \begin{cases} (w_j - z_i)^{-1} x_i u_j & \text{if} \quad z_i \ne w_j \\ \rho_{ij} & \text{if} \quad z_i = w_j. \end{cases}$$

As an exercise the reader can verify that

$$\Gamma A_\pi - A_\zeta \Gamma = B_\zeta C_\pi$$

(we remain that if $z_i = w_j$ then necessarily $x_i u_j = 0$, as one can easily check using the definitions of the pole vector and of the left null vector.) □

We now return to the general situation.

Some basic properties of the null-pole triples are collected in the following proposition.

PROPOSITION 3.1. *Let $(C, A_\pi;\ A_\zeta, B; \Gamma)$ be a null-pole triple of P with respect to the right half-plane. Then the following properties are valid:*

(i) *all the eigenvalues of A_π and of A_ζ are in the open right half-plane;*

(ii) *the pair (C, A_π) is observable: $\cap_{i\geq 0} \operatorname{Ker}(CA_\pi^i) = \{0\}$;*

(iii) *the pair (A_ζ, B) is controllable:*

$$\sum_{i\geq 0} Im(A_\zeta^i B) = \mathbb{C}^{n_5},$$

where $n_\zeta \times n_\zeta$ is the size of A_ζ;

(iv) *the Sylvester equation*

$$\Gamma A_\pi - A_\zeta \Gamma = B_\zeta C_\pi$$

is valid;

(v) *the rational matrix functions*

$$(zI - A_\zeta)^{-1} B P(z) u(z) \quad (\text{resp. } \ v(z) P(z)^{-1} C (zI - A_\pi)^{-1})$$

are analytic in the right half-plane whenever $u(z)$ (resp. $v(z)$) is a column (resp. row) vector function analytic in the right half-plane for which $P(z)u(z)$ (resp. $v(z)P(z)$) is analytic in the right half plane.

(vi) *there exist matrices $\widetilde{C}$ and $\widetilde{B}$ of suitable sizes such that the representation*

$$P(z)^{-1} = \widetilde{C}(zI - A_\zeta)^{-1} B + U(z)$$

and

$$P(z) = C(zI - A_\pi)^{-1} \widetilde{B} + V(z)$$

hold, where $U(z)$ and $V(z)$ are rational matrix function analytic in the right half-plane.

For the proof of various parts of Proposition 3.1 we refer to [BGR1, BGR3, GK].

Next, we introduce the notion of the null-pole subspace, or singular subspace, of P. By definition, the ***null-pole subspace*** $\mathcal{S}(P)$ consists of all n-dimensional vector rational functions of the form $P(z)x(z)$, where $x(z)$ is an arbitrary n-dimensional vector rational function without poles in $\overline{\Pi^+} \cup \{\infty\}$. This subspace was studied in [BGR2, BR2, BGR3, BCR, GK]. The description of the null-pole subspace in terms of the null-pole triple is given in the following theorem (proved in [BGR3]). It is convenient to use the notation $\mathcal{R}_n(\overline{\Pi^+} \cup \{\infty\})$ for the set of all n-dimensional rational vector functions without poles in $\overline{\Pi^+} \cup \infty$. We denote by $\operatorname{Res}_{z=z_0} f(z)$ the residue of rational matrix or vector function $f(z)$ at z_0 (here $z_0 \in \mathbb{C} \cup \{\infty\}$).

THEOREM 3.2. *Let $(C, A_\pi; A_\zeta, B; \Gamma)$ be a null-pole triple of P with respect to Π^+. Then the null-pole subspace $\mathcal{S}(P)$ is given by the formula*

$$\mathcal{S}(P) = \{C(zI - I_\pi)^{-1}x + h(z) : \; x \in \mathbf{C}^{n_\pi}, h \in \mathcal{R}_n(\overline{\Pi^+} \cup \{\infty\})$$

such that $\sum_{z_0 \in \Pi^+} \operatorname{Res}_{z=z_0}(zI - A_\zeta)^{-1}Bh(z) = \Gamma x\}.$

Here $n_\pi \times n_\pi$ is the size of A_π. Note that if (C, A_π) is vacuous ($n_\pi = 0$), i.e. $P(z)$ has no poles in Π^+, then the formula for $\mathcal{S}(P)$ simplifies to

$$\mathcal{S}(P) = \{h \in \mathcal{R}_n(\overline{\Pi^+} \cup \{\infty\}) : \sum_{z_0 \in \Pi^+} \operatorname{Res}_{z=z_0}(zI - A_\zeta)^{-1}Bh(z) = 0\}.$$

On the other hand, if (A_ζ, B) is vacuous (the size n_ζ of A_ζ is equal to 0), i.e., $P(z)$ has no zeros in Π^+, then $\mathcal{S}(P)$ collapses to

$$\mathcal{S}(P) = \{C(zI - A_\pi)^{-1}x : x \in \mathbf{C}^{n_\pi}\} + \mathcal{R}_n(\overline{\Pi^+} \cup \{\infty\}).$$

In the sequel we will use rational matrix functions with given null-pole triple (with respect to Π^+). The existence of such functions is ensured provided the conditions (i)–(iv) of Proposition 3.1 are satisfied. More precisely, given matrices $C, A_\pi, A_\zeta, B, \Gamma$ of suitable sizes which satisfy (i)–(iv) of Proposition 3.1, there exists a rational matrix function P with no poles and zeros on the imaginary axis and at infinity such that $(C, A_\pi; A_\zeta, B; \Gamma)$ is a null-pole triple of P with respect to Π^+. This statement follows from Theorem 2.3 in [GK] (see also [BR2]). Note that it applies also in the situations when one of the matrices A_π or A_ζ is empty (more precisely, it represents the linear transformation 0 acting on the zero-dimensional vector space). Such situations are interpreted as P having no poles (if A_π is empty) or no zeros (if A_ζ is empty) in the open right half plane.

We need also the solution of the general problem of determining a null-pole triple for the product WP from a null-pole triple for P and a minimal realization of the rational matrix function W. The result is as follows.

LEMMA 3.3. *Let $P(z)$ be $n \times n$ rational matrix function without poles and zeroes on the imaginary axis and at infinity. Suppose P has $(C_\pi, A_\pi; A_\zeta, B_\zeta; \Gamma)$ as a null-pole triple over Π^+ and $W(z)$ is rational $n \times n$ matrix function with no poles and no zeros in $\overline{\Pi^+} \cup \{\infty\}$ having the minimal realization*

$$W(z) = I + C(zI - A)^{-1}B. \tag{3.2}$$

Define matrices Q and L as the unique solutions of the Sylvester equations

$$QA_\pi - AQ = BC_\pi \tag{3.3}$$

and

$$L(A - BC) - A_\zeta L = B_\zeta C. \tag{3.4}$$

Then a null-pole triple for the product $W \cdot P$ over Π^+ is given by

$$(CQ + C_\pi, A_\pi; A_\zeta, LB + B_\zeta; LQ + \Gamma). \tag{3.5}$$

Note that (under the hypotheses of Lemma 3.3) each equation (3.3) and (3.4) has unique solution. Indeed, $\sigma(A_\pi) \cup \sigma(A_\zeta) \subset \Pi^+$ by the definition of a null-pole triple of P over Π^+. On the other hand, the minimality of realization (3.2) ensures (see, e.g., [BGK]) that the eigenvalues of A coincide with the poles of W, so $\sigma(A)$ is contained in the open left half-plane Π^-. Also, as a consequence of (3.2) one has a minimal realization for W^{-1} as well:

$$W(z)^{-1} = I - C(zI - (A - BC))^{-1}B,$$

hence the zeros of $W(z)$ (=poles of $W(z)^{-1}$) coincide with the eigenvalues of $A-BC$, and in particular $\sigma(A - BC) \subset \Pi^-$.

Proof. Without loss of generality we shall assume $P(\infty) = I$.

Next, we show that we can also assume that the zeros and poles of P in the left half-plane are disjoint from the zeros and poles of W. Indeed by Corollary 3.3.5 in [BGR3] (see also Lemma 1.3 of [BGR2] or Theorem 5.1 of [GK]) the null-pole triple (over Π^+) of a rational matrix function does not change if the function is postmultiplied by a rational matrix function without poles and zeros in $\overline{\Pi^+} \cup \{\infty\}$. Thus, one can replace P by a product $P\widetilde{P}$, where $\widetilde{P}$ is any rational matrix function with all its poles and zeros in the open left half-plane. It remains to show that $\widetilde{P}$ can be chosen in such a way that $P\widetilde{P}$ has all its poles and zeros away from the poles and zeros of W. To this end use the results of [GK] to construct a rational matrix function $R(z)$ having null-pole triple over Π^+ equal to $(C_\pi, A_\pi; A_\zeta, B_\zeta; \Gamma)$, having no poles or zeros on the imaginary axis (including infinity), and having its poles and zeros in Π^- disjoint from those of $W(z)$. By Lemma 1.3 of [BGR2] (see also Theorem 5.1 of [GK]), R has the form $R = P\widetilde{P}$ where $\widetilde{P}$ has all its poles and zeros in the left half plane. Thus without loss of generality, we may assume that the poles and zeros of P are disjoint from the poles and zeros of W.

We now return to the rational matrix function P. Let $(C_{\pi c}, A_{\pi c}; A_{\zeta c}, B_{\zeta c}; \Gamma_c)$ be a null-pole triple for P with respect to the open left half-plane Π^- (the definition of this notion is given in the same way as the null-pole triple for P over Π^+, by replacing Π^+ by Π^-). It follows from Theorem 4.2 in [GK] that P can be written in the form

$$P(z) = I + C_{\pi e}(zI - A_{\pi e})^{-1}\Gamma_e^{-1}B_{\zeta e}, \tag{3.6}$$

where

$$C_{\pi e} = [C_\pi,\ C_{\pi c}];\ A_{\pi e} = \begin{bmatrix} A_\pi & 0 \\ 0 & A_{\pi c} \end{bmatrix}; \qquad B_{\zeta e} = \begin{bmatrix} B_\zeta \\ B_{\zeta c} \end{bmatrix};$$

and

$$\Gamma_e = \begin{bmatrix} \Gamma & \Gamma_1 \\ \Gamma_2 & \Gamma_3 \end{bmatrix}$$

with Γ_1 and Γ_2 equal to the unique solutions of the Sylvester equations

$$\Gamma_1 A_{\pi c} - A_\zeta \Gamma_1 = B_\zeta C_{\pi c} \tag{3.7}$$

$$\Gamma_2 A_\pi - A_{\zeta c}\Gamma_2 = B_{\zeta c} C_\pi . \tag{3.8}$$

(Recall that the matrices $C_\pi, A_\pi, A_\zeta, B_\zeta$ and Γ are taken from the null-pole triple of P over Π^+ given in the statement of Lemma 3.3.) The matrix Γ_e turns out to be invertible, so (3.6) makes sense. Moreover, denoting

$$A_{\zeta e} = \begin{bmatrix} A_\zeta & 0 \\ 0 & A_{\zeta c} \end{bmatrix},$$

the Sylvester equation holds:

$$\Gamma_e A_{\pi e} - A_{\zeta e}\Gamma_e = B_{\zeta e} C_{\pi e}. \tag{3.9}$$

Since we have assumed that W and P have no common poles and no common zeros, the product WP has a minimal realization

$$W(z)P(z) = I + \underline{C}(zI - \underline{A})^{-1}\underline{B}, \tag{3.10}$$

where

$$\underline{A} = \begin{bmatrix} A & BC_{\pi e} \\ 0 & A_{\pi E} \end{bmatrix} \; ; \quad \underline{B} = \begin{bmatrix} B \\ \Gamma_e^{-1} B_{\zeta e} \end{bmatrix} \; ; \quad \underline{C} = [C, C_{\pi e}]$$

We calculate that $\underline{A}^\times \overset{\text{def}}{=} \underline{A} - \underline{BC}$ is given by

$$\underline{A}^\times = \begin{bmatrix} A - BC & 0 \\ -\Gamma_e^{-1} B_{\zeta e} C & A_{\pi e} - \Gamma_e^{-1} B_{\zeta e} C_{\pi e} \end{bmatrix}$$

and thus $(WP)^{-1}$ has a minimal realization

$$\left(W(z)P(z)\right)^{-1} = I - \underline{C}(zI - \underline{A}^\times)^{-1}\underline{B} \tag{3.11}$$

Writing in more detail, we have

$$\underline{C} = [C, C_\pi, C_{\pi c}]; \quad \underline{A} = \begin{bmatrix} A & BC_\pi & BC_{\pi c} \\ 0 & A_\pi & 0 \\ 0 & 0 & A_{\pi c} \end{bmatrix};$$

$$\begin{bmatrix} I & 0 & 0 \\ 0 & \Gamma & \Gamma_1 \\ 0 & \Gamma_2 & \Gamma_c \end{bmatrix} \underline{B} = \begin{bmatrix} B \\ B_\zeta \\ B_{\zeta c} \end{bmatrix};$$

$$\begin{bmatrix} I & 0 & 0 \\ 0 & \Gamma & \Gamma_1 \\ 0 & \Gamma_2 & \Gamma_c \end{bmatrix} \underline{A}^x = \begin{bmatrix} A-BC & 0 & 0 \\ -B_\varsigma C & A_\varsigma & 0 \\ -B_{\varsigma c}C & 0 & A_{\varsigma c} \end{bmatrix} \begin{bmatrix} I & 0 & 0 \\ 0 & \Gamma & \Gamma_1 \\ 0 & \Gamma_2 & \Gamma_c \end{bmatrix}.$$

We now construct a null-pole triple for WP over Π^+ using the definition. To this end we need the spectral $\underline{A}$-invariant subspace M corresponding to the eigenvalues of $\underline{A}$ in the open right half-plane, and we need also the Riesz projection Z onto the spectral $\underline{A}^x$-invariant subspace corresponding to the eigenvalues of $\underline{A}^x$ lying in Π^+.

We claim that

$$M = Im \begin{bmatrix} Q \\ I \\ 0 \end{bmatrix}$$

where Q is determined by (3.3). Indeed,

$$\underline{A} \begin{bmatrix} Q \\ I \\ 0 \end{bmatrix} = \begin{bmatrix} Q \\ I \\ 0 \end{bmatrix} A_\pi,$$

hence M is $\underline{A}$-invariant and $\underline{A}|M$ is similar to A_π. Since both $\sigma(A)$ and $\sigma(A_{\pi c})$ lie in the open left half plane, the dimensional considerations show that M is indeed the spectral $\underline{A}$-invariant subspace corresponding to the eigenvalues in Π^+.

Next, we will verify that the projection Z is given by

$$Z = \begin{bmatrix} 0 & 0 & 0 \\ \Delta_1 L & \Delta_1\Gamma & \Delta_1\Gamma_1 \\ \Delta_3 L & \Delta_3\Gamma & \Delta_3\Gamma_1 \end{bmatrix}, \tag{3.12}$$

where L is determined by (3.4) and

$$\begin{bmatrix} \Delta_1 & \Delta_2 \\ \Delta_3 & \Delta_4 \end{bmatrix} = \begin{bmatrix} \Gamma & \Gamma_1 \\ \Gamma_2 & \Gamma_c \end{bmatrix}^{-1}.$$

It is convenient to rewrite (3.12) in the form

$$Z = \begin{bmatrix} I & 0 & 0 \\ 0 & \Gamma & \Gamma_1 \\ 0 & \Gamma_2 & \Gamma_c \end{bmatrix}^{-1} \begin{bmatrix} 0 & 0 & 0 \\ L & I & 0 \\ 0 & 0 & 0 \end{bmatrix} \begin{bmatrix} I & 0 & 0 \\ 0 & \Gamma & \Gamma_1 \\ 0 & \Gamma_2 & \Gamma_c \end{bmatrix}$$

Using formula (3.11) we reduce the verification of (3.12) to the proof that

$$Z' \stackrel{\text{def}}{=} \begin{bmatrix} 0 & 0 & 0 \\ L & I & 0 \\ 0 & 0 & 0 \end{bmatrix}$$

is the Riesz projection on the spectral invariant subspace for the matrix

$$A' \stackrel{\text{def}}{=} \begin{bmatrix} A-BC & 0 & 0 \\ -B_\varsigma C & A_\varsigma & 0 \\ -B_{\varsigma c}C & 0 & A_{\varsigma c} \end{bmatrix}$$

corresponding to the eigenvalues in Π^+. Now

$$ImZ' = Im \begin{bmatrix} 0 \\ I \\ 0 \end{bmatrix} \quad ; \; KerZ' = Im \begin{bmatrix} I & 0 \\ -L & 0 \\ 0 & I \end{bmatrix}$$

and

$$A' \begin{bmatrix} 0 \\ I \\ 0 \end{bmatrix} = \begin{bmatrix} 0 \\ I \\ 0 \end{bmatrix} A_\zeta; \tag{3.13}$$

$$A' \begin{bmatrix} I & 0 \\ -L & 0 \\ 0 & I \end{bmatrix} = \begin{bmatrix} I & 0 \\ -L & 0 \\ 0 & I \end{bmatrix} \begin{bmatrix} A - BC & 0 \\ -B_{\zeta c}C & A_{\zeta c} \end{bmatrix}. \tag{3.14}$$

The equality (3.13) show that ImZ' is A'-invariant and the restriction of A' to ImZ' is similar to A_ζ, in particular, this restriction has all its eigenvalues in Π^+. Analogously, (3.14) shows that $\operatorname{Ker} Z'$ is A'-invariant and the restriction of A' to $\operatorname{Ker} Z'$ has all its eigenvalues in the open left half-plane. This proves our claim.

Finally, we are ready to finish the proof of Lemma 3.3. By definition,

$$(\underline{C}|_M, \underline{A}|_M; \underline{A}^\times|_{ImZ}, Z\underline{B}; Z|_M) \tag{3.15}$$

is a null-pole triple for WP with respect to Π^+. Choose a basis in M in the form

$$\left\{ \begin{bmatrix} Qe_i \\ e_i \\ 0 \end{bmatrix} \middle| e_1, \ldots, e_{n_\pi} \text{ is the standard basis in } \mathbb{C}^{n_\pi} \right\}$$

and a basis in ImZ in the form

$$\left\{ \begin{bmatrix} 0 \\ \Delta_1 f_j \\ \Delta_3 f_j \end{bmatrix} \middle| f_1, \ldots, f_{n_\zeta} \text{ is the standard basis in } \mathbb{C}^{n_\zeta} \right\}.$$

With this choice of bases, the above formulas for $\underline{C}, \underline{A}, \underline{A}^\times, \underline{B}, Z$, and M show that (3.15) actually coincides with (3.4). □

4. Interpolation characterization of stable sensitivities: the general case. As in Section 2, we consider here the characterization of all sensitivity functions $S = (I + PK)^{-1}$ associated with stabilizing compensators K. The rational matrix function P is assumed to be of square size ($n \times n$) and without zeros and poles on the imaginary axis (including infinity). However, here we drop the assumption that zeros and poles of P are simple, and allow the most general structure of zeros and poles of P. Thus, given P as above, we describe all rational matrix functions of the form $S = (I + PK)^{-1}$ where K is any rational matrix function such that the functions S, KS, SP and $(I + KP)^{-1}$ are all stable. This description is given by interpolation conditions expressed in terms of residues (or contour integrals). We denote by $\operatorname{Res}_{z=z_0} f(z)$ the residue of a rational matrix or vector function $f(z)$ at z_0 (here $z_0 \in \mathbb{C} \cup \{\infty\}$).

THEOREM 4.1. *Let P be $n \times n$ rational matrix function without poles and zeros on the imaginary axis (including infinity), and with null-pole triple $(C, A_\pi; A_\zeta, B; \Gamma)$ corresponding to the right half-plane.*

A stable rational matrix function S has the form $S = (I + PK)^{-1}$ for some rational matrix K such that SP, KS and $(I + KP)^{-1}$ are also stable if and only if $\det S \not\equiv 0$ *and the following equations are satisfied:*

$$\sum_{z_0 \in \Pi+} \mathrm{Res}_{z=z_0}\ S(z)C(zI - A_\pi)^{-1} = 0; \tag{4.1}$$

$$\sum_{z_0 \in \Pi+} \mathrm{Res}_{z=z_0}\ S(zI - A_\zeta)^{-1}BS(z) = B; \tag{4.2}$$

$$\sum_{z \in \Pi+} \mathrm{Res}_{z=z_0}\ S(zI - A_\zeta)^{-1}BS(z)C(zI - A_\pi)^{-1} = -\Gamma. \tag{4.3}$$

Proof. Suppose $S(z)$ is of the form $\big(I + P(z)K(z)\big)^{-1}$ and assume all four functions

$$S(z),\ S(z)P(z), -K(z)S(z),\ \big(I + K(z)P(z)\big)^{-1}$$

are analytic in $\overline{\Pi^+} \cup \{\infty\}$. Since S is the inverse of a rational matrix function, certainly $\det S \not\equiv 0$. It will be convenient for one analysis to rewrite the conditions (4.1)–(4.3) in the equivalent contour integral form:

$$\frac{1}{2\pi i}\int_\gamma S(z)C(zI - A_\pi)^{-1}dz = 0; \tag{4.1$'$}$$

$$\frac{1}{2\pi i}\int_\gamma (zI - A_\zeta)^{-1}BS(z)dz = B; \tag{4.2$'$}$$

$$\frac{1}{2\pi i}\int_\gamma (zI - A_\zeta)^{-1}BS(z)C(zI - A_\pi)^{-1}dz = -\Gamma \tag{4.3$'$}$$

Here γ is simple rectifiable contour in the right half-plane such that all eigenvalues of A_ζ and A_π are inside γ.

By Proposition 3.1 (vi) find matrices $\widetilde{C}$ and $\widetilde{B}$ such that the differences

$$V(z) = P(z)^{-1} - \widetilde{C}(zI - A_\zeta)^{-1}B$$

and

$$U(z) = P(z) - C(zI - A_\pi)^{-1}\widetilde{B}$$

are analytic in $\overline{\Pi^+} \cup \{\infty\}$. As $S(z)P(z)$ is analytic in $\overline{\Pi^+} \cup \{\infty\}$, it follows that so is also

$$S(z)C(zI - A_\pi)^{-1}\widetilde{B}.$$

Now for a fixed integer $k \geq 0$ we have

(4.4)

$$\frac{1}{2\pi i}\int\limits_{\gamma} S(z)C(zI-A_\pi)^{-1}A_\pi^k\widetilde{B}dz =$$

$$=\frac{1}{2\pi i}\int\limits_{\gamma} S(z)C(zI-A_\pi)^{-1}(A_\pi - zI)A_\pi^{k-1}\widetilde{B}dz+$$

$$+\frac{1}{2\pi i}\int\limits_{\gamma} S(z)C(zI-A_\pi)^{-1}zA_\pi^{k-1}\widetilde{B}dz = \frac{1}{2\pi i}\int\limits_{\gamma} S(z)C(zI-A_\pi)^{-1}zA_\pi^{k-1}\widetilde{B}dz,$$

and continuing in the manner we find that

$$\frac{1}{2\pi i}\int\limits_{\gamma} S(z)C(zI-A_\pi)^{-1}A_\pi^k\widetilde{B}dz = \frac{1}{2\pi i}\int\limits_{\gamma}(S(z)C(zI-A_\pi)^{-1}z^k\widetilde{B}dz = 0$$

So

$$\frac{1}{2\pi i}\int\limits_{\gamma} S(z)C(zI-A_\pi)^{-1}dz\cdot[\widetilde{B}, A_\pi\widetilde{B}, \ldots, A_\pi^{k-1}\widetilde{B}] = 0 \ ; \ k = 0, 1, \ldots,$$

and since the pair $(A_\pi, \widetilde{B})$ is easily seen to be controllable, the equality (4.1′) follows.

Next, we verify (4.2′). We have

$$\begin{aligned} -KS &= -K(I+PK)^{-1} = P^{-1}(I-(I+PK))(I+PK)^{-1} = \\ &= P^{-1}(S-I), \end{aligned} \tag{4.5}$$

and hence the function $P^{-1}(S-I)$ is analytic in $\overline{\Pi^+}\cup\{\infty\}$. So

$$\widetilde{C}(zI-A_\zeta)^{-1}B(S(z)-I)$$

is analytic $\overline{\Pi^+}\cup\{\infty\}$ as well. Arguing as in the verification of (4.1′) we conclude that

$$\frac{1}{2\pi i}\int\limits_{\gamma}(zI-A_\zeta)^{-1}B(S(z)-I)dz = 0,$$

which implies (4.2′).

To verify (4.3′), it will be convenient to denote by $\mathcal{R}_{n\times n}(\overline{\Pi^+}\cup\{\infty\})$ the set of all $n\times n$ rational matrix functions which are analytic in $\overline{\Pi^+}\cup\{\infty\}$. Write

$$(I+KP)^{-1} = P^{-1}(I+PK)^{-1}P = P^{-1}SP.$$

As $(I+KP)^{-1}$ is analytic in $\overline{\Pi^+}\cup\{\infty\}$, it follows that $P^{-1}(S-I)P$ is analytic in $\overline{\Pi^+}\cup\{\infty\}$ as well, and

$$-P+SP = (S-I)P \in P\mathcal{R}_{n\times n}(\overline{\Pi^+}\cup\{\infty\}).$$

Now

$$\begin{aligned} & - C(zI - A_\pi)^{-1}\widetilde{B} + S(z)C(zI - A_\pi)^{-1}\widetilde{B} = \\ & = -P(z) + U(z) + S(z)(P(z) - U(z)), \end{aligned}$$

where $U(z) \in \mathcal{R}_{n\times n}(\overline{\Pi^+} \cup \{\infty\})$. Thus

$$\begin{aligned} & - C(zI - A_\pi)^{-1}\widetilde{B} + S(z)C(zI - A_\pi)^{-1}\widetilde{B} = \\ & (-P(z) + S(z)P(z)) + (I - S(z))U(z), \end{aligned}$$

and both summands on the right-hand side belong to $P\mathcal{R}_{n\times n}(\overline{\Pi^+} \cup \{\infty\})$ (for the second summand we use the equality (4.5) that implies $(S-I) \in P\mathcal{R}_{n\times n}(\overline{\Pi^+}\cup\{\infty\})$. So $-C(zI - A_\pi)^{-1}\widetilde{B} + S(z)C(zI - A_\pi)^{-1}\widetilde{B} \in P\mathcal{R}_{n\times n}(\overline{\Pi^+} \cup \{\infty\})$. Using the controllability of $A_\pi, \widetilde{B}$ it is not difficult to deduce that for every $x \in \mathbb{C}^n$ we have

$$-C(zI - A_\pi)^{-1}x + S(z)C(zI - A_\pi)^{-1}x \in P\mathcal{R}_{n\times n}(\overline{\Pi^+} \cup \{\infty\}). \tag{4.6}$$

The subspace $P\mathcal{R}_{n\times n}(\overline{\Pi^+}\cup\{\infty\})$ is the singular subspace of P (introduced in Section 3). It follows from Theorem 3.2 that (4.6) is precisely equivalent to (4.3$'$).

Conversely, assume S is stable with $\det S \not\equiv 0$ and (4.1$'$) - (4.3$'$) are satisfied. If we define K by

$$K = P^{-1}(S^{-1} - I)$$

then S has the form $S = (I + PK)^{-1}$. Using the calculation (4.4) we find that

$$\frac{1}{2\pi i}\int_\gamma z^k S(z)C(zI - A_\pi)^{-1}\widetilde{B}dz = 0, \qquad k = 0, 1, \ldots,$$

and hence the function $S(z)C(zI - A_\pi)^{-1}\widetilde{B}$ is analytic in $\overline{\Pi^+} \cup \{\infty\}$. As the difference

$$P(z) - C(zI - A_\pi)^{-1}\widetilde{B}$$

is analytic in $\overline{\Pi^+} \cup \{\infty\}$ as well, we conclude that SP is stable. An analogous argument implies that $P^{-1}(S - I)$ is stable, and hence in view of (4.5) so is KS. Finally (4.3$'$) is equivalent to (4.6), which in turn implies (using the stability of $P^{-1}(S - I)$) that $-P + SP \in P\mathcal{R}_{n\times n}(\overline{\Pi^+} \cup \{\infty\})$. Consequently, $P^{-1}(I - S)P$ is analytic in $\overline{\Pi^+} \cup \{\infty\}$, and hence so is $(I + KP)^{-1} = P^{-1}SP$. □

We remark that the proof of Theorem 4.1 reveals that given the stability of S, the function SP is stable if and only (4.1) holds, and the function KS is stable if and only if (4.2) holds.

The problem (4.1)–(4.3) (or, equivalently, (4.1$'$)–(4.3$'$)) is precisely the two-sided Lagrange-Sylvester interpolation problem for rational matrix functions studied in [BGR3]. We quote a general result on this problem (adapted to the situation at hand) from this paper.

Given matrices $C_+, C_-, A_\pi, A_\zeta, B_+, B_-$ and Γ of sizes $M \times n_\pi, N \times n_\pi, n_\pi \times n_\pi, n_\zeta \times n_\zeta, n_\zeta \times M, n_\zeta \times N$ and $n_\zeta \times n_\pi$, respectively, such that all eigenvalues of

A_π and A_ζ are in Π^+. We consider the set of $M \times N$ rational matrix functions $W(z)$ with the properties that $W(z)$ has no poles in $\overline{\Pi^+} \cup \{\infty\}$ and the equations (the sums in (4.7), (4.8), (4.9) are taken over all $z_0 \in \Pi^+$)

$$\sum \operatorname{Res}_{z=z_0}(zI - A_\zeta)^{-1} B_+ W(z) = -B_-; \tag{4.7}$$

$$\sum \operatorname{Res}_{z=z_0} W(z) C_-(zI - A_\pi)^{-1} = C_+; \tag{4.8}$$

$$\sum \operatorname{Res}_{z=z_0}(zI - A_\zeta)^{-1} B_+ W(z) C_-(zI - A_\pi)^{-1} = \Gamma \tag{4.9}$$

are satisfied.

The equations (4.7)-(4.9) can be viewed as interpolation conditions for the unknown function W; thus, we shall speak of W as a solution of the interpolation problem (4.7)-(4.9), called the two-sided Lagrange-Sylvester interpolation problem.

Denote by $\mathcal{R}_{M\times N}(\overline{\Pi^+} \cup \{\infty\})$ the set of all $M \times N$ rational matrix functions with no poles in $\overline{\Pi^+} \cup \{\infty\}$.

THEOREM 4.2. *([BGR3]). Assume in addition that (C_-, A_π) is observable, (A_ζ, B_+) is controllable, and Γ satisfies the Sylvester equation*

$$\Gamma A_\pi - A_\zeta \Gamma = B_+ C_+ + B_- C_-. \tag{4.10}$$

Then there exists rational $M \times N$ matrix functions W with no poles in $\overline{\Pi^+} \cup \{\infty\}$ which are solutions of (4.7)-(4.9). Moreover, assume that $\theta = \begin{bmatrix} \theta_{11} & \theta_{12} \\ \theta_{21} & \theta_{22} \end{bmatrix}$ is any rational matrix function with no poles or zeros on the imaginary axis and at infinity, having the set

$$\left(\begin{bmatrix} C_+ \\ C_- \end{bmatrix}, A_\pi; A_\zeta, [B_+ B_-]; \Gamma \right)$$

as a null-pole triple over Π^+ and assume that ϕ^{-1} is a rational $M \times M$ matrix function without poles or zeros on the imaginary axis and at infinity, such that ϕ^{-1} has the set

$$(C_-, A_\pi; 0, 0; 0)$$

as a null-pole triple over Π^+(observe that ϕ^{-1} has no zeros in Π^+, hence ϕ has no poles in Π^+). Then $W \in \mathcal{R}_{M\times N}(\overline{\Pi^+} \cup \{\infty\})$ satisfies the problem (4.7)–(4.9) if and only if $W(z)$ has the following form: There exists rational matrix functions $Q_1 \in \mathcal{R}_{M\times N}(\overline{\Pi^+} \cup \{\infty\}$ and $Q_2 \in \mathcal{R}_{N\times N}(\overline{\Pi^+} \cup \{\infty\})$ for which the function

$$\phi(\theta_{21} Q_1 + \theta_{22} Q_2) \tag{4.11}$$

has no zeros and poles in $\overline{\Pi^+} \cup \{\infty\}$, and such that

$$W = (\theta_{11} Q_1 + \theta_{12} Q_2)(\theta_{21} Q_1 + \theta_{22} Q_2)^{-1}. \tag{4.12}$$

One can prove that the assumption on observability and controllability of pairs in Theorem 4.2 is made without loss of generality (i.e. the general two-sided Lagrange-Sylvester interpolation problem can be reduced to that problem with the additional observability and controllability assumptions). Another important observation is the equality (4.10) is necessary for solvability of (4.7)–(4.9)

Combining Theorems 4.1 and 4.2 we obtain the following result.

THEOREM 4.3. *Let P be as in Theorem 4.1 and assume that $S = (I + PK)^{-1}$ is stable and is associated with a stabilizing compensator K if and only if S has the form*

$$S = (\theta_{11}Q_1 + \theta_{12}Q_2)(\theta_{21}Q_1 + \theta_{22}Q_2)^{-1}.$$

Here

$$\theta = \begin{bmatrix} \theta_{11} & \theta_{12} \\ \theta_{21} & \theta_{22} \end{bmatrix}$$

is a fixed rational matrix function analytic and invertible on the imaginary axis and at infinity and having the set

$$\left(\begin{bmatrix} 0 \\ C \end{bmatrix}, A_\pi; A_\zeta, [B, -B], -\Gamma \right)$$

as a null-pole triple over Π^+, ϕ^{-1} is a fixed rational matrix function analytic and invertible on the imaginary axis and at infinity having $(C, A_\pi; 0, 0; 0)$ as its null-pole triple over Π^+, and Q_1, Q_2 are any stable rational matrix functions (of suitable sizes) such that $\phi(\theta_{21}Q_1 + \theta_{22}Q_2)$ has no poles and zeros in $\overline{\Pi^+} \cup \{\infty\}$.

Our next step will be to take into account the norm inequality $\|S\|_\infty \leq \mu$ (where μ is a given positive number) as well. This will be done (in Section 6 and 7) by means of a special choice of θ in Theorem 4.3.

5. The two-sided Nudelman interpolation problem. We state here general result on solutions of the two-sided Lagrange-Sylvester interpolation problem with additional metric constraints (the Nudelman interpolation problem).

Let

$$\omega = (C_+, C_-, A_\pi; A_\zeta, B_+, B_-; \Gamma) \tag{5.1}$$

be a set of given matrices as in Theorem 4.2, i.e. (C_-, A_π) is observable, (A_ζ, B_+) is controllable, and Γ satisfies the Sylvester equation

$$\Gamma A_\pi - A_\zeta \Gamma = B_+ C_+ + B_- C_-.$$

We assume further that the spectra of A_π and of A_ζ lie in the open right halfplane Π^+. We are looking for solutions of the following problem:
Define

$$\mathcal{BR}_{M\times N}(\Pi^+) = \{F \in \mathcal{R}_{M\times N}(\overline{\Pi^+} \cup \{\infty\}) : \sup_{z\in\Pi^+} \|F(z)\| < 1\}$$

Find $M \times N$ rational matrix functions $F(z)$ without poles in $\overline{\Pi^+} \cup \{\infty\}$ such that

$$F \in \mathcal{BR}_{M\times N}(\Pi^+) \tag{5.2}$$

and which satisfy the interpolation conditions

$$\sum_{z\in\Pi^+} \operatorname{Res}_{z=z_0} (zI - A_\zeta)^{-1} B_+ F(z) dz = -B_-; \tag{5.3}$$

$$\sum_{z\in\Pi+} \operatorname{Res}_{z=z_0} F(z)C_-(zI-A_\pi)^{-1}dz = C_+; \tag{5.4}$$

$$\sum_{z\in\Pi+} \operatorname{Res}_{z=z_0}(zI-A_\zeta)^{-1}B_+F(z)C_-(zI-A_\pi)^{-1}dz = \Gamma. \tag{5.5}$$

We call this problem the two-sided Nudelman interpolation problem, as it was first stated in the resolvent form in [N1, N2] (only conditions of type (5.3) or (5.4) were considered there).

To state the solution to this problem, we introduce the matrix $\Lambda(\omega)$ constructed from the interpolation data (5.1) as follows

$$\Lambda(\omega) = \begin{bmatrix} S_1 & \Gamma^* \\ \Gamma & S_2 \end{bmatrix} \tag{5.6}$$

where S_1 and S_2 are the unique solutions of the Lyapunov equations

$$S_1A_\pi + A_\pi^*\, S_1 = C_-^*C_- - C_+^*C_+ \tag{5.7}$$

$$S_2A_\zeta^* + A_\zeta\, S_2 = B_+B_+^* - B_-B_-^*. \tag{5.8}$$

The matrix $\Lambda(\omega)$ is nothing else but the Pick matrix for the Nevanlinna-Pick interpolation problem (which is a particular case of the two-sided Nudelman interpolation problem), see example later in this section.

Let $J = \begin{bmatrix} I_{n_\pi} & 0 \\ 0 & -I_{n_\zeta} \end{bmatrix}$ be the $(n_\pi + n_\zeta) \times (n_\pi + n_\zeta)$ signature matrix. An $(n_\pi + n_\zeta) \times (n_\pi + n_\zeta)$ rational matrix function $\theta(z)$ is called *J-unitary on the imaginary axis* if the equality

$$\left(\theta(-\overline{z})\right)^* J\, \theta(z) = J$$

holds for all $z \in \mathbb{C}$ such that both z and $-\overline{z}$ are not poles of $\theta(z)$. Assuming that the matrix $\Lambda(\omega)$ defined by (5.6) is invertible, we let $\theta(z) = \begin{bmatrix} \theta_{11}(z) & \theta_{12}(z) \\ \theta_{21}(z) & \theta_{22}(z) \end{bmatrix}$ be the rational matrix function which is J-unitary on the imaginary axis, has no zeros and poles on the imaginary axis and at infinity, and has

$$\left(\begin{bmatrix} C_+ \\ C_- \end{bmatrix}, A_\pi; A_\zeta, [B_+ B_-]; \Gamma\right)$$

as its null-pole triple with respect to Π^+. It turns out that such function is unique up to multiplication on the right by a constant J-unitary matrix, and is given (assuming $\theta(\infty) = I$) by the formula

$$\theta(z) = I + \begin{bmatrix} C_+ & -B_+^* \\ C_- & B_-^* \end{bmatrix} \begin{bmatrix} zI - A_\pi)^{-1} & 0 \\ 0 & (zI + A_\zeta)^{-1} \end{bmatrix} \Lambda(\omega)^{-1}.$$

$$\begin{bmatrix} -C_+^* & C_-^* \\ B_+ & B_- \end{bmatrix}$$

For the proof of this formula we refer the reader to [BGR5]; more information on realizations and factorizations of rational matrix function which are J-unitary on the imaginary axis is found in [AG]. The solution of the two-sided Nudelman interpolation problem is as follows.

THEOREM 5.1. *Let $w = (C_+, C_-, A_\pi; A_\zeta, B_+, B_-; \Gamma)$ be a set of matrices of appropriate sizes such that (C_-, A_π) is observable, (A_ζ, B_+) is controllable, the equation*

$$\Gamma A_\pi - A_\zeta \Gamma = B_+ C_+ + B_- C_- \tag{5.10}$$

is satisfied, and $\sigma(A_\pi) \cup \sigma(A_\zeta) \subset \Pi^+$. Define $\Lambda = \Lambda(w)$ by (5.6), and assume that Λ is invertible. Then there is $F \in \mathcal{BR}_{M\times N}(\Pi^+)$ satisfying the interpolation conditions (5.3)–(5.5) if and only if Λ is positive definite. If this is the case then all $F \in BR_{M\times N}(\Pi^+)$ satisfying the interpolation conditions (5.3)–(5.5) are given by the formula

$$F = (\theta_{11}G + \theta_{12})(\theta_{21}G + \theta_{22})^{-1}$$

where

$$\theta = \begin{bmatrix} \theta_{11} & \theta_{12} \\ \theta_{21} & \theta_{22} \end{bmatrix}$$

is given by (5.9) and G is an arbitrary function in $\mathcal{BR}_{M\times N}(\Pi^+)$.

The proof of Theorem 5.1 is found in [BGR5] (a complete exposition with full details of the two-sided Nudelman interpolation problem and related topics will be given in [BGR4]). For the case θ has no poles and only simple zeros in Π^+, a proof of Theorem 5.1 is given in [BGR1].

The two-sided Nudelman interpolation problem is equivalent to the bitangential Nevanlinna-Pick interpolation problem with arbitrarily high order interpolation conditions; special cases have been studied in [Fe1, Fe2, BR1, LA,K]. The work of [BH] handles a more general class of problems but in a less explicit form.

We conclude this section with the example when all poles and zeros of $\theta(z)$ are simple (cf Section 2 and Example 3.1).

Example 4.1. Let

$$B_+ = \begin{bmatrix} x_1^* \\ \vdots \\ x_n^* \end{bmatrix}, \quad B_- = -\begin{bmatrix} y_1^* \\ \vdots \\ y_n^* \end{bmatrix}$$

$$A_\zeta = \begin{bmatrix} z_1 & & 0 \\ & \vdots & \\ 0 & & z_n \end{bmatrix}$$

where $x_1, \cdots, x_n$ are nonzero vectors in $\mathbb{C}^M, y_1, \ldots, y_n$ are vectors in $\mathbb{C}^N$ and $z_1, \cdots, z_n$ are distinct points in Π^+. Also let C_-, A_π, C_+ be given by

$$C_- = [u_1, \cdots, u_m], \quad C_+ = [v_1, \cdots, v_m]$$

$$A_\pi = \begin{bmatrix} w_1 & & 0 \\ & \ddots & \\ 0 & & w_m \end{bmatrix}$$

where $u_1, \cdots, u_m$ are nonzero vectors in $\mathbb{C}^N, v_1, \ldots, v_m$ are vectors in $\mathbb{C}^M$ and $w_1, \ldots, w_m$ are distinct points in Π^+. Then the interpolation condition (5.3) amounts to

$$\text{(i)} \qquad x_j^* F(z_j) = y_j^* \quad , 1 \le j \le n$$

and (5.4) reduces to

$$\text{(ii)} \qquad F(w_j)u_j = v_j, \quad 1 \le j \le m.$$

If $\sigma(A_\zeta)$ and $\sigma(A_\pi)$ are disjoint (i.e. no z_i is also a w_j) then the condition (5.5) is automatic from (5.3) and (5.4) provided Γ satisfies the necessary consistency condition (5.10) (this observation applies in the general situation as well (Theorem 5.2.2 in [BGR3]), not only when all poles and zeros of $\theta(z)$ are simple). To make this explicit, note that if $z_i \neq w_j$ and F satisfies the interpolation conditions (5.3) and (5.4) then the (i,j)-th entry of (5.5) is

$$\begin{aligned}\Gamma_{ij} &= x_i^* F(z_i) u_j (z_i - w_j)^{-1} + (w_j - z_i)^{-1} x_i^* F(w_j) u_j \\ &= (z_i - w_j)^{-1} (y_i^* u_j - x_i^* v_j)\end{aligned}$$

On the other hand, if $z_i = w_j$, then the interpolation condition (5.5) collapses to

$$\text{(iii)} \qquad \rho_{ij} \overset{\text{def}}{=} x_i^* F'(z_i) u_j = \Gamma_{ij},$$

where Γ_{ij} is the (i,j) entry of Γ. The matrix $\Lambda = \Lambda(\omega)$ in this example is found as follows:

$$\Lambda = \begin{bmatrix} S_1 & \Gamma^* \\ \Gamma & S_2 \end{bmatrix}$$

where

$$\begin{aligned} S_1 &= \left[(w_j + \overline{w}_i)^{-1}(u_i^* u_j - v_i^* v_j)\right]_{1 \le i,j \le m}; \\ S_2 &= \left[(z_i + \overline{z}_j)^{-1}(x_i^* x_j - y_i^* y_j)\right]_{1 \le i,j \le m}; \\ \Gamma &= [\gamma_{ij}]_{1 \le i \le n; 1 \le j \le m} \end{aligned}$$

with

$$\begin{aligned} \gamma_{ij} &= (z_i - w_j)^{-1}(y_i^* u_j - x_i^* v_j) \qquad \text{if} \quad z_i \neq w_j \\ \gamma_{ij} &= \rho_{ij} \quad \text{if} \quad z_i = w_j. \end{aligned}$$

6. Back to the sensitivity minimization. We return now to the sensitivity minimization problem formulated in Section 1:

$$\text{(6.1)} \qquad \min_{K \text{ stabilizes } P} \|(I + PK)^{-1}\|_\infty.$$

First, we find out the minimum in (6.1):

THEOREM 6.1. *Let $P(z)$ be a rational $n \times n$ matrix function with no poles or zeros on the imaginary axis (including infinity) and let $\tau = (C, A_\pi; A_\zeta, B; \Gamma)$ be a null-pole triple for $P(z)$ over the open right half-plane Π^+. Then the number $\mu_{\inf}$ defined by $\mu_{\inf} = \{\mu : \|S(z)\|_\infty \leq \mu$ for some S of the form $S = (I + PK)^{-1}$, where K is a stabilizing compensator for $P\}$ is given by*

$$\mu_{\inf} = \|I + \Gamma_2^{-1/2}\Gamma\Gamma_1^{-1}\Gamma^*\Gamma_2^{-1/2}\|^{1/2} \tag{6.2}$$

where Γ_1 and Γ_2 are given by

$$\Gamma_1 A_\pi + A_\pi^* \Gamma_1 = C^* C; \tag{6.3}$$

$$\Gamma_2 A_\zeta^* + A_\zeta \Gamma_2 = BB^* \tag{6.4}$$

Using well-known inertia results (see, e.g., Theorem 13.1.4 in [LT]) it follows that the unique solutions Γ_1 and Γ_2 to (5.3) and (5.4) are positive definite. In particular Γ_1 and Γ_2 are invertible, and formula (6.2) makes sense.

If A_ζ is trivial (i.e. $n_\zeta = 0$ where $n_\zeta \times n_\zeta$ is the size of A_ζ) then (6.2) gives $\mu_{\inf} = 0$. This agrees with the result well known in the engineering literature that the optimal sensitivity for a minimal phase plant is zero in the limit.

Proof. We use Theorem 5.1. Note that S satisfies (4.1)–(4.3) together with $\|S\|_\infty \leq \mu$ if and only if $\mu^{-1}S$ satisfies (4.1) together with

$$\sum_{z_0 \in \Pi+} \operatorname{Res}_{z=z_0} (zI - A_\zeta)^{-1} B(\mu^{-1}S)(z)dz = \mu^{-1}B; \tag{6.5}$$

$$\sum_{z_0 \in \Pi+} \operatorname{Res}_{z=z_0} (zI - A_\zeta)^{-1} B(\mu^{-1}S)(z)C(zI - A_\pi)^{-1}dz = -\mu^{-1}\Gamma, \tag{6.6}$$

and $\|\mu^{-1}S\|_\infty \leq 1$. This is the type of problems considered in Theorem 5.1. Applying this theorem for $F = \mu^{-1}S$ we obtain that there exists a stable solution $\mu^{-1}S$ to (4.1), (6.5), (6.6) satisfying $\|\mu^{-1}S\|_\infty \leq 1$ if and only if the matrix

$$\Lambda_\mu = \begin{bmatrix} \Gamma_1 & -\mu^{-1}\Gamma^* \\ -\mu^{-1}\Gamma & (1 - \mu^{-2})\Gamma_2 \end{bmatrix} \tag{6.7}$$

is positive definite (here Γ_1 and Γ_2 are solutions of (6.3) and (6.4), respectively, and it is assumed in advance that μ is such that Λ_μ is invertible). Hence

$$\mu_{\inf} = \inf\{\mu : \Lambda_\mu \quad \text{is positive definite}\}.$$

Use the Schur complements:

$$\Lambda_\mu = X \begin{bmatrix} \Gamma_1 & 0 \\ 0 & \Gamma_2 - \mu^{-2}(\Gamma_2 + \Gamma\Gamma_1^{-1}\Gamma^*) \end{bmatrix} X^*,$$

where $X = \begin{bmatrix} I & 0 \\ -\mu^{-1}\Gamma\Gamma_1^{-1} & I \end{bmatrix}$. Thus Λ_μ is positive definite if and only if $\Gamma_2 - \mu^{-2}(\Gamma_2 + \Gamma\Gamma_1^{-1}\Gamma^*)$ is positive definite, that is to say if and only if

$$I + \Gamma_2^{-1/2}\Gamma\Gamma_1^{-1}\Gamma^*\Gamma_2^{-1/2} < \mu^2 I.$$

From this criterion the theorem follows. □

For the case where $\mu > \mu_{\inf}$ we see from the proof of Theorem 6.1 that the matrix Λ_μ given by (6.7) is positive definite (and in particular invertible). In this situation we define a rational $(2n \times 2n)$ matrix function $\theta_\mu(z) = \begin{bmatrix} \theta_{\mu 11}(z) & \theta_{\mu 12}(z) \\ \theta_{\mu 21}(z) & \theta_{\mu 22}(z) \end{bmatrix}$ by

(6.8)
$$\theta_\mu(z) = \begin{bmatrix} I & 0 \\ 0 & I \end{bmatrix}$$
$$+ \begin{bmatrix} 0 & \mu^{-1}B^* \\ C & B^* \end{bmatrix} \begin{bmatrix} (zI - A_\pi)^{-1} & 0 \\ 0 & (zI + A_\zeta^*)^{-1} \end{bmatrix} \Lambda_\mu^{-1} \begin{bmatrix} 0 & C^* \\ B & -\mu^{-1}B \end{bmatrix}$$

This function arises in the parametrization of all suboptional sensitivities as follows. Once one has explicit formulas for S, the stabilizing compensator K itself can be found as $K = P^{-1}(S^{-1} - I)$.

THEOREM 6.2. *Let $P(z), \tau = (C, A_\pi; A_\zeta, B_j\Gamma)$ and $\mu_{\inf}$ be as in Theorem 6.1 and suppose that $\mu > u_{\inf}$. Then a rational matrix function S satisfies the two conditions*

(i) $\|S\|_\infty \leq \mu$

and

(ii) *$S = (I + PK)^{-1}$ for some stabilizing compensator K for P if and only if $\det S \not\equiv 0$ and*

$$S = \mu(\theta_{\mu 11}G + \theta_{\mu 12})\,(\theta_{\mu 21}G + \theta_{\mu 22})^{-1}$$

where $\theta_{\mu ij}$ $(i, j = 1, 2)$ are given by (6.8) and where G is an arbitrary rational $n \times n$ matrix function with no poles in $\overline{\Pi^+} \cup \{\infty\}$ and such that $\|G\|_\infty \leq 1$.

The proof of Theorem 6.2 follows immediately by combining Theorem 5.1 and 4.1.

7. Sensitivity minimization with weights. In this section we extend the results of the previous section to the problem of weighted sensitivity minimization. For the weighted sensitivity minimization problem, besides the plant $P(z)$ and a null-pole triple $\tau = (C, A_\pi; A_\zeta, B; \Gamma)$ for $P(z)$ over Π^+, we assume that we are given also rational $n \times n$ matrix weight functions $W_1(z)$ and $W_2(z)$ (in practice generally taken to be scalar) along with minimal realizations

(7.1)
$$W_j(z) = I + C_j(zI - A_j)^{-1}B_j$$

for $j = 1, 2$. We also assume that both $W_1(z)$ and $W_2(z)$ have no zeros and no poles in $\Pi^+ \cup \{\infty\}$. The problem then is to minimize the norm of the weighted sensitivity $W_1(z)(I + P(z)K(z))^{-1}W_2(z)$ where K ranges over all stabilizing compensators for P.

We shall state and prove two main results: One is the characterization of the minimal weighted sensitivity norm; the other is parametrization of all suboptimal solutions in terms of a linear fractional map.

To describe the results we need to introduce more matrices. Given the null-pole triple τ over Π^+ for $P(z)$ and minimal realizations for $W_j(z)$ as in (7.1) for $j = 1, 2$, define matrices Q_2, E_1 and E_2 by

$$Q_2 A_\pi - (A_2 - B_2 C_2) Q_2 = B_2 C \tag{7.2}$$

$$E_1(A_1 - B_1 C_1) - A_\zeta E_1 = B C_1 \tag{7.3}$$

$$E_2 A_2 - A_\zeta E_2 = B C_2 \tag{7.4}$$

Note that by assumption the spectra of A_π and A_ζ are in the open right half plane while the spectra of $A_2 - B_2 C_2$, $A_1 - B_1 C_1$ and A_2 are in the open left half plane, so the Sylvester equations (7.2), (7.3) and (7.4) have unique solutions Q_2, E_2. Similarly we may define matrices Γ_1, Γ_{2a} and Γ_{2b} as the unique solutions of the Lyapunov equations

$$\Gamma_1 A_\pi + A_\pi^* \Gamma_1 = (C^* - Q_2^* C_2^*)(C - C_2 Q_2) \tag{7.5}$$

$$\Gamma_{2a} A_\zeta^* + A_\zeta \Gamma_{2a} = (E_1 B_1 + B)(B_1^* E_1^* + B^*) \tag{7.6}$$

$$\Gamma_{2b} A_\zeta^* + A_\zeta \Gamma_{2b} = (E_2 B_2 - B)(B_2^* E_2^* - B^*). \tag{7.7}$$

It is not difficult to verify that Γ_1 is positive definite. Indeed, in view of Theorem 13.1.4 of [LT] we have only to check that $(C - C_2 Q_2, A_\pi)$ is an observable pair. Let $M = \cap_{i \geq 0} \operatorname{Ker}(C - C_2 Q_2) A_\pi^i$. The subspace M is clearly A_π-invariant, and equality (7.2) gives

$$Q_2 A_\pi x = A_2 Q_2 x\ , \quad x \in M. \tag{7.8}$$

As $\sigma(A_{\pi|M}) \cap \sigma(A_2) = \emptyset$, we obtain from (7.8) that $Q_2 x = 0$ for all $x \in M$. Now using (7.2) it is not difficult to see that $C_2 Q_2 A_\pi^i x$, where $x \in M$, has the form

$$C_2 Q_2 A_\pi^i x = \sum_{j=0}^{i-1} X_{ji} C A_\pi^j x + Y_i Q_2 x \qquad (i \geq 1)$$

for some matrices X_{ji} and Y_i. So for $x \in M$ we have

$$\begin{aligned} C A_\pi^i &= (C - C_2 Q_2) A_\pi^i x + C_2 Q_2 A_\pi^i x \\ &= C_2 Q_2 A_\pi^i x = \sum_{j=0}^{i-1} X_{ji} C A_\pi^j x + Y_i Q_2 x = \\ &= \sum_{j=0}^{i-1} X_{ji} C A_\pi^j x \qquad , \quad i = 1, 2, \ldots \end{aligned}$$

By induction on i it is now evident that every $x \in M$ belongs to $\cap_{i \geq 0} \operatorname{Ker}(C A_\pi^i)$. Using the observability of (C, A_π) we conclude that $M = \{0\}$, as required. We can now give a computation of the minimal weighted sensitivity norm.

THEOREM 7. *Let $P(z)$ be a rational $n \times n$ matrix function having $\tau = (C, A_\pi; A_\zeta, B; \Gamma)$ as a null-pole triple over the open right half plane Π^+ and having no zeros or poles on the imaginary line and at infinity. Let W_1 and W_2 be rational $n \times n$ matrix weighting functions having no poles or zeros in $\overline{\Pi^+} \cup \{\infty\}$ and with minimal realizations (7.1). Then the infimum $\mu_{\inf}$ of the set of positive real numbers μ for which there exists a matrix function S such that*

(i) $S = (I + PK)^{-1}$ *for a stabilizing compensator K for P*

and

(ii) $\|W_1 S W_2\|_\infty \leq \mu$
is given by

$$\mu_{\inf} = \inf\{\mu > 0 : \Gamma_{2a} - \mu^{-2}\Gamma_{2b} - \mu^{-2}(\Gamma - E_2Q_2)\Gamma_1^{-1}(\Gamma - E_2Q_2)^*$$

is positive definite$\}$.

For $\mu > \mu_{\inf}$ we can also obtain a linear fractional parametrization of the set of all weighted sensitivities associated with stabilizing compensators and having norm at most μ. To do this we introduce the Hermitian matrix Λ_μ by

$$\Lambda_\mu = \begin{bmatrix} \Gamma_1 & -\mu^{-1}Q_2^*E_2^* - \mu^{-1}\Gamma^* \\ -\mu^{-1}E_2Q_2 - \mu^{-1}\Gamma & \Gamma_{2a} - \mu^{-2}\Gamma_{2b} \end{bmatrix} \tag{7.9}$$

where $\Gamma_1, \Gamma_{2a}, \Gamma_{2b}, Q_2, E_2$ are defined by (7.2)–(7.7). If $\mu > \mu_{\inf}$, then the matrix Λ_μ is positive definite. For these values of μ we introduce a rational $2n \times 2n$ matrix function $\theta_\mu(z)$ by
(7.10)

$$\theta_\mu(z) = \begin{bmatrix} \theta_{\mu 11}(z) & \theta_{\mu 12}(z) \\ \theta_{\mu 21}(z) & \theta_{\mu 22}(z) \end{bmatrix} = \begin{bmatrix} I & 0 \\ 0 & I \end{bmatrix} + \begin{bmatrix} 0 & -B_1^*E_1^* - B^* \\ -C_2Q_2 + C & \mu^{-1}B_2^*E_2^* - \mu^{-1}B^* \end{bmatrix}$$
$$\cdot \begin{bmatrix} (zI - A_\pi)^{-1} & 0 \\ 0 & (zI + A_\zeta^*)^{-1} \end{bmatrix} \Lambda_\mu^{-1} \begin{bmatrix} 0 & -Q_2^*C_2^* + C^* \\ E_1B_1 + B & \mu^{-1}E_2B_2 - \mu^{-1}B \end{bmatrix}$$

Here is our second main result, on the linear fractional parametrization of suboptimal solutions.

THEOREM 7.2. *Let $P(z), \tau = (C, A_\pi; A_\zeta, B; \Gamma)$, W_j $(j = 1, 2)$ and $\mu_{\inf}$ be as in Theorem 7.1 and let the matrices $Q_2, E_1, E_2, \Gamma_1, \Gamma_{2a}, \Gamma_{2b}$ and Λ_μ be given by (7.2) (7.7) and (7.9) where $\mu > \mu_{\inf}$. Then a matrix function $\widetilde{S}$ satisfies the two conditions*

(i) $\|\widetilde{S}\|_\infty \leq \mu$

and

(ii) $\widetilde{S} = W_1(I + PK)^{-1}W_2$ *for some stabilizing compensator K for P if and only if* $\det \widetilde{S} \not\equiv 0$ *and*

$$\widetilde{S} = \mu(\theta_{\mu 11}G + \theta_{\mu 12})(\theta_{\mu 21}G + \theta_{\mu 22})^{-1}$$

where $\theta_{\mu ij}$ $(i, j = 1, 2)$ are given by (7.10) and where G is an arbitrary $n \times n$ rational matrix function with no poles in $\overline{\Pi^+} \cup \{\infty\}$ and with $\|G\|_\infty \leq 1$.

The rest of this section is devoted to the proofs of Theorems 7.1 and 7.2. The idea of the proofs is to reduce weighted sensitivity optimization problem to the non-weighted one which was treated in Section 6. To make this idea work we have to calculate how the various components of the proofs of Theorems 6.1 and 6.2 behave under the transformation $S \longrightarrow W_1SW_2$. We start with the two-sided Lagrange-Sylvester interpolation problem (4.7)–(4.9) the solution of which is given in Theorem 4.2.

Let be given set of interpolation data

$$\omega = (C_-, C_+, A_\pi; A_\zeta, B_+, B_-; \Gamma) \tag{7.11}$$

where the matrices $C_-, C_+, A_\pi, A_\zeta, B_+, B_-, \Gamma$ are of sizes $N \times n_\pi, M \times n_\pi, n_\pi \times n_\pi, n_\zeta \times n_\zeta, n_\zeta \times M, n_\zeta \times N$, $n_\zeta \times n_\pi$, respectively, with the following properties:

(α) all eigenvalues of A_π and A_ζ are in Π^+;

(β) the pair (C_-, A_π) is observable, and the pair (A_ζ, B_+) is controllable;

(γ) the Sylvester equation

$$\Gamma A_\pi - A_\zeta \Gamma = B_+C_+ + B_-C_-$$

is satisfied.

Under these circumstances there exists an $(M+N) \times (M+N)$ rational matrix function $L(z)$ which is analytic and invertible on the imaginary axis and at infinity and having

$$\left(\begin{bmatrix} C_+ \\ C_- \end{bmatrix}, A_\pi;\ A_\zeta, [B_+ B_-]; \Gamma \right) \tag{7.12}$$

as its null-pole triple with respect to Π^+ (Theorem 2.3 in [GK]; see also [BR2]); note that such $L(z)$ is not unique. We say that $L(z)$ is *associated* with the interpolation data (7.11).

LEMMA 7.3. *Let $L(z)$ be a rational matrix function associated with (7.11). Given two rational matrix functions W_1 and W_2 of sizes $M \times M$ and $N \times N$, respectively, without poles and zeros in $\overline{\Pi^+} \cup \{\infty\}$, let*

$$\left(\begin{bmatrix} \widetilde{C}_+ \\ \widetilde{C}_- \end{bmatrix}, \widetilde{A}_\pi; \widetilde{A}_\zeta, [\widetilde{B}_+ \widetilde{B}_-]; \widetilde{\Gamma} \right) \tag{7.13}$$

be a null-pole triple over Π^+ for the product $\begin{bmatrix} W_1(z) & 0 \\ 0 & W_2^{-1}(z) \end{bmatrix} L(z)$. Then a stable $M \times N$ rational matrix function $F(z)$ satisfies the interpolation conditions

(i) $\displaystyle \frac{1}{2\pi i} \int_\gamma F(z)C_-(zI - A_\pi)^{-1} dz = C_+$

(ii) $\displaystyle \frac{1}{2\pi i} \int_\gamma (zI - A_\zeta)^{-1} B_+ F(z) dz = -B_-$

and

(iii) $\displaystyle \frac{1}{2\pi i}\int_\gamma (zI - A_\zeta)^{-1}B_+F(z)C_-(zI - A_\pi)^{-1}dz = \Gamma$

if and only if $\widetilde{F} = W_1FW_2$ *satisfies*

(iv) $\displaystyle \frac{1}{2\pi i}\int_\gamma \widetilde{F}(z)\widetilde{C}_-(zI - \widetilde{A}_\pi)^{-1}dz = \widetilde{C}_+$

(v) $\displaystyle \frac{1}{2\pi i}\int_\gamma (zI - \widetilde{A}_\zeta)^{-1}\widetilde{B}_+\widetilde{F}(z)dz = -\widetilde{B}_-$

(vi) $\displaystyle \frac{1}{2\pi i}\int_\gamma (zI - \widetilde{A}_\zeta)^{-1}\widetilde{B}_+\widetilde{F}(z)\widetilde{C}_-(zI - \widetilde{A}_\pi)^{-1}dz = \widetilde{\Gamma}.$

Here γ is a simple closed contour in the right half-plane such that $A_\pi, A_\zeta, \widetilde{A}_\pi$ and $\widetilde{A}_\zeta$ all have spectra inside γ.

Proof. For the proof we need one more construction and result from [BGR3].

Let ψ and ϕ be rational matrix functions without poles in $\overline{\Pi^+} \cup \{\infty\}$ of sizes $M \times M$ and $N \times N$, respectively, such that $(C_-, A_\pi; 0, 0; 0)$ is a null-pole triple of ϕ^{-1} over Π^+ and $(0, 0; A_\zeta, B_+; 0)$ is a null-pole triple of ψ over Π^+ (it is assumed also that ϕ and ψ have no zeros on the imaginary axis and at infinity).

Let stable K be any interpolating rational matrix function satisfying (i), (ii) and (iii). Then by Theorem 5.2.3 in [BGR3] the interpolation conditions (i), (ii) and (iii) for stable F are equivalent to F having the form

$$F = K + \psi Q \phi \tag{7.14}$$

for some stable Q. Introduce the function

$$L = \begin{bmatrix} \psi & K\phi^{-1} \\ 0 & \phi^{-1} \end{bmatrix}$$

It is easy to see that the formula (7.14) gives precisely the same functions as the formula (4.12) (with $\theta_{11} = \psi$, $\theta_{12} = K\phi^{-1}, \theta_{21} = 0, \theta_{22} = \phi^{-1}$ and $Q_1Q_2^{-1} = Q$). It follows from Lemma 5.2.5 in [BGR3] that (7.12) is a null-pole triple over Π^+ for L. Next note that stable F satisfies (7.14) if and only if $\widetilde{F} = W_1FW_2$ has the form

$$\widetilde{F} = \widetilde{K} + \widetilde{\psi}Q\widetilde{\phi}$$

for a stable Q, where

$$\widetilde{K} = W_1KW_2; \widetilde{\psi} = W_1\psi; \widetilde{\phi} = \phi W_2.$$

When we form the associated matrix function $\widetilde{L} = \begin{bmatrix} \widetilde{\psi} & \widetilde{K}\widetilde{\phi}^{-1} \\ 0 & \widetilde{\phi}^{-1} \end{bmatrix}$ from this transformed data, the result is

$$\widetilde{L} = \begin{bmatrix} W_1 & 0 \\ 0 & W_2^{-1} \end{bmatrix} L.$$

It remains to apply Theorem 4.2 again. □

We are now in a position to prove Theorems 7.1 and 7.2.

Proof of Theorem 7.1. By Theorem 4.1 a rational $n \times n$ matrix function S has the form $S = (I + PK)^{-1}$ for a stabilizing compensator K for P if and only if S is stable and satisfies the interpolation conditions (4.1)-(4.3).

We intend to use Lemma 7.3 to construct the interpolation conditions for the function $W_1(\mu^{-1}S)W_2$, here $\mu > 0$ is a parameter. So we need to compute a null-pole triple for $\widetilde{L} \stackrel{\text{def}}{=} \begin{bmatrix} W_1 & 0 \\ 0 & W_2^{-1} \end{bmatrix} \cdot L(z)$, where $L(z)$ has

$$\left(\begin{bmatrix} 0 \\ C \end{bmatrix}, A_\pi; A_\zeta, [B, -\mu^{-1}B]; -\mu^{-1}\Gamma \right)$$

as a null-pole triple over Π^+. As the realizations (7.1) for W_1 and W_2 are minimal, the realization

$$\begin{bmatrix} W_1(z) & 0 \\ 0 & W_2^{-1}(z) \end{bmatrix} = I + \begin{bmatrix} C_1 & 0 \\ 0 & -C_2 \end{bmatrix} \left(zI - \begin{bmatrix} A_1 & 0 \\ 0 & A_2 - B_2C_2 \end{bmatrix} \right)^{-1} \begin{bmatrix} B_1 & 0 \\ 0 & B_2 \end{bmatrix}$$

is minimal as well. We apply Lemma 3.3 to find a null-pole triple for $\widetilde{L}$ over Π^+; so define matrices Q and E which are solutions to the equations

$$QA_\pi - \begin{bmatrix} A_1 & 0 \\ 0 & A_2 - B_2C_2 \end{bmatrix} Q = \begin{bmatrix} B_1 & 0 \\ 0 & B_2 \end{bmatrix} \begin{bmatrix} 0 \\ C \end{bmatrix} \tag{7.15}$$

and

$$E \begin{bmatrix} A_1 - B_1C_1 & 0 \\ 0 & A_2 \end{bmatrix} - A_\zeta E = [B, -\mu^{-1}B] \begin{bmatrix} C_1 & 0 \\ 0 & -C_2 \end{bmatrix} \tag{7.16}$$

Write Q as $\begin{bmatrix} Q_1 \\ Q_2 \end{bmatrix}$ and E as $[E_{1\mu}, E_{2\mu}]$ We deduce from (7.15) that $Q_1 = 0$ and Q_2 is given by (7.2), and from (7.16) we deduce that $E_{1\mu} = E_1$, where E_1 is given by (7.3), and $E_{2\mu} = \mu^{-1}E_2$, where E_2 is given by (7.4). By Lemma 3.3 we conclude that a null-pole triple for $\widetilde{L}$ over Π^+ is given by

$$\left(\begin{bmatrix} C_1 & 0 \\ 0 & -C_2 \end{bmatrix} \begin{bmatrix} 0 \\ Q_2 \end{bmatrix} + \begin{bmatrix} 0 \\ C \end{bmatrix}, A_\pi; \right.$$

$$\left. A_\zeta, [E_1, \mu^{-1}E_2] \begin{bmatrix} B_1 & 0 \\ 0 & B_2 \end{bmatrix} + [B, -\mu^{-1}B]; [E_1, \mu^{-1}E_2] \begin{bmatrix} 0 \\ Q_2 \end{bmatrix} - \mu^{-1}\Gamma \right)$$

$$= \left(\begin{bmatrix} 0 \\ -C_2Q_2 + C \end{bmatrix}, A_\pi;\ A_\zeta, [E_1B_1 + B, \mu^{-1}E_2B_2 - \mu^{-1}B]; \right.$$

$$\left. -\mu^{-1}E_2Q_2 - \mu^{-1}\Gamma \right).$$

We conclude by Lemma 7.3 that S is a stable sensitivity for P if and only if the function $F = \mu^{-1}W_1SW_2$ is stable and satisfies the interpolation conditions

$$\sum_{z_0\in\Pi+} \mathrm{Res}_{z=z_0} F(z)(-C_2Q_2+C)(zI-A_\pi)^{-1} = 0;$$
$$\sum_{z_0\in\Pi+} \mathrm{Res}_{z=z_0}(zI-A_\zeta)^{-1}(E_1B_1+B)F(z) = -\mu^{-1}E_2B_2+\mu^{-1}B;$$
$$\sum_{z_0\in\Pi+} \mathrm{Res}_{z=z_0}(zI-A_\zeta)^{-1}(E_1B_1+B)F(z)(-C_2Q_2+C)(zI-A_\pi)^{-1} = -\mu^{-1}E_2Q_2-\mu$$

Next, plug these data into Theorem 5.1 to compute the matrix (5.6) for this interpolation problem; the result is

$$\Lambda_\mu = \begin{bmatrix} \Gamma_{1\mu} & -\mu^{-1}(Q_2^*E_2^*E_2+\Gamma^*) \\ -\mu^{-1}(E_2Q_2+\Gamma) & \Gamma_{2\mu} \end{bmatrix} \tag{7.17}$$

where

$$\Gamma_{1\mu}A_\pi + A_\pi^*\Gamma_{1\mu} = (C^*-Q_2^*C_2^*)(C-C_2Q_2)$$
$$\Gamma_{2\mu}A_\zeta^* + A_\zeta\Gamma_{2\mu} = (E_1B_1+B)(B_1^*E_1^*+B^*)$$
$$-(\mu^{-1}E_2B_2-\mu^{-1}B)(\mu^{-1}B_2^*E_2^*-\mu^{-1}B^*)$$

Thus $\Gamma_{1\mu} = \Gamma_1$ is independent of μ and is given by (7.5), and $\Gamma_{2\mu}$ can be written as

$$\Gamma_{2\mu} = \Gamma_{2a} - \mu^{-1}\Gamma_{2b}$$

where Γ_{2a} and Γ_{2b} are given by (7.6) and (7.7), respectively. Plugging these expressions into (7.17) we see that Λ_μ is also as in (7.9). To test when Λ_μ is positive definite, since Γ_1 is positive definite it suffices to check when the Schur complement

$$\Gamma_{2a} - \mu^{-2}\Gamma_{2b} - \mu^{-2}(\Gamma - P_2Q_2)\Gamma_1^{-1}(\Gamma - P_2Q_2)^*$$

of Γ_1 is positive definite. By Theorem 5.1 this yields the formula for $\mu_{\inf}$ in theorem 7.1 and complete the proof of Theorem 7.1.

Proof of Theorem 7.2. This is just a continuation of the proof of Theorem 7.1. Observe that $\theta_\mu(z)$ given by (7.10) is a J-unitary on the imaginary axis having

$$\left(\begin{bmatrix} 0 \\ -C_2Q_2+C \end{bmatrix}, A_\pi; A_\zeta, [E_1B_1+B, \mu^{-1}E_2B_2-\mu^{-1}B]; -\mu^{-1}E_2Q_2-\mu^{-1}\Gamma\right)$$

as its null-pole triple over Π^+ (see formula (5.9)). Now Theorem 7.2 follows by combining Theorem 5.1 and Lemma 5.3. □

REFERENCES

[AG] D. Alpay, D. Gohberg, *Unitary rational matrix functions and orthogonal matrix polynomials*, to appear in Operator Theory: Advanced and Applications.

[BCR] J.A. Ball, N. Cohen, A.C.M. Ran, *Inverse spectral problems for regular improper rational matrix functions*, Topics in Interpolation Theory of Rational Matrix-valued Functions (ed. I. Gohberg), Birkhaüser, Basel (1988), pp. 123–173.

[BGR1] J.A. Ball, I. Gohberg, L. Rodman, *Realization and Interpolation of rational of rational matrix functions*, Operator Theory: Advances and Applications, 33 (1988), 1–72.

[BGR2] ——————, *Minimal factorizations of meromorphic matrix functions in terms of local data*, Integral Equations and Operator Theory 10 (1987), 309–348.

[BGR3] ——————, *Two-sided Lagrange-Sylvester interpolation problems for rational matrix functions*, submitted to Proceedings of the AMS Summer Research Institute in Operator Theory and Operators Algebras, Durham, New Hampshire (1988).

[BGR4] ——————, *Interpolation problems for rational matrix functions*, monograph in preparation.

[BGR5] ——————, *Two-sided Nudelman interpolation problem for rational matrix functions*, submitted.

[BH] J.A. Ball and J.W. Helton, *A Beurling-Lax Theorem for the Lie group $U(m,n)$ which contains most classical interpolation*, J. Operator Theory 9 (1983), 107–142.

[BL] J. A. Ball and D.W. Luse, *Sensitivity minimization as a Nevanlinna-Pick interpolation problem*, in: Modelling, Robustness and Sensitivity Reduction in Control Systems, ed. R.F. Curtain, NATO ASI, Springer-Verlag (1987), 451–462,.

[BR1] J.A. Ball and A.C.M. Ran, *Local inverse spectral problems for rational matrix functions*, Integral Equations and Operator Theory 10 (1987), 349–415.

[BR2] ——————, *Global inverse spectral problems for rational matrix functions*, Linear Algebra and Applications 86 (1987), 237–282.

[BGK] H. Bart, I. Gohberg and M.A. Kaashoek, *Minimal Factorization of Matrix and Operator Functions*, Birkhaüser, Basel (1979).

[DFGK] J. Doyle, B.A. Francis, K. Glover, P. Khargonekar, *State-space solutions to standard H^2 and H^∞ control problems*, proceedings of the American Control Conference (1988).

[F1] I.I. Fedchina, *Tangential Nevanlinna-Pick problem with multiple points*, Akad. Nauk Armjan. SSR Dokl 61 (1975), 214–218 [in Russian].

[F2] ——————, *Description of solutions of the tangential Nevanlinna-Pick problem*, Akad. Nauk Armjan. SSR Dokl. 60 (1975), 37–42 [in Russian].

[Fr] B. Francis, *A course in H_∞ Control Theory*, Lecture Notes in Control and Information Sciences #88, Springer-Verlag (1987).

[GK] I. Gohberg, M.A. Kaashoek, *An inverse spectral problem for rational matrix functions and minimal divisibility*, Integral Equations and Operator Theory 10 (1987), 437–465.

[GKLR] I. Gohberg, M.A. Kaashoek, L. Lerer and L. Rodman, *Minimal divisors of rational matrix functions with prescribed zero and pole structure*, in: Topics in Operator Theory, Systems and Networks (eds. H. Dym, I. Gohberg), Operator Theory: Advances and Applications, 12 (1984), Birkhaüser Verlag, Basel, pp. 241–275.

[K] H. Kimura, *Directional interpolation in the state space*, Systems and Control Letters 10 (1988), 317–324.

[LA] D.J.N. Limebeer and B.D.O. Anderson, *An interpolation theory approach to H^∞ controller degree bounds*, Linear Algebra Appl., (to appear).

[LT] P. Lancaster, M. Tismenetsky, *The theory of matrices with applications*, Academic Press (1985).

[N1] A.A. Nudelman, *On a new problem of moment problem type*, Soviet Math. Doklady 18 (1977), 507–510 [Doklady Akademii Nauk SSSR (1977)].

[N2] ——————, *A generalization of classical interpolation problems*, Soviet Math Doklady 23 (1981), 125–128 [Doklady Akademii Nauk SSSR (1981)].

[R] A.C.M. Ran, *State space formulas for a model matching problem*, preprint.

[YBL] D.C. Youla, J. Bongiorno, Y. Lu, *Single loop feedback stabilization of linear multivariable dynamic plants*, Automatica 10 (1974), 151–170.

UNIFORM BOUNDED INPUT – BOUNDED OUTPUT STABILIZATION OF NONLINEAR SYSTEMS*

CHRISTOPHER I. BYRNES† AND ALBERTO ISIDORI‡

Dedicated to our friend and teacher
Roger Brockett
on the occasion of his fiftieth birthday

1. Introduction. Achieving some form of closed-loop stability for nonlinear feedback systems has been the object of intense research for many decades. Yet, despite remarkable advances using for example an operator-theoretic point of view to obtain generalizations of various classical graphical stability criteria (see e.g. [1]–[4]), even the basic problem of feedback stabilization about an equilibrium was recently called "by far the most important open problem...in nonlinear control" in the consensus report [5] of the IEEE Santa Clara meeting on future directions in systems and control. Following closely on the heels of the encouraging development of exact linearization techniques using nonlinear feedback and coordinate transformations (see [6]–[9]), research on nonlinear feedback stabilization has attracted considerable interest during the past five years. In fact, there are now some general methods for deriving feedback laws locally and globally stabilizing some rather broad classes of nonlinear affine control systems

$$\begin{aligned} \dot{x} &= f(x) + g(x)u \\ y &= h(x) \end{aligned} \tag{1.1}$$

In this paper, we present an extension of some of these results, yielding solutions to the problem of uniform BIBO stabilization of systems (1.1) using smooth state feedback laws of the standard form

$$u = \alpha(x) + \beta(x)v \tag{1.2}$$

Since our results will also imply asymptotic stability, according to Milnor [10] we may take, without loss of generality, our state space to be diffeomorphic with $\mathcal{R}^n$. Matters being so, we also assume that (1.1) has an equilibrium at 0 and adopt the standard Euclidean norm to express our estimates. In this setting, by uniform BIBO stability we will mean that there exist positive constants c_1, c_2 such that the closed-loop trajectories $x(t)$ satisfy

$$\|x\|_\infty \leq c_1\|v\|_\infty + c_1\|x_0\| \tag{1.3}$$

*Research supported in part by grants from AFOSR, NSF and Ministero della Publica Instruzione.

†Department of Systems Science and Mathematics, Washington University, St. Louis, MO 63131 USA.

‡Istituto di Informatica e Sistemistica, Universita di Roma – "La Sapienza" Via Eudosiana 18, 00184 Roma, Italy.

If we take $x_0 = 0$, this is stronger than but related to the classical definition of uniform BIBO stability for linear time-varying systems (see [11]). If $x_0 \neq 0$ is fixed (1.3) takes the form

$$\|x\|_\infty \leq \alpha \|v\|_\infty + \beta$$

which plays a central role in a general treatment of the small gain theorem, which of course implies certain uniform BIBO stabilization results as special cases. Our first main BIBO stabilization result establishes (1.3) for a broad class of nonlinear systems (1.1), roughly speaking, for nonlinear minimum phase systems having relative degree one. In fact we show more, there exists a feedback law (1.2) so that the closed-loop trajectories satisfy

$$\|x(t)\| \leq c_1 \|v\|_\infty + 0(e^{-\gamma t}\|x_0\|) \tag{1.4}$$

Estimates such as (1.4) are familiar in the linear case and hence must hold for systems which are globally linearizable by feedback transformations (1.2). Our first result, however, does not require, and in fact is quite far from, feedback linearizability. Our second result obtains (1.4) Estimates such as (1.4) are familiar in the linear case and hence must hold for systems which are globally linearizable by feedback transformations (1.2). Our first result, however, does not require, and in fact is quite far from, feedback linearizability. Our second result obtains (1.4) for systems having relative degree $r, 1 \leq r \leq n$, but which satisfy an additional involutivity hypothesis. For the extreme cases $r = 1$ and $r = n$ this condition is vacuous, the case $r = n$ corresponding however to systems which are feedback linearizable. In the intermediate cases, this condition is equivalent to the existence of a special normal form used by several other authors in feedback stabilization schemes. We conjecture that this condition is in fact not necessary and could be supplemented by weaker hypotheses.

In section 2 we give a careful discussion of various notions of BIBO stability. Section 3 provides preliminary material on recent work on feedback stabilization, in particular stating results (Theorem 3.3) on asymptotic stabilization for compact sets of initial data and on BIBO stabilization for small inputs. Section 4 beings with an application of Theorem 3.3, proving global asymptotic null controllability for certain nonlinear minimum phase systems. The main result, Theorem 4.2, provides a partial converse to this "open-loop" result, by showing existence of globally asymptotically stabilizing smooth feedback laws for certain nonlinear systems. In section 5, we apply these results to obtain the results on uniform BIBO stabilization discussed above.

We want to thank Giorgio Picci for very insightful and helpful conversations and remarks.

2. BIBO Stability for Nonlinear Systems. In some of some of the early treatments of BIBO stability for linear systems, there was confusion between the property that bounded inputs would yield boundded outputs and the property that the corresponding map between the appropriate L^∞ space be continuous. In the operator theoretic treatments of stability for linear and for nonlinear systems this question was treated very elegantly. However, for nonlinear systems such as (1.1) there emerges yet another distinction: Continuity at $u(t) \equiv 0$ versus continuity as a globally defined mapping on L^∞. Actually, both concepts are important, sometimes in different contexts, and in this section we delineate three definitions reflecting what we feel are useful variations on this theme. Throughout we will concentrate on the mapping from control functions $u(\cdot)$ to the corresponding state trajectory $x(\cdot)$.

DEFINITION 2.1. The system (1.1) is bounded input-bounded output stable for small inputs and initial data x_0 provided for all $\epsilon > 0$ there exists $\delta > 0$ such that

$$\|u\|_\infty < \delta \Rightarrow \|x(t)\| < \epsilon \quad \text{for } t \gg 0. \tag{2.1}$$

We remark that this notion is closely related to the notion of uniform BIBO stability for time-varying linear systems (see [11]) except that we do not require $x_0 = 0$. For this reason, we cannot of course ask that $\|x\|_\infty < \epsilon$, as is required in the standard definition which is equivalent to continuity of the input-state trajectory map at $u = 0$. This property can be checked under various conditions (as in [12]) related to Volterra series representations of $x(t)$ in terms of the control $u(t)$, see especially Remark 3.4. On the other hand, continuity at $u = 0$ does not provide for an analysis of the system response to large inputs, e.g. to large but bounded disturbances. For this reason, we introduce a second notion of BIBO stability.

DEFINITION 2.2. The system (1.1) is uniformly BIBO stable provided there exist constants $c_1, c_2 > 0$ such that

$$\|x\|_\infty \le c_1\|u\|_\infty + c_2\|x_0\|_\infty \tag{2.2}$$

If one were to fix the initial condition x_0, the estimate (2.2) would take the form

$$\|x\|_\infty \le \alpha\|u\|_\infty + \beta \tag{2.3}$$

which is the condition used in [4] to establish the small-gain theorem [3]. Of course, Definitions 2.1–2.2 could be modified, as in [4], by adding a constant in order to accommodate nonsmooth nonlinearities such as limiters, etc. On the other hand, for smooth f, g and h in (1.1) we are able in section 5 to derive feedback laws achieving uniform BIBO stability in the stronger sense of Definition 2.2.

Definitions 2.1–2.2 express forms of continuity, at $u = 0$, for equilibrium solution $x(t) \equiv 0$ of (1.1). One would certainly want a notion of BIBO stability for systems which are stable or which have been stabilized about attractors more general than equilibria (see e.g. [13]–[16]), where the origin does not play a special role.

DEFINITION 2.3. The system (1.1) is BIBO stable for initial data x_0 provided bounded, measurable inputs produce bounded, measurable state trajectories and the corresponding mapping

$$T_{x_0} : L^\infty((0,\infty); \mathcal{R}^m) \to L^\infty((0,\infty); \mathcal{R}^n)$$

is continuous. the system is globally BIBO stable provided the mapping

$$T : \mathcal{R}^n \times L^\infty((0,\infty); R^m) \to L^\infty((0,\infty); \mathcal{R}^n),$$

defined via $T(x_0, u) = T_{x_0}(u)$, is continuous.

While our main results are concerned with uniform BIBO 'stability, we expect to have more to say about global BIBO stability in the future paper.

3. Recent Results in Feedback Stabilization. In this section we present some preliminary results on nonlinear feedback stabilization which motivate, or are closely related to, the feedback design techniques we shall use in section 4 for global stabilization of certain nonlinear systems. We begin with a discussion of the local case.

It has long been a central problem of significant importance in nonlinear control theory to design a state feedback law, $u = \alpha(x)$, which renders the origin of the system

$$\dot{x} = f(x) + g(x)u, \ \ f(0) = 0 \tag{3.1}$$

locally (or globally) asymptotically stable. Here, $x \in \mathcal{R}^n$ and $g(x) = \big(g_1(x), \ldots, g_m(x)\big)$ and $u = \mathrm{col}(u_1, \ldots, u_m)$, but, without significant loss of generality or complexity, we may focus on the case $m = 1$. If the linearization of (3.1) near 0 is controllable, then the local stabilization problem is of course trivial so that most current research has focused on the uncontrollable case, with special attention devoted to the "critical" case. We note that (3.1) can still be locally controllable as a nonlinear system, but it is now known that for nonlinear systems stabilizability and properties such as local controllability or local reachability are quite distinct, in contrast to the linear case (see [17]–[19]). In particular, [19] shows that even commonly used models for rigid spacecraft can be controllable in every desired sense yet may not be locally asymptotically stabilizable. One of the techniques used in [19] is a necessary condition discovered by Brockett [18], see also [20] for a degree theoretic criterion. The paper [18] also discusses sufficient conditions for local asymptotic stabilization and is now part of a sizable literature on feedback stabilization which is quite divorced from nonlinear controllability and, in particular, from nonlinear feedback linearizability.

In [18], Brockett showed, in terms of a "finite gain" hypotheses, that the existence of both an invariant manifold on which the closed-loop flow was asymptotically stable and a certain candidate Lyapunov function would imply local asymptotic stabilizability. As developed in Aeyels [21] and Crouch-Irving [22], the finite gain hypotheses can be significantly streamlined using center manifold methods, but the question of existence of suitable invariant manifolds remains a crucial open problem. By the implicit function theorem, any invariant manifold must be defined by (independent) smooth constraints

$$h_1(x) = \cdots = h_l(x) = 0 \tag{3.2}$$

while the invariance condition is simply that

$$\langle dh_i, f\rangle = -\langle dh_i, g\rangle u \qquad \text{wherever (3.2) holds.} \tag{3.3}$$

Beginning in 1984, the authors developed a general procedure for designing stabilizing feedback laws for input-output systems (1.1) which, in this context, interprets the constraints (3.2) as being either given by or implied by the output constraint

$$y(t) = 0 \tag{3.4}$$

while the control laws satisfying (3.3) are derived using a nonlinear form of Silverman's Algorithm [23]–[25], [13]–[16]. We stress that these methods can, of course, be used in two settings: Where the output y represents a measured variable or where no output functions is specified, so that a choice of y can be viewed as a design option similar, for example, to sensor location and design. In more detail, denoting by $D_X H$ the directional derivative of the function H in the direction determined by the vector field X, recall [26] that the relative degree, r of (1.1) at x_0 is defined by

$$\begin{aligned} &D_g h(x) = \cdots = D_g D_f^{r-2} h(x) = 0 \qquad \text{for } x \text{ near } x_0 \\ &D_g D_f^{r-1} h(x_0) 0. \end{aligned} \tag{3.5}$$

If r is finite and independent of x_0, we say (1.1) has uniform relative degree r at x_0. In this case, repeated differentiation along the trajectories of (1.1) shows that the quantities

$$\begin{aligned} y_1 = y \qquad &= h \\ y_r = y^{(r-1)} \quad &= D_f^{r-1} h \end{aligned} \tag{3.6}$$

are state variable with independent differentials $dy_1, \ldots, dy_r$. Assuming, without loss of generality, that $h(0) = 0$, we observe that the constraint (3.4) implies the constraint

$$y_1(x) = \cdots = y_r(x) = 0 \tag{3.7}$$

which defines a smooth submanifold, compare (3.2). Moreover, the invariance conditions (3.3) take the simple form

$$D_f y_i = -D_g y_i \cdot u \tag{3.8}$$

which are trivially satisfied when $i = 1, \ldots, r-1$, while for $i = r$(3.8) takes the form

$$u = \left(D_g D_f^{r-1} h\right)^{-1} D_f^r h \tag{3.8$'$}$$

The closed-loop system (1.1)–(3.8)$'$ leaves the constraint submanifold Z^* defined by (3.7) invariant and therefore defines an autonomous dynamical system

$$\dot{z} = (f + gu)(z), \quad z \epsilon Z^* \tag{3.9}$$

which we refer to as the *zero dynamics*, for reasons which will become clear shortly.

REMARK 3.1. It is possible to define zero dynamics more generally, i.e. without strong relative degree or equilibrium assumptions, see [27]–[28], [13] for alternative treatments which have evolved from the local, scalar case to the global, multivariable case, as well as for algorithms for computing Z^* and the zero dynamics vector field.

Our first result concerns local asymptotic stabilization about 0 when the system (1.1) is "locally minimum phase;" i.e. when the zero dynamics is locally asymptotically stable. This includes the critical case.

THEOREM 3.2 [29]. *If (1.1) has locally asymptotically stable zero dynamics, then there exists a state feedback law $u = \alpha(x)$, for which the closed-loop system is locally asymptotically stable.*

If the system has uniform relative degree r, one can be much more explicit about the design of stabilizing control laws and about the global asymptotic behavior of the closed-loop system. In this case, one can augment the partial local coordinate system (3.6) by including functions $z_1, \ldots z_{n-r}$ satisfying

(i) $\dim sp\{dz_1, \ldots, dz_{n-r}, dy_1, \ldots, dy_r\} = n$

(ii) $\langle dz_i, g\rangle = 0$.

thereby obtaining the following normal form for (1.1)

$$\begin{aligned} \dot{z} &= f_0(z, y_i) \\ \dot{y}_1 &= y_2 \\ &\vdots \\ \dot{y}_{r-1} &= y_r \\ \dot{y}_r &= D_f^r h(z, y_i) + D_g D_f^{-1} h(z, y)_i)u \end{aligned} \tag{3.10}$$

We note that in this local normal form, the constraint $y(t) = 0$ yields the constraint (3.7) and hence an explicit, but local form, of the zero dynamics

$$\dot{z} = f_0(z, 0) \tag{3.11}$$

In order for (3.10), and hence (3.11), to be globally valid we make the following assumption

(**H1**) The vector fields $g, ad_f g, \ldots, ad_f^{r-1} g$ are complete.

In [13] it is shown that (H1) and the condition that the level set (3.7) be connected is sufficient for the existence of a global change of coordinates to normal form. Connectivity will follow, for example, whenever the zero dynamics is globally asymptotically stable. In the (invertible) linear case (3.7) defines a subspace, viz. $V^*(\operatorname{Ker} C)$, and (3.11) is an autonomous linear system whose spectrum coincide with the transmission zeroes (see e.g. [30]). For this reason we shall say that (1.1) is (exponentially) minimum phase provided the zero dynamics is globally (exponentially) asymptotically stable.

We now want to derive a refinement of the feedback law/constraint law (3.8) which will stabilize minimum phase systems (1.1) having uniform relative degree r for large sets of initial data. For stabilization "in the large", we will need an auxiliary "nonspeaking" condition

(H2) There exists a normal form representation for (1.1) such that, for any compact set V of points (z_1, z_2, η, ξ) in $Z^* \times Z^* \times \mathbf{R}^r \times \mathbf{R}^r$ there exists constants $c(v), \alpha(v)$ with $c(v) > 0$ and $0 \leq \alpha(v) < 1$ such that

$$\|f_0(z_1, K^{j-1}\eta_j) - f_0(z_2, K^{j-1}\xi_j)\| < c(v)K^{\alpha(v)}(\|z_1 - z_2\| + \|\eta - \xi\|) \tag{H2}$$

We note that since we allow c and α to depend on v, the nonpeaking condition does not follow from a global Lipschitz condition on $f_0(z, \xi)$. We also note that weaker conditions also suffice for our stabilization results, e.g. under the two conditions

NP(1) $\|f_0(z, K^{j-1}\xi_j) - f_0(z, 0)\| \leq c(v)K^{\alpha(v)}\|\xi\|$

NP(2) $\|f_0(z_1, K^{j-1}\xi_j) - f_0(z_2, K^{j-1}\xi_j)\| \leq c(v)K^{\alpha(v)}\|z_1 - z_2\|$

Theorem 3.3 will also hold.

If

$$p(s) = s^r + c_1 s^{r-1} + \cdots + c_r$$

and k is a positive constant, we set

$$p_k(s) = s^r + c_1 k s^{r-1} + \cdots + c_r k^r.$$

We consider refinements of the basic feedback law (3.8)'

$$u = -\left(D_h D_f^{r-1} h\right)^{-1} (p_k(D_f)h) \tag{3.12}$$

$$u = -\left(D_h D_f^{r-1} h\right)^{-1} (p_k(D_f)h + v) \tag{3.13}$$

In [13]–[16], (see also [23] we have proved the following stabilization results for compact sets of initial data using fixed point methods generalizing the basic arguments in center manifold theory.

THEOREM 3.3. *Suppose (1.1) has uniform relative degree r, is minimum phase and satisfies hypotheses (H1)-(H2). Moreover, suppose $p(s)$ is a Hurwitz polynomial of degree r and choose $U \subset \Re^n$ to be any relatively compact set of initial data.*

(i) *There exists a k_U such that for any $k > k_U$ the feedback law (3.12) drives all closed-loop trajectories initialized in U asymptotically to 0.*

(ii) *If (1.1) is exponentially stable, there exists a k_U such that for any $k > k_U$ the closed-loop system (1.1)–(3.13) is BIBO stable for small inputs, for arbitrary initial data in U.*

REMARK 3.4. The result on BIBO stabilization for small inputs follows from exponential stabilization on a compact set U of initial data. In particular, the spectrum of the Jacobian of the closed-loop drift vectorfield at 0 lies in the left-half plane. For $U = \{0\}$, one can then derive BIBO stability for small inputs from a Volterra series argument, under various additional hypotheses. For example, Brockett [12] has shown that if f, g have a complex analytic extension to $\mathbf{C}^n$, then the Volterra series with $x_0 = \{0\}$ has a non-zero radius of convergence on $L^\infty((0,\infty), \mathcal{R}^m)$ thereby implying BIBO stability for small inputs.

REMARK 3.5. In [15]–[16], BIBO stabilization for small inputs is proved by a different method. It is shown that the system, driven by a sufficiently small input, has a zero dynamics with a compact attractor V. Then, part (i) of Theorem 3.3 is established in the non-equilibrium case, giving positive results concerning stabilization about an attractor. This result promises to be considerable independent interest.

REMARK 3.6. In section 4 under an additional hypothesis which is superfluous in the cases $r = 1, r = n$ we can obtain Theorem 3.3 for $U = \mathcal{R}^n$, i.e. we obtain global stabilization in part (i), and improve part (ii) by obtaining uniform BIBO stabilization.

REMARK 3.7. Setting $k = 1$ in (3.12), we obtain explicit formulae for a family of locally asymptotically stabilizing laws. Under relative degree hypotheses, similar results were obtained for the local case, apparently independently, by Marino [31] using singular perturbation/highgain methods. Local results have also been obtained using these methods by Khalil-Saberi [32] for a more restrictive class of systems, assumed to be in the normal form

$$\begin{aligned}
\dot{z} &= f_0(z, y_1) \\
\dot{y} &= y_2 \\
&\vdots \\
\dot{y}_{r-1} &= y_r \\
\dot{y}_r &= f_1(z, y_i) + g_1(z, y_i)u, \;\; g(0,0) \neq 0.
\end{aligned} \tag{3.14}$$

While for linear systems one can always assume that only y_1 influences as the first equation, this is however a nontrivial assumption in the nonlinear case if $1 < r < n$, see section 4. Assumptions concerning the existence of this restricted normal form also plays an important role in [33], which derives a local result closely related to Theorem 3.2 in the spirit of Brockett [18] and Aeyels [21] using the existence of a suitable invariant submanifold (3.2) in lieu of conditions on the zero constraint manifold (3.7). We remark, however, that if the zero dynamics (or slow dynamics) is not critical, the local case follows trivially from a linear argument, while there exist counterexamples to high output gain stabilization, even locally, in the critical case [34]. This is in sharp contrast to the linear case.

4. Null Controllability and Global Feedback Stabilization. Our first observation is based on the fact that the compact set of initial data in Theorem

3.3(i) is arbitrary; i.e., for all initial data $x_0 \in \mathcal{R}^n$ there exists an open loop control $u(t)$ such that the corresponding state trajectory satisfies

$$\lim_{t \to \infty} x(t) = 0$$

PROPOSITION 4.1. *Any minimum phase system with uniform relative degree r, satisfying (H1)-(H2), is globally asymptotically null controllable.*

REMARK 4.2. An inspection of the formal form (2.3) shows that this property has very little to do with controllability or reachability of (2.1) or with Lie brackets, accessibility rank conditions, etc. One interpretation of this is by analogy with the Kalman normal form; i.e., the high frequency (or y_i) dynamics in (2.3) are of course controllable while the zero dynamics must contain all the "uncontrollable modes" (see [34]).

These observations lead to the question of whether minimum phase systems can in fact be globally stabilized by some fixed choice of a globally defined state feedback law, as has recently been shown in the relative degree one case [34]. A recent unpublished example due to H.J. Sussman shows that this is not the case, so that additional hypotheses - such as (H2) - are required. In order to give a partial solution to this problem, we will need to fix some notation concerning the special normal form (3.14) discussed in connection with local stabilization, see Remark 3.7. We note that a necessary condition for (1.1) to be expressible in the form (3.14) is:

(**H3**) The distribution $\Delta_r = sp\{g, ad_f g, \ldots, ad_f^{r-1} g\}$ has constant dimension r and is involutive.

REMARK 4.3. Condition (H3) is, of course, vacuous in case $r = 1$. It is also vacuous if $r = n$ which implies feedback linearizability of (1.1). Our next result shows that in these two extreme cases, as well as in intermediate cases $1 < r < n$ when (H3) is satisfied, there exists globally stabilizing state feedback laws, vastly generalizing the results discussed in section 3. We note that (H3) implies the nonpeaking condition (H2).

THEOREM 4.4. *Suppose the system (1.1) is minimum phase, has uniform relative degree r and satisfies hypotheses (H1), (H2). Then, there exists a smooth, globally defined state feedback law for which the closed-loop system is globally asymptotically stable. Furthermore, if (1.1) is exponentially minimum phase, the closed-loop system is also exponentially stable.*

Proof. We first consider the relative degree one case. Following [34], we refine the normal form (2.3) by expanding $f_0(z, y)$ using the remainder theorem

$$\begin{aligned} \dot{z} &= f_0(z,0) + yR(zy) \\ \dot{y} &= D_f h(zy) + D_g h(z,y) \cdot u \end{aligned} \tag{4.1}$$

Let $V(z)$ be a Lyapunov function for the zero dynamics and construct the feedback law

$$u = (D_g h^{-1}(-D_f h - y - \langle \operatorname{grad} V, R(z,y)\rangle) \tag{4.2}$$

Using the Lyapunov function $W(z,y) = V(z) + \frac{1}{2}y^2$ it is clear that (4.1)–(4.2) is globally asymptotically stable. Moreover, if the zero dynamics is exponentially stable, exponential stability of (4.2)–(4.3) is immediate.

Now consider the case $r \geq 1$. According to ([13]), Proposition 5.4), (H3) implies that there is a global change of coordinates to the refined normal form (3.14). We shall prove that there exists a new output so that the system takes the form (3.14), but with relative degree $r = 1$, so that the construction of a globally stabilizing feedback law follows inductively from the case $r = 1$.

Explicitly, consider

$$\tilde{y}_2 = y_1 + y_2 + \langle \operatorname{grad} V, R(z,y)\rangle$$

where V and R are defined as above. After a nonsingular change of coordinates, (3.14) takes the form

$$\begin{aligned} \dot{z} &= f_0(z,0) + y_1 R(z,y_1) \\ \dot{y}_1 &= \tilde{y}_2 - y_1 - \langle \operatorname{grad} V(z), R(z,y_1)\rangle \\ \dot{\tilde{y}}_2 &= \tilde{y}_3 \\ \vdots &= \\ \dot{\tilde{y}}_r &= f_1(z,y_1,\tilde{y}_i) + g_1(z,y_1,\tilde{y}_i)\cdot u \end{aligned} \tag{4.3}$$

With $\tilde{y}_2$ as an output, (4.3) has relative degree $r-1$ with zero dynamics

$$\begin{aligned} \dot{z} &= f_0(z,0) + y_1 R(z,y_1) \\ \dot{y}_1 &= -y_1 - \langle \operatorname{grad} V(z), R(z,y_1)\rangle \end{aligned} \tag{4.4}$$

coinciding with the globally asymptotically stable closed-loop system considered in our analysis of the case $r = 1$.

□

5. Uniform BIBO Stabilization. We now apply the global stabilization results obtained in section 4 to derive our main results on BIBO stabilization of certain classes of scalar input-scalar output nonlinear systems, see however Remark 5.3. These include positive results for uniform BIBO stabilization of exponentially minimum phase systems having relative degree one and of systems with relative degree n. To say a system has uniform relative degree n is to say it is globally feedback linearizable, so that uniform BIBO stabilization follows immediately from the linear case. Such systems are well-known to be very special if $n \geq 3$, comprising a set of infinite codimension in the Whitney on C^∞-topology [35]. On the other extreme, exponentially minimum phase systems with uniform relative degree one, satisfying (H1), contain an open set of nonlinear systems with a complete control field g, in the Whitney C^∞-topology, and also contains an open set of unstable systems. In the intermediate cases, $1 < r < n$, we derive uniform BIBO stabilization results under some auxiliary involutivity hypotheses which in fact we conjecture could be replaced by weaker conditions satisfied by open sets of unstable systems.

THEOREM 5.1. *Suppose (1.1) is an exponentially minimum phase system having uniform relative degree 1 and satisfying hypotheses (H1). There exist state feedback laws*

$$u = \alpha(x) + \beta(x)v$$

such that the closed-loop system, with input $v(t)$, is uniformly BIBO stable for arbitrary initial data in s^n. In fact there exist constants $c, \gamma > 0$ such that

$$\|x(t)\| \leq c\|v(t)\|_\infty + 0(e^{-\gamma t}\|x_0\|) \tag{5.1}$$

REMARK 5.2. Of course (5.1) implies the required estimate (2.2) for uniform BIBO stability but it also implies exponential stability of the unforced closed-loop system. If one relaxes (5.1) to an estimate of the form

$$\|x(t)\| \leq c\|v(t)\|_\infty + 0(\gamma(t)\|x_0\|) \tag{5.2}$$

where $\gamma(t) \to 0$ as $t \to \infty$, then one obtains only global asymptotic stabilization, yet one still has uniform BIBO stability. Estimates such as (5.2) can be obtained for minimum phase systems with critically stable zero dynamics, but requires more technical hypotheses concerning (algebraic) rates of decay for critically stable systems, Lyapunov functions, etc. For example, for the open-loop system

$$\begin{aligned} \dot{z} &= -z^3 + y \\ \dot{y} &= u \end{aligned} \tag{5.3}$$

and the feedback law

$$u = -y^3 - z + y^2 v \tag{5.4}$$

the closed-loop trajectories satisfy (5.2). We prefer, however, stating our principal results in the exponentially minimum phase case where the hypotheses are more intuitive and clear.

REMARK 5.2. As the proof of Theorem 4.4 for the case $r = 1$ (see [34] shows, Theorem 5.1 remains valid in the MIMO case, mutatis mutandis. Indeed, following the analysis in [13]–[15], the only new ingredient occurs in the derivation of the normal form (3.10) where we require that the distribution, span $\{g_1, \dots g_m\}$, be involutive. As is well-known, this condition is not restrictive if we replace memoryless state feedback by compensation, since this distribution can always be rendered involutive by adding one integrator per input channel.

Theorem 5.1 is itself a corollary of the next result.

THEOREM 5.4. *Suppose (2.1) is an exponentially minimum phase system having strong relative degree r, and satisfying hypotheses (H1), (H3). There exists a state feedback law*

$$u = \alpha(x) + \beta(x)v$$

such that the closed-loop system, with input $v(t)$, is uniformly BIBO stable for arbitrary initial data. In fact there exist constants $c, \gamma > 0$ such (5.1) holds for all closed-loop trajectories.

Proof. By Theorem 4.4, there exists a state feedback law $u = \alpha(x)$ such that the closed-loop system

$$\dot{x} = F(x) \tag{5.5}$$

is globally exponentially stable. In particular, there exists ([36] p. 273) a Lyapunov function $V(x)$ and positive constants $\alpha_1, i = 1, \ldots, 4$ such that

$$\alpha_1 \|x\|^2 \leq V((x) \leq \alpha_2 \|x\|^2 \tag{5.6a}$$

$$\dot{V} = \langle \operatorname{grad} V, F \rangle \leq -\alpha_3 \|x\|^2 \tag{5.6b}$$

$$\| \operatorname{grad} V \| \leq \alpha_4 \|x\| \tag{5.6c}$$

Therefore, implementing

$$u = \alpha(x) + \beta(x)V, \qquad \beta(x) = D_g D_f^{r-1} h(x)^{-1} \tag{5.7}$$

in the normal form (3.14) we obtain the controlled closed-loop system

$$\dot{x} = F(x) + bv, \qquad b = e_n \epsilon \mathcal{R}^n \tag{5.8}$$

Computing $\dot{V}$ along trajectories of (5.8) one obtains

$$\dot{V} \leq -\alpha_3 \|x\|^2 + \alpha_4 \|x\| \; \|v\|_\infty \tag{5.9}$$

Choosing $c_1 > \alpha_3/\alpha_4$, one sees that $\dot{V} < 0$ for $\|x\| \leq c_1 \; \|v\|_\infty$; i.e. the closed ball $B(0, c_1 \|v\|_\infty)$ contains an attractor for (5.8), see also [16]. In particular, if $x_0 \in B(0, c_1 \|v\|_\infty)$, then

$$\|x(t)\| \leq c_1 \|v\|_\infty, \qquad t \geq 0. \tag{5.10}$$

If, on the other hand, $\|x_0\| > c_1 \|v\|_\infty$, then for $t \ll \infty$, $\dot{V}(x,t))$ is decreasing. In fact, for $x(t) \notin B(0, c_1 \|v\|_\infty)$, in the light of (5.6a), we have

$$\begin{aligned} \dot{V} &\leq -\left(\alpha_3 + \frac{c_1}{\alpha_4}\right) \|x\|^2 \\ &\leq - - \beta V \; , \; \beta > 0 \end{aligned}$$

so that

$$\alpha_1 \|x(t)\|^2 \leq V(x(t)) \leq e^{-\beta t} V(x, (0)) \leq \alpha_2 e^{-\beta t} \|x(0)\|^2 \tag{5.11}$$

In particular, there exist positive constants c_2, γ such that

$$\|x(t)\| \leq c_2 e^{-\gamma t} \|x(0)\| \tag{5.12}$$

whenever $x(t) \notin B(0, c_1 \|v\|_\infty)$. Adding (5.10) to (5.12) yields the desired estimate (5.1).

REFERENCES

[1] I.W. Sandberg, *An observation concerning the application of the contraction mapping fixed-point theorem and a result concerning the norm-boundedness of solutions of nonlinear functional equations*, Bell Sys. Tech. J. 44 (1965), 1809–1812.

[2] G. Zames, *On the input-output stability of nonlinear time-varying feedback systems, Parts I and II*, IEEE Trans. Aut. Control, AC-11 (1966), 228–238 and AC-11 (1966), 465–477.

[3] J.C. Willems, *The Analysis of Feedback Systems*, MIT Press, Cambridge (1971).

[4] C.A. Desoer and M. Vidyasagar, *Feedback Systems: Input-Output Properties*, Academic Press, N.Y. (1975).

[5] *Challenges to Control: A Collective View*, IEEE Trans. Aut. Control, AC-32 (1987), 275–285.

[6] R.W. Brockett, *Feedback Invariants for Nonlinear System*, VIII IFAC Cong., Helsinki (1978), 1115–1120.

[7] R. Sommer, *Control Design for Multivariable Nonlinear Time Varying System*, Int. J. Control 31 (1980), 883–891.

[8] B. Jakubczyk and W. Respondek, *On Linearization of Control System*, Bull. Acad Polon. Sci. Ser. Sci. Math, 28 (1980), 517–522.

[9] L.R. Hunt, R. Su and G. Meyer, *Global Transformation of Nonlinear System*, IEEE Trans. Aut. Control, AC 28 (1983), 24–31.

[10] J.W. Milnor, *Differentiable Topology*, in Lectures in Modern Mathematics, II (T. Saaty, ed) J. Wiley, N.Y. (1964).

[11] R.W. Brockett, *Finite Dimensional Linear Systems*, J. Wiley and Sons, N.Y. (1970).

[12] R.W. Brockett, *Convergence of Volterra series on infinite intervals and bilinear approximations*, Nonlinear Systems and Their Applications (P. Lakshmikanthan, ed.), Academic Press, N.Y. (1977), 39–46.

[13] C.I. Byrnes and A. Isidori, *The Analysis and Design of Nonlinear Feedback Systems I: Zero Dynamics and Global Normal Forms*, submitted to IEEE Trans. Aut. Control.

[14] C.I. Byrnes and A. Isidori, *The Analysis and Design of Nonlinear Feedback Systems II: Global Stabilization of Minimum Phase Systems*, submitted to IEEE Trans. Aut. Control.

[15] C.I. Byrnes and A. Isidori, *Nonlinear Disturbance Decoupling with Stability*, Proc. of 26th IEEE Conf. on Dec. and Control, Los Angeles (1987).

[16] C.I. Byrnes and A. Isidori, *Feedback Stabilization About Attractors and the Problem of Asymptotic disturbance Rejection*, Proc. of 27th IEEE Conf. on Dec. and Control, Austin (1988).

[17] E. D. Sontag and H.J. Sussman, *Remarks on Continuous Feedback*, Proc. of 19th IEEE Conf. on Dec. and Control, Albuquerque (1980).

[18] R.W. Brockett, *Asymptotic stability and feedback stabilization*, Differential Geometric Control Theory (R.W. Brocket, R.S. Millman and H. Sussman, eds.) Birkhauser (1983).

[19] C.I. Byrnes and A. Isidori, *Attitude Stabilization of Rigid Spacecraft*, (to appear in Automatica).

[20] M.A. Krasnosel'ski and P.P. Zabreiko, *Geometric Methods of Nonlinear Analysis*, Springer—Verlag (1984).

[21] D. Aeyels, *Stabilization of a class of nonlinear systems by a smooth feedback control*, systems and Control Letters 5 (1985), 289–294.

[22] P.E. Crouch and M. Irving, *On Sufficient Conditions for Local Asymptotic Stability of Nonlinear Systems whose Linearization is Uncontrollable*, Control Theory Centre Report No. 114, Univ. of Warwick (1983).

[23] C.I. Byrnes and A. Isidori, *A Frequency Domain Philosophy for Nonlinear Systems, with Applications to Stabilization and to Adaptive Control*, Proc. of 23rd IEEE Conf. on Dec. and Control, Las Vegas (1984) 1569–1573.

[24] C.I. Byrnes and A. Isidori, *Global Feedback Stabilization of Nonlinear Minimum Phase Systems*, Proc. of 24th IEEE Conf. on Dec. and Control, Ft. Lauderdale (1985), 1031–1037.

[25] C.I. Byrnes and A. Isidori, *Heuristics for Nonlinear Control*, Modelling and Adaptive Control (C.I. Byrnes and A. Kurzhansky, eds.) Lecture Notes in Inf. and Control 105, Springer–Verlag, Heidelberg (1988).

[26] R. Hirschorn, *Invertibility of Nonlinear Control System*, SIAM J. Control and Optimiz 17 (1979), 289–297.

[27] A. Isidori and A.J. Krener, *Nonlinear zero distributions*, Proc. of 19th IEEE Conf. on Dec. and Control (1980).

[28] A. Isidori and C.H. Moog, *On the nonlinear equivalent of the notion of transmission zeroes*, Modelling and Adaptive Control (C.I. Byrnes and A. Kurzhansky, eds.) Lecture Notes in Inf. and Control, 105, Springer–Verlag, Heidelberg (1988).

[29] C.I. Byrnes and A. Isidori, *Local Stabilization of Minimum Phase Nonlinear Systems*, Sys. and Control Letters (1988), 9–17.

[30] W.M. Wonham, 2nd Ed., Springer–Verlag, NY (1979), *Linear multivariable control: A Geometric Approach*.

[31] R. Marino, *Nonlinear Compensation by High Gain Feedback*, Int. J. Control, 42 (1985), 1369–1385.

[32] H. Khalil and A. Saberi, *Adaptive stabilization of a class of nonlinear systems using high-gain feedback*, IEEE Trans. Aut. Control AC-32 (1987).

[33] R. Marino, *Feedback stabilization of single-input nonlinear systems*, Systems and Control Letters 10 (1988), 201–206.

[34] C.I. Byrnes and A. Isidori, *New results and examples in nonlinear feedback stabilization*, Systems and Control Letters, 12 (1989), 437–442.

[35] K. Tchon, *On some applications to transversality theory to system theory*, Systems and Control Letters 4 (1984), 149–156.

[36] W. Hahn, *Stability of Motion*, Springer–Verlag (1967).

OPERATOR THEORETIC METHODS IN THE CONTROL OF DISTRIBUTED AND NONLINEAR SYSTEMS*

CIPRIAN FOIAS† AND ALLEN TANNENBAUM‡

Abstract. In this paper we discuss some applications of operator theory (in particular dilation theory and the commutant lifting theorem) to some problems in robust control, i.e. the control of systems with parameter uncertainty. We base our work here on our previous papers [7], [12], [13].

1. INTRODUCTION. In the past several years, mainly because of the burst of activity in the area of H^∞ design theory, there has evolved a major interest in the employment of operator theoretic methods in systems and control. In particular, there has been a great deal of research in the uses of interpolation and dilation techniques in this context. The point of the present paper is to show how these methods may be used to solve a very general case of the four block problem in H^∞ design valid for a large class of distributed, i.e., infinite dimensional systems and even a nonlinear generalization of the sensitivity minimization problem.

In the first part of this paper we study the spectral properties of certain "four block operators" from control and systems engineering. These operators naturally appear in the most general H^∞ synthesis problems and also have a number of intriguing mathematical properties in the sense that they are natural extensions of both the Hankel and Toeplitz operators. For this reason they fit into the skew Toeplitz framework developed in [7]. For the full details of our arguments and details about the skew Toeplitz theory applied to this problem we refer the reader to [12]. Here we will just consider the four block problem for single input / single output systems. See also the monograph of Francis [15], and the references therein for more details about the engineering aspects of this research area.

We now will give a precise mathematical statement of the "four block problem." Let $w, f, g, h, \in H^\infty$, where w, f, g, h are rational and m is nonconstant inner. (All of our Hardy spaces will be defined on the unit disc D in the standard way.)

Set

$$\mu := \inf\{\| \begin{bmatrix} w - mq & f \\ g & h \end{bmatrix} \|_\infty : q \in H^\infty\}. \tag{1}$$

Then the four block problem amounts to calculating the quantity μ. Note that for $f = g = h = 0$, this reduces to the classical Nehari problem. In this paper following [9], we will identify μ as the norm of a certain "four block operator" (see Section 2 for the precise definition), and then in Sections 3 and 4 give an explicit

*This research was supported in part by grants from the Research Fund of Indiana University, Department of Energy DE-FG02-86ER25020, NSF (ECS-8704047), NSF (DMS-8811084), and the Air Force Office of Scientific Research AFOSR-88-0020

†Department of Mathematics Indiana University Bloomington, Indiana 47405

‡Department of Electrical Engineering University of Minnesota 123 Church Street SE Minneapolis, Minnesota 55455 and Technion, Israel Institute of Technology Haifa, Israel

determinantal formula for its computation. The techniques given here are based on the previous work in [7], [11], [12], [14].

The second part of this paper will be concerned with nonlinear extensions of the H^∞ design theory. In the papers [3], [4] an extension of the commutant lifting theorem to a local nonlinear setting was given, together with a discussion of how this result could be used to develop a design procedure for nonlinear systems. In the present paper, we continue this line of research with a constructive extension of the linear H^∞ theory to nonlinear systems. We should note that our colleagues Joe Ball and Bill Helton [6] have developed a completely different novel approach to this problem based on a nonlinear version of Ball-Helton theory.

In the theory presented below, we will consider majorizable input/output operators (see Section 6 for the precise definition). In particular, these operators are analytic in a ball around the origin in a complex Hilbert space, and it turns out that it is possible to express each n-linear term of the Taylor expansion of such an operator as a linear operator on a certain tensor space. (Our class of operators also include Volterra series of fading memory [8]). This allows us to iteratively apply the classical commutant lifting theorem in designing a compensator. (The general technique we call the "iterative commutant lifting procedure." See Section 9.) For single input/single output (SISO) systems, this leads to the construction of a compensator which is optimal relative to a certain sensitivity function which will be defined below. Moreover in complete generality (i.e. for multiple input/multiple ouput (MIMO) systems), our procedure will ameliorate (in the sense of our nonlinear weighted sensitivity criterion), any given design. We note that for linear systems, our method reduces to the standard H^∞ design technique as discussed for example in [15].

In developing the present theory, we have had to extend some of the skew Toeplitz techniques of [7] and [12] to linear operators defined on certain tensor spaces. This has lead to several novel results in computational operator theory, and for example provides a way of iteratively constructing the nonlinear intertwining dilation of the nonlinear commutant lifting theorem considered in [3] and [4].

This research was supported in part by grants from the Research Fund of Indiana University, Department of Energy DE-FG02-86ER25020, NSF (DMS-8811084, ECS-8704047), and the Air Force Office of Scientific Research (AFOSR-88-0020).

2. THE FOUR BLOCK OPERATOR. In this section we will precisely define the *four block operator* which will be the major object of study in this paper. We will not give complete proofs for the various results in this section, and so for all the details we refer the reader to [9] and [12]. We use the notation of the Introduction. Moreover, we let $H(m) := H^2 \ominus mH^2$, $L(m) := L^2 \ominus mH^2$, and we let $P_{H(m)} : H^2 \to H(m)$, $P_{L(m)} : L^2 \to L(m)$ denote the corresponding orthogonal projections. Let $S : H^2 \to H^2$ denote unilateral shift, $T : H(m) \to H(m)$ the compression of S, and let $U : L^2 \to L^2$ denote bilateral shift, with $T(m) : L(m) \to L(m)$ the compression of U. Next for $w, f, g, h \in H^\infty$ rational, we set

$$A := \begin{bmatrix} P_{L(m)}w(S) & P_{L(m)}f(U) \\ g(S) & h(U) \end{bmatrix}. \tag{7}$$

Note that

$$A = \begin{bmatrix} w(T)P_{H(m)} & f(T(m))P_{L}(m) \\ g(S) & h(U) \end{bmatrix} \tag{8}$$

(Clearly $A : H^2 \oplus L^2 \to L(m) \oplus L^2$.)

PROPOSITION (2.1). *Notation as above. Then* $||A|| = \mu$.

Proof. Use the commutant lifting theorem. (See [9] and [12].) □

Thus in order to solve the four block problem we are required to compute the norm of the operator A. This we will show how to do in the next two sections. In order to do this however, we will first need to identify the essential norm of A (denoted by $||A||_e$). We are using the standard notation from operator theory as, for example, given in [17], [21]. In particular σ_e will denote the essential spectrum, and $A(\overline{D})$ will stand for the set of analytic functions on D which are continuous on the closed disc $\overline{D}$. We can now state the following result whose proof we refer the reader to [12]:

THEOREM (2.2). *Notation as above. Let* $w, f, g, h \in A(\overline{D})$, *and set*

$$\alpha := \max\{|| \begin{bmatrix} w(\zeta) & f(\zeta) \\ g(\zeta) & h(\zeta) \end{bmatrix} || : \zeta \in \sigma_e(T)\} \tag{9}$$

$$\beta := \max\{|| \begin{bmatrix} 0 & 0 \\ g(\zeta) & h(\zeta) \end{bmatrix} || : \zeta \in \partial D\} \tag{10}$$

$$\gamma := \sup\{|| \begin{bmatrix} f(\zeta) \\ h(\zeta) \end{bmatrix} || : \zeta \in \partial D\}.$$

Then

$$||A||_e = \max(\alpha, \beta, \gamma). \tag{12}$$

3. Invertibility of skew Toeplitz operators. In this section, we will study the invertibility of certain skew Toeplitz operators as considered in [7] which occur as basic elements in our procedure for computing the norm and singular values of the four block operator. More precisely, we will show that the calculation of the singular values of the four block operator A amounts to expliciting inverting two ordinary Toeplitz operators, and essentially inverting an associated skew Toeplitz operator. The Fredholm conditions on the invertibility of the skew Toeplitz operator (which is essentially invertible), and the coupling between the various systems (expressed as "matching conditions"; see Section 4) constitutes a certain linear system of equations called the *singular system* which allows one to determine the invertibility of A.

Using the notation of Section 2, we let $\rho > \max(\alpha, \beta, \gamma)$. Note that when $||A|| > ||A||_e$, $||A||^2$ is an eigenvalue of AA^*. By slight abuse of notation, ζ will denote a complex variable as well as an element of ∂D (the unit circle). The context will always make the meaning clear. Of course, if $\zeta \in \partial D$, then $\overline{\zeta} = 1/\zeta$.

As in the Introduction, we take w, f, g, h to be rational, and thus we can write $w = a/q$, $f = b/q$, $g = c/q$, $h = d/q$, where a, b, c, d, q are polynomials of degree $\leq n$. Then we have that

$$A := \begin{bmatrix} P_{L(m)}(\frac{a}{q})(S) & P_{L(m)}(\frac{b}{q})(U) \\ (\frac{c}{q})(S) & (\frac{d}{q})(U) \end{bmatrix}. \tag{13}$$

Clearly ρ^2 is an eigenvalue of AA^* if and only if

$$\begin{bmatrix} \rho^2 q(T(m))q(T(m))^* & 0 \\ 0 & \rho^2 q(U)q(U)^* \end{bmatrix} \begin{bmatrix} u \\ v \end{bmatrix} - \tag{14}$$

$$\begin{bmatrix} P_{L(m)}a(S) & P_{L(m)}b(U) \\ c(S) & d(U) \end{bmatrix} \begin{bmatrix} a(S)^*P & Pc(U)^* \\ b(U)^*P_L(m) & d(U)^* \end{bmatrix} \begin{bmatrix} u \\ v \end{bmatrix} = 0$$

for some non-zero

$$\begin{bmatrix} u \\ v \end{bmatrix} \in L(m) \oplus L^2$$

and where $P : L^2 \to H^2$ denotes orthogonal projection.

Next set

$$u_+ := Pu, \;\; u_- := (I - P)u$$

and

$$v_+ := Pv, \;\; v_- := (I - P)v, \;\; v_{++} := (I - P_{H(m)})v.$$

Then we can write (14) equivalently as

(14a)

$$\begin{bmatrix} \rho^2 q(T(m))q(T(m))^* - b(T(m))b(T(m))^* & -b(T(m))P_{L(m)}d(U)^* \\ -d(U)b(T(m))^* & \rho^2 q(U)q(U)^* - d(U)d(U)^* \end{bmatrix} \begin{bmatrix} u \\ v \end{bmatrix} -$$

$$\begin{bmatrix} a(T)a(T)^* & a(T)P_{H(m)}c(S)^* \\ c(S)a(T)^* & c(S)c(S)^* \end{bmatrix} \begin{bmatrix} u_+ \\ v_+ \end{bmatrix} = 0.$$

Set $V := U^*|L^2 \ominus H^2$. Then if we apply $(I - P)$ to both rows of (14a), we see that the basic block operator applied to

$$\begin{bmatrix} u_- \\ v_- \end{bmatrix}$$

is

(15a)

$$C_- \begin{bmatrix} u_- \\ v_- \end{bmatrix} := \begin{bmatrix} \rho^2 q(V^*)q_*(V) - b(V^*)b_*(V) & -b(V^*)d_*(V) \\ -d(V^*)b_*(V) & \rho^2 q(V^*)q_*(V) - d(V^*)d_*(V) \end{bmatrix} \begin{bmatrix} u_- \\ v_- \end{bmatrix}.$$

Next applying $(I - P_H(m))$ to both rows of (14a), we see that the basic operator applied to v_{++} is

$$C_{++}\overline{m}v_{++} := P\{(\rho^2|q|^2 - |c|^2 - |d|^2)\overline{m}v_{++}\}. \tag{15b}$$

Finally, applying $P_H(m)$ to (14a), we derive that the basic operator applied to

$$\begin{bmatrix} u_+ \\ P_H(m)v_+ \end{bmatrix}$$

is

$$C_+ \begin{bmatrix} u_+ \\ P_H(m)v_+ \end{bmatrix} :=$$
$$\begin{bmatrix} \rho^2 q(T)q(T)^* - b(T)b(T)^* - a(T)a(T)^* & -b(T)P_{H(m)}d(S)^* \\ -d(T)b(T)^* & \rho^2 q(T)q(T)^* - d(T)d(T)^* - c(T)c(T)^* \end{bmatrix} \begin{bmatrix} u_+ \\ P_H(m)v_+ \end{bmatrix}.$$

The operators C_-, C_{++}, C_+ are all skew Toeplitz (see [1]). We will now show how to invert C_- and C_{++} under the assumption $\rho > \|A\|_e$. The essential inversion of C_+ can be handled exactly as in [7], and we will postpone the details of this for Section 4.

We start with C_- . Namely, let $f_-, g_- \in L^2 \ominus H^2$, and consider the equation

$$C_- \begin{bmatrix} u_- \\ v_- \end{bmatrix} = \begin{bmatrix} f_- \\ g_- \end{bmatrix}. \tag{16a}$$

But (16a) is equivalent to

$$\begin{bmatrix} \rho^2|q|^2 - |b|^2 & -b\bar{d} \\ -d\bar{b} & \rho^2|q|^2 - |d|^2 \end{bmatrix} \begin{bmatrix} u_- \\ v_- \end{bmatrix} = \begin{bmatrix} f_- \\ g_- \end{bmatrix} + \sum_{j=0}^{n-1} \zeta^j \begin{bmatrix} x_j \\ y_j \end{bmatrix} \tag{16b}$$

where

$$\begin{bmatrix} x_j \\ y_j \end{bmatrix} \in \mathbf{C}^2$$

$(0 \leq j \leq n-1)$ are to be determined.

Put $q_o(\zeta^{-1}) := \zeta^{-n} q(\zeta)$. If we multiply (16b) by ζ^{-n}, we get that (with all the polynomials in ζ^{-1})

$$\begin{bmatrix} \rho^2 q_o q_* - b_o b_* & -b_o d_* \\ -d_o b_* & \rho^2 q_o q_* - d_o d_* \end{bmatrix} \begin{bmatrix} u_- \\ v_- \end{bmatrix} = \begin{bmatrix} \zeta^{-n} f_- \\ \zeta^{-n} g_- \end{bmatrix} + \sum_{j=0}^{n-1} \zeta^{j-n} \begin{bmatrix} x_j \\ y_j \end{bmatrix}. \tag{17}$$

Now by definition, $\rho > \|A\|_e$, and so using (3.2), we see that $\rho^2|q|^2 - |b|^2 - |d|^2 > 0$, and hence we can write

$$\det \left\{ \begin{bmatrix} \rho^2 q\bar{q} & 0 \\ 0 & \rho^2 q\bar{q} \end{bmatrix} - \begin{bmatrix} \bar{b} \\ \bar{d} \end{bmatrix} [bd] \right\} = \overline{\Delta}\Delta. \tag{18}$$

Thus we see that

$$\zeta^{-n} \det \left\{ \begin{bmatrix} \rho^2 q\bar{q} & 0 \\ 0 & \rho^2 q\bar{q} \end{bmatrix} \begin{bmatrix} \bar{b} \\ \bar{d} \end{bmatrix} [bd] \right\} = \Delta_o(\zeta^{-1})\Delta_*(\zeta^{-1}).$$

We now make the following *assumption of genericity*

$$\Delta_o \Delta_* \quad \textit{has distinct roots all of which are non-zero.} \tag{19}$$

Below we will discuss how to remove (19). However if we assume (19) for the present, it is easy to see that $\Delta_o \Delta_*$ has $2n$ distinct roots $\overline{z}_1, \cdots, \overline{z}_{2n}$ in D, and $2n$ distinct zeros $1/z_1, \cdots, 1/z_{2n}$ in the complement of $\overline{D}$. Set

$$\hat{D}(\zeta^{-1}) := \begin{bmatrix} \rho^2 q_o q_* - b_o b_* & -b_o d_* \\ -d_o b_* & \rho^2 q_o q_* - d_o d_* \end{bmatrix}$$

and let $\hat{D}^{ad}(\zeta^{-1})$ denote the algebraic adjoint of $\hat{D}(\zeta^{-1})$. Then if we apply $\hat{D}^{ad}(\zeta^{-1})$ to both sides of (17), we get

$$\Delta_o \Delta_* \begin{bmatrix} u_- \\ v_- \end{bmatrix} = \hat{D}^{ad}(\zeta^{-1}) \begin{bmatrix} \zeta^{-n} f_- \\ \zeta^{-n} g_- \end{bmatrix} + \hat{D}^{ad}(\zeta^{-1}) \sum_{j=0}^{n-1} \zeta^{j-n} \begin{bmatrix} x_j \\ y_j \end{bmatrix}. \tag{19a}$$

Hence plugging the $1/z_k$ into the last expression, we derive that

$$\widehat{D}^{ad}(z_k) \begin{bmatrix} z_k^n f_-(z_k) \\ z_k^n g_-(z_k) \end{bmatrix} + \widehat{D}^{ad}(z_k) \sum_{j=0}^{n-1} z_k^{-j+n} \begin{bmatrix} x_j \\ y_j \end{bmatrix} = 0 \tag{20}$$

for $k = 1, \cdots, 2n$.

Next note that

$$\Delta_o \Delta_* = \rho^2 q_o q_* (\rho^2 q_o q_* - b_o b_* - d_o d_*).$$

Let $1/z_1, \cdots, 1/z_n$ be such that

$$q_o q_*(z_k) = 0$$

for $1 \leq k \leq n$, and $1/z_{n+1}, \cdots, 1/z_{2n}$ be such that

$$(\rho^2 q_o q_* - b_o b_* - d_o d_*)(z_{n+k}) = 0$$

for $1 \leq k \leq n$. We can now state the following (the proofs of the following results can all be found in [12]):

PROPOSITION (3.1). *With the above notation, and under assumption (19), we have that the $x_j, y_j \in \mathbf{C}$, $0 \leq j \leq n-1$, are uniquely defined by*

$$\begin{bmatrix} x_o \\ \cdot \\ \cdot \\ \cdot \\ x_{n-1} \\ y_o \\ \cdot \\ \cdot \\ \cdot \\ y_{n-1} \end{bmatrix} = E^{-1} \hat{E} \begin{bmatrix} z_1^n f_-(z_1) \\ z_1^n g_-(z_1) \\ \cdot \\ \cdot \\ z_{2n}^n f_-(z_{2n}) \\ z_{2n}^n g_-(z_{2n}) \end{bmatrix}$$

where

$$\hat{E} := \begin{bmatrix} E_1 & 0_{(n,2n)} \\ 0_{(n,2n)} & E_2 \end{bmatrix}$$

for

$$E_1 := \begin{bmatrix} d_o(z_1) & -b_o(z_1) & 0 & 0 & \cdot & 0 & 0 \\ 0 & 0 & d_o(z_2) & -b_o(z_2) & \cdot & 0 & 0 \\ 0 & 0 & 0 & 0 & \cdot & 0 & 0 \\ \cdot & \cdot & \cdot & \cdot & \cdot & \cdot & \cdot \\ 0 & 0 & \cdot & 0 & \cdot & d_o(z_n) & -b_o(z_n) \end{bmatrix},$$

$$E_2 := \begin{bmatrix} b_*(z_{n+1}) & d_*(z_{n+1}) & 0 & 0 & \cdot & 0 & 0 \\ 0 & 0 & b_*(z_{n+2}) & d_*(z_{n+2}) & \cdot & 0 & 0 \\ 0 & 0 & 0 & 0 & 0 & \cdot & 0 & 0 \\ \cdot & \cdot & \cdot & \cdot & \cdot & \cdot & \cdot \\ 0 & 0 & 0 & 0 & \cdot & b_*(z_{2n}) & d_*(z_{2n}) \end{bmatrix},$$

($0_{(n,2n)}$ *denotes the* $n \times 2n$ *matrix all of whose entries are* 0), *and where*

$$E := \begin{bmatrix} \operatorname{diag}(d_o(z_1), \cdots, d_o(z_n))V_1 & \operatorname{diag}(-b_o(z_1), \cdots, -b_o(z_n))V_1 \\ \operatorname{diag}(b_*(z_{n+1}), \cdots, b_*(z_{2n}))V_2 & V_2 \operatorname{diag}(d_*(z_{n+1}), \cdots, d_*(z_{2n}))V_2 \end{bmatrix}$$

for

$$V_1 := \begin{bmatrix} 1 & z_1 & \cdot & \cdot & \cdot & z_1^{n-1} \\ \cdot & \cdot & \cdot & \cdot & \cdot & \cdot \\ \cdot & \cdot & \cdot & \cdot & \cdot & \cdot \\ \cdot & \cdot & \cdot & \cdot & \cdot & \cdot \\ 1 & z_n & \cdot & \cdot & \cdot & z_n^{n-1} \end{bmatrix}, V_2 := \begin{bmatrix} 1 & z_{n+1} & \cdot & \cdot & \cdot & z_{n+1}^{n-1} \\ \cdot & \cdot & \cdot & \cdot & \cdot & \cdot \\ \cdot & \cdot & \cdot & \cdot & \cdot & \cdot \\ \cdot & \cdot & \cdot & \cdot & \cdot & \cdot \\ 1 & z_{2n} & \cdot & \cdot & \cdot & z_{2n}^{n-1} \end{bmatrix}.$$

(Note that $\operatorname{diag}(a_1, \cdots, a_N)$ *denotes the* $N \times N$ *diagonal matrix with entries* $a_1, \cdots, a_N$ *on the diagonal.)*

Let $\hat{E}^{-1}E =: [e_{ij}]$ for $1 \le i \le 2n$, $1 \le j \le 4n$. Then we can now state (again see [12] for the details):

COROLLARY (3.2). *With the above notation, we have that*

(21)

$$\begin{bmatrix} u_- \\ v_- \end{bmatrix} =$$

$$\frac{1}{\Delta_o \Delta_*}\left\{\hat{D}^{ad}(\zeta^{-1})\begin{bmatrix} \zeta^{-n} f_- \\ \zeta^{-n} g_- \end{bmatrix} + \hat{D}^{ad}(\zeta^{-1}) \sum_{k=0}^{n-1} \zeta^{k-n} \begin{bmatrix} \sum_{j=1}^{2n} [e_{k+1,2j-1} z_j^n f_-(z_j) + e_{k+1,2j} z_j^n g_-(z_j)] \\ \sum_{j=1}^{2n} [e_{n+k+1,2j-1} z_j^n f_-(z_j) + e_{n+k+1,2j} z_j^n g_-(z_j)] \end{bmatrix}\right\}.$$

Now we consider the inverse of C_{++}. Since $\rho > \beta$, we have of course that C_{++} is invertible. For $p(\zeta)$ a polynomial of degree $\le n$, we let $\tilde{p}(\zeta) := \zeta^n \overline{p(\zeta)}$. We now make our second assumption of genericity that

(22) $\lambda(\zeta) := (\rho^2 \tilde{q} q - \tilde{c} c - \tilde{d} d)$ *has distinct nonzero roots all of which are nonzero.*

Note that $\lambda(\zeta)$ has n roots $\zeta_1, \cdots \zeta_n \in D$, and n roots $1/\bar{\zeta}_1, \cdots, 1/\bar{\zeta}_n$ which are in the complement of $\overline{D}$. We then have

PROPOSITION (3.3). *With assumption (22), if*

$$C_{++}(\bar{m}v_{++}) = f$$

for $f \in H^2$, then

$$v_{++} := m \begin{pmatrix} \zeta^n f - \sum_{j=1}^n \zeta_j \zeta^{n-j} \\ -(\rho^2 \tilde{q} q - \tilde{c} c - d\tilde{d}) \end{pmatrix}$$

where

$$\begin{bmatrix} \eta_1 \\ \cdot \\ \cdot \\ \cdot \\ \eta_n \end{bmatrix} = R_1^{-1} \begin{bmatrix} \zeta_1^{n-1} f(\zeta_1) \\ \cdot \\ \cdot \\ \cdot \\ \zeta_n^{n-1} f(\zeta_n) \end{bmatrix}, R_1 := \begin{bmatrix} \zeta_1^{n-1} & \cdot & \cdot & \cdot & 1 \\ \cdot & \cdot & \cdot & \cdot & \cdot \\ \cdot & \cdot & \cdot & \cdot & \cdot \\ \cdot & \cdot & \cdot & \cdot & \cdot \\ \zeta_n^{n-1} & \cdot & \cdot & \cdot & 1 \end{bmatrix}.$$

We will see in the next section how (3.2) and (3.3) will be used in determining the invertibility of A.

4. Singular system. We give in this section our procedure for computing the norm and singular values of the four block operator A. This will be accomplished through a certain associated linear system of equations which we call *the singular system.* We use the notation of the previous sections here. Moreover, the genericity assumptions (19) and (22) will be in force throughout this section. As above, we let $\rho > \max(\alpha, \beta, \gamma)$. For the proofs of the various computational lemmas in this section, we refer the interested reader to [12].

Once again, we work with the eigenvalue equation (14a) from Section 4. If as before, we set $V := U^*|L^2 \ominus H^2$, and apply $(I - P)$ to both rows of (14a), we get that

$$(24)\quad \begin{aligned} 0 = &\begin{bmatrix} \rho^2 q(V^*)q_*(V) - b(V^*)b_*(V) & -b(V^*)d_*(V) \\ -d(V^*)b_*(V) & \rho^2 q(V^*)q_*(V) - d(V^*)d_*(V) \end{bmatrix} \begin{bmatrix} u_- \\ v_- \end{bmatrix} + \\ &\begin{bmatrix} \rho^2 q(V^*)(I-P)q(T(m))^* & -b(V^*)(I-P)d(U)^* \\ -d(V^*)(I-P)b(T(m))^* & \rho^2 q(V^*)(I-P)q(U)^* \end{bmatrix} \begin{bmatrix} u_+ \\ v_+ \end{bmatrix} - \\ &\begin{bmatrix} b(V^*)(I-P)b(T(m))^* & 0 \\ 0 & d(V^*)(I-P)d(U)^* \end{bmatrix} \begin{bmatrix} u_+ \\ v_+ \end{bmatrix}. \end{aligned}$$

Next we write

$$u_+ = \sum_{j=0}^{\infty} u_j \zeta^j, v_+ = \sum_{j=0}^{\infty} v_j \zeta^j.$$

We will need the following lemma:

LEMMA (4.1). *Let*

$$\alpha(\zeta) := \sum_{j=0}^{n} \alpha_j \zeta^j, \beta(\zeta) := \sum_{j=0}^{n} \beta_j \zeta^j$$

be polynomials. Then

$$\alpha(V^*)(I-P)\beta(U)^*v_+ = \sum_{j=1}^{n}\sum_{i\geq k}^{n}\sum_{k\geq j}^{n}\bar{\beta}_i\alpha_{k-j}v_{i-k}\zeta^{-j}$$

and similarly for u_+. *(Note that since* $u_+ \in H(m)$, $\beta(U)^*u_+ = \beta_*(U^*)u_+ = \beta(T(m))^*u_+$.*)*

Using (4.1), we see that (24) becomes

(25)

$$0 = \begin{bmatrix} \rho^2 q(V^*)q_*(V) - b(V^*)b_*(V) & -b(V^*)d_*(V) \\ -d(V^*)b_*(V) & \rho^2 q(V^*)q_*(V) - d(V^*)d_*(V) \end{bmatrix} \begin{bmatrix} u_- \\ v_- \end{bmatrix} + \sum_{j=1}^{n}\zeta^{-j}\begin{bmatrix} M_j \\ N_j \end{bmatrix} w$$

where

$$w^t := [u_0 \cdots u_{n-1} v_0 \cdots v_{n-1}]$$

(w^t denotes the transpose of w), and

$$M_j := [\sum_{k\geq j}^{n}(\rho^2\bar{q}_k q_{k-j} - \bar{b}_k b_k - j)\cdots\sum_{k\geq j}^{n-j}(\rho^2\bar{q}_{k+n-j}q_{k-j} - \bar{b}_{k+n-j}b_{k-j})0_{j-1} - \sum_{k\geq j}^{n}\bar{d}_k b_{k-j}\cdots$$
$$\cdots - \sum_{k\geq j}^{n-j}\bar{d}_{k+n-j}b_{k-j}0_{j-1}]$$

$$N_j := [-\sum_{k\geq j}^{n}\bar{b}_k d_{k-j}\cdots - \sum_{k\geq j}^{n} -j\bar{b}_{k+n-j}d_{k-j}0_j - 1\sum_{k\geq j}^{n}(\rho^2\bar{q}_k q_{k-j} - \bar{d}_k d_{k-j})\ldots$$
$$\cdots\sum_{k\geq j}^{n-j}(\rho^2\bar{q}_{k+n-j}q_{k-j} - \bar{d}_{k+n-j}d_{k-j})0_{j-1}].$$

(Note that 0_{j-1} denotes the $j-1$ row matrix all of whose entries are 0.)

We now want to solve (25) for

$$\begin{bmatrix} u_- \\ v_- \end{bmatrix}.$$

But this we can do immediately from (3.2). Indeed, using the notation from (3.1) and (3.2), we set

$$\begin{bmatrix} \hat{M}(\zeta^{-1}) \\ \hat{N}(\zeta^{-1}) \end{bmatrix} :=$$

$$\frac{1}{\Delta_o\Delta_*}\{-\hat{D}^{ad}(\zeta^{-1})\sum_{j=1}^{n}\zeta^{-j-n}\begin{bmatrix} M_j \\ N_j \end{bmatrix} + \hat{D}^{ad}(\zeta^{-1})\sum_{k=0}^{n-1}\zeta^{k-n}\begin{bmatrix} \sum_{j=1}^{2n}[e_{k+1,2j-1}z_j^{-n}M_j + e_{k+1,2j}z_j^{-n}N_j] \\ \sum_{j=1}^{2n}[e_{n+k+1,2j-1}z_j^{-n}M_j + e_{n+k+1,2j}z_j^{-n}N_j] \end{bmatrix}$$

LEMMA (4.2). *With the above notation,*

$$\begin{bmatrix} u_- \\ v_- \end{bmatrix} = \begin{bmatrix} \hat{M} \\ \hat{N} \end{bmatrix} w.$$

We now play a similar, but slightly more complicated game for

$$\begin{bmatrix} u_+ \\ P_{H(m)}v_+ \end{bmatrix}$$

and $(I - P_{H(m)})v_+$.

In order to do this, we first set

$$Q_1 := \begin{bmatrix} \rho^2 q(T(m))q(T(m))^* - b(T(m))b(T(m))^* & -b(T(m))P_{L(m)}d(U)^* \\ -d(U)b(T(m))^* & \rho^2 q(U)q(U)^* - d(U)d(U)^* \end{bmatrix}$$

and

$$\begin{bmatrix} \hat{M}_1 \\ \hat{N}_1 \end{bmatrix} := Q_1 \begin{bmatrix} \hat{M} \\ \hat{N} \end{bmatrix}.$$

(Notice that the entries of $\hat{M}_1$ are in $L(m)$, and those of $\hat{N}_1$ are in L^2.) Then we can express (14a) as

$$(26) \qquad 0 = \left\{ Q_1 - \begin{bmatrix} a(T)a(T)^* & a(T)P_{H(m)}c(S)^* \\ c(S)a(T)^* & c(S)c(S)^* \end{bmatrix} \right\} \begin{bmatrix} u_+ \\ v_+ \end{bmatrix} + \begin{bmatrix} \hat{M}_1(\zeta) \\ \hat{N}_1(\zeta) \end{bmatrix} w.$$

As above, it is very easy to give a closed form expression for

$$\begin{bmatrix} \hat{M}_1 \\ \hat{N}_1 \end{bmatrix}.$$

Indeed, this follows from the following:

LEMMA (4.3). *Let α and β be as in (4.1). Let*

$$p(\zeta^{-1}) = \sum_{i=1}^{\infty} p_{-i} \in L^2 \ominus H^2.$$

Then

$$\alpha(T(m))P_{L(m)}\beta(U)^* p(\zeta^{-1}) = J_\alpha^\beta \begin{bmatrix} p_{-1} \\ p_{-2} \\ \cdot \\ \cdot \\ \cdot \\ p_{-n} \\ \cdot \\ \cdot \end{bmatrix}$$

where J_α^β is the row matrix whose j^{th} $(j = 1, 2, \ldots)$ element is

$$\sum_{k=0}^{n}\sum_{i=0}^{n} \alpha_i \bar{\beta}_k \zeta^{i-j-k} - m \sum_{k=0}^{n} \sum_{j+k \le i \le n} \sum_{l=0}^{i-j-k} \alpha_i \bar{\beta}_k \bar{m}_l \zeta^{i-j-k-l}.$$

(ii) We have

$$\alpha(U)\beta(U)^* p(\zeta^{-1}) = K_\alpha^\beta \begin{bmatrix} p_{-1} \\ p_{-2} \\ \cdot \\ \cdot \\ \cdot \\ p_{-n} \\ \cdot \\ \cdot \end{bmatrix}$$

where K_α^β is the row matrix with j^{th} ($j = 1, 2, \ldots$) element equal to $\sum_{k=0}^n \sum_{i=0}^n \bar{\beta}_k \alpha_i \zeta^{i-j-k}$.
(iii) Finally, we have the following equalities:

$$\alpha(T(m))(I-P)\beta(U)^* u_+ = F_\alpha^\beta \begin{bmatrix} u_0 \\ \cdot \\ \cdot \\ \cdot \\ u_{n-1} \end{bmatrix}$$

where F_α^β is the $1 \times n$ row matrix whose $(j+1)^{st}$ ($j = 0, \cdots, n-1$) entry is

$$\sum_{k=0}^{n} \sum_{1 \le i \le n-j} \alpha_k \bar{\beta}_{i+j} \zeta^{k-i} - m \sum_{k \ge i} \sum_{1 \le i \le n-j} \sum_{l=0}^{k-i} \alpha_k \bar{\beta}_i + j \bar{m}_l \zeta^{k-i-l}$$

and

$$\alpha(U)(I-P)\beta(U)^* u_+ = \hat{F}_\alpha^\beta \begin{bmatrix} u_0 \\ \cdot \\ \cdot \\ \cdot \\ u_{n-1} \end{bmatrix}$$

where $\hat{F}_\alpha^\beta$ is the $1 \times n$ row matrix with $(j+1)^{st}$ element ($j = 0, \cdots, n-1$)

$$\sum_{k=0}^{n} \sum_{1 \le i \le n-j} \alpha_k \bar{\beta}_{i+j} \zeta^{k-i}.$$

Thus from (4.3), we see that

$$\begin{bmatrix} \hat{M}_1 \\ \hat{N}_1 \end{bmatrix} = \begin{bmatrix} (\rho^2 J_q^q - J_b^b) M_- - J_b^d N_- \\ -K_d^b M_- + (\rho^2 K_q^q - K_d^d) N_- \end{bmatrix}$$

where

$$M_- := \begin{bmatrix} M_{-1} \\ \cdot \\ \cdot \\ \cdot \\ M_{-n} \\ \cdot \\ \cdot \\ \cdot \end{bmatrix}, N_- := \begin{bmatrix} N_{-1} \cdot \\ \cdot \\ \cdot \\ N_{-n} \\ \cdot \\ \cdot \\ \cdot \end{bmatrix}$$

for $\hat{M} = \sum_{i=1}^{\infty} M_{-i}\zeta^{-i}$, $\hat{N} = \sum_{i=1}^{\infty} N_{-i}\zeta^{-i}$. Consequently, from (26) and (4.3) we have

$$0 = Q_2 \begin{bmatrix} u_+ \\ v_+ \end{bmatrix} + \begin{bmatrix} \hat{M}_2 \\ \hat{N}_2 \end{bmatrix} w \tag{27}$$

where

$$Q_2 := \begin{bmatrix} \rho^2 q(T)q(T)^* - b(T)b(T)^* - a(T)a(T)^* & -b(T)P_{H(m)}d(S)^* \\ -d(S)b(T)^* & \rho^2 q(S)q(S)^* - d(S)d(S)^* - c(S)c(S)^* \end{bmatrix}$$

and

$$\begin{bmatrix} \hat{M}_2 \\ \hat{N}_2 \end{bmatrix} := \begin{bmatrix} \hat{M}_1 \\ \hat{N}_1 \end{bmatrix} + \begin{bmatrix} \rho^2 F_q^q - F_b^b & -F_b^d \\ -\hat{F}_d^b & \rho^2 \hat{F}_q^q - \hat{F}_d^d \end{bmatrix}.$$

Notice that $Q_2|H(m) \oplus H(m) = C_+$ as defined in Section 3. One can check that the entries of $\hat{M}_2$ are in $H(m)$, while those of $\hat{N}_2$ are in H^2. In order to develop the expression in (27), we will need another computational result:

LEMMA (4.4). *(i) Let $f \in H^2, f = \sum_{i=0}^{\infty} f_i\zeta^i$, and let $\alpha(\zeta)$ be a polynomial of degree $\leq n$. Then*

$$\alpha(S)^* f = \bar{\alpha} f - M_\alpha(\zeta^{-1}) \begin{bmatrix} f_0 \\ \cdot \\ \cdot \\ \cdot \\ f_{n-1} \end{bmatrix}$$

where $M_\alpha(\zeta^{-1})$ is a $1 \times n$ row matrix with j^{th} element equal to

$$\sum_{i=j}^{n} \bar{\alpha}_i \zeta^{-i+j-1}.$$

(ii) Let $f \in H(m), \beta(\zeta) = \sum_{j=0}^{n} \beta_j\zeta^j$. Then

$$\beta(T)\alpha(S)^* f = \beta(T)\alpha(T)^* f = \beta\bar{\alpha} f - \hat{M}_{\alpha,\beta} \begin{bmatrix} f_0 \\ \cdot \\ \cdot \\ \cdot \\ f_{n-1} \end{bmatrix} - \hat{N}_{\alpha,\beta} \begin{bmatrix} f_{-1} \\ \cdot \\ \cdot \\ \cdot \\ f_{-n} \end{bmatrix}$$

where

$$\bar{m} f = \sum_{i=1}^{\infty} f_{-i}\zeta^{-i}$$

and

$$\hat{M}_{\alpha,\beta} := \beta M_\alpha - m \sum_{i \geq j+k} \sum_{k=1}^{n} \sum_{j=1}^{n} \bar{m}_k \beta_i M_{\alpha j} \zeta^{i-j-k}$$

for $M_\alpha(\zeta^{-1}) = \sum_{j=1}^n M_{\alpha j}\zeta^{-j}$, and

$$\hat{N}_{\alpha,\beta} := m[\sum_{i=0}^{n-1}\sum_{k\geq i+1}^{n} \beta_k\bar{\alpha}_i\zeta^{-i-1+k} \ldots \sum_{i=0}^{n-j}\sum_{k\geq i+j}^{n} \beta_k\bar{\alpha}_i\zeta^{-i-j+k} \ldots \bar{\alpha}_0\beta_n] \begin{bmatrix} f_{-1} \\ \cdot \\ \cdot \\ \cdot \\ f_{-n} \end{bmatrix}.$$

Proceeding now with our computations, since $u_+, P_{H(m)}v_+ \in H(m)$, we can write

$$\begin{aligned} \bar{m}u_+ &= u_{-1}\zeta^{-1} + \cdots + u_{-n}\zeta^{-n} + \cdots \\ \bar{m}P_{H(m)}v_+ &= v_{-1}\zeta^{-1} + \cdots + v_{-n}\zeta^{-n} + \cdots \\ P_{H(m)}v_+ &= v_{+0} + \cdots + v_{+,n-1}\zeta^{n-1} + \cdots \end{aligned}$$

Set

$$w_-^t := [u_-1 \cdots u_{-n}v_{-1} \cdots v_{-n}]$$

and

$$w_+^t = [v_{+,0} \cdots v_{+,n-1}]$$

where as before w_-^t, w_+^t denote the transposes of w_-, w_+ respectively. Using (4.4), we can now express (27) as

$$0 = Q_2 \begin{bmatrix} u_+ \\ P_{H(m)}v_+ \end{bmatrix} + Q_2 \begin{bmatrix} 0 \\ (I - P_{H(m)})v_+ \end{bmatrix} + \begin{bmatrix} \hat{M}_2 \\ \hat{N}_2 \end{bmatrix} w \tag{28}$$

or using the above discussion as

(28a)

$$0 = \begin{bmatrix} \rho^2|q|^2 - |a|^2 - |b|^2 & -b\bar{d} \\ -d\bar{b} & \rho^2|q|^2 - |c|^2 - |d|^2 \end{bmatrix} \begin{bmatrix} u_+ \\ P_{H(m)}v_+ \end{bmatrix} +$$

$$\begin{bmatrix} -b(T)P_{H(m)}d(S)^* \\ \rho^2 q(S)q(S)^* - c(S)c(S)^* - d(S)d(S)^* \end{bmatrix} (I - P_{H(m)})v_+ + J_1 w + J_2 w_- + J_3 w_+$$

where

$$\begin{aligned} J_1 &:= \begin{bmatrix} \hat{M}_2 \\ \hat{N}_2 \end{bmatrix} + \begin{bmatrix} \hat{M}_{q,b,a} & 0_n \\ dM_b & 0_n \end{bmatrix} \\ J_2 &:= \begin{bmatrix} \hat{N}_{q,b,a} & \hat{N}_{d,b} \\ 0_n & 0_n \end{bmatrix} \\ J_3 &:= \begin{bmatrix} \hat{M}_{d,b} \\ -\rho^2 qM_q + cM_c + dM_d \end{bmatrix} \end{aligned}$$

for

$$\begin{aligned} \hat{M}_{q,b,a} &:= \hat{M}_{b,b} + \hat{M}_{a,a} - \rho^2\hat{M}_{q,q} \\ \hat{N}_{q,b,a} &:= \hat{N}_{b,b} + \hat{N}_{a,a} - \rho^2\hat{N}_{q,q}. \end{aligned}$$

From the second equation of (28), we see that

$$0 = -d(S)b(T)^*u_+ + (\rho^2 q(S)q(S)^* - c(S)c(S)^* - d(S)d(S)^*) \tag{29}$$
$$((I - P_{H(m)})v_+ + P_{H(m)}v_+) + \hat{N}_2 w.$$

If we apply $(I - P_{H(m)})$ to (29), we obtain

$$\begin{aligned} 0 = &-(I - P_{H(m)})d(S)b(T)^*u_+ + \\ &(I - P_{H(m)})(\rho^2 q(S)q(S)^* - c(S)c(S)^* - d(S)d(S)^*)((I - P_{H(m)})v_+) + \\ &(I - P_{H(m)})(\rho^2 q(S)q(S)^* - c(S)c(S)^* - d(S)d(S)^*)(P_{H(m)}v_+) + \\ &(I - P_{H(m)})\hat{N}_2 w. \end{aligned} \tag{30}$$

Now from (4.4), notice that we have

$$-(I - P_{H(m)})d(S)b(T)^*u_+ = dM_d \begin{bmatrix} u_0 \\ \cdot \\ \cdot \\ \cdot \\ u_{n-1} \end{bmatrix} - \hat{M}_{b,d} \begin{bmatrix} u_0 \\ \cdot \\ \cdot \\ \cdot \\ u_{n-1} \end{bmatrix} - \hat{N}_{b,d} \begin{bmatrix} u_{-1} \\ \cdot \\ \cdot \\ \cdot \\ u_{-n} \end{bmatrix}$$

We set $\hat{L}_{b,d} := dM_d - \hat{M}_{b,d}$. Similarly, we have that

$$(I - P_{H(m)})(\rho^2 q(S)q(S)^* - c(S)c(S)^* - d(S)d(S)^*)P_{H(m)}v_+ = T_{qcd}w_+ + [0_n \;\; U_{q,c,d}]w_-$$

where

$$\begin{aligned} T_{q,c,d} &:= -\rho^2 \hat{L}_{q,q} + \hat{L}_{c,c} + \hat{L}_{d,d} \\ U_{q,c,d} &:= \rho^2 \hat{N}_{q,q} - \hat{N}_{c,c} - \hat{N}_{d,d} \end{aligned}$$

and 0_n denotes the $1 \times n$ zero matrix. Thus (30) becomes

$$\begin{aligned} 0 = &[\hat{L}_{b,d} \; 0_n]w + [-\hat{N}_{b,d} \; U_{q,c,d}]w_- + T_{q,c,d}w_+ + \\ &(I - P_{H(m)})(\rho^2 q(S)q(S)^* - c(S)c(S)^* - d(S)d(S)^*)((I - P_{H(m)})v_+) + (I - P_H(m))\hat{N}_2 w \end{aligned}$$

Set

$$\begin{aligned} U_1 &:= [\hat{L}_{b,d} \; 0_n] + (I - P_H(m))\hat{N}_2 \\ U_2 &:= [-\hat{N}_{b,d} \; U_{q,c,d}] \\ U_3 &:= T_{q,c,d} \end{aligned}$$

Note that the U_i have entries in mH^2 for $i = 1, 2, 3$, the explicit form of which follows directly from the above. Now as in Section 3, denote $v_{++} := (I - P_{H(m)})v_+$. If we let $v_{++} = \sum_{j=0}^{\infty} v_{++}, j\zeta^j$, and

$$w_{++}^t := [v_{++,0} \ldots v_{++,n-1}],$$

we have that

$$w_{++} = I_o w - w_+$$

where

$$I_o := [\, 0_{(n)} \quad I \,]$$

for $0_{(n)}$ the $n \times n$ matrix all of whose entries are 0, and I the $n \times n$ identity matrix. Thus we can write (31) as

(32a)

$$0 = U_1 w + U_2 w_- + U_3 w_+ + (I - P_{H(m)})(\rho^2 q(S) q(S)^* - c(S)c(S)^* - d(S)d(S)^*)v_{++}.$$

Now from (4.4) we have that

$$\alpha(S)\alpha(S)^* v_{++} = |\alpha|^2 v_{++} - \alpha M_\alpha w_{++}.$$

Therefore

$$(I - P_{H(m)})\alpha(S)\alpha(S)^* v_{++} = mP(|\alpha|^2 \bar{m} v_{++}) + m\hat{L}_\alpha w_{++}$$

where

$$m\hat{L}_\alpha := (P_{H(m)} - I)\alpha M_\alpha = -mP\bar{m}\alpha M_\alpha.$$

If we now let $L'_{q,c,d} := \rho^2 \hat{L}_q - \hat{L}_c - \hat{L}_d$, then (32a) is equivalent to

$$0 = U_1 w + U_2 w_- + U_3 w_+ + mP\{(\rho^2|q|^2 - |c|^2 - |d|^2)\bar{m} v_{++}\} + mL'_{qcd} w_{++}.$$

Finally setting $\hat{U}_i := \bar{m} U_i$, $i = 1, 2, 3$, the latter equality becomes equivalent to

(32b) $\quad 0 = \hat{U}_1 w + \hat{U}_2 w_- + \hat{U}_3 w_+ + P\{(\rho^2|q|^2 - |c|^2 - |d|^2)\bar{m} v_{++}\} + L'_{qcd} w_{++}.$

But now in order to find v_{++}, it is clear that we must invert C_{++}. Indeed, let

$$U'_1 := \zeta^n \hat{U}_1 + \zeta^n L'_{q,c,d} I_o,$$
$$U'_2 := \zeta^n \hat{U}_2,$$
$$U'_3 := \zeta^n \hat{U}_3 - \zeta^n L'_{q,c,d},$$

and for R_1 as in Section 3, set

$$R_1^{-1} R_2 := \begin{bmatrix} R_{11} w + R_{21} w_- + R_{31} w_+ \\ \cdot \\ \cdot \\ \cdot \\ R_{1n} w + R_{2n} w_- + R_{3n} w_+ \end{bmatrix}$$

where

$$R_2 := \begin{bmatrix} U'_1(\zeta_1) w + U'_2(\zeta_1) w_- + U'_3(\zeta_1) w_+ \\ \cdot \\ \cdot \\ \cdot \\ U'_1(\zeta_n) w + U'_2(\zeta_n) w_- + U'_3(\zeta_n) w_+ \end{bmatrix}.$$

Finally, we denote for $i = 1, 2, 3$

$$A_i := m \left(\frac{U'_i - \sum_{j=1}^n R_{ij} \zeta^{n-j}}{-(\rho^2 \tilde{q} q - \tilde{c} c - \tilde{d} d)} \right).$$

We can now state:

LEMMA (4.5). *With the above notation, we have*

$$v_{++} = A_1 w + A_2 w_- + A_3 w_+ \tag{33}$$

i.e., v_{++} *is determined by* w, w_-, *and* w_+.

We now return to (28a). First express

$$A_i := \sum_{k\geq 0} A_{i,k}\zeta^k$$

for $i = 1, 2, 3$, and then plug (33) into (28a). We get that
(34)

$$0 = \begin{bmatrix} \rho^2|q|^2 - |a|^2 - |b|^2 & -b\bar{d} \\ -d\bar{b} & \rho^2|q|^2 - |c|^2 - |d|^2 \end{bmatrix} \begin{bmatrix} u_+ \\ P_{H(m)}v_+ \end{bmatrix} + J_1' w + J_2' w_- + J_3' w_+$$

where

$$J_i' := J_i + \begin{bmatrix} -b(T)P_{H(m)}d(S)^* \\ q(S)q(S)^* - c(S)c(S)^* - d(S)d(S)^* \end{bmatrix} A_i =$$

$$J_i + \begin{bmatrix} b\bar{d}A_i - mP(\bar{m}b\bar{d}A_i) - (bM_d - mP(\bar{m}bM_d)) \begin{bmatrix} A_i, 0 \\ \cdot \\ \cdot \\ \cdot \\ A_{i,n-1} \end{bmatrix} \\ \rho^2|q|^2 - |c|^2 - |d|^2 - (\rho^2 qM_q - cM_c - dM_d) \begin{bmatrix} A_i, 0 \\ \cdot \\ \cdot \\ \cdot \\ A_{i,n-1} \end{bmatrix} \end{bmatrix}$$

for $i = 1, 2, 3$.

If we multiply (34) by ζ^n, we get that

$$0 = V(\zeta) \begin{bmatrix} u_+ \\ P_{H(m)}v_+ \end{bmatrix} + J_1'\zeta^n w + J_2'\zeta^n w_- + J_3'\zeta^n w_+ \tag{34a}$$

where

$$V(\zeta) := \begin{bmatrix} \rho^2\tilde{q}q - a\tilde{a} - b\tilde{b} & -b\tilde{d} \\ -d\tilde{b} & \rho^2\tilde{q}q - \tilde{c}c - \tilde{d}d \end{bmatrix}.$$

With $d(\zeta) := \det V(\zeta)$, we make our final assumption of genericity:

$$d(\zeta) \text{ has distinct roots all of which are non-zero.}$$

We now have:

LEMMA (4.6). *Under assumption (35), $d(\zeta)$ has r zeros $\alpha_1, \cdots, \alpha_r \in D$, r zeros $1/\bar{\alpha}_1, \cdots, 1/\bar{\alpha}_r$, and $4n - r$ zeros $\alpha_{2r+1}, \cdots, \alpha_{4n} \in \partial D \backslash \sigma(T)$.*

Let $V^{ad}(\zeta)$ denote the algebraic adjoint of $V(\zeta)$, i.e. $V^{ad}(\zeta)V(\zeta) = d(\zeta)I$. Set for $i = 1, 2, 3$, $\hat{J}_i := -V^{ad}J_i'\zeta^n$. Then multiplying (34a) by $V^a d(\zeta)$, we see

$$d(\zeta)\begin{bmatrix} u_+ \\ P_{H(m)}v_+ \end{bmatrix} = \hat{J}_1 w + \hat{J}_2 w_- + \hat{J}_3 w_+. \tag{36}$$

Next since $P_{H(m)}u_+$, $v_+ \in H(m)$, they must be analytic in a neighborhood of $\partial D\sigma(T)$ as well as D. Thus using (4.6), from (36) we get

$$0 = \hat{J}_1(\alpha_i)w + \hat{J}_2(\alpha_i)w_- + \hat{J}_3(\alpha_i)w_+ \tag{37}$$

$1 \le i \le r$, $2r + 1 \le i \le 4n$. Multiplying (36) by $\bar{\zeta}^{4n}\bar{m}$, we have

$$\bar{\zeta}^{4n}d(\zeta)\begin{bmatrix} \bar{m}u_+ \\ \bar{m}P_{H(m)}v_+ \end{bmatrix} = \bar{\zeta}^{4n}\bar{m}\hat{J}_1 w + \bar{\zeta}^{4n}\bar{m}\hat{J}_2 w_- + \bar{\zeta}^{4n}\bar{m}\hat{J}_3 w_+. \tag{38}$$

Note that (38) admits an analytic extension to the complement of $\overline{D}$, i.e. all the functions are analytic in $1/\zeta$. Set

$$J_i^o(1/\zeta) := (1/\zeta)^{4n}\hat{J}_i(\zeta)$$

for $i = 1, 2, 3$, and

$$d^o(1/\zeta) := (1/\zeta)^{4n}d(\zeta).$$

Then (38) can be expressed equivalently as

$$d_o(1/\zeta)\begin{bmatrix} (\bar{m}u_+)(1/\zeta) \\ (\bar{m}P_{H(m)}v_+)(1/\zeta) \end{bmatrix} = \overline{m(1/\bar{\zeta})}\{J_1^o w + J_2^o w_- + J_3^o w_+\}. \tag{38a}$$

If we now plug the $1/\bar{\alpha}_i$ into (38a), $1 \le i \le r$, we get

$$0 = \overline{m(\alpha_i)}\{J_1^o(\bar{\alpha}_i)w + J_2^o(\bar{\alpha}_i)w_- + J_3^o(\bar{\alpha}_i)w_+\} \tag{39}$$

for $1 \le i \le r$.

Note that from (37) and (39), we have $8n$ equations in $5n$ unknowns. In order to complete our *singular system* we will now need some "matching equations".

Namely, from (36) we see that

$$\begin{bmatrix} \bar{m}u_+ \\ \bar{m}P_{H(m)}v_+ \end{bmatrix} = J_{o1}(\zeta^{-1})w + J_{o2}(\zeta^{-1})w_- + J_{o3}(\zeta^{-1})w_+$$

where $J_{oi}(\zeta^{-1}) := (I - P)\bar{m}\hat{J}_i d^-1$ for $i = 1, 2, 3$. Write

$$J_{oi} = \begin{bmatrix} J_{oi}^1 \\ J_{oi}^2 \end{bmatrix}$$

for $i = 1, 2, 3$, where the J^l_{oi} are $1 \times 2n$ matrices $i = 1, 2$, $l = 1, 2$, and the J^{3l}_o are $1 \times n$ matrices for $l = 1, 2$. We also can express

$$J^l_{oi}(\zeta^{-1}) = \sum_{k \geq 1} J_{oi,k} \zeta^{-k}$$

for $l = 1, 2$, and $i = 1, 2, 3$. Thus from (40), we have

$$\begin{bmatrix} \sum_{k \geq 1} u_{-k} \zeta^{-k} \\ \sum_{k \geq 1} v_{-k} \zeta^{-k} \end{bmatrix} = \begin{bmatrix} \sum_{k \geq 1} (J^1_{o1,k} w + J^1_{o2,k} w_- + J^1_{o3,k} w_+) \zeta^{-k} \\ \sum_{k \geq 1} (J^2_{o1,k} w + J_o 2, k^2 w_- + J^2_{o3,k} w_+) \zeta^{-k} \end{bmatrix}.$$

Matching the first n coefficients, we have

$$\begin{aligned} u_{-k} &= J^1_{o1,k} w + J^1_{o2,k} w_- + J^1_{o3,k} w_+ \\ v_- k &= J^2_{o1,k} w + J^2_{o2,k} w_- + J^2_{o3,k} w_+ \end{aligned} \tag{42}$$

for $k = 1, \ldots, n$.

Now multiplying (36) by $d(\zeta)^{-1}$ and adding the resulting expression to (33), we derive

$$\begin{bmatrix} u_+ \\ v_+ \end{bmatrix} = A^o_1 w + A^o_2 w_- + A^o_3 w_+ \tag{43}$$

where

$$A^o_i = \hat{J}_i d^{-1} + \begin{bmatrix} 0_{2n} \\ A_i \end{bmatrix}$$

for $i = 1, 2$, and

$$A^o_3 = \hat{J}_3 d^{-1} + \begin{bmatrix} 0_n \\ A_3 \end{bmatrix}.$$

As before, we may express

$$A^o_1 w + A^o_2 w_- + A^o_3 w_+ = \begin{bmatrix} \sum_{k \geq 0} (A^{o1}_{1,k} w + A^{o1}_{2,k} w_- + A^{o1}_{3,k} w_+) \zeta^k \\ \sum_{k \geq 0} (A^{o2}_{1,k} w + A^{o2}_{2,k} w_- + A^{o2}_{3,k} w_+) \zeta^k \end{bmatrix}.$$

Matching the first n coefficients, we have

$$\begin{aligned} u_k &= A^{o1}_{1,k} w + A^{o1}_{2,k} w_- + A^{o1}_{3,k} w_+ \\ v_k &= A^{o2}_{1,k} w + A^{o2}_{2,k} w_- + A^{o2}_{3,k} w_+ \end{aligned}$$

for $0 \leq k \leq n - 1$. Finally, matching the coefficients for

$$v_{++} = A_1 w + A_2 w_- + A_3 w_+ = \sum_{k \geq 0} (A_{1,k} w + A_{2,k} w_- + A_{3,k} w_+) \zeta^k$$

we get that,

$$v_{+k} = A_{1,k} w + A_{2,k} w_- + A_{3,k} w_+ \tag{45}$$

for $0 \leq k \leq n-1$.

From (37), (39), (42), (44), (45) we can now define the following singular system for w, w_-, w_+ :

$$0 = \hat{J}_1(\alpha_i)w + \hat{J}_2(\alpha_i)w_- + \hat{J}_3(\alpha_i)w_+ \tag{46a}$$

$1 \leq i \leq r$, $2r+1 \leq i \leq 4n$,

$$0 = \overline{m(\alpha_i)}J_1^o(\alpha_i)w + \overline{m(\alpha_i)}J_2^o(\bar{\alpha}_i)w_- + \overline{m(\alpha_i)}J_3^o(\bar{\alpha}_i)w_+ \tag{46b}$$

$1 \leq i \leq r$,

$$0 = \Lambda_1 w + (\Lambda_2 - I)w_- + \Lambda_3 w_+ \tag{46c}$$

$$0 = (\Theta_1 - I)w + \Theta_2 w_- + \Theta_3 w_+ \tag{46d}$$

$$0 = \Psi_1 w + \Psi_2 w_- + (\Psi_3 - I)w_+ \tag{46e}$$

where

$$\Lambda_i := \begin{bmatrix} J^1_{oi,1} \\ \cdot \\ \cdot \\ \cdot \\ J^1_{oi,n} \\ J^2_{oi,1} \\ \cdot \\ \cdot \\ \cdot \\ J^2_{oi,n} \end{bmatrix}, \Theta_i := \begin{bmatrix} A^{o1}_{i,0}\cdot \\ \cdot \\ \cdot \\ A^{o1}_{i,0} \\ A^{o2}_{i,0} \\ \cdot \\ \cdot \\ \cdot \\ A^{o2}_{i,n-1} \end{bmatrix}$$

$i = 1, 2, 3$,

$$\Psi_i := \begin{bmatrix} A_{i0} \\ \cdot \\ \cdot \\ \cdot \\ A_{in-1} \end{bmatrix}$$

for $i = 1, 2, 3$.

Notice the system of equations (46) (the singular system) consists of $13n$ equations in $5n$ unknowns. It is easy to see that if $w = w_- = 0$, and $w_+ = 0$, then

$$\begin{bmatrix} u \\ v \end{bmatrix} = 0.$$

Set

$$\Omega := \begin{bmatrix} \hat{J}_1(\alpha_1) & \hat{J}_2(\alpha_1) & \hat{J}_3(\alpha_1) \\ \vdots & \vdots & \vdots \\ \hat{J}_1(\alpha_r) & \hat{J}_2(\alpha_r) & \hat{J}_3(\alpha_r) \\ \hat{J}_1(\alpha_{2r+1}) & \hat{J}_2(\alpha_{2r+1}) & \hat{J}_3(\alpha_{2r+1}) \\ \vdots & \vdots & \vdots \\ \hat{J}_1(\alpha_{4n}) & \hat{J}_2(\alpha_{4n}) & \hat{J}_3(\alpha_{4n}) \\ \overline{m(\alpha_1)}J_1^o(\bar{\alpha}_1) & \overline{m(\alpha_1)}J_2^o(\bar{\alpha}_1) & \overline{m(\alpha_1)}J_3^o(\bar{\alpha}_1) \\ \cdots & & \\ \cdots & & \\ \cdots & & \\ \overline{m(\alpha_r)}J_1^o(\bar{\alpha}_r) & \overline{m(\alpha_r)}J_2^o(\bar{\alpha}_r) & \overline{m(\alpha_r)}J_3^o(\bar{\alpha}_r) \\ \Lambda_1 & (\Lambda_2 - I) & \Lambda_3 \\ (\Theta_1 - I) & \Theta_2 & \Theta_3 \\ \Psi_1 & \Psi_2 & (\Psi_3 - I) \end{bmatrix}$$

We can at long last state the main result of this paper:

THEOREM (4.7). *With the above notation, and with the genericity assumptions (19), (22), (35), ρ is a singular value of the four block operator A if and only if Ω has rank $< 5n$.*

Proof. Immediate from the above discussion. □

Remark. We can remove the assumptions of genericity via a limiting argument of the kind which we gave in [14].

Remark. The above procedure also gives a way of computing the optimal compensator in a given four block problem. Indeed, from the above determinantal formula one can compute the Schmidt pair ψ, η corresponding to the singular value $s := \|A\|$ when $s > \|A\|_e$. We will indicate how one derives the optimal interpolant (and thus the opimal compensator) from these Schmidt vectors. In order to do this, notice

$$A\psi = s\eta.$$

Thus, there exists $q_{opt} \in H^\infty$ with

$$(w - q_{opt})\psi_1 + f\psi_2 = s\eta_1$$

$$g\psi_1 + h\psi_2 = s_k\eta_2$$

where

$$\psi = \begin{bmatrix} \psi_1 \\ \psi_2 \end{bmatrix}, \quad \eta = \begin{bmatrix} \eta_1 \\ \eta_2 \end{bmatrix}.$$

One can show (see [25]), that $\psi_1 \neq 0$, so that

$$q_{opt} = w - \frac{s\eta_1 - f\psi_2}{\psi_1}.$$

Note from q_{opt}, we can derive the corresponding optimal controller in a given systems design problem. See also [25] for an extension of the theory of [1] (valid for the Hankel operator) to the singular values of the four block operator and their relationship to more general interpolation and distance problems.

5. Analytic Mappings on Hilbert Space. We now would like to begin our discussion of the nonlinear generalization of the above H^∞ linear theory. In order to do this, we will need to first discuss a few standard results about analytic mappings on Hilbert spaces. We are essentially following the treatments of [3], [4], [5], and [8] to which the reader may refer for all of the details. In particular, input/output operators which admit Volterra expansions are special cases of the operators which we study here. See [8], [16], [20].

Let G and H denote complex Hilbert spaces. Set

$$B_{r_o}(G) := \{g \in G : \|g\| < r_o\}$$

(the open ball of radius r_o in G about the origin). Then we say that a mapping $\phi : B_{r_o}(G) \to H$ is *analytic* if the complex function $(z_1, \ldots, z_n) \to \langle \phi(z_1 g_1 + \cdots + z_n g_n), h\rangle$ is analytic in a neighborhood of $(1, 1, \ldots, 1) \in \mathbf{C}^n$ as a function of the complex variables $z_1, \ldots, z_n$ for all $g_1, \ldots, g_n \in G$ such that $\|g_1 + \cdots + g_n\| < r_o$, for all $h \in H$, and for all $n > 0$. (Note that we denote the Hilbert space norms in G and H by $\| \quad \|$ and the inner products by $<,>$.)

We will now assume that $\phi(0) = 0$. It is easy to see that if $\phi : B_{r_o}(G) \to H$ is analytic, then ϕ admits a convergent Taylor series expansion, i.e.

$$\phi(g) = \phi_1(g) + \phi_2(g, g) + \cdots + \phi_n(g, \cdots, g) + \cdots$$

where $\phi_n : G \times \cdots \times G \to H$ is an n-linear map. Clearly, without loss of generality we may assume that the n-linear map $(g_1, \cdots, g_n) \to \phi(g_1, \cdots, g_n)$ is symmetric in the arguments $g_1, \cdots, g_n$. This assumption will be made throughout this paper for the various analytic maps which we consider. For ϕ a Volterra series, ϕ_n is basically the nth-Volterra kernel.

Now set

$$\hat{\phi}_n(g_1 \otimes \cdots \otimes g_n) := \phi_n(g_1, \cdots, g_n).$$

Then $\hat{\phi}_n$ extends in a unique manner to a dense set of $G^{\otimes n} := G \otimes \cdots \otimes G$ (tensor product taken n times). Notice by $G^{\otimes n}$ we mean the Hilbert space completion of the algebraic tensor product of the G's. Clearly if $\hat{\phi}_n$ has finite norm on this dense set, then $\hat{\phi}_n$ extends by continuity to a bounded linear operator $\hat{\phi}_n : G^{\otimes n} \to H$. By abuse of notation, we will set $\phi_n := \hat{\phi}_n$.

We now conclude this section with two key definitions.

DEFINITIONS (5.1). (i) Notation as above. By a *majorizing sequence* for the holomorphic map ϕ, we mean a positive sequence of numbers α_n $n = 1, 2, \ldots$ such that $\|\phi_n\| < \alpha_n$ for $n \geq 1$. Suppose that $\rho := \limsup \alpha_n{}^1/n < \infty$. Then it is completely standard ([8]) that the Taylor series expansion of ϕ converges at least on the ball $B_r(G)$ of radius $r = 1/\rho$.
(ii) If ϕ admits a majorizing sequence as in (i), then we will say that ϕ is *majorizable*.

We will see in the next section that a very important class of input/output operators from systems and control theory are in point of fact majorizable.

6. Operators with Fading Memory. In this section, we will show that perhaps the most natural class of input/output operators from the systems standpoint are majorizable. Moreover for this class of operators we will even derive *a priori* majorizing sequence. We begin with the following key definition:

DEFINITION (6.1). An analytic map $\phi : B_{r_o}(G) \to H$, $\phi(0) = 0$ has *fading memory* if its nonlinear part $\phi - \phi'(0)$ admits a factorization

$$\phi - \phi'(0) = \hat{\phi}_o W$$

where $\hat{\phi}$ is an analytic defined in some neighborhood of $0 \in G$, and W is a linear Hilbert-Schmidt operator. (In this case, one can assume that there exists an orthonormal basis of eigenvectors for W in G, $\{e_k\}, k = 1, 2, \ldots$ such that $We_k = \lambda_k e_k$ with

$$\|W\|_2^2 := \sum_{k=1}^{\infty} |\lambda_k|^2 < \infty.$$

$\|W\|_2$ is called the *Hilbert-Schmidt norm* of W.)

Remark (6.2). System-theoretically fading memory input/output operators have the property that any two input signals which are close in the recent past but not necessarily close in the remote past will yield present outputs which are close. For more details about this important class of operators see [8].

For fading memory operators, we can construct an explicit majorizing sequence:

LEMMA (6.3). *Let $\phi : B_{r_o}(G) \to H, \phi(0) = 0$, have fading memory. Suppose moreover that if we write*

$$\phi - \phi'(0) = \hat{\phi}_o W$$

as in (6.1), then $\hat{\phi} : B_{r_1}(G) \to B_{r_2}(H)$. Then the sequence

$$\alpha_1 := \|\phi'(0)\|$$

$$\alpha_n := \frac{r_2 e^n \|W\|_2^n}{r_1^n}$$

for $n \geq 2$, is a majorizing sequence for ϕ.

Proof. See [4], lemma (3.5). □

In what follows, we will assume that all of the input/output operators we consider are causal and have fading memory. An interesting and useful property of fading memory operators is the following:

PROPOSITION (6.4). *Notation and hypotheses as in (6.3). Then each ϕ_n (regarded as a linear operator on $G^{\otimes n}$) is compact for $n \geq 2$.*

Proof. See (3.5) of [13]. □

7. Control Theoretic Preliminaries. We start here with the control problem definition. First, we will need to consider the precise kind of input/output operator we will be considering. See [3], [4] for closely related discussions. We will assume that all of the operators we consider are causal and majorizable. Throughout this paper $H^2(\mathbf{C}^k)$ will denote the standard Hardy space of $\mathbf{C}^k$-valued functions on the unit circle (k may be infinite, i.e., in this case $\mathbf{C}^k$ is replaced by h^2, the space of one-sided square summable sequences). We now make the following definition:

DEFINITION (7.1). Let $S : H^2(\mathbf{C}^k) \to H^2(\mathbf{C}^k)$ denote the canonical unilateral right shift. Then we say an input/output operator ϕ is *locally stable* if it is causal and majorizable, $\phi(0) = 0$, and if there exists an $r > 0$ such that $\phi : B_r(H^2(\mathbf{C}^k)) \to H^2(\mathbf{C}^k)$ with $S\phi = \phi \circ S$ on $B_r(H^2(\mathbf{C}^k))$. We set

$$C_l := \{\text{space of locally stable operators}\}.$$

Since the theory we are considering is local, the notion of local stability is sufficient for all of the applications we have in mind. The interested reader can compare this notion, with the more global notions of stability as for example discussed in [6].

The theory we are about to give holds for all plants which admit coprime locally stable factorizations. However, for simplicity we will assume that our plant is also locally stable. Accordingly, let P, W denote locally stable operators, with W invertible. In a typical feedback system [24], P represents the plant, and W the weight or filter on the set of disturbances whose energy is bounded by 1. Now we say that the feedback compensator C *locally stabilizes* the closed loop if the operators $(I + P \circ C)^{-1}$ and $C \circ (I + P \circ C)^{-1}$ are well-defined and locally stable. By a result of [2], C locally stabilizes the closed loop if and only if

$$C = \hat{q} \circ (I - P \circ \hat{q})^{-1} \tag{47}$$

for some $\hat{q} \in C_l$. Notice then that the weighted sensitivity (see [15] and [24] for all the relevant engineering definitions and motivation), $(I + P \circ C)^{-1} \circ W$ can be written as $W - P \circ q$, where $q := \hat{q} \circ W$. (Since W is invertible, the data q and $\hat{q}$ are equivalent.) In this context, we will call such a q, a *compensating parameter*. Note that from the compensating parameter q, we get a locally stabilizing compensator C via the formula (47).

The problem we would like to solve here, is a version of the classical disturbance attenuation problem of [15], [24]. This of course corresponds to the "minimization" of the "sensitivity" $W - P \circ q$ taken over all locally stable q. In order to formulate a precise mathematical problem, we need to say in what sense we want to minimize $W - P \circ q$. This we will do in the next section where we will propose a notion of "sensitivity minimization" which we seems quite natural to analytic input/output operators.

Notice that the linear 4-block problem discussed above is a straightforward generalization of the linear version of the sensitivity minimization problem. Indeed using the notation of the Introduction, we get sensitivity minimization *(1-block problem)* for $f = g = h = 0$.

8. Sensitivity Function. In this section we define a fundamental object, namely a nonlinear version of *sensitivity.* We will see that while the optimal H^∞ sensitivity is a real number in the linear case, the measure of performance which seems to be more natural in this nonlinear setting is a certain function defined in a real interval.

In order to define our notion of sensitivity, we will first have to partially order germs of analytic mappings. All of the input/output operators here will be locally stable. We also follow here our convention that for given $\phi \in C_l$, ϕ_n will denote the bounded linear map on the tensor space $(H^2(\mathbf{C}^k))^{\otimes n}$ associated to the n-linear part of ϕ which we also denote by ϕ_n (and which we always assume without loss of generality is symmetric in its arguments). The context will always make the meaning of ϕ_n clear.

We can now state the following key definitions:

DEFINITIONS (8.1). (i) For $W, P, q \in C_l$ (W is the weight, P the plant, and q the compensating parameter), we define the *sensitivity function* $S(q)$,

$$S(q)(\rho) := \sum_{n=1} \rho^n \|(W - P \circ q)_n\|$$

for all $\rho > 0$ such that the sum converges. Notice that for fixed P and W, for each $q \in C_l$, we get an associated sensitivity function.
(ii) We write $S(q) \lesssim S(\tilde{q})$, if there exists a $\rho_o > 0$ such that $S(q)(\rho) \leq S(\tilde{q})(\rho)$ for all $\rho \in [0, \rho_o]$. If $S(q) \lesssim S(q)$ and $S(\tilde{q}) \lesssim S(q)$, we write $S(q) \cong S(\tilde{q})$. This means that $S(q)(\rho) = S(\tilde{q})(\rho)$ for all $\rho > 0$ sufficiently small, i.e. $S(q)$ and $S(\tilde{q})$ are equal as germs of functions.
(iii) If $S(q) \lesssim S(\tilde{q})$, but $S(\tilde{q}) \not\lesssim S(q)$, we will say that q *ameliorates* $\tilde{q}$. Note that this means $S(q)(\rho) < S(\tilde{q})(\rho)$ for all $\rho > 0$ sufficiently small.

Now with (8.1), we can define a notion of "optimality" relative to the sensitivity function:

DEFINITIONS (8.2). (i) $q_o \in C_l$ is called *optimal* if $S(q_o) \lesssim S(q)$ for all $q \in C_l$.
(ii) We say $q \in C_l$ is "optimal with respect to its n-th term" q_n, if for every n-linear $\hat{q}_n \in C_l$, we have

$$S(q_1 + \cdots + q_{n-1} + q_n + q_{n+1} \ldots) \lesssim S(q_1 + \cdots + q_{n-1} + \hat{q}_n + q_{n+1} + \ldots).$$

If $q \in C_l$ is optimal with respect to all of its terms, then we say that it is "partially optimal."

Clearly, if q is optimal, then it is partially optimal, but the converse may not hold. Notice moreover that if ϕ is a Volterra series, then our definition of sensitivity measures in a precise sense the amplication of energy of each Volterra kernel on signals whose energy is bounded by a given ρ. For this reason, it appears that in this context the definition (8.1) of the sensitivity function $S(q)$ seems physically natural. In the next section, we will discuss a procedure for constructing partially optimal compensating parameters, and then in Section 10 we will show how this procedure leads to the construction of optimal compensating parameters for SISO systems. Of course, from formula (47) above, one can derive the corresponding partially optimal (resp., optimal) compensator from the partially optimal (resp., optimal) compensating parameter.

9. Iterative Commutant Lifting Method. In this section, we discuss the main construction of this paper from which we will derive both partially optimal and optimal compensators relative to the sensitivity function given in (8.1) above. As before, P will denote the plant, and W the weighting operator, both of which we assume are locally stable. As in the linear case, we always suppose that P_1 is an isometry, i.e. P_1 is **inner.** In order to state our results, we will need a few preliminary remarks and to set-up some notation. We refer the interested reader to [13] for the precise proofs of the various results in this section.

We begin by noting the following key relationship:

$$(W - P \circ q)_k = W_k - \sum_{1 \leq j \leq k} \sum_{i_1 + \ldots + i_j = k} P_j(q_{i_1} \otimes \cdots \otimes q_{i_j})$$

Note that once again for ϕ of fading memory, ϕ_n denotes the n-linear part of ϕ, as well as the associated linear operator on the appropriate tensor space.

We are now ready to formulate the *iterative commutant lifting procedure.* Let $\Pi : H^2(\mathbf{C}^k) \to H^2(\mathbf{C}^k) \ominus P_1 H^2(\mathbf{C}^k)$ denote orthogonal projection. Using the linear commutant lifting theorem (CLT) (see [21] for the details), we may choose q_1 such that

$$\|W_1 - P_1 q_1\| = \|\Pi W_1\|.$$

Now given this q_1, we choose (using CLT) q_2 such that

$$\|W_2 - P_2(q_1 \otimes q_1) - P_1 q_2\| = \|\Pi(W_2 - P_2(q_1 \otimes q_1))\|.$$

Inductively, given $q_1, \ldots, q_{n-1}$, set

$$A_n := (W_n - \sum_{2 \leq j \leq n} \sum_{i_1 + \ldots + i_j = n} P_j(q_{i_1} \otimes \cdots \otimes q_{i_j}))$$

for $n \geq 2$. Then from the CLT, we may choose q_n such that

$$\|A_n - P_1 q_n\| = \|\Pi A_n\|. \tag{48}$$

We now come to the key point on the convergence of the iterative commutant lifting method.

PROPOSITION (9.1). *With the above notation, let* $q^{(1)} := q_1 + q_2 + \ldots$. *Then* $q^{(1)} \in C_l$.

Note that given any $q \in C_l$, we can apply the iterative commutant lifting procedure to $W - P \circ q$. Now set

$$S_\Pi(q)(\rho) := \sum_{n=1} \rho^n \|\Pi(W - P \circ q)_n\|.$$

Clearly, $S_\Pi(q) \leq S(q)$ (as functions). We can now state the following result whose proof is immediate from the above discussion:

PROPOSITION (9.2). *Given* $q \in C_l$, *there exists* $\tilde{q} \in C_l$, *such that* $S(\tilde{q}) \equiv S_\Pi(q)$. *Moreover* $\tilde{q}$ *may be constructed from the iterated commutant lifting procedure.*

Moreover, we easily have the following result:

PROPOSITION (9.3). q *is partially optimal if and only if* $S(q) \cong S_\Pi(q)$.

We can now summarize the above discussion with the following:

THEOREM (9.4). *For given* P *and* W *as above, any* $q \in C_l$ *is either partially optimal or can be ameliorated by a partially optimal compensating parameter.*

Proof. Immediate from (9.1), (9.2), and (9.3). □

It is important to emphasize that a partially optimal compensating parameter need not be optimal in the sense of (9.2)(i). Basically, what we have shown here is that using the iterated commutant lifting procedure, we can ameliorate any given design. The question of optimality will be considered in the next section.

Optimal Compensators. In this section we will derive our main results about optimal compensators. Basically, we will show that in the single input / single output setting, the iterated commutant lifting procedure leads to an optimal design. We begin with the following:

THEOREM (10.1). *There exist optimal compensators.*

Proof. See (7.1) of [13]. □

For the construction of the optimal compensator in Theorem (10.3) below, we will need one more technical result. Accordingly, we will need to set-up a bit more notation. First set $H^2 := H^2(\mathbf{C})$, and $H^\infty := H^\infty(\mathbf{C})$ (the space of bounded analytic complex-valued functions on the unit disc). Let $m \in H^\infty$ be a nonconstant inner function, let $\Pi_1 : H^2 \to H^2 \ominus mH^2 =: H(m)$ denote orthogonal projection, and set $T := \Pi_1 S|H(m)$, where S is the canonical unilateral shift on H^2. (T is the compressed shift.) For H a complex separable Hilbert space, let $S_\infty : H \to H$ denote a unilateral shift, i.e. an isometric operator with no unitary part. This means that $S_\infty^{*n} \to 0$ for all $h \in H$ as $n \to \infty$. (See [17] and [21].) We can now state the following generalization of a nice result which appears in [20]:

LEMMA (10.2). *Notation as above. Let* $A : H \to H^2 \ominus mH^2$ *be a bounded linear operator which attains its norm, i.e. such that there exists* $h_o \in H$ *with* $\|Ah_o\| = \|A\|\|h_o\| \neq 0$. *Suppose moreover that*

$$AS_\infty = TA.$$

Then there exists a unique minimal intertwining dilation B *of* A, *i.e. an operator* $B : H \to H^2$ *such that* $BS_\infty = SB$, $||A|| = ||B||$, *and* $\Pi_1 B = A$.

Proof. See (7.2) of [13]. □

We now come to the main result of this section:

THEOREM (10.3). *Let* W *and* P *be single SISO locally stable operators, with* W *the weight and* P *the plant. Suppose that* ΠW_j *is compact for* $j \geq 1$ *and* ΠP_k *is compact for* $k \geq 2$. *(*$\Pi : H^2 \to H^2 \ominus P_1 H^2$ *denotes orthogonal projection.) Let* q_{opt} *be a partially optimal compensating parameter as constructed by the iterated commutant lifting procedure. Then* q_{opt} *is optimal.*

Proof. First of all, since ΠW_1 attains its norm, from (10.2) we have that the optimal q_1 constructed relative to W_1 and P_1 is unique. (Actually in this special case since we are working in H^2, this follows from [20].) Now from our above hypotheses, each ΠA_k is compact for $k \geq 2$, and hence each ΠA_k attains its norm. Therefore by (10.2) each optimal q_k constructed by the iterated commutant lifting procedure is unique. Theorem (10.3) now follows immediately from (9.4). □

COROLLARY (10.4). *Let* P *and* W *be locally stable and SISO, with linear part* P_1 *rational. Then the partially optimal compensating parameter* q_{opt} *constructed by the iterated commutant lifting procedure is optimal.*

Proof. Indeed, since P_1 is SISO rational (recall that we also always assume that P_1 is inner), $H^2 \ominus P_1 H^2$ is finite dimensional, and so we are done by (10.3). □

Remarks (10.5). (10.4) gives a constructive procedure for finding the optimal compensator under the given hypotheses. Indeed, when P_1 is SISO rational, the iterative commutant lifting procedure can be reduced to ***finite dimensional matrix calculations***. In a subsequent paper, we plan to show that when the hypotheses of (10.3) are satisfied, the skew Toeplitz theory of [7] provides an algorithmic design procedure for distributed nonlinear systems as well.

REFERENCES

[1] V. M. ADAMJAN, D. Z. AROV, AND M. G. KREIN, *Infinite Hankel matrices and generalized problems of Caratheodory-Fejer and F. Riesz*, Functional Anal. Appl., 2 (1968), pp. 1–18.

[2] V. ANANTHARAM AND C. DESOER, *On the stabilization of nonlinear systems*, IEEE Trans. Automatic Control, AC-29 (1984), pp. 569–573.

[3] J. BALL, C. FOIAS, J. W. HELTON, AND A. TANNENBAUM, *On a local nonlinear commutant lifting theorem*, Indiana J. Mathematics, 36 (1987), pp. 693–709.

[4] J. BALL, C. FOIAS, J. W. HELTON, AND A. TANNENBAUM, *Nonlinear interpolation theory in* H^∞, in *Modelling, Robustness, and Sensitivity in Control Systems*, (edited by Ruth Curtain), NATO-ASI Series, Springer-Verlag, New York, 1987.

[5] J. BALL, C. FOIAS, J. W. HELTON, AND A. TANNENBAUM, *A Poincare-Dulac approach to a nonlinear Beurling-Lax-Halmos theorem*, Journal of Math. Anal. and Applications (to appear).

[6] J. BALL AND J. W. HELTON, *Sensitivity bandwidth optimization for nonlinear feedback systems*, Technical Report, Department of Mathematics, University of California at San Diego, 1988.

[7] H. BERCOVICI, C. FOIAS, AND A. TANNENBAUM, *On skew Toeplitz operators, I*, Operator Theory: Advances and Applications, 29 (1988), pp. 21–44.

[8] S. BOYD AND L. CHUA, *Fading memory and the problem of approximating nonlinear operators with Volterra series*, IEEE Trans. Circuits and Systems, CAS-32 (1985), pp. 1150–1161.

[9] A. FEINTUCH AND B. FRANCIS, *Uniformly optimal control of linear systems*, Automatica, 21 (1986), pp. 563-574.

[10] C. FOIAS AND A. FRAZHO, *On the Schur representation in the commutant lifting theorem, I*, Operator Theory: Advances and Applications, 18 (1986), pp. 207–217.

[11] C. FOIAS AND A. TANNENBAUM, ON THE NEHARI PROBLEM FOR A CERTAIN CLASS OF L^∞ FUNCTIONS APPEARING IN CONTROL THEORY, J. Functional Analysis, 74 (1987), pp. 146–159.

[12] C. FOIAS AND A. TANNENBAUM, *On the four block problem, II: the singular system,*, Integral Equations and Operator Theory, 11 (1988), pp. 726–767.

[13] C. FOIAS AND A. TANNENBAUM, *Weighted optimization theory for nonlinear systems*, SIAM J. on Control and Optimization (to appear).

[14] C. FOIAS, A. TANNENBAUM, AND G. ZAMES, *Some explicit formula for the singular values of certain Hankel operators with factorizable symbol*, SIAM J. on Math. Analysis, 19 (1988), pp. 1081–1089.

[15] B. FRANCIS, *A Course in, H^∞ Control Theory*, McGraw-Hill, New York, 1981.

[16] E. HILLE AND R. S. PHILLIPS, *Functional Analysis and Semigroups*, AMS Colloquium Publications XXIII, Providence, 1957.

[17] N. K. NIKOLSKII, *Treatise on the Shift Operator*, Springer, New York, 1986.

[18] W. J. RUGH, *Nonlinear System Theory: the Volterra/Wiener Approach*, Johns Hopkins Univ. Press, Baltimore, 1981.

[19] D. SARASON, *Generalized Interpolation in H^∞*, Transactions of the AMS, 127 (1967), pp. 179–203.

[20] D. SARASON,, *Function Theory on the Unit Circle*, Lecture Notes, Virginia Polytechnic Institute, 1978.

[21] B. SZ.-NAGY AND C. FOIAS, *Harmonic Analysis of Operators on Hilbert Space*, North-Holland Publishing Company, Amsterdam, 1970.

[22] N. WIENER, *Nonlinear Problems in Random Theory*, Technology Press of M.I.T., Cambridge, Mass., 1958.

[23] N. YOUNG, *The Nevanlinna-Pick problem for matrix-valued functions*, J. Operator Theory, 15 (1986), pp. 239–265.

[24] G. ZAMES, *Feedback and optimal sensitivity: model reference transformations, multiplicative seminorms, and approximate inverses*, IEEE Trans. Auto. Control, AC-26 (1981), pp. 301–320.

[25] C. FOIAS AND A. TANNENBAUM, *On the singular values of the four block operator and certain generalized interpolation problems*, Integral Equations and Operator Theory (to appear).

SNIPPETS OF $\mathcal{H}_\infty$ CONTROL THEORY

BRUCE FRANCIS*

1. Introduction. The subject of $\mathcal{H}_\infty$ control, now over a decade old, has several main parts: the mathematical theory of $\mathcal{H}_\infty$-optimization; extensions from the standard linear time-invariant framework to more general settings; methods for computing optimal controllers and associated software; results on design constraints and achievable performance, including tradeoffs between conflicting objectives; case studies of control system designs. In the past there have been several plenary talks on the big picture [Zames, 1979, 1988; Francis, 1986], a few expository papers [Francis and Doyle, 1987; Safonov *et al.*, 1987], a critical analysis of the value of $\mathcal{H}_\infty$ optimization as a design methodology [Freudenberg and Looze, 1986], a chapter in a research monograph [Vidyasagar, 1985], a set of lecture notes [Helton, 1987], and a published set of course notes [Francis, 1987].

This paper is directed at non-experts and its goal is to pique their interest. Four illustrative fragments from the subject are extracted for presentation, namely:

(i) The problem of robust stabilization with an uncertain plant gain (also called the optimal gain margin problem). Posed and solved by Tannenbaum (1980, 1982), this is one of the earliest problems in the subject and contains a nice blend of control theory and mathematics: feedback stability theory, conformal mapping, and the Nehari problem (or Nevanlinna-Pick interpolation).

(ii) $\mathcal{H}_\infty$-optimal state-feedback control. Although $\mathcal{H}_\infty$-optimal control started out in the framework of input-output models, i.e., linear operators, state-space methods were gradually invoked for the purpose of computation. Now, work from the 1960s on linear-quadratic optimal control and differential games is having application in $\mathcal{H}_\infty$-optimal control.

(iii) A distance formula for the model-matching problem. The model-matching problem is to minimize the $\mathcal{H}_\infty$-norm of $T_1 - T_2QT_3$, where the T_is are given $\mathcal{H}_\infty$-matrices and Q is sought in $\mathcal{H}_\infty$. The objective is to find a formula for the minimum norm. For the scalar-valued case there is the famous formula of Sarason (1967).

(iv) Tradeoffs in classical control theory. Plant uncertainty places a limit on high performance of a feedback control system. Some analysis methods are available, but so far a complete synthesis theory is not.

The notation is fairly standard: $\mathcal{H}_2$ and $\mathcal{H}_\infty$ are Hardy spaces with respect to the right half-plane; $\mathcal{L}_2$ and $\mathcal{L}_\infty$ are Lebesgue spaces with respect to the imaginary axis; prefix $\mathcal{R}$ stands for real-rational.

*Department of Electrical Engineering, University of Toronto, Toronto, Canada M5S 1A4

2. Robust stability for gain uncertainty. In general, a property of a control system, such as its stability, is said to be *robust* if it is preserved under plant perturbation, of a type which would be made precise in any given context. A control system must be robust because of unavoidable modeling errors. In this section we look at the problem of stabilizing a plant with an uncertain multiplicative real constant [Tannenbaum, 1980, 1982].

Consider the setup in the following figure:

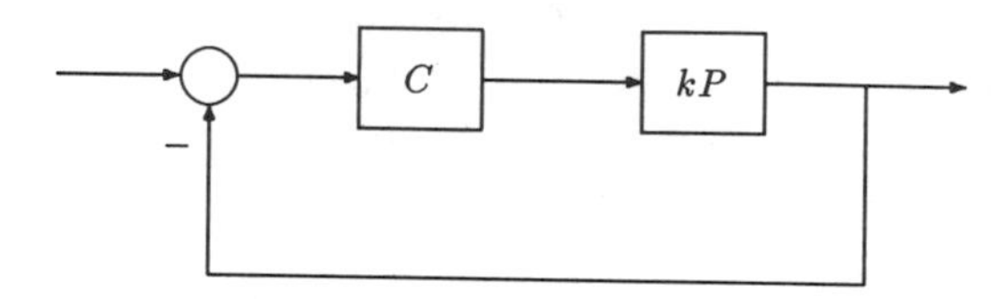

Here P and C are real-rational transfer functions for, respectively, the nominal plant and the controller. We'll assume P is strictly proper and C is proper. It is supposed that the real gain k is known only to the extent that it belongs to an interval, $[k_0, k_1]$, with $0 < k_0 < k_1 < \infty$. By absorbing a scaling factor into P, we can assume without loss of generality that $k_0 = 1$.

Consider the question, Does there exist a C which stabilizes kP for all k in $[1, k_1]$? Such a controller is robustly stabilizing for the specified uncertainty model. The answer is yes if k_1 is small enough, by continuity. But in general there is an upper bound on k_1 for an affirmative answer. The least upper bound will be denoted $k_{\max}$. The objective is to find a formula for $k_{\max}$ and then a procedure for computing a robustly stabilizing controller when $k_1 < k_{\max}$.

Bring in a factorization of P, $P = N/M$, where N and M are coprime functions in the ring $\mathcal{RH}_\infty$. Because of this coprimeness, there exist functions X and Y in $\mathcal{RH}_\infty$ satisfying the "corona equation"

$$MX + NY = 1$$

Now let α denote the distance in $\mathcal{H}_\infty$ from NY to the subspace of functions of the form NMQ, $Q \in \mathcal{H}_\infty$, i.e.,

$$\alpha := \text{dist}\,(NY, NM\mathcal{H}_\infty)$$

THEOREM 1. *There exists a C stabilizing kP for all $1 \le k \le k_1$ iff*

$$\frac{\sqrt{k_1}+1}{\sqrt{k_1}-1} > \alpha$$

Noting that $(\sqrt{k_1}+1)/(\sqrt{k_1}-1)$ is monotonically decreasing, we conclude that

$$k_{\max} = \begin{cases} \left(\frac{\alpha+1}{\alpha-1}\right)^2, & \alpha > 1 \\ \infty, & \alpha \le 1 \end{cases}$$

Proof. The proof is in several steps. First, define two subsets of the complex plane:

$$\mathbf{H} := \{\ \mathrm{Re}\ s \geq 0\} \cup \{\infty\}$$
$$\mathbf{G} := \{s \in \mathbf{C} : s \notin [a, \infty)\}, \quad a := \frac{1}{k_1 - 1}$$

Claim 1. There exists a C which stabilizes kP for all $1 \leq k \leq k_1$ iff there exists F in $\mathcal{RH}_\infty$ such that

$$F\mathbf{H} \subset \mathbf{G}, N|F, M|(1+F)$$

(division in $\mathcal{H}_\infty$).

Parametrize all C's stabilizing P [Vidyasagar, 1985]:

$$C = (Y - MQ)/(X + NQ), \quad Q \in \mathcal{RH}_\infty$$

Then for $k > 1$, C stabilizes kP

$$\begin{aligned}
&\Leftrightarrow M(X + NQ) + kN(Y - MQ) \text{ is a unit of } \mathcal{RH}_\infty \\
&\Leftrightarrow MX + kNY - (k-1)NMQ \text{ is a unit of } \mathcal{RH}_\infty \\
&\Leftrightarrow 1 + (k-1)(NY - NMQ) \text{ is a unit of } \mathcal{RH}_\infty \\
&\Leftrightarrow \frac{1}{k-1} + (NY - NMQ) \text{ is a unit of } \mathcal{RH}_\infty \\
&\Leftrightarrow -(NY - NMQ)\mathbf{H} \subset \{s \in \mathbf{C} : s \neq \frac{1}{k-1}\}
\end{aligned}$$

Hence C stabilizes kP for all $1 \leq k \leq k_1$ iff

$$-(NY - MNQ)\mathbf{H} \subset \mathbf{G}$$

Thus there exists C stabilizing kP for all $1 \leq k \leq k_1$

$$\begin{aligned}
&\Leftrightarrow (\exists Q \in \mathcal{RH}_\infty) \ -(NY - MNQ)\mathbf{H} \subset \mathbf{G} \\
&\Leftrightarrow (\exists F \in \mathcal{RH}_\infty) F\mathbf{H} \subset \mathbf{G}, N|F, M|(1+F) \quad \Box
\end{aligned}$$

Now bring in a conformal mapping from $\mathbf{G}$ onto the open unit disk:

$$\Phi(s) := \frac{1 - \sqrt{1 - a^{-1}s}}{1 + \sqrt{1 - a^{-1}s}}$$

Claim 2.

$$\begin{aligned}
&F \in \mathcal{RH}_\infty, F\mathbf{H} \subset \mathbf{G}, N|F, M|(1+F), G := [\Phi(-1)]^{-1}\Phi F \\
&\Rightarrow G \in \mathcal{H}_\infty, \|G\|_\infty < |\Phi(-1)|^{-1}, N|G, M|(G-1)
\end{aligned}$$

The first two properties of G follow from the first two of F. The proof that $N|G$ is illustrated like this. Suppose z is a zero of N in $\mathbf{H}$ of multiplicity, say, 2. Then $N|F$ implies

$$F(z) = 0, F'(z) = 0$$

Thus

$$\begin{aligned} G(z) &= [\Phi(-1)]^{-1}\Phi(F(z)) = [\Phi(-1)]^{-1}\Phi(0) = 0 \\ G'(z) &= [\Phi(-1)]^{-1}\frac{d\Phi}{dF}(F(z))F'(z) = [\Phi(-1)]^{-1}\frac{d\Phi}{dF}(0) \times 0 = 0 \end{aligned}$$

so z is a zero of G of multiplicity at least 2. Doing this for all zeros of N in $\mathbf{H}$ shows that $N^{-1}G \in \mathcal{H}_\infty$. Similarly for the condition $M|(G-1)$. □

Claim 3.

$$\begin{aligned} &G \in \mathcal{RH}_\infty, \|G\|_\infty < |\Phi(-1)|^{-1}, N|G, M|(G-1), F := \Phi^{-1}[\Phi(-1)G] \\ &\Rightarrow F \in \mathcal{RH}_\infty, F\mathbf{H} \subset \mathbf{G}, N|F, M|(1+F) \end{aligned}$$

The proof is like the previous one except we note that if G is real-rational, then so is F. This can be observed by inverting Φ:

$$F = a\left[1 - \left(\frac{1-\Phi(-1)G}{1+\Phi(-1)G}\right)^2\right] \qquad \square$$

Now for the proof of necessity in the theorem. By Claims 1 and 2

$$(\exists G \in \mathcal{H}_\infty)\|G\|_\infty < |\Phi(-1)|^{-1}, N|G, M|(G-1)$$

Define Z via $G = NY + NMZ$. It is claimed that $Z \in \mathcal{H}_\infty$. Indeed,

$$\begin{aligned} NMZ &= G - NY, N|G \Rightarrow MZ \in \mathcal{H}_\infty \\ NMZ &= G - 1 + MX, M|(G-1) \Rightarrow NZ \in \mathcal{H}_\infty \end{aligned}$$

and

$$MX + NY = 1 \Rightarrow Z = XMZ + YNZ \in \mathcal{H}_\infty$$

Thus we have

$$\begin{aligned} &\text{dist}\,(NY, NM\mathcal{H}_\infty) \\ &\le \|NY + NMZ\|_\infty \\ &= \|G\|_\infty \\ &< |\Phi(-1)|^{-1} = \frac{\sqrt{k_1}+1}{\sqrt{k_1}-1} \end{aligned}$$

Finally, sufficiency. We omit the proof that the distances from NY to $NM\mathcal{H}_\infty$ and $NM\mathcal{RH}_\infty$ are equal. Then

$$\text{dist}\,(NY, NM\mathcal{RH}_\infty) < |\Phi(-1)|^{-1}$$

so there exists Z in $\mathcal{RH}_\infty$ such that

$$\|NY + NMZ\|_\infty < |\Phi(-1)|^{-1}$$

Set $G := NY + NMZ$. Then

$$G \in \mathcal{RH}_\infty, \|G\|_\infty < |\Phi(-1)|^{-1}, N|G, M|(G-1)$$

By Claim 3 then Claim 1, the conclusion follows. □

How is α to be computed? Assume P has no poles on the imaginary axis and at least one pole in Re $s > 0$. (If P is stable then $\alpha = 1$.) Bring in inner-outer factorizations

$$N = N_i N_o, M = M_i M_o$$

Thus $M_o^{-1} \in \mathcal{RH}_\infty$ and we have

$$\alpha = \text{dist }(N_i N_o Y, N_i N_o M_i M_o \mathcal{H}_\infty) = \text{dist }(N_o Y, N_o M_i \mathcal{H}_\infty)$$

However, it can be shown that the distances from $N_o Y$ to $N_o M_i \mathcal{H}_\infty$ and $M_i \mathcal{H}_\infty$ are equal. Therefore

$$\alpha = \text{dist }(N_o Y, M_i \mathcal{H}_\infty)$$

so α equals the norm of the Hankel operator with symbol $N_o Y M_i^{-1}$.

Finally, the proof of sufficiency leads to the following procedure for computing a C stabilizing kP for all $1 \leq k \leq k_1$, where $k_1 < k_{\max}$:

Step 1 Get functions N, M, X, Y in $\mathcal{RH}_\infty$ such that

$$P = N/M, \quad MX + NY = 1$$

Step 2 Find Z in $\mathcal{RH}_\infty$ such that

$$\|NY + NMZ\|_\infty < \frac{\sqrt{k_1} + 1}{\sqrt{k_1} - 1}$$

Step 3 Set $G = NY + NMZ$.

Step 4 Set $F = \Phi^{-1}[\Phi(-1)G]$.

Step 5 Solve $-NY + NMQ = F$ for Q.

Step 6 Set $C = (Y - MQ)/(X + NQ)$.

For more on robust stability, see Chapter 7 of [Vidyasagar, 1985] and the references therein.

3. State-space methods. $\mathcal{H}_\infty$-control theory began in the input-output framework of operators and complex functions [Zames, 1981]. Indeed, one of Zames' main theses is that input-output models are more appropriate than state models in many circumstances because uncertainty is often more readily accommodated in the former. However, computers must be fed a finite amount of data, so at some point the model must be represented by finitely many numbers. Because of the wealth of experience in computing with state-space models, it is natural to start with them, by getting a realization if necessary.

We begin with the computation of $\mathcal{H}_\infty$-norm. Let $G(s)$ be real-rational, strictly proper, and stable (analytic in the closed right half-plane). So $G \in \mathcal{RH}_\infty$. Instead of computing $\|G\|_\infty$ directly, we seek a test for the inequality $\|G\|_\infty \leq \gamma$. Obviously, by testing this inequality for γ in a suitable finite set, we can estimate $\|G\|_\infty$ to any desired accuracy. By scaling, it suffices to test the inequality $\|G\|_\infty \leq 1$.

Let (A, B, C) be a realization, i.e.,

$$G(s) = C(s - A)^{-1}B$$

with A stable (all eigenvalues in the open left half-plane). The following, Lemma 5 in [Willems, 1971], provides a (nonconstructive) test in terms of solvability of a Riccati equation:

Suppose (A, B, C) is minimal. Then $\|G\|_\infty \leq 1$ iff $\exists X$ such that

$$\begin{aligned} &A'X + XA + XBB'X + C'C = 0 \\ &\sigma(A + BB'X) \subset \{s : \text{Re } s \leq 0\} \end{aligned}$$

The necessity part of this result is Theorem 2, Section 25, in [Brockett, 1970]. Let's restate the result in terms of the Hamiltonian matrix

$$H := \begin{bmatrix} A & BB' \\ -C'C & -A' \end{bmatrix}$$

Let A be $n \times n$; then H is $2n \times 2n$.

Suppose (A, B, C) is minimal. Then $\|G\|_\infty \leq 1$ iff $\exists$ a subspace $\mathcal{V}$ of $\mathbf{R}^{2n}$ with the three properties

$$\begin{aligned} &\mathcal{V} \textit{ and } Im \begin{bmatrix} 0 \\ I \end{bmatrix} \textit{ are complementary} \\ &\mathcal{V} \textit{ is invariant under } H \\ &\sigma(H|\mathcal{V}) \subset \{s : Re\ s \leq 0\} \end{aligned}$$

The first property means that $\mathcal{V}$ has the form

$$\mathcal{V} = \text{Im}\begin{bmatrix} I \\ X \end{bmatrix}$$

Then the second property is equivalent to the Riccati equation.

It turns out to be simpler (in principle) to test the strict inequality [Boyd *et al.*, 1988]:

LEMMA 1. $\|G\|_\infty < 1$ *iff* H *has no imaginary axis eigenvalues.*

(Note that minimality is unnecessary.) It's convenient to write a transfer matrix as

$$\left[\begin{array}{c|c} A & B \\ \hline C & D \end{array}\right] := D + C(s-A)^{-1}B$$

and to define $G^{\sim}(s) := G(-s)'$.

Proof. We have

$$(I - G^{\sim}G)^{-1}(s) = \left[\begin{array}{cc|c} A & BB' & B \\ -C'C & -A' & 0 \\ \hline 0 & B' & I \end{array}\right]$$

so H is the A-matrix of $(I - G^{\sim}G)^{-1}$. It is easy to check that the pair

$$H, \quad \begin{bmatrix} B \\ 0 \end{bmatrix}$$

has no uncontrollable modes on the imaginary axis. Nor has the pair

$$[0 \quad B'], \quad H$$

any unobservable modes there. Thus, H has no eigenvalues on the imaginary axis iff $(I-G^{\sim}G)^{-1}$ has no poles there, i.e., $(I-G^{\sim}G)^{-1} \in \mathcal{RL}_\infty$. So it suffices to prove that

$$\|G\|_\infty < 1 \Leftrightarrow (I - G^{\sim}G)^{-1} \in \mathcal{RL}_\infty$$

If $\|G\|_\infty < 1$, then

$$I - G(j\omega)^*G(j\omega) > 0, \quad \forall\omega$$

and hence $(I - G^{\sim}G)^{-1} \in \mathcal{RL}_\infty$.

Conversely, if $\|G\|_\infty \geq 1$, then $\sigma_{max}[G(j\omega)] = 1$ for some ω, i.e., 1 is an eigenvalue of $G(j\omega)^*G(j\omega)$, so $I - G(j\omega)^*G(j\omega)$ is singular. □

Now we turn to $\mathcal{H}_\infty$-optimal control via state-feedback. Consider the system

$$\dot{x} = Ax + Bu + Ev, \quad x(0) = 0$$

with control input u and disturbance input v. The control law is restricted to be state-feedback, $u = Fx$, and stabilizing, and the cost function is taken to be the usual quadratic one except maximized over all v's in the unit ball of $\mathcal{L}_2[0, \infty)$:

$$\sup_{\|v\|_2 \le 1} \int_0^\infty x(t)'Qx(t) + u(t)'u(t)\; dt$$

We'll assume $Q \ge 0$. Then there exists the factorization

$$\begin{bmatrix} Q & 0 \\ 0 & I \end{bmatrix} = \begin{bmatrix} C' \\ D' \end{bmatrix} [C \quad D]$$

Notice that $D'C = 0$ and $D'D = I$. Define

$$z := Cx + Du$$

to get the equivalent expression for the cost function

$$\sup_{\|v\|_2 \le 1} \|z\|_2$$

This supremum equals the $\mathcal{H}_\infty$-norm of the transfer matrix T_{zv} from v to z. In this way we are led to the *$\mathcal{H}_\infty$-state-feedback problem*:

$$\begin{aligned} &\dot{x} = Ax + Bu + Ev, \quad x(0) = 0 \\ &z = Cx + Du \\ &u = Fx \\ &\text{minimize } \{\|T_{zv}\|_\infty : A + BF \text{ is stable}\} \end{aligned}$$

The infimal cost is

$$\alpha := \inf_F \{\|T_{zv}\|_\infty : A + BF \text{ is stable}\}$$

Along with $D'C = 0$, $D'D = I$, we shall assume in addition that (A, B) is stabilizable and (C, A) is detectable. As with $\mathcal{H}_\infty$-norm computation, it is difficult to compute α directly. Instead, one tests if $\alpha < \gamma$ for several values of γ. Again, by scaling if necessary, we may as well assume $\gamma = 1$.

The above problem has the structure of a game: there are two players, u and v, v trying to maximize $\|z\|_2$ subject to the constraint $\|v\|_2 \le 1$, u trying to minimize this maximum subject to the constraint $u = Fx$. Let's drop the constraint on v and consider the associated cost function

$$\begin{aligned} J(u,v) &:= \|z\|_2^2 - \|v\|_2^2 \\ &= \int_0^\infty x'C'Cx + u'u - v'v\; dt \end{aligned}$$

Early references for such linear-quadratic differential games are [Medanic, 1967], [Rhodes and Luenberger, 1969], and [Mageirou, 1976]. There it is shown that the relevant Hamiltonian matrix is

$$H_\infty := \begin{bmatrix} A & EE' - BB' \\ -C'C & -A' \end{bmatrix}$$

The solution of our $\mathcal{H}_\infty$-state-feedback problem is indeed in terms of this matrix, but we need some preliminaries.

Consider the $2n \times 2n$ Hamiltonian matrix

$$H := \begin{bmatrix} A & R \\ Q & -A' \end{bmatrix}$$

with Q and R symmetric. Assume H has no eigenvalues on the imaginary axis. Then it must have n in Re $s < 0$ and n in Re $s > 0$. Consider the two spectral subspaces $\mathcal{X}_-(H)$ and $\mathcal{X}_+(H)$: the former is the span of all (real) generalized eigenvectors corresponding to eigenvalues in Re $s < 0$; the latter, to eigenvalues in Re $s > 0$. In this case $\mathcal{X}_-(H)$ and $\mathcal{X}_+(H)$ both have dimension n. Finding a basis for $\mathcal{X}_-(H)$, stacking the basis vectors up to form a matrix, and partitioning the matrix, we get

$$\mathcal{X}_-(H) = \text{Im} \begin{bmatrix} X_1 \\ X_2 \end{bmatrix}$$

where $X_1, X_2 \in \mathbb{R}^{n \times n}$. If X_1 is nonsingular, i.e., if the two subspaces

$$\mathcal{X}_-(H), \quad \text{Im} \begin{bmatrix} 0 \\ I \end{bmatrix}$$

are complementary, we can set $X := X_2 X_1^{-1}$ to get

$$\mathcal{X}_-(H) = \text{Im} \begin{bmatrix} I \\ X \end{bmatrix}$$

Then X is uniquely determined by H, i.e., $H \mapsto X$ is a function, which will be denoted *Ric*; thus, $X = \mathit{Ric}(H)$. The domain of *Ric*, denoted *dom*(*Ric*), thus consists of Hamiltonian matrices H with two properties, namely, H has no eigenvalues on the imaginary axis and the two subspaces

$$\mathcal{X}_-(H), \quad \text{Im} \begin{bmatrix} 0 \\ I \end{bmatrix}$$

are complementary. It follows that X satisfies the Riccati equation

$$A'X + XA + XRX - Q = 0$$

THEOREM 2. $\alpha < 1$ *only if* $H_\infty \in \mathit{dom}(\mathit{Ric})$ *and* $\mathit{Ric}(H_\infty) > 0$. *Conversely, if these two conditions hold, the control law*

$$u = F_\infty x, \; F_\infty := -B'X_\infty, \; X_\infty := \mathit{Ric}(\mathcal{H}_\infty)$$

yields $\|T_{zv}\|_\infty < 1$.

Proof of sufficiency We first show that $A+BF_\infty$ is stable. The Riccati equation for X_∞ can be re-arranged to give

$$(A+BF_\infty)'X_\infty + X_\infty(A+BF_\infty) + (C+DF_\infty)'(C+DF_\infty) + X_\infty EE'X_\infty = 0$$

This is a Lyapunov equation with X_∞ positive definite. Furthermore, the pair

$$A+BF_\infty, (C+DF_\infty)'(C+DF_\infty) + X_\infty EE'X_\infty$$

is detectable since (C, A) is. Thus, $A+BF_\infty$ is stable.

The controlled system is

$$\begin{aligned} \dot{x} &= (A+BF_\infty)x + Ev \\ z &= (C+DF_\infty)x \end{aligned}$$

so

$$T_{zv}(s) = \left[\begin{array}{c|c} A+BF_\infty & E \\ \hline C+DF_\infty & 0 \end{array}\right]$$

To confirm that $\|T_{zv}\|_\infty < 1$, according to Lemma 1 we should check that

$$H := \begin{bmatrix} A+BF_\infty & EE' \\ -(C+DF_\infty)'(C+DF_\infty) & -(A+BF_\infty)' \end{bmatrix}$$

has no imaginary axis eigenvalues. Now

$$H = H_\infty + \begin{bmatrix} I \\ X_\infty \end{bmatrix} BB' \begin{bmatrix} -X_\infty & I \end{bmatrix}$$

so $H = H_\infty$ on the subspace $\mathcal{X}_-(H_\infty)$, and hence $H|\mathcal{X}_-(H_\infty)$ is stable. By symmetry of its spectrum, H has no imaginary axis eigenvalues. □

For the proof of necessity, which is much harder, see [Doyle *et al.*, 1988]. There are many other papers on state-space methods and $\mathcal{H}_\infty$-optimal control; a few are [Ball and Cohen, 1987], [Ball and Ran, 1987], [Bernstein and Haddad, 1988], [Glover and Doyle, 1988], [Khargonekar *et al.*, 1988], [Limebeer and Hung, 1987], and [Petersen, 1987].

4. A distance formula for the model-matching problem. Standard practice in computing $\mathcal{H}_\infty$-optimal controllers is to do it in two steps: first compute the minimum value of the cost function; then compute the optimal controller. The first step can be reduced to a distance problem, find the distance from one operator to a subspace of operators. This section illustrates how this goes.

Let T_1, T_2, and T_3 be matrices in $\mathcal{H}_\infty$. In the $\mathcal{H}_\infty$ context the model-matching problem is to find a fourth matrix Q in $\mathcal{H}_\infty$ to minimize the $\mathcal{H}_\infty$-norm of the difference $T_1 - T_2QT_3$. The model-matching error is

$$\alpha := \inf_{Q \in \mathcal{H}_\infty} \|T_1 - T_2QT_3\|_\infty$$

Our goal in this section is a formula for α in terms of the original data, T_1, T_2, T_3.

For a matrix G in $\mathcal{H}_\infty$, define $G^\sim(s) := G(-\bar{s})^*$, so that $G^\sim = G^*$ on the imaginary axis.

We may as well assume T_2 is inner ($T_2^\sim T_2 = I$); otherwise, do an inner-outer factorization of T_2 and absorb the outer factor into Q. Similarly, assume T_3 is co-inner ($T_3 T_3^\sim = I$).

The simplest case is when all T_is are of dimension 1×1. Then we may as well suppose in addition that $T_3 = 1$. It's a basic fact that $\|T_1 - T_2 Q\|_\infty$ equals the norm of the multiplication operator

$$f \mapsto (T_1 - T_2 Q) f : \mathcal{H}_2 \to \mathcal{H}_2$$

Let $\mathcal{K}$ denote the orthogonal complement of $T_2 \mathcal{H}_2$ in $\mathcal{H}_2$ and let Π be the orthogonal projection from $\mathcal{H}_2$ onto $\mathcal{K}$. Then for every f in $\mathcal{H}_2$

$$\|(T_1 - T_2 Q) f\|_2 \geq \|\Pi (T_1 - T_2 Q) f\|_2 = \|\Pi T_1 f\|_2$$

By defining Ξ to be the operator

$$f \mapsto \Pi T_1 f : \mathcal{H}_2 \to \mathcal{K}$$

and sup-ing over f we get that

$$\|T_1 - T_2 Q\|_\infty \geq \|\Xi\|$$

Now inf-ing over Q gives $\alpha \geq \|\Xi\|$. In fact equality holds: $\alpha = \|\Xi\|$. This is Sarason's famous formula [Sarason, 1967], a special case of the commutant lifting theorem.

Now back to the general case. Fix Q in $\mathcal{H}_\infty$. It's easy to check that the matrix

$$V := \begin{bmatrix} T_2^\sim \\ I - T_2 T_2^\sim \end{bmatrix}$$

has the property $V^\sim V = I$, which implies that

$$\|T_1 - T_2 Q T_3\|_\infty = \left\| \begin{bmatrix} T_2^\sim \\ I - T_2 T_2^\sim \end{bmatrix} (T_1 - T_2 Q T_3) \right\|_\infty$$

Similarly for the matrix

$$[\, T_3^\sim \quad I - T_3^\sim T_3 \,]$$

Thus we have

$$\|T_1 - T_2 Q T_3\|_\infty = \left\| \begin{bmatrix} T_2^\sim \\ I - T_2 T_2^\sim \end{bmatrix} (T_1 - T_2 Q T_3) [\, T_3^\sim \quad I - T_3^\sim T_3 \,] \right\|_\infty$$

Define

$$R := \begin{bmatrix} T_2^\sim \\ I - T_2 T_2^\sim \end{bmatrix} T_1 [\, T_3^\sim \quad I - T_3^\sim T_3 \,]$$

and note that

$$\begin{bmatrix} T_2^{\sim} \\ I - T_2 T_2^{\sim} \end{bmatrix} T_2 = \begin{bmatrix} I \\ 0 \end{bmatrix}$$

and

$$T_3 \, [\, T_3^{\sim} \quad I - T_3^{\sim} T_3 \,] = [\, I \quad 0 \,]$$

We conclude that

$$\|T_1 - T_2 Q T_3\|_\infty = \left\| R - \begin{bmatrix} Q & 0 \\ 0 & 0 \end{bmatrix} \right\|_\infty$$

The matrix R induces a multiplication operator

$$f \mapsto Rf : \mathcal{L}_2 \oplus \mathcal{L}_2 \to \mathcal{L}_2 \oplus \mathcal{L}_2$$

Let Π denote the orthogonal projection from $\mathcal{L}_2 \oplus \mathcal{L}_2$ onto $\mathcal{H}_2^\perp \oplus \mathcal{L}_2$ and define Ξ to be the operator

$$f \mapsto \Pi R f : \mathcal{H}_2 \oplus \mathcal{L}_2 \to \mathcal{H}_2^\perp \oplus \mathcal{L}_2$$

It should be plain from the above argument that $\alpha \geq \|\Xi\|$. Again, equality holds, $\alpha = \|\Xi\|$, a formula due to Feintuch and Francis (1986).

There's an equivalent formula due to Young (1986) which is easier to state. Define two subspaces, $\mathcal{X}$ and $\mathcal{Y}$, of $\mathcal{H}_2$,

$$\mathcal{X} := \{f \in \mathcal{L}_2 : T_3 f \in \mathcal{H}_2\}$$
$$\mathcal{Y} := \mathcal{L}_2 \ominus T_2 \mathcal{H}_2$$

and the operator Ξ,

$$f \mapsto (\text{orthogonal projection of } T_1 f \text{ onto } \mathcal{Y}) : \mathcal{X} \to \mathcal{Y}$$

Then $\alpha = \|\Xi\|$. This is a true generalization of Sarason's formula.

There has been recent research on the spectrum of operators like the ones labelled Ξ above. See [Bercovici *et al.*, 1988], [Foias and Tannenbaum, 1988], [Jonckheere and Juang, 1987], and [Zames and Mitter, 1988].

5. Tradeoffs in classical control theory. Classical control theory treats controller design in the frequency-domain. Actually, prior to 1980 the content of the theory was mainly a collection of graphical techniques and rules-of-thumb. Gain and phase plots versus frequency were modified using various tricks, but what it was all for was rarely stated precisely. Reading books on the subject, even books written in the 1980s, one seeks in vain for a clear problem statement. One of Zames' goals in [Zames, 1981] was to rigorize classical control theory. An equally influential source for current thinking is the paper by Doyle and Stein (1981), which clearly stated a robust control problem for a specified uncertainty model. This section is adapted from [Doyle, 1988].

We begin by looking at a typical performance specification, namely, tracking a reference signal, and see how to quantify it in terms of a weighted $\mathcal{H}_\infty$-norm bound. The relevant block diagram is

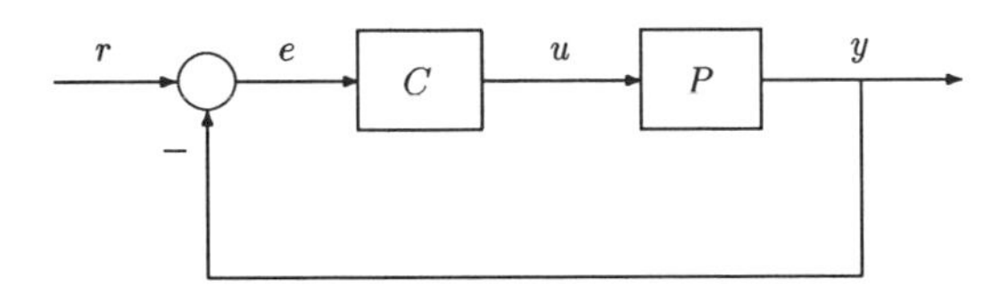

where r is the reference input, u the plant input, y the plant output, and e the tracking error. The transfer functions P and C are assumed to be real-rational and proper, with P strictly proper. Let L denote the loop transfer function, $L := PC$. The transfer function from reference input r to tracking error e is

$$S := \frac{1}{1+L}$$

also called the *sensitivity function.* In the analysis to follow, it will always be assumed that the feedback system is internally stable, so $S \in \mathcal{RH}_\infty$. Observe that since L is strictly proper, $S(j\infty) = 1$.

Now we have to decide on a performance specification, a measure of goodness of tracking. This decision depends on two things: what we know about r and what measure we choose to assign to the tracking error. Usually, r is not known in advance. Rather, a set of possible r's is known, or at least postulated for the purpose of design. Let's begin with an idealized scenario: suppose r can be any sinusoid of amplitude ≤ 1 and we want e to have amplitude $< \epsilon$. Then the performance specification can be expressed succinctly as

$$\|S\|_\infty < \epsilon$$

Here we used the fact that the maximum amplitude of e equals the $\mathcal{H}_\infty$-norm of the transfer function. Or if we define the (trivial, in this case) weighting function $W(s) = 1/\epsilon$, then the performance specification is $\|WS\|_\infty < 1$.

The situation becomes more realistic, and more interesting, with a frequency-dependent weighting function. Assume $W(s)$ is real-rational, stable, and proper. Let's consider four scenarios giving rise to an $\mathcal{H}_\infty$-norm bound on WS.

(i) Suppose the family of reference inputs is all signals of the form $r = Wr_{pf}$, where r_{pf}, a pre-filtered input, is any sinusoid of amplitude ≤ 1. Thus the set of r's consists of sinusoids with frequency-dependent amplitudes. Then the maximum amplitude of e equals $\|WS\|_\infty$.

(ii) Recall that

$$\|r\|_2^2 = \frac{1}{2\pi} \int_{-\infty}^{\infty} |r(j\omega)|^2 \, d\omega$$

and that $\|r\|_2^2$ is a measure of the energy of r. Thus, we may think of $|r(j\omega)|^2$ as *energy spectral density*, or energy spectrum. Suppose the set of all r's is

$$\{r : r = Wr_{pf}, \|r_{pf}\|_2 \leq 1\}$$

i.e.,

$$\left\{r : \frac{1}{2\pi} \int_{-\infty}^{\infty} |r(j\omega)/W(j\omega)|^2 \, d\omega \leq 1\right\}$$

Thus, r has an energy constraint and its energy spectrum is weighted by $1/|W(j\omega)|^2$. For example, if W were a bandpass filter, then the energy spectrum of r would be confined to the passband. More generally, W could be used to shape the energy spectrum of the expected class of reference inputs.

Now suppose the tracking error measure is the $\mathcal{H}_2$-norm of e. Then

$$\sup_r \|e\|_2 = \sup\{\|SWr_{pf}\|_2 : \|r_{pf}\|_2 \leq 1\} = \|WS\|_\infty$$

so $\|WS\|_\infty < 1$ means $\|e\|_2 < 1$ for all r's in the above set.

(iii) This scenario is like the previous one except for signals of finite power. The power-norm of r is defined to be

$$\left(\lim_{T\to\infty} \frac{1}{2T} \int_{-T}^{T} |r(t)|^2 \, dt\right)^{1/2}$$

It's a fact that $\|WS\|_\infty$ equals the supremum power-norm of e over all r_{pf} of power-norm ≤ 1. So W could be used to shape the power spectrum of the expected class of r's.

(iv) In several applications designers have acquired through experience desired shapes for the Bode magnitude plot of S. In particular, suppose good performance is known to be achieved if the plot of $|S(j\omega)|$ lies under some curve. We could rewrite this as

$$|S(j\omega)| < |W(j\omega)|^{-1} \quad \forall\omega$$

or in other words $\|WS\|_\infty < 1$.

There is a nice graphical interpretation of the norm bound $\|WS\|_\infty < 1$. Note that

$$\begin{aligned}
&\|WS\|_\infty < 1 \\
\Leftrightarrow &\left|\frac{W(j\omega)}{1 + L(j\omega)}\right| < 1 \quad \forall\omega \\
\Leftrightarrow &|W(j\omega)| < |1 + L(j\omega)| \quad \forall\omega
\end{aligned}$$

The last inequality says that, at every frequency, the point $L(j\omega)$ on the Nyquist plot lies outside the circle with center -1, radius $|W(j\omega)|$.

To recap, it has been argued that a reasonable performance criterion for the tracking problem is $\|WS\|_\infty < 1$. (See also [Zames, 1981].)

Now we turn to the issue of robust stability. Common measures of robust stability are gain and phase margin, but these alone are inadequate. A better measure is the distance from the critical point to the nearest point on the Nyquist plot of L, that is, $1/\|S\|_\infty$. Although better, this measure is not entirely adequate either because it contains no frequency information.

To get a satisfactory characterization of robust stability we shall focus on a specific uncertainty model, multiplicative perturbations. Suppose the nominal plant transfer function is P and consider perturbed plant transfer functions of the form $(1+\Delta W)P$. Here W is a fixed weighting function, assumed stable and proper, and Δ is a variable real-rational transfer function, not necessarily stable, having two properties:

(i) P and $(1+\Delta W)P$ have the same number of poles in $\mathrm{Re}\, s \geq 0$

(ii) $\|\Delta\|_\infty \leq 1$

Note that (ii) implies that Δ has no poles on the imaginary axis. In what follows, all Δ's are assumed to be of this sort.

The idea behind this uncertainty model is that ΔW is the relative plant perturbation:

$$\frac{(1+\Delta W)P - P}{P} = \Delta W$$

Hence

$$\left| \frac{[1+\Delta(j\omega)W(j\omega)]P(j\omega) - P(j\omega)}{P(j\omega)} \right| \leq |W(j\omega)| \quad \forall \omega$$

so $|W(j\omega)|$ provides the uncertainty profile. Typically, $|W(j\omega)|$ is a monotonically increasing function of ω — uncertainty increases with increasing frequency. The main purpose of Δ is to account for phase uncertainty and to act as a scaling factor on the magnitude of the perturbation, i.e., $|\Delta|$ varies between 0 and 1.

Assume the nominal feedback system, i.e., with $\Delta = 0$, is internally stable. When is internal stability a robust property, i.e., when is internal stability preserved for all Δ? Define the *complementary sensitivity function*

$$T := 1 - S = \frac{L}{1+L}$$

The following is an instance of the small-gain theorem.

LEMMA 2. [Doyle and Stein, 1981] *Internal stability is robust iff* $\|WT\|_\infty < 1$.

The condition $\|WT\|_\infty < 1$ also has a nice graphical interpretation. Draw the Nyquist plot of L. At each frequency ω, draw the closed disk with centre $L(j\omega)$, radius $|W(j\omega)L(j\omega)|$. The union of these disks forms a tube. Then $\|WT\|_\infty < 1$ iff the critical point, -1, lies outside the tube.

Finally, robust performance — the preservation of tracking and internal stability under plant perturbation. Now we need two weighting functions, W_1 and W_2, both stable and proper. Then the *nominal performance* condition is $\|W_1 S\|_\infty < 1$ and the *robust stability* condition, with respect to multiplicative perturbations of the form $(1+\Delta W_2)P$, is $\|W_2 T\|_\infty < 1$.

When P is perturbed to $(1+\Delta W_2)P$, S is perturbed to

$$\frac{1}{1+(1+\Delta W_2)L} = \frac{S}{1+\Delta W_2 T}$$

Clearly the *robust performance* condition should be

$$\|W_2 T\|_\infty < 1 \text{ and } \left\|\frac{W_1 S}{1+\Delta W_2 T}\right\|_\infty < 1 \quad \forall \Delta$$

(Remember that Δ is implicitly assumed to satisfy the two conditions above.) The next lemma gives a test for robust performance. The function

$$s \mapsto |W_1(s)S(s)| + |W_2(s)T(s)|$$

is denoted $|W_1 S| + |W_2 T|$.

LEMMA 3. *A necessary and sufficient condition for robust performance is*

$$\| |W_1 S| + |W_2 T| \|_\infty < 1 \tag{1}$$

Proof. (Sufficiency) Assume (1), or equivalently

$$\|W_2 T\|_\infty < 1 \text{ and } \left\|\frac{|W_1 S|}{1-|W_2 T|}\right\|_\infty < 1 \tag{2}$$

Fix Δ. In what follows, functions are evaluated at an arbitrary point $j\omega$, but this is suppressed to simplify notation. We have

$$1 = |1+\Delta W_2 T - \Delta W_2 T| \le |1+\Delta W_2 T| + |W_2 T|$$

and therefore

$$1 - |W_2 T| \le |1+\Delta W_2 T|$$

This implies that

$$\left\|\frac{|W_1 S|}{1-|W_2 T|}\right\|_\infty \ge \left\|\frac{W_1 S}{1+\Delta W_2 T}\right\|_\infty$$

This and (2) yield

$$\left\|\frac{W_1 S}{1+\Delta W_2 T}\right\|_\infty < 1$$

(Necessity) Assume

$$\|W_2 T\|_\infty < 1 \text{ and } \left\|\frac{W_1 S}{1+\Delta W_2 T}\right\|_\infty < 1 \quad \forall \Delta \tag{3}$$

Pick a frequency ω where

$$\frac{|W_1 S|}{1 - |W_2 T|}$$

is maximum. Now pick Δ so that

$$1 - |W_2 T| = |1 + \Delta W_2 T|$$

(The idea here is that $\Delta(j\omega)$ should rotate $W_2(j\omega)T(j\omega)$ so that $\Delta(j\omega)W_2(j\omega)T(j\omega)$ is negative real.) Now we have

$$\begin{aligned} \left\| \frac{|W_1 S|}{1 - |W_2 T|} \right\|_\infty &= \frac{|W_1 S|}{1 - |W_2 T|} \\ &= \frac{|W_1 S|}{|1 + \Delta W_2 T|} \\ &\leq \left\| \frac{W_1 S}{1 + \Delta W_2 T} \right\|_\infty \end{aligned}$$

So from this and (3) there follows (2). □

Test (1) also has a nice graphical interpretation. For each frequency ω, construct two closed disks: one with centre -1, radius $|W_1(j\omega)|$; the other with centre $L(j\omega)$, radius $|W_2(j\omega)L(j\omega)|$. Then (1) holds iff for each ω these two disks are disjoint.

The robust performance condition says that the robust performance level 1 is achieved. More generally, let's say that robust performance level α is achieved if

$$\|W_2 T\|_\infty < 1 \text{ and } \left\| \frac{W_1 S}{1 + \Delta W_2 T} \right\|_\infty < \alpha \quad \forall \Delta$$

We can find out how small α can be by applying a scaling technique in Lemma 3. The procedure is as follows:

Step 1 Check that $\|W_2 T\|_\infty < 1$. (If not, then no robust performance level is achieved.)

Step 2 Decrease α while

$$\| |\alpha^{-1} W_1 S| + |W_2 T| \|_\infty < 1$$

Step 3 Stop at α_{min} when

$$\| |\alpha_{min}^{-1} W_1 S| + |W_2 T| \|_\infty = 1$$

Then robust performance level α is achieved iff $\alpha > \alpha_{min}$.

Alternatively, we may wish to know how large the uncertainty can be while the robust performance condition holds. To do this, we scale the uncertainty level, i.e., we allow Δ to satisfy $\|\Delta\|_\infty < ta$, together with the other condition that P and $(1 + \Delta W_2)P$ have the same number of poles in Re $s \geq 0$. Application of Lemma

2 shows that internal stability is robust iff $\|taW_2T\|_\infty < 1$. Let's say that the uncertainty level β is permissible if

$$\|\beta W_2T\|_\infty < 1 \text{ and } \left\|\frac{W_1S}{1+\Delta W_2T}\right\|_\infty < 1 \quad \forall\Delta$$

The procedure for finding the largest β is as follows:

Step 1 Check that $\|W_1S\|_\infty < 1$.

Step 2 Increase β while

$$\||W_1S| + |\beta W_2T|\|_\infty < 1$$

Step 3 Stop at β_{max} when

$$\||W_1S| + |\beta_{max} W_2T|\|_\infty = 1$$

Then the uncertainty level β is permissible iff $\beta < \beta_{max}$.

Inequality (1) is an analysis test for a given controller. It is a shocking state of affairs that constructive necessary and sufficient conditions for the existence of a controller so that (1) holds are as yet unknown. Luckily, there is a constructive sufficient condition which is frequently adequate in applications, namely,

$$\|(|W_1S|^2 + |W_2T|^2)^{1/2}\|_\infty < 1/\sqrt{2}$$

The left-hand side equals the $\mathcal{H}_\infty$-norm of the 2×1 matrix

$$\begin{bmatrix} W_1S \\ W_2T \end{bmatrix}$$

For much deeper work on the subject of this section, see [Freudenberg and Looze, 1988], [Helton, 1985], and [Packard, 1988].

Acknowledgements. On the subjects in this paper I have profited greatly from discussions with John Doyle, Abie Feintuch, Keith Glover, Pramod Khargonekar, Allen Tannenbaum, and George Zames.

REFERENCES

[1] Ball, J.A. and N. Cohen, *Sensitivity minimization in an $\mathcal{H}_\infty$ norm: parametrization of all suboptimal solutions*, Int. J. Control, 46, pp. 785-816. (1987).

[2] Ball, J.A. and A.C.M. Ran, *Optimal Hankel norm model reductions and Wiener-Hopf factorizations I: the canonical case*, SIAM J. Control Opt., 25, pp. 362-382 (1987).

[3] Bercovici, H., C. Foias, and A. Tannenbaum, *On skew Toeplitz operators, I*, Topics in Operator Theory and Interpolation, Operator Theory: Advances and Applications, 29 (1988), pp. 21–44.

[4] Bernstein, D.S. and W.M. Haddad, *LQG control with an $\mathcal{H}_\infty$ performance bound*, Technical report, Government Aerospace Systems Division, Melbourne, FL. (1988).

[5] Boyd, S., V. Balakrishnan, and P. Kabamba, *On computing the $\mathcal{H}_\infty$-norm of a transfer matrix*, Math. Control, Signals, and Systems (1988) (to appear).

[6] Brockett, R.W., *Finite Dimensional Linear Systems*, Wiley, New York, 1970.

[7] Doyle, J.C., *Course notes for EE104, Dept. Elect. Eng.*, Caltech (1988).

[8] DOYLE, J.C. AND G. STEIN, *Multivariable feedback design: concepts for a classical modern synthesis*, IEEE Trans. Auto. Control, AC-26 (1981), pp. 4–16.

[9] DOYLE, J.C., K. GLOVER, P. KHARGONEKAR, AND B. FRANCIS, *State-space solutions to standard $\mathcal{H}_2$ and $\mathcal{H}_\infty$ control problems*, submitted for publication. (1988).

[10] FEINTUCH, A. AND B. FRANCIS, *Uniformly optimal control of linear systems*, Automatica, 21, (1986), pp. 563-574.

[11] FOIAS, C. AND A. TANNENBAUM, *On the four-block problem, parts I and II*, Integral Equations and Operator Theory (1988) (to appear).

[12] FRANCIS, B.A., *A guide to $\mathcal{H}_\infty$ control theory*, Proc. Workshop on Modelling, Robustness and Sensitivity Reduction in Control Systems, Groningen (1987).

[13] FRANCIS, B.A., *A Course in $\mathcal{H}_\infty$ Control Theory*, in *Lecture Notes in Control and Information Sciences, vol. 88*, Springer-Verlag, 1987.

[14] FRANCIS, B.A. AND J.C. DOYLE, *Linear control theory with an $\mathcal{H}_\infty$ optimality criterion*, SIAM J. Control Opt., 25 (1987), pp. 815–844.

[15] FREUDENBERG, J. AND D. LOOZE, *An analysis of $\mathcal{H}_\infty$-optimization design methods*, IEEE Trans. Auto. Control, AC-31 (1986), pp. 194–200.

[16] FREUDENBERG, J. AND D. LOOZE, *Frequency Domain Properties of Scalar and Multivariable Feedback Systems*, in *Lecture Notes in Control and Information Sciences, 104*, Springer-Verlag, 1988.

[17] GLOVER, K. AND J. DOYLE, *State-space formulae for all stabilizing controllers that satisfy an $\mathcal{H}_\infty$-norm bound and relations to risk sensitivity*, submitted for publication (1988).

[18] HELTON, J.W., *Worst case analysis in the frequency-domain: an $\mathcal{H}_\infty$ approach to control*, IEEE Trans. Auto. Control, AC-30 pp. 1154-1170.

[19] HELTON, J.W., *Operator Theory, Analytic Functions, Matrices, and Electrical Engineering*, CBMS no. 68, AMS (1987).

[20] JONCKHEERE, E.A. AND J.-C. JUANG, *Fast computation of achievable feedback performance in mixed sensitivity $\mathcal{H}_\infty$ design*, IEEE Trans. Auto. Control, AC-32, pp. 896–906.

[21] KHARGONEKAR, P.P., I.R. PETERSEN, AND M.A. ROTEA, *$\mathcal{H}_\infty$- optimal control with state-feedback*, IEEE Trans. on Auto. Control (1988) (to appear).

[22] LIMEBEER, D.J.N. AND Y.S. HUNG, *An analysis of pole-zero cancellations in $\mathcal{H}_\infty$-optimal control problems of the first kind*, SIAM J. Control Opt., 25 (1987), pp. 1457–1493.

[23] MAGEIROU, E.F., *Values and strategies for infinite time linear quadratic games*, IEEE Trans. Auto. Control, AC-21 (1976), pp. 547–550.

[24] MEDANIC, J., *Bounds on the performance index and the Riccati equation in differential games*, IEEE Trans. Auto. Control, vol AC-12 (1967), pp. 613–614.

[25] PACKARD, A.K. PH.D. THESIS IN ENGINEERING, UNIVERSITY OF CALIFORNIA AT BERKELEY..

[26] PETERSEN, I.R., *Disturbance attenuation and $\mathcal{H}_\infty$ optimization: a design method based on the algebraic Riccati equation*, IEEE Trans. Auto. Contr., vol. AC-32 (1987), pp. 427–429.

[27] RHODES, I.B. AND D.G. LUENBERGER, *Differential games with imperfect state information*, IEEE Trans. Auto. Control, vol. AC-14 (1969), pp. 29–38.

[28] SAFONOV, M.G., E. JONCKHEERE, M. VERMA, AND D. LIMEBEER, *Synthesis of positive-real multivariable feedback systems*, Int. J. Control, vol. 45 (1987), pp. 817–842.

[29] SARASON, D., *Generalized interpolation in $\mathcal{H}_\infty$*, Trans. AMS., vol. 127 (1967), pp. 179–203.

[30] TANNENBAUM, A., *Feedback stabilization of linear dynamical plants with uncertainty in the gain factor*, Int. J. Control, vol. 32 (1980), pp. 1–16.

[31] TANNENBAUM, A., *Modified Nevanlinna-Pick interpolation of linear plants with uncertainty in the gain factor*, Int. J. Control, vol. 36 (1982), pp. 331–336.

[32] VIDYASAGAR, M., *Control System Synthesis: a Factorization Approach*, MIT Press., 1985.

[33] WILLEMS, J.C., *Least-squares stationary optimal control and the algebraic Riccati equation*, IEEE Trans. Auto. Control, vol. AC-16 (1971), pp. 621–634.

[34] YOUNG, N., *The Nevanlinna-Pick problem for matrix-valued functions*, J. Operator Theory, vol. 15, (1986), pp. 239–265.

[35] ZAMES, G., *Optimal sensitivity and feedback: weighted seminorms, approximate inverses, and plant invariant schemes*, Proc. Allerton Conf. (1979).

[36] ZAMES, G., *Feedback and optimal sensitivity: model reference transformations, multiplicative seminorms, and approximate inverses*, IEEE Trans. Auto. Control, vol. AC-26 (1981), pp. 301–320.

[37] ZAMES, G., *Feedback organizations, learning and complexity in $\mathcal{H}_\infty$*, Proc. American Control Conf. (1988).

[38] ZAMES, G. AND S.K. MITTER, *A note on essential spectra and norms of mixed Hankel-Toeplitz operators*, Systems & Control Letters, vol. 10 (1988), pp. 159–166.

NONGAUSSIAN LINEAR FILTERING, IDENTIFICATION OF LINEAR SYSTEMS, AND THE SYMPLECTIC GROUP*

MICHIEL HAZEWINKEL†

Abstract. Consider stochastic linear dynamical systems, $dx = Axdt + Bdw, dy = Cxdt + dv, y(0) = 0, x(0)$ a given initial random variable independent of the standard independent Wiener noise processes w, ν. The matrices A, B, C are supposed to be constant. In this paper I consider two problems. For the first one A, B and C are supposed known and the question is how to calculate the conditional probability density of x at time t given the observations $y(s), 0 \leq s \leq t$ in the case that $x(0)$ is not necessarily gaussian. (In the gaussian case the answer is given by the Kalman-Bucy filter). The second problem concerns identification, i.e. the A, B, C are unknown (but assumed constant so that $dA = 0, dB = 0, dC = 0$), and one wants to calculate the joint conditional probability density at time t of (x, A, B, C), again given the observations $y(s), 0 \leq s \leq t$. The methods used rely on Wei-Norman theory, the Duncan-Mortensen-Zakai equation and a "real form" of the Segal-Shale-Weil representation of the symplectic group $Sp_n(\mathbf{R})$.

Key words. nongaussian distribution, identification, non-linear filtering, DMZ equation, Duncan-Mortensen-Zakai equation, propagation of nongaussian initials, Wei-Norman theory, Segal-Shale-Weil representation, reference probability approach, unnormalized density, Kalman-Bucy filter, Lie algebra approach to nonlinear filtering.

AMS(MOS) subject classifications. 93E11, 93B30, 17B99, 93C10, 93B35, 93E12

1. Introduction. Consider a general nonlinear filtering problem of the following type:

$$dx = f(x)dt + G(x)dw, \qquad x \in \mathbf{R}^n,\ w \in \mathbf{R}^m \tag{1.1}$$

$$dy = h(x)dt + d\nu, \qquad y \in \mathbf{R}^p,\ \nu \in \mathbf{R}^p \tag{1.2}$$

where f, G, h are vector and matrix valued functions of the appropriate dimensions, and the w, ν are standard Wiener processes independent of each other and also independent of the initial random variable $x(0)$. One takes $y(0) = 0$.

The general non-linear filtering problem is this setting asks for (effective) ways to calculate and/or approximate the conditional density $\pi(x,t)$ of x given the observations $y(s), 0 \leq s \leq t$; i.e. $\pi(x,t)$ is the density of $\hat{x} := (t)E[x(t)|y(s), 0 \leq s \leq t]$ the conditional expectation of the state $x(t)$.

One approach to this problem proceeds via the so called DMZ equation which is an equation of a rather nice form for an unnormalized version $\rho(x,t)$ of $\pi(x,t)$. Here unnormalized means that $\rho(x,t) = r(t)\pi(x,t)$ for some function $r(t)$ of time alone. A

*Various subselections of the material in this article have formed the subject of various talks at different conferences; e.g. the 2nd conference on the road–vehicle system in Torino in June 1987, the 24–th Winter school on theoretical physics in Karpacz in January 1988, the 3rd meeting of the Bellman continuum in Valbonne in June 1988, and the special program on signal processing of the IMA in Minneapolis in the summer of 1988, the present one. As a result this article may also appear in the proceedings of these meetings.

†Centre for Mathematics and Computer Science P.O. Box 4079, 1009 AB Amsterdam, The Netherlands.

capsule description of this approach is given in section 2 below. Using this approach was strongly advocated by BROCKET and MITTER (cf. e.g. their contributions in [6]), and initially the approach had a number of nontrivial successes, both in terms of positive and negative results (cf. e.g. the surveys [9] and [4]). Subsequently, the approach became less popular; perhaps because a number of rather formidable mathematical problems arose, and because the number of systems to which the theory can be directly applied appears to be quite small. Cf [4] for a discussion of some aspects of these two points.

It is the purpose of this paper to apply this approach to two problems concerning linear systems, which do not fall within the compass of the usual Kalman-Bucy linear filtering theory. More precisely, consider a linear stochastic dynamical system

$$dx = Axdt + Bdw, \quad x \in \mathbf{R}^n, w \in \mathbf{R}^m \tag{1.3}$$

$$dy = Cxdt + dv, \quad y \in \mathbf{R}^p, v \in \mathbf{R}^p \tag{1.4}$$

where the A, B, C are matrices of the appropriate sizes. The first problem I want to consider is the filtering of (1.3) - (1.4) in the case that the initial condition $x(0)$ is a non-gaussian random variable. The second problem concerns the identification of (1.3) - (1.4); i.e. one assumes that the matrices A, B, C are constant but unknown and it is desired to calculate the conditional density $\pi(x, A, B, C, t)$ of the (enlarged) state (x, A, B, C) at time t. Technically this means that one adds to (1.3) - (1.4) the equations

$$dA = 0, \quad dB = 0, \quad dC = 0 \tag{1.5}$$

and one considers the filtering problem for the nonlinear system (1.3) - (1.5). Strictly speaking this problem is not well posed. Simply because A, B, C can not be uniquely identified on the basis of the observations alone. In the DMZ equation approach this shows up only at the very end in the form that $\rho(x, A, B, C, t)$ will be degenerate in the sense that $\rho(Sx, SAS^{-1}, SB, CS^{-1}, t) = \rho(x, A, B, C, t)$ for all constant invertible real matrices S. As a result the normalization factor $\int \rho(x, A, B, C, t)dxdAdBdC$ does not exist, and in fact $\pi(x, A, B, C, t)$ is also degenerate. One gets rid of this by passing to the quotient space (finite moduli space) $\{(x, A, B, C)\}/GL_n(\mathbf{R})$ for the action just given and/or by considering (local) canonical forms. The normalization factor can be calculated by integrating over this quotient space.

Besides the DMZ-equation, already mentioned, the tools used to tackle the two problems described above are Wei-Norman theory and something which could be called a real form of the Segal-Shale-Weil representation of the symplectic Lie group $Sp_n(\mathbf{R})$. These two topics are discussed in sections 3 and 4 below.

2. The DMZ Approach to Nonlinear Filtering. Consider again the general nonlinear system (1.1) - (1.2). These stochastic differential equations are to be considered as Ito equations. Let $\pi(x, t)$ be the probability density of $E[x(t)|y(s), 0 \leq s \leq t]$, the conditional expectation of $x(t)$. (Given sufficiently nice f, G and h it can be shown that $\pi(x,t)$ exists.) Then the Duncan-Mortensen-Zakai result [1, 10, 12] is that there exists an unnormalized version $\rho(x,t)$ of $\pi(x,t)$, i.e. $\rho(x,t) = r(t)\pi(x,t)$, which satisfies an evolution equation

$$d\rho = \mathcal{L}\rho dt + \sum h_k \rho dy_k(t), \quad \rho(x,0) = \psi(x) \tag{2.1}$$

where $\psi(x)$ is the distribution of the initial random variable $x(0)$ and where $\mathcal{L}$ is the second-order partial differential operator

$$\mathcal{L}\phi = \frac{1}{2}\sum_{i,j} \frac{\partial^2}{\partial x_i \partial x_j}(GG^T)_{ij}\phi - \sum_i \frac{\partial}{\partial x_i} f_i \phi - \frac{1}{2}\sum_k h_k^2 \phi. \tag{2.2}$$

Here $h_k, y_k(t), f_i$ are components of $h, y(t)$ and f respectively and $(GG^T)_{ij}$ is the (i,j)-entry of the product GG^T of the matrix G and its transpose.

Equation (2.1) is a Fisk-Stratonovič stochastic differential equation. The corresponding Ito differential equation is obtained by removing the $-\frac{1}{2}\sum h_k^2\phi$ term from (2.2).

As it stands (2.1) is a stochastic partial differential equation. However the transformation

$$\tilde{\rho}(x,t) = \exp\left(\sum h_k(x) y_k(t)\right) \rho(x,t) \tag{2.3}$$

turns it into the equation

$$d\tilde{\rho} = \left(\mathcal{L}\tilde{\rho} + \sum \mathcal{L}_i \tilde{\rho} y_i + \frac{1}{2}\sum \mathcal{L}_{i,j}\tilde{\rho} y_i y_j\right) dt \tag{2.4}$$

where $\mathcal{L}_i$ is the operator commutator $\mathcal{L}_i = [h_i, \mathcal{L}] = h_i\mathcal{L} - \mathcal{L}h_i$ and $\mathcal{L}_{ij} = [h_i, [h_j, \mathcal{L}]]$. Cf. [4] for more details. In (2.3) I have explicitly indicated the dependence of the various quantities on x, t to stress that here $h(x)$ should simply be seen as a known function of x and not as the time function $h(x(t))$. Equation (2.4) does not involve the derivatives dy_i any more; it makes sense for all possible paths $y(t)$, and can be regarded as a family of PDE parametrized by the possible observation paths $y(t)$. Thus there is a robust version of (2.1) and we can work with (2.1) as a parametrized family of PDE parametrized by the $y(t)$. Note that knowledge of $\tilde{\rho}(x,t)$ (and $y(t)$) immediately gives $\rho(x,t)$ and that the conditional expectation of any function $\phi(x(t))$ of the state at time t can be calculated by

$$E[\phi(x(t))|y(s), 0 \le s \le t] = \left(\int \rho(x,t)dx\right)^{-1} \int \phi(x)\rho(x,t)dx \tag{2.5}$$

Possibly the simplest example of a filtering problem is provided by one-dimensional Wiener noise linearly observed:

$$dx = dw, \quad x, w \in \mathbf{R} \tag{2.6}$$

$$dy = x dt + dv, \quad y, v \in \mathbf{R}. \tag{2.7}$$

In this case the corresponding DMZ equation is

$$d\rho = \left(\frac{1}{2}\frac{d^2}{dx^2} - \frac{1}{2}x^2\right)\rho dt + x\rho dy \tag{2.8}$$

an Euclidean Schrödinger equation for a forced harmonic oscillator.

3. Wei-Norman Theory. Wei-Norman theory is (for instance) concerned with solving partial differential equations of the form

$$\frac{\partial p}{\partial t} = u_1 A_1 \rho + \cdots + u_m A_m \rho \tag{3.1}$$

where the $A_i, i = 1, \ldots, m$ are linear partial differential operators in the space variables $x_1, \ldots, x_n$, and the u_i, $i = 1, \ldots, m$ are given functions of time, in terms of solutions of the simpler equations

$$\frac{\partial \rho}{\partial t} = A_i \rho, i = 1, \ldots, m \tag{3.2}$$

which we write as

$$\rho(x,t) = e^{A_i t}\psi(x), \psi(x) = \rho(x,0). \tag{3.3}$$

Originally, the theory was developed for the finite dimensional case, i.e. for systems of ordinary differential equations

$$\dot{z} = u_1 A_1 z + \cdots + u_m A_m z \tag{3.4}$$

where $z \in \mathbf{R}^k$, and the A_i are $k \times k$ matrices. Both in the finite dimensional case (3.4) and the infinite dimensional case (3.1) it is well known that besides in the given directions $A_1\rho, \ldots A_m\rho$, the to be determined function or vector can also move (infinitesimally) in the directions given by the commutators $[A_i, A_j]\rho = (A_i A_j - A_j A_i)\rho$, and in the directions given by repeated commutators $[[A_i, A_j], A_k]$, $[[A_i, A_j], [A_k, A_l]]$, etc. etc.

Let $\text{Lie}(A_1, \ldots, A_m)$ be the Lie algebra of operators generated by the operators $A_1, \ldots A_m$. This is the smallest vector space L of operators containing $A_1, \ldots A_n$ and such that if $A, B \in L$ then also $[A, B] := AB - BA \in L$. In the finite dimensional case (3.4) L is always finite dimensional, a subvector space of $gl_k(\mathbf{R})$, the vector space (Lie algebra) of all $k \times k$ matrices. In the infinite dimensional case the Lie algebra generated by the operators $A_1, \ldots A_m$ in (3.1) can easily be infinite dimensional and it often is; also in the cases coming from filtering problems via the DMZ equation. Cf. [5] for a number of examples.

This is the essential difference between (3.1) and (3.4). Accordingly, here I shall assume that the Lie algebra $L = \text{Lie}(A_1, \ldots, A_m)$ generated by the operators $A_1, \ldots, A_m$ in (3.1) is finite dimensional. For a discussion of various infinite dimensional versions of Wei-Norman theory cf. [4]. Hence, granting this finite dimensionality property, by setting, if necessary, some of the $u_i(t)$ equal to zero, and by combining other $u_j(t)$ in the case of linear dependence among the operators on the RHS of (3.1), without loss of generality, we can assume that we are dealing with an equation

$$\frac{\partial \rho}{\partial t} = u_1 A_1 \rho + \cdots + u_n A_n \rho \tag{3.5}$$

with the additional property that

$$[A_i, A_j] = \sum_k \gamma_{ij}^k A_k; \quad i,j = 1,\dots,n \tag{3.6}$$

for suitable real constants γ_{ij}^k; $i,j,k = 1,\dots,n$.

The central idea of Wei-Norman theory is now to try for a solution of the form

$$\rho(t) = e^{g_1(t)A_1} d^{g_2(t)A_2} \dots e^{g_n(t)A_n} \psi \tag{3.7}$$

where the g_i are still to be determined functions of time. The next step is to insert the Ansatz (3.7) into (3.5), to obtain

$$\begin{aligned} \dot{\rho} &= \dot{g}_1 A_1 e^{g_1 A_1} \dots e^{g_n A_n} \psi + e^{g_1 A_1} \dot{g}_2 A_2 e^{g_2 A_2} \dots e^{g_n A_n} \psi + \cdots \\ &+ e^{g_1 A} \dots e^{g_{n-1} A_{n-1}} \dot{g}_n A_n e^{g_n A_n} \psi \end{aligned} \tag{3.8}$$

Now, for $i = 2,\dots,n$ insert a term

$$e^{-g_{i-1}A_{i-1}} \dots e^{-g_1 A_1} e^{g_1 A_1} \dots e^{g_{i-1}A_{i-1}}$$

just behind $\dot{g}_i A_i$ in the i-th term of (3.8). Then use the adjoint representation formula

$$e^A B e^{-A} = B + [A,B] + \frac{1}{2!}[A,[A,B]] + \frac{1}{3!}[A,[A,[A,B]]] + \cdots \tag{3.9}$$

and (3.6) repeatedly, and use the linear independence of the $A_1,\dots,A_n$ to obtain a system of ordinary differential equations for the $g_1,\dots g_n$ (with initial conditions $g_1(0) = 0 = g_2(0) = \cdots = g_n(0)$).

These equations are always solvable for small time. However they may not be solvable for all time, meaning that finite escape time phenomena can occur.

Let's consider an example, viz. the example afforded by the DMZ equation (2.8). One calculates that

$$\begin{aligned} &[\frac{1}{2}\frac{d^2}{dx^2} - \frac{1}{2}x^2, x] = \frac{d}{dx}, \quad [\frac{1}{2}\frac{d^2}{dx^2} - \frac{1}{2}x^2, \frac{d}{dx}] = x \\ &[\frac{d}{dx}, x] = 1, \quad [A,1] = 0 \end{aligned}$$

where A is any linear combination of the four operators $\frac{1}{2}\frac{d^2}{dx^2} - \frac{1}{2}x^2, x, \frac{d}{dx}, 1$. Applying the recipe sketched above to the equation

$$\dot{\rho} = (\frac{1}{2}\frac{d^2}{dx^2} - \frac{1}{2}x^2)\rho + x\rho u(t) + \frac{d}{dx}\rho 0 + 1\rho 0 \tag{3.10}$$

one finds the equations

$$\begin{aligned} &\dot{g}_1 = 0, \quad \cosh(g_1)\dot{g}_2 + \sinh(g_1)\dot{g}_3 = u(t), \\ &\sinh(g_1)\dot{g}_2 + \cosh(g_1)\dot{g}_3 = 0, \quad \dot{g}_4 = \dot{g}_3 g_2 \end{aligned} \tag{3.11}$$

which are solvable for all time.

This fact and the form of the resulting equations: straightforward quadratures and one set of linear equations $B(t)g = b(t)$, with $B(t), b(t)$ known and $B(t)$ invertible, is typical for the case that the Lie algebra $L = \oplus \mathbf{R}A_i$ spanned by the $A_1, \ldots, A_n$ is solvable. This means the following. Let $[L, L]$ be the subvector space of L spanned by all the operators of the form $[A, B], A, B \in L$. It is easily seen that this is again a Lie algebra. Inductively, let $L^{(n)} = [L, L^{(n-1)}]$ be the subvector space of L spanned by all operators of the form $[A, B], A \in L, B \in L^{(n-1)}, L^{(0)} = L$. These are all sub Lie algebras of L.

The Lie algebra of L is called nilpotent if $L^{(n)} = 0$ for n large enough. It is called solvable if $[L, L]$ is nilpotent. The phenomenon alluded to above, i.e. solvability of the Wei-Norman equations for all time, always happens in case L is solvable [11]. (And it is no accident that these algebras have been called solvable. Though this is not the result which gave them that name.)

Note that the DMZ equation (2.1) corresponding to a nonlinear filtering problem (1.1) - (1.2) is of the type (3.1) (with $u_h(t) = dy_k(t)$). Thus the Lie Algebra generated by the operators $\mathcal{L}, h_1(x), \ldots, h_p(x)$ occuring in (2.1) clearly has much to say about how difficult the filtering problem is. This Lie algebra is called the *estimation Lie algebra* of the system (1.1) - (1.2) and it can be used to prove a variety of positive and negative results about the filtering problem [4,5,9].

4. The Segal-Shale-Weil Representation and a 'Real Form'. Let J be the standard symplectic matrix $J = \begin{pmatrix} 0 & I_n \\ -I_n & 0 \end{pmatrix}$, where I_n is the $n \times n$ unit matrix. Consider the vector space of $2n \times 2n$ real matrices defined by

$$sp_n(\mathbf{R}) = \{M : JM + M^T J = 0\}. \tag{4.1}$$

Writing M as a 2×2 block matrix, $M = \begin{pmatrix} A & B \\ C & D \end{pmatrix}$, the conditions on the $n \times n$ blocks A, B, C, D become

$$B^T = B, \quad C^T = C, \quad D = -A^T, \tag{4.2}$$

As we shall see shortly below this set of matrices occurs naturally for filtering problems coming from linear systems (1.1) - (1.2).

The corresponding Lie group to $Sp_n(\mathbf{R})$ is the group of invertible $2n \times 2n$ matrices defined by

$$Sp_n(\mathbf{R}) = \{S \in \mathbf{R}^{2n \times 2n} : S^T J S = J\} \tag{4.3}$$

(This is a *group* of matrices in that if $S_1, S_2 \in Sp_n(\mathbf{R})$ then also $S_1 S_2 \in Sp_n(\mathbf{R})$ and $S_1^{-1} \in Sp_n(\mathbf{R})$ as is easily verified.)

There is a famous representation of $Sp_n(\mathbf{R})$ (or more precisely of its two-fold covering group $\tilde{S}p_n(\mathbf{R})$) in the Hilbert space $L^2(\mathbf{R}^n)$, called the Segal-Shale-Weil representation or the oscillator representation; cf. [8]. Here the word 'representation' means that to each $S \in Sp_n(\mathbf{R})$ there is associated a unitary operator U_S such that $U_{S_1S_2} = U_{S_1}U_{S_2}$ for all $S_1, S_2 \in Sp_n(\mathbf{R})$.

For the purposes of this paper a modification of it is of importance. It can be described as follows by explicit operators associated to certain specific kinds of elements of $Sp_n(\mathbf{R})$:

(i) Let P be a symmetric $n \times n$ matrix; then to the element

$$\begin{pmatrix} I & P \\ 0 & I \end{pmatrix} \in Sp_n(\mathbf{R})$$

there is associated the operator

$$f(x) \mapsto \exp(x^T Px) f(x)$$

(ii) Let $A \in GL_n(\mathbf{R})$ be an invertible $n \times n$ matrix. Then to the element

$$\begin{pmatrix} A & o \\ 0 & (A^{-1})^T \end{pmatrix} \in Sp_n(\mathbf{R})$$

there is associated the operator

$$f(x) \mapsto |\det(A)|^{1/2} f(A^T x)$$

(iii) Let Q be a symmetric $n \times n$ matrix. Then to the element

$$\begin{pmatrix} I & 0 \\ Q & I \end{pmatrix} \in Sp_n(\mathbf{R})$$

there is associated the operator

$$f(x) \mapsto \mathfrak{F}^{-1}(\exp(x^T Qx)\mathfrak{F} f(x))$$

where $\mathfrak{F}$ denotes the Fourier transform.

(The operator corresponding to the element

$$\begin{pmatrix} 0 & I \\ -I & 0 \end{pmatrix} \in Sp_n(\mathbf{R})$$

is the fact the Fourier transform itself).

Except for one snag to be discussed below, this suffices to describe the operator which should be associated to any element $S \in Sp_n(\mathbf{R})$. Indeed let

$$S = \begin{pmatrix} S_1 & S_2 \\ S_3 & S_4 \end{pmatrix} \in Sp_n(\mathbf{R}); \tag{4.4}$$

then there is an $s > 0, s \in \mathbf{R}$ such that $S_1 + sS_2$ is invertible and we have a factorisation

$$\begin{pmatrix} S_1 & S_2 \\ S_3 & S_4 \end{pmatrix} = \begin{pmatrix} 1 & 0 \\ (S_3 + sS_4)(S_1 + sS_2)^{-1} & I \end{pmatrix} \begin{pmatrix} I & S_2(S_1 + sS_2)^T \\ 0 & I \end{pmatrix} \times$$
$$\times \begin{pmatrix} S_1 + sS_2 & 0 \\ 0 & (S_1^T + sS_2^T)^{-1} \end{pmatrix} \begin{pmatrix} I & 0 \\ -s & I \end{pmatrix} \tag{4.5}$$

(It is easily verified that all four factors on the right are in fact in $Sp_n(\mathbf{R})$).

Now assign to the operator S the product of the four operators corresponding to the factors on the RHS of (4.5) according to the recipe (i) - (iii) given above. There is a conceivable second snag here in that it seems a priori possible that different factorisations could give different operators. This in fact does not happen precisely because the 'representation' described by (i) - (iii) is a 'real form' of the oscillator representation $Sp_n(\mathbf{R}) \hookrightarrow$ Aut $(L^2(\mathbf{R}^n))$. The relation between the oscillator representation and (i) - (iii) above is given by the substitution $x_k \mapsto \sqrt{i}x_k$ where $i = \sqrt{-1}$. (The possible sign ambiguity which could come from the fact that the oscillator representation is really a representation of the covering $\tilde{S}p_n(\mathbf{R})$ rather than $Sp_n(\mathbf{R})$ itself also seems not to happen; it would in any case be irrelevant for the applications discussed below.)

It remains to discuss the first snag mentioned just above (4.4) and why the words 'representation' and 'real form' above have been placed in quotation marks. The trouble lies in part (iii) of the recipe. Taking a Fourier transform and then multiplying with a quadratic exponential may well take one out of the class of functions which are inverse Fourier transformable. Another way to see this is to observe that the operator described in (iii) assigns to a function ψ the value in $t = 1$ of the solution of the evolution equation

$$\frac{\partial \rho}{\partial t} = ((\frac{\partial}{\partial x})^T Q \frac{\partial}{\partial x})\rho, \quad \rho(x,0) = \psi(x) \tag{4.6}$$

and if Q is not nonnegative definite this involves anti–diffusion components for which the solution at $t = 1$ may not exist. Additionally, – but this is really the same snag – applying recipe (i) to a function may well result in a function that is not Fourier transformable.

What we have in fact is not a representation of all of $Sp_n(\mathbf{R})$ but only a representation of a certain sub-semi-group cone in $Sp_n(\mathbf{R})$.

For the applications to be described below this means that we must be careful to take factorisations such that applying the various operators successively continues to make sense. The factorisation (4.5) does not seem optimal in that respect and we shall for the special elements of $Sp_n(\mathbf{R})$ which come from filtering problems use a different one.

Incidently, one says that two structures over $\mathbf{R}$ are real forms of one another if after tensoring with $\mathbf{C}$ (= extending scalars to $\mathbf{C}$) they become isomorphic (over $\mathbf{C}$). It is in this sense that the 'representation' described by the recipe (i) - (iii) is a 'real form' of the oscillator representation.

5. Propagation of Non-Gaussian Initials. Now, finally, after all this preparation, consider a known linear dynamical system

$$dx = Axdt + Bdw, \quad Cxdt + dv, \quad x \in \mathbf{R}^n, \quad w \in \mathbf{R}^m, \quad y, v \in \mathbf{R}^p. \tag{5.1}$$

with a known, not necessarily Gaussian, initial random variable $x(0)$ with probability distribution $\psi(x)$.

The DMZ equation in this case is as follows

$$d\rho = \mathcal{L}\rho dt + \sum_{j=1}^{p}(Cx)_j dy_j(t) \tag{5.2}$$

where $(Cx)_j$ is the j-th component of the p-vector Cx. The operator $\mathcal{L}$ in this case has the form

$$\mathcal{L} = \frac{1}{2}\sum_{i,j}(BB^T)_{i,j}\frac{\partial^2}{\partial x_i \partial x_j} - \sum_{i,j} A_{ji}x_j\frac{\partial}{\partial x_i} - Tr(A) - \frac{1}{2}\sum_j (Cx)_j^2 \tag{5.3}$$

Taking brackets of the multiplication operators $(Cx)_j$ with $\mathcal{L}$ yields a linear combination of the operators

$$x_1, \ldots, x_n; \frac{\partial}{\partial x_1}, \ldots, \frac{\partial}{\partial x_n}; 1. \tag{5.4}$$

This is a straightforward calculation to check. Moreover, the bracket (= commutator product) of $\mathcal{L}$ with any of the operators in (5.4) again yields a linear combination of the operators listed in (5.4). It follows that for linear stochastic dynamical systems (5.1) the associated estimation Lie algebra (= the Lie algebra generated by $\mathcal{L}, (Cx)_1, \ldots, (Cx)_p$) is always solvable of dimension $\leq 2n+2$.

As a matter of fact it is quite simple to prove that the system (5.1) is completely reachable and completely observable if and only if the dimension of the estimation Lie algebra is precisely $2n+2$ so that a basis of the algebra is formed by the $(2n+1)$ operators of (5.4) and $\mathcal{L}$ itself.

In all cases Wei-Norman theory is applicable (working perhaps with a slightly larger Lie algebra than strictly necessarily makes no real difference).

Thus we can calculate effectively the solutions of the unnormalized density equation (5.2) provided we have good ways of calculating the expressions.

$$e^{t\mathcal{L}}\psi, \; e^{tx_i}\psi, \; e^{t\frac{\partial}{\partial x_i}}\psi, \; e^t\psi \tag{5.5}$$

for arbitrary initial data ψ. The last three expressions of (5.5) cause absolutely zero difficulties $(\exp(t\frac{\partial}{\partial x_i})\psi = \psi(x_1, \ldots, x_{i-1}, x_i + t, x_{i+1}, \ldots, x_n))$. Thus it remains to calculate the $e^{t\mathcal{L}}\psi$ where $\mathcal{L}$ is an operator of the form (5.3). It is at this point that the business of the Segal-Shale-Weil representation of the previous section comes in. As a matter of fact, the Segal-Shale-Weil representation itself, not the 'real form' described in section 4 above, is a representation of the Lie algebra spanned by the operators

$$(5.6)\qquad ix_kx_j,\quad x_k\frac{\partial}{\partial x_j}+\frac{1}{2}\delta_{k,j},\quad i\frac{\partial^2}{\partial x_k\partial x_j},\quad i=\sqrt{-1}$$

and apart from multiples of the identity (which hardly matter) and the occurence of $\sqrt{-1}$ these are the constituents of the operators $\mathcal{L}$ in (5.3). It is to remove the factors $\sqrt{-1}$ that we have to go to a real form. Cf. [3] for more details on the Segal-Shale-Weil representation itself, and what it, and its real form, have to do with Kalman-Bucy filters.

It is convenient not to have to worry about multiples of the identity. To this end note that if $\mathcal{L}'=\mathcal{L}+aI$ then $\exp(t\mathcal{L}')\psi=\exp(ta)\exp(t\mathcal{L})\psi$, so that neglecting multiples of the identity indeed matters hardly.

The first observation is now that, modulo multiples of the identity operator, if $\mathcal{L}$ and $\mathcal{L}'$ are two operators of the form (5.3) then their commutator difference $[\mathcal{L},\mathcal{L}']=\mathcal{L}\mathcal{L}'-\mathcal{L}'\mathcal{L}$ is again of the same form. (To make this exact replace $\mathcal{L}$ in (5.3) by $\mathcal{L}+\frac{1}{2}Tr(A)$ and similarly for $\mathcal{L}'$.) Thus these operators actually form a finite dimensional Lie algebra and this is, of course, the symplectic Lie algebra $sp_n(\mathbf{R})$. The correspondence is given by assigning to $\mathcal{L}(=\mathcal{L}(A,B,C))$ the $2n\times 2n$ matrix

$$(5.7)\qquad \mathcal{L}(A,B,C)\to\begin{pmatrix}-A^T & -C^TC\\ -BB^T & A\end{pmatrix}$$

(If you want to be finicky it is the operator $\mathcal{L}(A,B,C)+\frac{1}{2}Tr(A)$ which corresponds to the matrix on the right of (5.7).)

In terms of a basis on the left and right side the correspondence (i.e. the isomorphism of Lie algebras) is given as follows. Let E_{ij} be the $n\times n$ matrix with a 1 in spot (i,j) and zero everywhere else. Then

$$(5.8)\qquad \frac{\partial^2}{\partial x_i\partial x_j}\mapsto\begin{pmatrix}0 & 0\\ -E_{ij}-E_{ji} & 0\end{pmatrix}$$

$$(5.9)\qquad x_i\frac{\partial}{\partial x_j}+\frac{1}{2}\delta_{i,j}\mapsto\begin{pmatrix}E_{ij} & 0\\ 0 & -E_{ji}\end{pmatrix}$$

$$(5.10)\qquad x_ix_j\mapsto\begin{pmatrix}0 & E_{ij}+E_{ji}\\ 0 & 0\end{pmatrix}$$

It is now straightforward to check that this does indeed define an isomorphism of Lie algebras from the Lie algebra of all operators $\mathcal{L}(A,B,C)+\frac{1}{2}Tr(A)$ where $\mathcal{L}$ is as in (5.3) and the algebra $sp_n(\mathbf{R})$ described and discussed in section 4 above. For example one has

$$(5.11)\qquad \left[\frac{\partial^2}{\partial x_1\partial x_2},x_2x_3\right]=x_3\frac{\partial}{\partial x_1}$$

which fits perfectly with

$$\left[\begin{pmatrix} 0 & 0 \\ -E_{12}-E_{21} & 0 \end{pmatrix}, \begin{pmatrix} 0 & E_{23}+E_{32} \\ 0 & 0 \end{pmatrix}\right] = \begin{pmatrix} E_{31} & 0 \\ 0 & -E_{13} \end{pmatrix} \tag{5.12}$$

It is precisely the correspondence (5.8) - (5.10) or, modulo multiples of the identity, (5.7), plus the fact that 'real form' described in section 4 of the SSW representation is precisely the way to remove the $\sqrt{-1}$ factors, plus, again, the fact that the SSW is really a representation, which makes it possible to use finite dimensional calculations to obtain expressions for

$$\exp(t(\pounds(A,B,C)+\frac{1}{2}Tr(A))\psi \tag{5.13}$$

for arbitrary initial conditions.

Basically the recipe is as follows. Take $\pounds(A,B,C)+\frac{1}{2}Tr(A)$. Let $M \in sp_n(\mathbf{R})$ be its associated matrix as defined by the RHS of (5.7). Calculate $\exp(tM)=S(t)$. Write $S(t)$ as a product of matrices as in (i), (ii), (iii) in section 4. Apply successively the operators associated to the factors. The result, if defined, will be an expression for (5.13). One factorisation which can be used is that of (4.5) above. It does not, however, seem to be very optimal and it is difficult to show that everything is well defined.

It is better and more efficient to use a preliminary reduction. Consider the algebraic Riccati equation

$$A^TP+PA-PBB^TP+C^TC=0 \tag{5.14}$$

determined by the triple of matrices (A,B,C). It is easy to check that for any solution P one has

$$\begin{pmatrix} I & -P \\ 0 & I \end{pmatrix} M \begin{pmatrix} I & P \\ 0 & I \end{pmatrix} = \begin{pmatrix} -\tilde{A}^T & 0 \\ -BB^T & \tilde{A} \end{pmatrix} \tag{5.15}$$

where $\tilde{A}=A-BB^TP$. Given this it becomes useful to know when (5.14) has a solution and to know some properties of the solutions. These will also be important for the next section. In fact the function $rc(A,B,C)$ that assigns to the triple (A,B,C) under suitable conditions the unique positive definite solution of (5.14) is important enough to be considered a standard named function which should be available in accurate tabulated form much as say the Airy function or Bessel functions. I know of no such tables. The symbol 'rc' of course stands for Riccati.

Let A^* be the adjoint of the complex $n\times n$ matrix A, i.e., the conjugated transpose of A, so, if A is real, $A^*=A^T$. Consider the equation (algebraic Riccati equation)

$$A^*P+PA=PBB^*P-C^*C \tag{5.16}$$

(Here A is an $n\times n$ matrix, B an $n\times m$ matrix, C an $p\times n$ matrix.) Some facts about (5.16) are then as follows:

(5.17) If (A,B) is stabilizable, i.e. if there exists an F such that $A-BF$ has all eigenvalues with negative real part, then there is a solution of (5.16) which is positive semidefinite ($P\geq 0$) (and for this solution $\tilde{A}=A-BB^*P$ is stable).

(So in particular if (A, B) is completely reachable there is a solution of (5.14).)

(5.18) Suppose (5.16) has a solution $P \geq 0$ and suppose that (A, C) is completely observable. Then P is the only nonnegative definite solution of (5.16) and $P > 0$.

(5.19) If (A, B, C) is *co* and *cr* then there is a unique $P > 0$ which solves (5.16).

This last property is the essential one for this section. For the next one we need something better. Let $L^{co,cr}_{m,n,p}(\mathbf{R})$ be the space of all triples of real matrices (A, B, C) such that (A, B) is completely reachable and (A, C) is completely observable. Let $rc(A, B, C) := P$ be the unique solution P of (5.16) such that $P > 0$ (the matrix P is positive definite and selfadjoint). Then

(5.20) The function $rc(A, B, C)$ from $L^{co,cr}_{m,n,p}(\mathbf{R})$ to the space of selfadjoint matrices is real analytic (and so in particular C^∞ (= smooth))

Moreover

(5.21) $rc(TAT^{-1}, TB, CT^{-1}) = (T^*)^{-1} rc(A, B, C) T^{-1}$

(5.22) $rc(-A^*, \pm C^*, \pm B^*) = rc(A, B, C)^{-1}$

Property (5.21) is important in section 6; more precisely it will be important when these results are really implemented for multi-input multi-output systems. The point is that the matrices (A, B, C) are not determinable from the observations alone, simply because the systems (A, B, C) and (TAT^{-1}, TB, CT^{-1}) for $T \in GL_n(\mathbf{R})$ produce exactly the same input-output behavior . For completely reachable and completely observable systems this is also the only indeterminacy. Property (5.21) guarantees that the whole analysis of these two section 5 and 6 'descends' to the moduli space (quotient manifold) $L^{co,cr}_{m,n,p}(\mathbf{R})/GL_n(\mathbf{R})$.

Having all this available it is tempting (and natural) to play the trick embodied by (5.15) again, this time using conjugation by a 2×2 block matrix with identities on the diagonal, a zero in the upper right hand corner and a Riccati equation solution Q in the lower left hand corner. This, however, is no particular good because this will introduce both the two factors

$$\begin{pmatrix} I & 0 \\ -Q & I \end{pmatrix}, \begin{pmatrix} I & o \\ Q & I \end{pmatrix}$$

in the factorisation of $S(t) = \exp(tM)$, and at least one will cause difficulties with inverse and direct Fourier transforms; cf. part (iii) of the recipe of section 4.

Instead, writing

$$\exp\left(t\begin{pmatrix} -\tilde{A}^T & 0 \\ -BB^T & \tilde{A} \end{pmatrix}\right) = \begin{pmatrix} \exp(-t\tilde{A}^T) & 0 \\ -R & \exp(t\tilde{A}) \end{pmatrix} \tag{5.23}$$

one uses the factorisation

$$\exp\left(t\begin{pmatrix} -\tilde{A}^T & 0 \\ -BB^T & \tilde{A} \end{pmatrix}\right) = \begin{pmatrix} I & 0 \\ -R\exp(t\tilde{A}^T) & I \end{pmatrix}\begin{pmatrix} \exp(-t\tilde{A}^T) & 0 \\ 0 & \exp(t\tilde{A}) \end{pmatrix} \tag{5.24}$$

giving the following total factorisation for $S(t) = \exp(tM)$
(5.25)

$$S(t) = \begin{pmatrix} I & P \\ 0 & I \end{pmatrix} \begin{pmatrix} I & 0 \\ -R\exp(t\tilde{A}^T) & I \end{pmatrix} \begin{pmatrix} \exp(-t\tilde{A}^T) & 0 \\ 0 & \exp(t\tilde{A}) \end{pmatrix} \begin{pmatrix} I & -P \\ 0 & I \end{pmatrix}$$

Except for possibly the second factor on the right hand side of (5.25) applying the recipe of section 4 is a total triviality.

As to that second factor observe that

$$\frac{d}{dt}\exp\left(t\begin{pmatrix} -\tilde{A}^T & 0 \\ -BB^T & \tilde{A} \end{pmatrix}\right) = \begin{pmatrix} -\exp(-t\tilde{A})A^T & 0 \\ -\frac{d}{dt}R & \exp(t\tilde{A})\tilde{A} \end{pmatrix} \tag{5.26}$$
$$= \begin{pmatrix} \exp(-t\tilde{A}^T) & 0 \\ -R & \exp(t\tilde{A}) \end{pmatrix} \begin{pmatrix} -\tilde{A}^T & 0 \\ -BB^T & \tilde{A} \end{pmatrix}$$

from which it follows that

$$\frac{dR}{dt} = -R\tilde{A}^T + \exp(\tilde{t}A)BB^T. \tag{5.27}$$

As a result
(5.28)

$$\frac{d}{dt}(R\exp(t\tilde{A}^T)) = -R\tilde{A}^T\exp(t\tilde{A}^T) + \exp(t\tilde{A})BB^T\exp(t\tilde{A}^T) + R\tilde{A}^T\exp(t\tilde{A}^T)$$
$$= \exp(t\tilde{A})BB^T\exp(t\tilde{A}^T) \geq 0$$

and it follows that

$$R\exp(t\tilde{A}^T) \geq 0 \quad \text{all } t \tag{5.29}$$

which means that applying part (iii) of the recipe of section 4 (= part (iii) of the definition of the real form of the SSW representation) just involves solving a diffusion equation (no anti diffusion component); or, in other words that the inverse Fourier transformation involved will exist. Note also that if the initial condition ψ is Fourier transformable then, if P is nonnegative definite, the result of applying the parts of the recipe corresponding to the third and fourth factors on the RHS of 5.25 will still be a Fourier transformable function.

This concludes the description of the algorithm for propagating non-gaussian initial densities.

6. Identification. Given all that has been said above, this section can be mercifully short. The problem is the following. Given a linear system

$$dx = Axdt + Bdw, \quad dy = Cxdt + dv \tag{6.1}$$

with *unknown* A, B, C, but constant A, B, C, we want to calculate the joint conditional density (given the observations $y(s), 0 \leq s \leq t$) for A, B, C, x. This can be approached as a nonlinear filtering problem by adding the equations

$$dA = 0, \quad dB = 0, \quad dC = 0 \tag{6.2}$$

or, more precisely, the equations stating (locally) that the free parameters remaining after specifying a local canonical form are constant but unknown. More generally one has the same setup and problem when, say, part of the parameters of (A, B, C) are known (or, generalizing a bit more, imperfectly known).

The approach, of course, will be to calculate the DMZ unnormalized version of the conditional density $\rho(x, A, B, t)$ given the observations $y(s), 0 \leq s \leq t$. Writing down the DMZ equation for the system (6.1) - (6.2) gives

$$d\rho = \mathcal{L}\rho dt + \sum_{j=1}^{p}(Cx)_j dy_j(t) \tag{6.3}$$

with $\mathcal{L}$ given by (5.3); i.e. exactly the same equation as occurred in section 5 for the case of known A, B, C. And, indeed the only difference is that in section 5 the A, B, C are known, while (6.3) should be seen as a family of equations parametrized by (the unknown parameters in) the A, B, C. Thus if $\rho(x, t|A, B, C)$ denotes the solution of (5.2) and $\rho(x, A, B, C, t)$ denotes the solution of (6.3) then

$$\rho(x, t|A, B, C) = \rho(x, A, B, C, t) \tag{6.4}$$

Now the bank of Kalman-Bucy filters for $\hat{x}$ parametrized by $(A, B, C) \in L^{co,cr}_{m,n,p}$ gives the probability density

$$\pi(x, t|A, B, C) = r(t, A, B, C)^{-1}\rho(x, t|A, B, C) \tag{6.5}$$

so that the normalization factor $r(t, A, B, C)$ can be calculated as $\int \rho(x, t, A, B, C)dx$. By Bayes

$$\pi(x, A, B, C, t) = \pi(x, t|A, B, C)\pi(A, B, C, t) \tag{6.6}$$

so that the normalization factor $r(t, A, B, C)$ is, so to speak, precisely equal to the difference between the solution of the DMZ equation (6.3) (or (5.2)) and the bank of Kalman filters producing $\pi(x, t|A, B, C)$, i.e., the marginal conditional density

$$\pi(A, B, C, t) = \int \pi(x, A, B, C, t)dx = \int \rho(x, A, B, C, t)dx / \int \rho(x, A, B, C, t)dxdAdBdC \tag{6.7}$$

is obtainable from the unnormalized version of the bank of Kalman-Bucy filters parametrized by (A, B, C). Given the relations between this bank of filters described in [13] and briefly recalled in section 7 below, this may offer further opportunities.

Be that as it may the marginal density $\pi(A, B, C, t)$ which up to a normalization factor is equal to $\int \rho(x, A, B, C, t)dx$ can be effectively calculated by the procedure of section 5 above with the only difference that $P = rc(A, B, C)$ now has to be treated as a function. Once $\pi(A, B, C, t)$ (or in various cases some unnormalized version $\rho(A, B, C, t)$) is available a host of well known techniques such as maximum likelyhood become available.

If it is possible (as it will be in many cases) to work with a $\rho(A, B, C, t) = r(t)\pi(A, B, C, t)$ there is no (immediate) need to descent to the quotient manifold $L^{co,cr}_{m,n,p}(\mathbf{R})/GL_n(\mathbf{R})$.

7. On the Relation Between the 'Real Form' of the SSW Representation and the Kalman-Bucy Filter. We have seen that the essential difficulty in obtaining the (unnormalized) conditional density $\rho(x,t)$ lies in 'solving' $\exp(t\mathcal{L})\psi$ where $\mathcal{L}$ is the second order differential operator (5.3). Now $\mathcal{L}$ corresponds in a fundamental way with the $2n \times 2n$ matrix

$$\begin{pmatrix} -A^T & -C^TC \\ -BB^T & A \end{pmatrix} \tag{7.1}$$

Not very surprisingly this matrix in turn is very much related to the matrix Riccati equation part of the Kalman-Bucy filter. Indeed, consider the matrix differential equation

$$\begin{pmatrix} \dot{X} \\ \dot{Y} \end{pmatrix} = \begin{pmatrix} -A^T & -C^TC \\ -BB^T & A \end{pmatrix} \begin{pmatrix} X \\ Y \end{pmatrix} \tag{7.2}$$

and, assuming that $X(t)$ is invertible, let

$$-P = YX^{-1}. \tag{7.3}$$

Then

$$\begin{aligned} \dot{P} &= -\dot{Y}X^{-1} + YX^{-1}\dot{X}X^{-1} = (+BB^TX - AY)X^{-1} + YX^{-1}(-A^TX - C^TCY)X^{-1} \\ &= +BB^T + AP + PA^T - PC^TCP \end{aligned}$$

which is the covariance equation of the Kalman-Bucy filter.

REFERENCES

[1] T.E. Duncan, *Probability densities for diffusion processes with applications to nonlinear filtering theory*, PhD thesis, Stanford (1967).

[2] M. Hazewinkel, *(Fine) moduli (spaces) for linear systems; what are they and what are they good for.*, In: C.I. Byrnes, C.F. Martin (eds), Geometric methods for linear system theory (Harvard, June 1979), Reidel (1980), pp. 125-193.

[3] M. Hazewinkel, *The linear systems Lie algebra, the Segal-Shale-Weil representation and all Kalman-Bucy filters*, J. Syst. Th. & Math. Sci., 5 (1985), pp. 94-106.

[4] M. Hazewinkel, *Lectures on linear and nonlinear filtering*, In: W. Schiehlen, W. Wedig (eds), Analysis and estimation of stochastic mechanical systems, CISM course June 1987, Springer (Wien) (1988), pp. 103-135.

[5] M. Hazewinkel, S.I. Marcus, *On Lie algebras and finite dimensional filtering*, Stochastics, 7 (1982), pp. 29-62.

[6] M. Hazewinkel, J.C. Willems (eds), *Stochastic systems: the mathematics of filtering and identification and applications*, Reidel (1981).

[7] O. Hijab, *Stabilization of control systems*, Springer (1987).

[8] R.E. Howe, *On the role of the Heisenberg group in harmonic analysis*, Bull. Amer. Math. Soc., 3 (1980), pp. 821-844.

[9] S.I. Marcus, *Algebraic and geometric methods in nonlinear filtering*, SIAM J. Control and Opt., 22 (1984), pp. 817-844.

[10] R.E. Mortensen, *Optimal control of continuous time stochastic systems*, PhD thesis, Berkeley (1966).

[11] J. Wei, E. Norman, *On the global representation of the solutions of linear differential equations as products of exponentials*, Proc. Amer. Math. Soc, 15 (1964), pp. 327-334.

[12] M. Zakai, *On the optimal filtering of diffusion processes*, Z. Wahrsch. verw. Geb., 11 (1969), pp. 230-243.

APPROXIMATION OF HANKEL OPERATORS: TRUNCATION ERROR IN AN H^∞ DESIGN METHOD*

J.W. HELTON† AND N.J. YOUNG‡

Introduction. Many fashionable methods of designing filters, particularly in the context of control systems, make use of theorems giving the best H^∞ approximation to an L^∞ function in terms of singular vectors of a Hankel operator: such results may be found in [AAK], [C], [G]. Instances are the "H^∞ control" solutions to the problems of sensitivity minimization and robust stabilization [F]. Another example is the H^∞ disc method, which is applied to control problems in [H1] and to gain equalization in circuits in [H2]. To implement these methods it is necessary to represent Hankel operators numerically, a task for which several methods have been proposed [GR, SB, Y3, T, G1]. Two of these have gained currency. The most straightforward has been developed by Trefethen [T], and christened by him the *Caratheodory-Fejér method.* Mathematically a Hankel operator, which acts between two infinite-dimensional Hilbert spaces, can be represented by an infinite matrix

$$\begin{bmatrix} G_{-1} & G_{-2} & G_{-3} & \dots \\ G_{-2} & G_{-3} & G_{-4} & \dots \\ \cdot & \cdot & \cdot & \dots \end{bmatrix} \tag{1}$$

where

$$G(z) \sim \sum_{n=-\infty}^{\infty} G_n z^n \tag{2}$$

is the Fourier series of an L^∞ function G on the unit circle $\partial\mathbb{D}$. G is called a *symbol* of the operator. The Caratheodory-Fejér method simply approximates the Hankel operator by taking the $N \times N$ truncation of the matrix (1) for some large integer N.

The second popular method uses a state space representation. The requisite formulae were derived by Glover [G1], and numerous authors have reported favorably on their implementation [F]. However, the state space method is not directly comparable with the Caratheodory-Fejér method as it has a different starting point: it presupposes that a state space model of the symbol G (or "plant") is available. In some applications (for instance, the H^∞ disc method) the data of the problem (i.e. G) are supplied as function values on a grid, and so Glover's representation can only

*This research was partially supported by the Institute for Mathematics and its Applications, Minneapolis, with funds provided by the National Science Foundation and also by the Science and Engineering Research Council (UK) and Office of Naval Research (USA).

†Department of Mathematics, University of California, San Diego, La Jolla, CA 92093

‡Department of Mathematics, Lancaster University, Lancaster LA1 4 YF, England

be used after a preliminary stage of identification and realization. This adds to the computational burden and introduces error. The Caratheodory-Fejér method replaces the identification/realization stage by some procedure for extracting Fourier coefficients from data - typically the fast Fourer transform.

In this paper we propose and analyze a hybrid method which is close in spirit to the Caratheodory-Fejér method but does also involve an element of identification. The reason for our innovation is that in practice the Caratheodory-Fejér has proved inadequate for the design application under study. We can illustrate simply why this should be so by looking at the case of rational functions G. If all the unstable poles of G are well away from the unit circle then the negative Fourier coefficients of G decay quickly to zero. In this case a modestly-sized truncation of (1) affords a reasonable approximation to the true Hankel operator, and the Caratheodory-Fejér method works well. However poles close to the circle, manifesting themselves in "spikes" in the graph of $G(e^{i\theta})$, give rise to slowly decaying Fourier coefficients. In such circumstances it will require an inordinately large truncation of the Hankel matrix (1) to achieve a satisfactory representation. We propose a modified form of truncation which is intended to cope better with rapidly-varying G.

To address the problem of bad poles of the symbol G (or slow modes of the plant, in system-theoretic language) we begin with an identification stage. From the data we derive estimates $\beta_1, \ldots, \beta_m$ for the true poles $\alpha_1, \ldots, \alpha_m$ of G. Using the β_j we construct, for $N \in \mathbb{N}$, an $m + N$-dimensional subspace E of H^2 with the property that if the β_j are close to the α_j and N is large, then the restriction to E of the Hankel operator H_G is a good approximation to H_G for the purpose of computing the best H^∞ approximation to G.

There are two ways to assess the potential of our algorithm - numerical experimentation and mathematical analysis. The subject of this paper is a truncation error analysis. Our main result (Theorem 3) takes a fixed rational G and estimates the L^∞-distance from the true best H^∞ approximation $\widehat{G}$ of G to the function $\widetilde{G}$ computed by our modified truncation algorithm. We show:

Suppose that the highest singular value of the Hankel operator corresponding to G is simple. Suppose further that the only poles of G in the annulus $\{z \in \mathbb{C} : r < |z| < 1\}$ are $\alpha_1, \ldots, \alpha_m$. Then there exist $A, B > 0$ such that if $\widetilde{G}$ is the computed approximation obtained from the modified truncation algorithm based on $\beta_1, \ldots, \beta_m$ and $N \in \mathbb{N}$ then, for N sufficiently large, and for β_j sufficiently close to α_j,

$$\|\widehat{G} - \widetilde{G}\|_\infty \leq A \sum_{j=1}^{m} |\alpha_j - \beta_j| + Br^{N/2}.$$

This provides some evidence in favor of modified truncation as a way of computing with Hankel operators. Further evidence comes from numerical testing in joint work with P.G. Spain [HSY]. We have tried the method with the "candidate poles" β_j at varying distances from the true poles α_j. Good guesses for the β_j improve substantially on the Caratheodory-Fejér method. The closer the α_j to the circle, the better the β_j must be guessed for the algorithm to work well. This is consistent with the estimate in Theorem 3; it is unfortunate, but we believe it is intrinsic to the problem*

In the event that $\alpha_1, \ldots, \alpha_m$ are *all* the unstable poles of G and $\beta_j = \alpha_j$ (the poles are known exactly) Theorem 3 shows that the truncation error is zero. One can view this as the case of perfect knowledge, or alternatively as a truly rational plant generating data uncontaminated by noise, from which the denominator of the plant may be found precisely. Computational methods for this case have been given before [SB, Y3, GR]: the case of limited or noisy data seems to us to be at least as important. In particular it occurs in the H^∞ disc method. For our purpose there is no point in trying to guess *all* the unstable poles of G: poles of magnitude less than, say, 0.9 cause no difficulties for the Caratheodory-Fejér method, and so we handle them in the same way. This is the reason for the integer N in our theorem above. The poles close to the unit circle are the ones which cause trouble, but they are also, paradoxically, the ones which are most easily detected from the inspection of $G(e^{i\theta})$.

The modified truncation approach will only be advantageous in conjunction with a good way of tracking the bad poles of G directly from data (i.e. of finding reasonable approximations β_j to the true poles of G near the unit circle). This will be the subject of a future article [HSY]. Even quite crude tracking methods enable us to improve substantially on Caratheodory-Fejér. This is not surprising, as the latter is equivalent to choosing the β_j all to be zero.

For some applications it will be more important to obtain near-optimal performance rather than to get close to the true optimum (in other words, to find $\widetilde{G} \in H^\infty$ such that $\|G - \widetilde{G}\|_\infty$ is near its minimum rather than such that $\|\widehat{G} - \widetilde{G}\|_\infty$ is small). Nevertheless the issue of representing Hankel operators satisfactorily for computation remains, and the present results may show the way for convergence theorems for related computational tasks.

1. Hankel operators and their restrictions. We consider the Nehari problem corresponding to discrete-time systems. That is, we suppose given a function G in the space L^∞ of essentially bounded Lebesgue measurable complex-valued functions on the unit circle $\partial\mathbb{D}$. The problem is to find a function $\widehat{G} \in H^\infty$, the space of bounded analytical functions in the open unit disc $\mathbb{D}$, such that the L^∞ norm $\|G - \widehat{G}\|_\infty$ is minimized. One of the Adamyan-Arov-Krein theorems gives the solution in terms of the Hankel operator $H_G : H^2 \to L^2 \ominus H^2$ defined by

$$H_G x = P_-(Gx), \qquad x \in H^2.$$

Here L^2 is the space of square-integrable functions on $\partial\mathbb{D}$ with its usual inner product and H^2 is the Hardy space on the unit disc (See [F] or [Ho]). P_- is the

*It would be very interesting to compare our estimates here to comparable estimates which one could do on Glover's algorithm. Suppose the state equations input to Glover's algorithm are not perfectly determined; we have nominal values for the state space operators and are given bounds δ_j on the error in λ_j, the system eigenvalues near the circle. How much error does this produce in the solution $\hat{c}$ to the approximation problem? It would be interesting to compute this error and compare it to that in Corollary 2. That would give an idea of how our algorithm would compare to Glover's if his state space coefficients are actually estimated from data on a grid.

orthogonal projection operator from L^2 to $L^2 \ominus H^2$. We say that ν is a *maximizing vector* for an operator T if $\nu \neq 0$ and $\|T\nu\| = \|T\| \, \|\nu\|$. The AAK theorem in question asserts that if H_G has a maximizing vector (and so in particular if G is continuous on $\partial\mathbb{D}$) then the best approximation $\widehat{G}$ is unique and is given by

$$\widehat{G} = G - \frac{H_G \nu}{\nu} \tag{1.1}$$

for any maximizing vector ν of H_G. An exposition of this theorem may be found in [Y1].

The *standard bases* in $H^2, L^2 \ominus H^2$ are the orthonormal sequences $1, z, z^2, \ldots$ and $z^{-1}, z^{-2}, z^{-3}, \ldots$, respectively, or when expressed in terms of their restrictions to $\partial\mathbb{D}, 1, e^{i\theta}, e^{i2\theta}, \ldots$ and $e^{-i\theta}, e^{-i2\theta}, \ldots$. It is easy to check that the matrix of H_G with respect to these bases is given by (1)–(2). Finding a maximizing vector is a singular value problem: to solve this problem computationally we need to represent H_G, or at least an approximation to H_G, by means of a finite matrix. The Caratheodory-Fejér method works with the matrix of the compression of H_G acting from span $\{1, z, \ldots, z^{N-1}\}$ to span $\{z^{-1}, z^{-2}, \ldots, z^{-N}\}$. This is the first approximation one would think of, but since the two subspaces are not tuned in any way to the characteristics of G it is not likely to be the best one can do. A simple example will make this plain.

Consider the function

$$G(z) = \frac{1}{z - a}$$

where $a \in \mathbb{D}$. It is easy to show that H_G is a rank 1 operator. Its cokernel is

$$(\operatorname{Ker} H_G)^{\perp} = \text{ span } \{(1 - \overline{a}z)^{-1}\}.$$

Since the maximizing vectors of any operator lie in its cokernel we should ideally like to use the restriction of H_G to a space containing $(\operatorname{Ker} H_G)^{\perp}$: we could then find the maximizing vector ν without any truncation error. Failing this we should look for a finite-dimensional restriction to a subspace which is in some way close to $(\operatorname{Ker} H_G)^{\perp}$. In the example we can readily calculate that the cosine of the angle between the 1-dimensional space $(\operatorname{Ker} H_G)^{\perp}$ and span $\{1, z, \ldots, z^{N-1}\}$ is $1 - |a|^{2N}$, so that if $|a|$ is very close to 1 the spaces are nearly perpendicular until N becomes large. Experience confirms that it requires enormously large truncations to get satisfactory results if G has a pole of modulus, say, 0.98.

We can do better by attempting to guess the space $(\operatorname{Ker} H_G)^{\perp}$ and working with the restriction of H_G to this guess. Now $(\operatorname{Ker} H_G)^{\perp}$ is determined by the poles of G. That is, if G is expressible in the form f/p where $f \in H^\infty$ and P is a polynomial whose zeros lie in $\mathbb{D}$ then $(\operatorname{Ker} H_G)^{\perp} \subseteq H^2 \ominus pH^2$. For if $x \in pH^2$, say $x = py$ with $y \in H^2$, then

$$H_G x = P_-(Gpy) = P_-(fy) = 0,$$

i.e. $x \in \operatorname{Ker} H_G$. This motivates the algorithm which we call the

Modified truncation algorithm.

1. Guess a polynomial q whose zeros are close to the poles of G in $\mathbb{D}$ close to $\partial\mathbb{D}$.
2. Pick a positive integer N.
3. Form
   (a) the restriction operator

$$\dot{H}_G = H_G | H^2 \ominus z^N q H^2$$

   (b) the compression operator

$$\ddot{H}_G = P_{(1/z^N q)H^2 \ominus H^2} \quad H_G | H^2 \ominus z^N q H^2$$

4. Compute a maximizing vector ν of $\dot{H}_G$ or $\ddot{H}_G$ and form
   (a) $\widetilde{G} = G - \dfrac{\dot{H}_G \nu}{\nu}$, or
   (b) $\widetilde{G} = G - \dfrac{\ddot{H}_G \nu}{\nu}$

Then $\widetilde{G}$ is an approximation to the closest $\widehat{G}$ in H^∞ to G with respect to the L^∞ norm.

While this algorithm is stated at a high level, the (b) track of it is finite–dimensional and so can be implemented. One performs Gram-Schmidt orthogonalization to the basis

$$1,\ z,\ z^2, \dots, z^{N-1}, \frac{1}{1-\overline{\beta}_1 z}, \dots, \frac{1}{1-\overline{\beta}_m z}$$

to obtain an orthonormal basis for $H^2 \ominus z^N q H^2$, and thereby obtains a matrix representation of $\dot{H}_G$. The details will be published separately.

If the zeros of q are close to the zeros of p, then it is plausible to expect that $H^2 \ominus qH^2$ should come close to containing $(\mathrm{Ker}\ H_G)^\perp$, and hence that the restriction $\dot{H}_G$ and compression $\ddot{H}_G$ should be satisfactory approximations to H_G for the purpose of computing $\widehat{G}$. Our goal is to give these imprecise notions a solid analytical foundation. We consider only the restriction procedure, since we believe that it gives a good indication of what happens with compression as well. The key problem is

Error analysis problem

Given $G \in L^\infty$ of the form f/p, where $f \in H^\infty$ and p is a polynomial whose zeros lie in $\mathbb{D}$, and given a polynomial q whose zeros are close to those of p, estimate the error (in the L^∞ norm) incurred in using $H_G | H^2 \ominus qH^2$ instead of H_G to compute the best H^∞ approximation $\widehat{G}$ to G.

In other words, we seek a bound for $\|\widehat{G} - \widetilde{G}\|_\infty$ where

$$\widetilde{G} = G - \frac{H_G \nu}{\nu}$$

and ν is a maximizing vector for $H_G | H^2 \ominus qH^2$.

We shall obtain such a bound, albeit one involving quantities which cannot be computed explicitly. Our result is thus a convergence theorem rather than a true error estimate.

2. Singular values of projection operators. One ingredient of our analysis is to show that if p is close to q then the spaces $H^2 \ominus pH^2$ and $H^2 \ominus qH^2$ are so close that any maximizing vector for $H_G | H^2 \ominus qH^2$ is near-maximizing for $H_G | H^2 \ominus pH^2$. The relevant sense of closeness of the two spaces is that the orthogonal projection

$$P_{H^2 \ominus pH^2} | H^2 \ominus qH^2$$

from $H^2 \ominus qH^2$ onto $H^2 \ominus pH^2$ have all singular values close to 1.

THEOREM 1. *Let $n \geq m \geq 1$ and let*

$$p(z) = (z - \alpha_1)(z - \alpha_2) \cdots (z - \alpha_m),$$
$$q(z) = (z - \beta_1)(z - \beta_2) \cdots (z - \beta_n)$$

where $\alpha_1, \ldots, \alpha_m, \beta_1, \ldots, \beta_n \in \mathbf{D}$. Then

$$\underline{\sigma}(P_{H^2 \ominus pH^2} | H^2 \ominus qH^2)^2 \geq 1 - \left\{ \sum_{i=1}^{m} \left| \frac{\alpha_j - \beta_j}{1 - \overline{\beta}_i \alpha_i} \right| \right\}^2$$

where $\underline{\sigma}(.)$ denotes the smallest singular value of an operator.

Proof. For brevity let

$$R = P_{H^2 \ominus pH^2} | H^2 \ominus qH^2.$$

By definition $\underline{\sigma}(R)$ is the smallest eigenvalue of $(RR^*)^{\frac{1}{2}}$. Let

$$v(z) = \prod_{i=1}^{n} \frac{z - \beta_i}{1 - \overline{\beta}_i z}$$

and let

$$P : H^2 \to H^2 \ominus pH^2, \qquad Q : H^2 \to H^2 \ominus qH^2$$

be orthogonal projection operators, so that P^*, Q^* are the corresponding injections into H^2. Then $R = PQ^*$ and so

$$RR^* = PQ^*QP^*.$$

Now Q^*Q is the Hermitian projection on H^2 with range $H^2 \ominus qH^2$, and we can give an alternative expression for this operator in terms of v. Let S denote the forward shift operator on H^2 (i.e. $(Sf)(z) = zf(z)$), so that $v(S)$ is multiplication by the inner function v and is an isometry. Thus $v(S)^*v(S) = I$ and so $v(S)v(S)^*$ is a Hermitian projection with range $vH^2 = qH^2$. Thus $I - v(S)v(S)^*$ is the Hermitian projection with range $H^2 \ominus qH^2$, i.e.

$$Q^*Q = I - v(S)v(S)^*,$$

and so

$$RR^* = P(I - v(S)v(S)^*)P^*.$$

PP^* is the identity operator on $H^2 \ominus pH^2$, and since $H^2 \ominus pH^2$ is invariant under $v(S)^*$ we have $PP^*v(S)^*P^* = v(S)^*P^*$ and so

$$RR^* = I_{H^2 \ominus pH^2} - Pv(S)P^*Pv(S)^*P^*.$$

Now

$$Pv(S)P^* = v(PSP^*) = v(S_p)$$

where

$$S_p = PSP^*$$

is the compression of S to $H^2 \ominus pH^2$, known as the Sarason operator after [S]. Hence

$$R^*R = I - v(S_p)v(S_p)^* \in \mathcal{L}(H^2 \ominus pH^2).$$

It follows that

$$\underline{\sigma}(R)^2 = 1 - \|v(S_p)\|^2 \tag{2.2}$$

To estimate $\|v(S_p)\|$ we use an ingenious observation due to the late Constantin Apostol in an oral communication. This useful result was published (with his consent) in [PY] and extended in [Y2]. The proof is so short we repeat it here.

LEMMA 1. *Let $T_1, \ldots, T_m$ be $m \times m$ upper triangular matrices with $\|T_i\| \leq 1$, $1 \leq i \leq m$. Let the ith diagonal entry of T_i have modulus $r_i, 1 \leq i \leq m$. Then*

$$\|T_1 T_2 \ldots T_m\| \leq \sum_{i=1}^{m} r_i.$$

Proof. The assertion is true when $m = 1$. Suppose it holds for $m - 1$. Partition T_i as

$$T_i = \begin{bmatrix} R_i & * \\ 0 & \lambda_i \end{bmatrix},$$

so that $|\lambda_m| = r_m$. Note that

$$T_1 T_2 \ldots T_m = \begin{bmatrix} R_1 \ldots R_{m-1} & 0 \\ 0 & 0 \end{bmatrix} T_m + T_1 \ldots T_{m-1} \begin{bmatrix} 0 & 0 \\ 0 & \lambda_m \end{bmatrix}.$$

On taking norms and using the triangle inequality we have

$$\begin{aligned}\|T_1 \dots T_m\| &\le \|R_1 \dots R_{m-1}\| \, \|T_m\| + \|T_1 \dots T_{m-1}\| \, \|\lambda_m| \\ &\le \sum_1^{m-1} r_i.1 + 1.r_m = \sum_1^m r_i,\end{aligned}$$

establishing the result by induction. In fact the proof shows that the inequality even holds with the operator norm replaced by the trace norm.

To use the lemma we write

$$v(S_p) = T_1 \dots T_n$$

where

$$T_i = (S_p - \beta_i I)(I - \overline{\beta}_i S_p)^{-1}.$$

It is easy to check that the eigenvalues of S_p are the zeros of p, so that for a suitable choice of orthonormal basis in $H^2 \ominus pH^2$ S_p will be represented by an upper triangular matrix with diagonal entries $\alpha_1, \dots, \alpha_m$. Thus T_i is also upper triangular, and so Apostol's Lemma gives

$$\|v(S_p)\| \le \|T_1 \dots T_m\| \le \sum_{i=1}^m \left| \frac{\alpha_i - \beta_i}{1 - \overline{\beta}_i \alpha_i} \right|$$

Combining this inequality with (2.2) gives the desired result (2.1).

LEMMA 2. *Let p, q be as in Theorem 1 and let $G \in p^{-1} H^\infty$. Then*

$$\|H_G | H^2 \ominus qH^2\| \ge \underline{\sigma}(P_{H^2 \ominus pH^2} | H^2 \ominus \; qH^2) \, \|H_G\|.$$

Proof. Since H_G has finite rank it has a maximizing vector $x \neq 0$. Necessarily $x \in (\text{Ker } H_G)^\perp \subseteq H^2 \ominus pH^2$. $P_{H^2 \ominus pH^2} | H^2 \ominus qH^2$ maps its domain surjectively onto its range, and hence there exists $y \neq 0$ in $H^2 \ominus qH^2$ such that $Py = x$ (P being as in the proof of Lemma 1). Then

$$\begin{aligned}\|H_G y\| = \|H_G P y\| = \|H_G x\| = \|H_G\| \, \|x\| \\ = \|H_G y\| \, \|Py\| \ge \underline{\sigma}(P_{H^2 \ominus pH^2} | H^2 \ominus qH^2) \, \|H_G\| \, \|y\|.\end{aligned}$$

3. Computing with near-maximizing vectors. The upshot of the preceding section is that, for polynomials q sufficiently close to the denominator of G, $\|H_G | H^2 \ominus qH^2\|$ is close to $\|H_G\|$. It follows that a maximizing vector for the former operator is nearly maximizing for H_G. We shall make this observation precise and examine its implications for the purpose of computing $\widehat{G}$. Let us say that a vector x is *ϵ-maximizing* for an operator T if $x \neq 0$ and

$$\|Tx\| \ge (1 - \epsilon) \, \|T\| \, \|x\|.$$

LEMMA 3. *Let p, q and G be as in Lemma 2 and write*

$$M = \sum_{i=1}^{m} \left| \frac{\alpha_i - \beta_i}{1 - \overline{\beta}_i \alpha_i} \right|.$$

Let ν be a maximizing vector for $H_G | H^2 \ominus qH^2$. Then ν is ϵ-maximizing for H_G, where

$$\epsilon = 1 - \surd(1 - M^2).$$

Proof. Using Theorem 1 and Lemma 2 we have $\nu \neq 0$ and

$$\begin{aligned} \|H_G \nu\| &= \|H_G | H^2 \ominus qH^2\| \; \|\nu\| \\ &\geq \underline{\sigma}(P_{H^2 \ominus qH^2} | H^2 \ominus pH^2) \; \|H_G\| \; \|\nu\| \\ &\geq \surd(1 - M^2) \|H_G\| \; \|\nu\| \\ &= (1 - \epsilon) \|H_G\| \; \|\nu\| \end{aligned}$$

where $1 - \epsilon = \surd(1 - M^2)$.

LEMMA 4. *Let $G \epsilon L^\infty, 0 < \epsilon < 1$ and let ν be an ϵ-maximizing vector for H_G. Let $\widehat{G}$ be a best approximation in H^∞ to G. Then*

$$(G - \widehat{G})\nu = H_G \nu + \eta$$

where $\eta \epsilon H^2$ and

$$\|\eta\| \leq \surd(2\epsilon - \epsilon^2) \|H_G\| \; \|v\|. \tag{3.1}$$

Proof. Let $P_+ : L^2 \rightarrow H^2$ be the orthogonal projection operator. We take

$$\eta = P_+((G - \widehat{G})\nu).$$

Then

$$\begin{aligned} (G - \widehat{G})\nu &= P_-(G - \widehat{G})\nu + (G - \widehat{G})\nu \\ &= H_G \nu + \eta. \end{aligned}$$

The important point is that η has small norm. We may assume $\|\nu\| = 1$. Then

$$\begin{aligned} (1 - \epsilon)^2 \|H_G\|^2 &\leq \|H_G \nu\|^2 = \|P_-(G - \widehat{G})\nu\|^2 \\ &\leq \|(G - \widehat{G})\nu\|^2 \leq \|G - \widehat{G}\|_\infty^2 \|\nu\|^2 = \|H_G\|^2, \end{aligned}$$

the last step by Nehari's theorem. Thus

$$\|(G - \widehat{G})\nu\|^2 - \|P_-(G - \widehat{G})\nu\|^2 \leq (1 - (1 - \epsilon)^2) \|H_G\|^2.$$

That is,

$$\|\eta\|^2 \leq (2\epsilon - \epsilon^2) \|H_G\|^2.$$

Lemma 4 gives us an L^2 bound on the error vector η, whereas we are aiming for an L^∞ bound. To succeed we must pin η down to a finite-dimensional subspace of H^2 for which we can estimate the constant relating the two norms.

Recall from [AAK] that $G - \widehat{G}$ has constant modulus $\|H_G\|$ a.e. on the unit circle. Thus, in the case that H_G has finite rank, when $G - \widehat{G}$ is rational, we can write

$$G - \widehat{G} = \|H_G\| \theta \overline{\varphi} \tag{3.2}$$

for some relatively prime finite Blaschke products θ and φ.

LEMMA 5. *With the assumptions of Lemma 4, suppose further that* $v \in H^2 \ominus qH^2$ *for some polynomial* q *and that* $G - \widehat{G}$ *is expressible in the form (3.2). Then* η *in Lemma 4 lies in the space* $H^2 \ominus \theta q H^2$.

Proof. We have

$$\begin{aligned}(G-\widehat{G})\nu &= \|H_G\|\theta\overline{\varphi}\nu \\ &\in \theta\overline{\varphi}(H^2 \ominus qH^2) \\ &= \overline{\varphi}(\theta H^2 \ominus \theta q H^2) \\ &\subseteq \overline{\varphi}(H^2 \ominus \theta q H^2).\end{aligned}$$

Hence

$$\begin{aligned}\eta = P_+(G-\widehat{G})\nu \;&\in\; P_+\overline{\varphi}(H^2 \ominus \theta q H^2) \\ &= \varphi(S)^*(H^2 \ominus \theta q H^2)\end{aligned}$$

where, as before, S is the forward shift on H^2. Since $H^2 \ominus \theta q H^2$ is invariant under S^* it follows that $\eta \in H^2 \ominus \theta q H^2$.

LEMMA 6. *Let* $\rho_1, \ldots, \rho_k \in \mathbf{D}$ *and let*

$$\chi(z) = \prod_{j=1}^{k} \frac{z-\rho_j}{1-\overline{\rho}_j z}.$$

For any $x \in H^2 \ominus \chi H^2, x \in H^\infty$ *and*

$$\|x\|_\infty \leq \sum_{j=1}^{k} \left(\frac{1+|\rho_j|}{1-|\rho_j|}\right)^{\frac{1}{2}} \|x\|$$

Proof. Let

$$K(z,w) = \frac{1-\chi(\overline{w})\chi(z)}{1-\overline{w}z}, \qquad z, w \in \mathbf{D}.$$

K is the reproducing kernel for $H^2 \ominus \chi H^2$ - that is $K(.,w) \in H^2 \ominus \chi H^2$ for all $w \in \mathbf{D}$ and, for any $x \in H^2 \ominus \chi H^2, w \in \mathbf{D}$,

$$x(w) = (x, K(.,w)).$$

By the Cauchy-Schwarz inequality,

$$|x(w)| \leq \|x\| \; \|K(.,w)\|_{H^2} \tag{3.3}$$

Let

$$\chi_j(z) = \frac{z-\rho_j}{1-\overline{\rho}_j z}, \qquad 1 \leq j \leq k,$$

so that $\chi = \chi_1 \dots \chi_k$. Fixing $w \in \mathbf{D}$ and regarding $z \in \partial\mathbf{D}$ as the independent variable we have

$$\begin{aligned}\|K(z,w)\| &= \left\| \frac{1-\chi(\overline{w})\chi(z)}{1-\overline{w}z} \right\| \\ &= \left\| \frac{\chi(z)-\chi(w)}{z-w} \right\|.\end{aligned}$$

Now

$$\begin{aligned}\frac{\chi(z)-\chi(w)}{z-w} &= \chi_1 \cdots \chi_{k-1}(z) \frac{\chi_k(z)-\chi_k(w)}{z-w} + \chi_1 \cdots \chi_{k-2}(z) \frac{\chi_{k-1}(z)-\chi_{k-1}(w)}{z-w} \chi_k(w) \\ &\quad + \dots + \frac{\chi_1(z)-\chi_1(w)}{z-w} \chi_2 \cdots \chi_k(w).\end{aligned}$$

It follows that

$$\begin{aligned}\|K(.,w)\| &\leq \sum_{j=1}^{k} \left\| \frac{\chi_j(z)-\chi_j(w)}{z-w} \right\| \\ &= \sum_{j=1}^{k} \frac{1-|\rho_j|^2}{|1-\overline{\rho}_j w|} \left\| \frac{1}{1-\overline{\rho}_j z} \right\| \\ &= \sum_{j=1}^{k} \frac{1-|\rho_j\|^2}{|1-\overline{\rho}_j w|} \cdot \frac{1}{(1-|\rho_j|^2)^{\frac{1}{2}}} \\ &= \sum_{j=1}^{k} \frac{(1-|\rho_j\|^2)^{\frac{1}{2}}}{|1-\overline{\rho}_j w|} \\ &\leq \sum_{j=1}^{k} \frac{(1-|\rho_j|^2)^{\frac{1}{2}}}{|1-\rho_j|} \\ &= \sum_{j=1}^{k} \left(\frac{1+|\rho_j|}{1-|\rho_j|} \right)^{\frac{1}{2}}\end{aligned}$$

Combining this inequality with (3.3) we have

$$|x(w)| \leq \sum_{j=1}^{k} \left(\frac{1+|\rho_j|}{1-|\rho_j|} \right)^{\frac{1}{2}} \cdot \quad \|x\|.$$

As this holds for all $w \in \mathbf{D}$, the Lemma follows.

Lemmas 3 to 6 taken together suffice to give an L^∞ bound on the error vector $\eta = (G-\widehat{G})\nu - H_G\nu$ where ν is maximizing for $[H_G|H^2 \ominus qH^2]$ and q is close to the denominator of G. Indeed, by Lemma 3 ν is ϵ-maximizing, where $\epsilon = 1-\surd(1-M^2)$, and so by Lemma 4

$$\begin{aligned} \|\eta\| &\leq \sqrt{1-(1-\epsilon)^2}\ \|H_G\|\ \|\nu\|. \\ &= M\|H_G\|\ \|\nu\|. \end{aligned} \tag{3.4}$$

Lemma 5 tells us that $\eta \in H^2 \ominus \theta q H^2$, where $G - \widehat{G}$ is expressible in the form (3.2), and so Lemma 6 enables us to convert (3.4) to an L^∞-estimate:

$$\|\eta\|_\infty \leq \left(\sum_{j=1}^{m} \left(\frac{1+|\beta_j|}{1-|\beta_j|}\right)^{\frac{1}{2}} + \sum_{j=1}^{k} \left(\frac{1+|\rho_j|}{1-|\rho_j|}\right)^{\frac{1}{2}} \right) M\|H_G\|\ \|\nu\| \tag{3.5}$$

where β_j are the zeros of q and the ρ_j are the zeros of θ. The ρ_j's are unknown in the situation we envisage, but the inequality is enough to give us a rate of convergence.

4. A convergence theorem.

THEOREM 2. *Let $G \in L^\infty$ be such that $(z-\alpha_1)(z-\alpha_2)\ldots(z-\alpha_m)G(z) \in H^\infty$ for some $\alpha_1, \ldots \alpha_m \in \mathbb{D}$ and suppose that $\|H_G\|$ is a simple singular value of H_G. There exist a constant $K > 0$ and neighborhoods U_i of α_i, $1 \leq i \leq m$, with the following property. If $\beta_i \in U_i$, $1 \leq i \leq m$, and v is a maximizing vector of*

$$H_G | H^2 \ominus (z-\beta_1)(z-\beta_2)\ldots(z-\beta_m)H^2$$

then ν is bounded away from zero on $\partial\mathbb{D}$ and the function $\widetilde{G}$ defined by

$$\widetilde{G} = G - \frac{H_G \nu}{\nu}$$

satisfies

$$\|\widetilde{G} - \widehat{G}\|_\infty \leq K \sum_{j=1}^{m} |\alpha_j - \beta_j|$$

where $\widehat{G}$ is the best approximation in H^∞ to G with respect to the L^∞ norm.

A constant K is given explicitly in (4.9).

Proof. We may assume ν is always chosen to be a unit vector. Section 3 (inequality (3.5)) shows that, for any choice of $\beta_1, \ldots, \beta_m \in \mathbb{D}$

$$(\widetilde{G} - \widehat{G})\nu = \eta \tag{4.1}$$

where

$$\|\eta\|_\infty \leq \left(\sum_{j=1}^{m} \left(\frac{1+|\beta_j|}{1-|\beta_j|}\right)^{\frac{1}{2}} + \sum_{j=1}^{k} \left(\frac{1+|\rho_j|}{1-|\rho_j|}\right)^{\frac{1}{2}} \right) \times \sum_{j=1}^{m} \left| \frac{\alpha_j - \beta_j}{1-\overline{\beta}_j \alpha_j} \right| \cdot \|H_G\|$$

where $\rho_1, \ldots, \rho_k$ are the zeros (with repetition according to multiplicity) of the function θ occurring in the expression

$$G - \widehat{G} = \|H_G\| \theta \overline{\varphi}$$

with θ, φ coprime finite Blaschke products (the assumption that a polynomial multiple of G is in H^∞ entails that $G - \widehat{G}$ is rational, and since it has constant modulus it does have such an expression). It follows that $\rho_1, \ldots, \rho_k$ are determined by G and are independent of the β_j. As $\beta_j \to \alpha_j$ we have

$$\sum_{j=1}^{m} \left(\frac{1 + |\beta_j|}{1 - |\beta_j|} \right)^{\frac{1}{2}} \longrightarrow \sum_{j=1}^{m} \left(\frac{1 + |\alpha_j|}{1 - |\alpha_j|} \right)^{\frac{1}{2}}$$

and

$$1 - \overline{\beta}_j \alpha_j \to 1 - |\alpha_j|^2.$$

It follows that, for β_j sufficiently close to α_j, $1 \le j \le k$, we have

$$\|\eta\|_\infty \le A \sum_{j=1}^{k} |\alpha_j - \beta_j| \tag{4.2}$$

with

(4.3)

$$A = 2 \left(\sum_{j=1}^{m} \left(\frac{1 + |\alpha_j|}{1 - |\alpha_j|} \right)^{\frac{1}{2}} + \sum_{j=1}^{k} \left(\frac{1 + |\rho_j|}{1 - |\rho_j|} \right)^{\frac{1}{2}} \right) \cdot \left\{ \min_{1 \le j \le m} \; (1 - |\alpha_j|^2) \right\}^{-1} \|H_G\|.$$

Dividing by ν in (4.1) gives

$$\begin{aligned} \|\widetilde{G} - \widehat{G}\|_\infty &= \|\eta/\nu\|_\infty \\ &\le \|\eta\|_\infty \; \{\inf_{|z|=1} |\nu(z)|\}^{-1} \end{aligned} \tag{4.4}$$

To show that this gives a finite bound we need the singular value condition.

LEMMA 7. *Let T be a compact operator between Hilbert spaces and let $\|T\|$ be a simple singular value of T. For any ϵ, $0 < \epsilon < 1$, there exists $\delta > 0$ such that whenever x is a δ-maximizing unit vector for T, there is a unit maximizing vector x' of T such that $\|x - x'\| < \epsilon$. Indeed we may take $\delta = K\epsilon^2$ where $K > 0$ is a constant depending only on T.*

Proof. Let E be the 1-dimensional space of maximizing vectors for T. With respect to the direct decompositions $E \oplus E^\perp, TE \oplus (TE)^\perp$ of the domain and codomain of T respectively we can write

$$T \sim \begin{bmatrix} \|T\| & 0 \\ 0 & T_1 \end{bmatrix}$$

where $\|T_1\| < \|T\|$. Consider a δ-maximizing unit vector

$$x = \begin{bmatrix} x_0 \\ x_1 \end{bmatrix}$$

for T. Then

$$\|Tx\|^2 \geq (1-\delta)^2 \|T\|^2,$$

i.e.

$$\|T\|^2 |x_0|^2 + \|T_1 x_1\|^2 \geq (1-\delta)^2 \|T\|^2,$$

and hence

$$\|T\|^2 |x_0|^2 + \|T_1\|^2 + \|T_1\|^2 (1 - |x_0|^2) \geq (1 - 2\delta) \|T\|^2.$$

On rearranging we obtain

$$|x_0|^2 \geq 1 - D\delta$$

where

$$D = \frac{2\|T\|^2}{\|T\|^2 - \|T_1\|^2} > 0.$$

A *fortiori*

$$|x_0| \geq 1 - D\delta$$

as long as $\delta \leq 1/D$.

Let x' be the unit maximizing vector

$$x' = \begin{bmatrix} x_0 \\ 0 \end{bmatrix} \Big/ |x_0|.$$

Then

$$\begin{aligned} \|x - x'\|^2 &= (|x_0| - 1)^2 + \|x_1\|^2 \\ &= (|x_0| - 1)^2 + 1 - |x_0|^2 \\ &= 2(1 - |x_0|) \\ &\leq 2D\delta \end{aligned}$$

for $\delta \leq 1/D$. Hence we may take $\delta = \epsilon^2/(2D)$.

We resume the demonstration that $|\nu|$ is bounded away from zero. It is a well-known result of deLeeuw-Rudin, cf [G], that, when $\|H_G\|$ is a simple singular value of H_G, any maximizing vector of H_G is non-zero everywhere. Such maximizing vectors lie in $H^2 \ominus pH^2$, where

$$p(z) = (z - \alpha_1) \dots (z - \alpha_m),$$

and so are rational. Hence there exists $c > 0$ such that

$$|x(z)| \geq c \tag{4.5}$$

for all $z \in \partial\mathbb{D}$ and all unit maximizing vectors x of H_G.

Next observe that if ν is maximizing for $H_G|H^2 \ominus qH^2$, where

$$q(z) = (z - \beta_1)\dots(z - \beta_m), \tag{4.6}$$

and ν' is any maximizing vector for H_G then $\nu' \in H^2 \ominus qH^2, \nu' \in H^2 \ominus pH^2$ and so $\nu - \nu' \in H^2 \ominus pqH^2$. Hence, by Lemma 6

$$\|\nu - \nu'\|_\infty \leq \left(\sum_{j=1}^{m} \left(\frac{1 + |\alpha_j|}{1 - |\alpha_j|} \right)^{\frac{1}{2}} + \sum_{j=1}^{m} \left(\frac{1 + |\beta_j|}{1 - |\beta_j|} \right)^{\frac{1}{2}} \right) \|\nu - \nu'\|$$

Hence, for β_j sufficiently close to α_j,

$$\|\nu - \nu'\|_\infty \leq L\|\nu - \nu'\| \tag{4.7}$$

where

$$L = 3 \sum_{j=1}^{m} \left(\frac{1 + |\alpha_j|}{1 - |\alpha_j|} \right)^{\frac{1}{2}}$$

By Lemma 7 there exists $\delta > 0$ such that, if ν is a δ-maximizing unit vector for H_G then there exists a unit maximizing vector ν' for H_G such that

$$\|\nu - \nu'\| < \frac{c}{2L} \, .$$

By (4.7) there are neighborhoods V_j of $\alpha_j, 1 \leq j \leq m$, such that, if in addition $\beta_j \in V_j$ and $\nu \in H^2 \ominus qH^2$ with q given by (4.6) then

$$\|\nu - \nu'\|_\infty < L\|\nu - \nu'\| < \frac{c}{2} \, .$$

In view of (4.6) this implies that

$$|v(z)| \geq \frac{c}{2}, \qquad \text{all} \quad z \in \partial\mathbb{D}. \tag{4.8}$$

Now by Lemma 3 there exist neighborhoods W_j of $\alpha_j, 1 \leq j \leq m$, such that if $\beta_j \in W_j$ and ν is maximizing for $H_G|H^2 \ominus qH^2$ then ν is δ-maximizing for H_G: we simply have to make the $\alpha_j - \beta_j$ so small that

$$1 - \sqrt{\left\{ 1 - \left(\sum_{j=1}^{m} \left| \frac{\alpha_j - \beta_j}{1 - \overline{\beta}_j \alpha_j} \right| \right)^2 \right\}} < \delta$$

Thus, for β_j sufficiently close to α_j (4.8) is satisfied and so (4.4) shows that

$$\|\widetilde{G} - \widehat{G}\|_\infty \leq \frac{2}{c} \, \|\eta\|_\infty.$$

It follows from (4.2) that

$$\|\widetilde{G} - \widehat{G}\|_\infty \le \frac{2A}{c} \sum_{j=1}^{m} |\alpha_j - \beta_j|$$

whenever β_j is sufficiently close to $\alpha_j, 1 \le j \le m$. Thus the theorem holds with

(4.9)

$$\begin{aligned} K &= \frac{2A}{c} \\ &= \frac{4}{c}\left(\sum_{j=1}^{m}\left(\frac{1+|\alpha_j|}{1-|\alpha_j|}\right)^{\frac{1}{2}} + \sum_{j=1}^{k}\left(\frac{1+|\rho_j|}{1-|\rho_j|}\right)^{\frac{1}{2}}\right) \cdot \left\{\min_{1\le j\le m} (1-|\alpha_j|^2)\right\}^{-1} \|H_G\| \end{aligned}$$

where c is the inf of $|\nu|$ over $\partial\mathbb{D}$ for a unit maximizing vector ν for H_G and $\rho_1, \ldots, \rho_k$ are the zeros in $\mathbb{D}$ of the rational function $G - \widehat{G}$.

5. An example. In Section 1 we considered the function

$$G(z) = \frac{1}{z-a}$$

where $a \in \mathbb{D}$ is very close to the unit circle. This example is easy to analyse by hand but nevertheless illustrates the issues we are addressing. We have pointed out that the Caratheodory-Fejér method does not work well in such a case: let us investigate the alternatives. It is easy to show that the exact best H^∞ approximation to G is the constant function

$$\widehat{G}(z) = \frac{\overline{a}}{1-|a|^2}\ .$$

Suppose that the pole is a mis-identified as $a + \delta a \in \mathbb{D}$, with δa small. If we carry out best analytical approximation exactly, by whatsoever method, on the "nominal plant" $(z - a - \delta a)^{-1}$ we shall obtain a computed best approximant

$$\widehat{G}(z) = \frac{\overline{a} + \overline{\delta a}}{1 - |a+\delta a|^2}\ .$$

For a real the error is approximately

$$\frac{d}{da}\,\frac{a}{1-a^2}\cdot \delta a = \frac{\delta a}{2(1-a)^2}.$$

On the other hand, if we use modified truncation and succeed in representing $H_G|H^2 \ominus (z - a - \delta a)H^2$ exactly then an easy calculation shows that we obtain error approximately $\delta a/4(1-a)^2$. Thus by using the data twice we have halved the error! Of course, in neither case is the idealization realistic: state space methods, for example, typically involve solving Lyapunov equations which become ill-conditioned in the presence of poles close to the unit circle.

More generally, let us analyse what happens when we guess m candidates poles $\beta_1, \ldots, \beta_m$ in ignorance that G has only one pole. Write

$$v(z) = \prod_{j=1}^{m} \frac{z - \beta_j}{1 - \overline{\beta}_j z}$$

Since $(\text{Ker } H_G)^{\perp} = \text{span } \{(1 - \overline{a}z)^{-1}\}$, a maximizing vector for $H_G | H^2 \ominus vH^2$ is the projection on this space of $(1 - \overline{a}z)^{-1}$, i.e. the vector

$$v(z) = \frac{1 - v(\overline{a})v(z)}{1 - \overline{a}z}.$$

Let $b = v(a)$. Then

$$H_G \nu(z) = P_- \frac{1}{z - a} \cdot \frac{1 - v(\overline{a})v(z)}{1 - \overline{a}z}.$$

$$- \frac{1 - |b|^2}{1 - |a|^2} \cdot \frac{1}{z - a},$$

so that the computed approximation is

$$\begin{aligned} \widetilde{G}(z) &= \frac{1}{z - a} - \frac{1 - |b|^2}{1 - |a|^2} \cdot \frac{1}{z - a} \cdot \frac{1 - \overline{a}z}{1 - \overline{b}v(z)} \\ &= \frac{1}{(1 - |a|^2)(1 - \overline{b}v(z))} \left[\overline{a} - \overline{b} \frac{(1 - |a|)^2 v(z) - (1 - \overline{a}z)b}{z - a} \right]. \end{aligned}$$

We observe that here $\widetilde{G}$ does belong to H^∞, since $|b| < 1$. If one or more candidate poles β_j are close to α then b will be small and so $\widetilde{G}$ will be close to $\widehat{G}$, as expected. We can also make b small by taking v to have high degree, compensating for poor pole tracking by brute force. In particular we could take $v(z) = z^N$ for large N: this would be a half way state to the Caratheory-Fejér method, in which the Hankel matrix is approximated by its $\infty \times N$ truncation.

6. Convergence of the modified truncation algorithm.

THEOREM 3. *Let $G \in L^\infty$ be rational and suppose that $\|H_G\|$ is a simple singular value of H_G. Let $0 < r < 1$ and let $\alpha_1, \ldots, \alpha_m$ be the poles of G which lie in the annulus $\{z : r \le |z| < 1\}$. Let $\widehat{G}$ be the closest function in H^∞ to G with respect to the L^∞ norm. Then there exist constants $A, B > 0$ and a positive integer N_0 with the following property. For each $N \ge N_0$ there are neighborhoods U_j of $\alpha_j, 1 \le j \le m$, such that if $\beta_j \in U_j, 1 \le j \le m$, and if ν is a maximizing vector of*

$$H_G | H^2 \ominus z^N (z - \beta_1) \ldots (z - \beta_n) H^2$$

for some $n \ge m$ and $\beta_j \in \mathbb{D}, m < j \le n$, then ν is bounded away from zero on $\partial\mathbb{D}$ and the function $\widetilde{G}$ defined by

$$\widetilde{G} = G - \frac{H_G \nu}{\nu}$$

satisfies

$$\|\widetilde{G} - \widehat{G}\|_\infty \leq A \sum_{j=1}^{m} |\alpha_j - \beta_j| + Br^{N/2}.$$

Proof. Let

$$p(z) = \prod_{j=1}^{m} z - \alpha_j, \qquad q(z) = \prod_{j=1}^{n} z - \beta_j.$$

Let

$$T = H_G | H^2 \ominus z^N q H^2,$$

so that ν is maximizing for T. Let the Fourier series of pG be

$$(pG)(z) \;\sim\; \sum_{-\infty}^{\infty} c_k z^k.$$

Since pG is analytic in the annulus $\{z : r \leq |z| \leq 1\}$ it follows from Cauchy's integral formula that

$$c_{-k} = O(r^k), \qquad k \in \mathbb{N}.$$

Hence, if we write

$$g_N(z) = \sum_{k=-N}^{\infty} c_k z^k.$$

we have

$$\|pG - g_N\| \leq Cr^N$$

for some constant C. Thus, if

$$G_N = g_N / p,$$

we have

$$\text{(6.1)} \qquad \|G - G_N\|_\infty \leq Cr^N$$

for some (possibly different) $C > 0$, and furthermore

$$\text{(6.2)} \qquad z^N p G_N = z^N g_N = \sum_{k=0}^{\infty} c_{k-n} z^k \in H^\infty$$

We show that ν is near-maximizing for the operator

$$\widetilde{T} = H_{G_N} | H^2 \ominus z^N q H^2.$$

Indeed,

$$\begin{aligned} \|T - \widetilde{T}\| &= \|(H_G - H_{G_N}) | H^2 \ominus z^N q H^2\| \\ &\leq \|G - G_N\| \leq \|G - G_N\|_\infty \\ &\leq Cr^N. \end{aligned}$$

Since $\|T\nu\| = \|T\| \; \|\nu\|$ we have

$$\|\widetilde{T}\nu\| = \|T\nu + (\widetilde{T} - T)\nu\|$$
$$\geq (\|T\| - Cr^N)\|\nu\|$$
$$\geq (\|\widetilde{T}\| - 2Cr^N)\|\nu\| \tag{6.3}$$

Next we show that ν is near-maximizing for H_G. In view of (6.2) Lemma 2 shows that

$$\|\widetilde{T}\| \geq \underline{\sigma}(P_{H^2 \ominus pH^2}|H^2 \ominus qH^2)\|H_{G_N}\|.$$

By Theorem 1,

$$\|\widetilde{T}\| \geq \sqrt{1 - M^2} \; \|H_{G_N}\| \tag{6.4}$$

where

$$M = \sum_{j=1}^{m} \left| \frac{\alpha_j - \beta_j}{1 - \overline{\beta}_j \alpha_j} \right| . \tag{6.5}$$

From (6.1) we have

$$\|H_{G_N}\| \geq \|H_G\| - \|G - G_N\|_\infty$$
$$\geq \|H_G\| - Cr^N,$$

and so (6.4) yields

$$\|\widetilde{T}\| \geq \sqrt{1 - M^2} \; (\|H_G\| - Cr^N)$$
$$\geq \sqrt{1 - M^2} \; \|H_G\| - Cr^N.$$

Now we invoke (6.4) to obtain

$$\|H_{G_N}\nu\| = \|\widetilde{T}\nu\|$$
$$\geq (\|\widetilde{T}\| - 2Cr^N)\|\nu\|$$
$$\geq (\sqrt{1 - M^2} \; \|H_G\| - 3Cr^N) \; \|\nu\|. \tag{6.6}$$

Next it transpires that ν is near-maximizing for H_G. We have, using (6.6)

$$\|H_G\nu\| \geq \|H_{G_N}\nu\| - \|H_{G-G_N}\nu\|$$
$$\geq (\sqrt{1 - M^2} \; \|H_G\| - 4Cr^N) \; \|\nu\|.$$

That is, ν is ϵ-maximizing for H_G, where

$$1 - \epsilon = \sqrt{1 - M^2} \; - \frac{4Cr^N}{\|H_G\|} \tag{6.7}$$

Now we may show as in the proof of Theorem 2 that $(\widetilde{G}-\widehat{G})\nu$ is small in the L^2 norm. By Lemma 4,

$$(\widetilde{G}-\widehat{G})\nu=\eta \tag{6.8}$$

where $\eta = P_+(G-\widehat{G})\nu$ and

$$\|\eta\| \le \surd\{1-(1-\epsilon)^2\}\ \|H_G\|\ \|\nu\|.$$

Here

$$\begin{aligned}\surd\{1-(1-\epsilon)^2\} &= \sqrt{\left\{1-\left(1-M^2-\frac{\sqrt{1-M^2}\,8Cr^N}{\|H_G\|}+\frac{16C^2r^{2N}}{\|H_G\|^2}\right)\right\}}\\ &= \sqrt{\left\{M^2+\frac{\sqrt{1-M^2}\,8Cr^N}{\|H_G\|}-\frac{16C^2r^{2N}}{\|H_G\|^2}\right\}}\\ &\le \sqrt{\left\{M^2+\frac{8Cr^N}{\|H_G\|}\right\}}\\ &\le M+\left(\frac{8Cr^N}{\|H_G\|}\right)^{\frac{1}{2}}.\end{aligned}$$

Thus

$$\|\eta\| \le (M\|H_G\|+(8C\|H_G\|r^N)^{\frac{1}{2}})\ \|\nu\|.$$

Lemma 6 enables us to convert this to an L^∞ bound. Since G is rational, $G-\widehat{G}$ is also, and its zeros in $\mathbb{D}$ are $\rho_1,\ldots,\rho_k$ then we have

$$\begin{aligned}\|\eta\|_\infty &\le \sum_{j=1}^{k}\left(\frac{1+|\rho_j|}{1-|\rho_j|}\right)^{\frac{1}{2}}\ \|\eta\|\\ &\le \sum_{j=1}^{k}\left(\frac{1+|\rho_j|}{1-|\rho_j|}\right)^{\frac{1}{2}}\ (M\|H_G\|+(8C\|H_G\|r^N)^{\frac{1}{2}})\|\nu\| \end{aligned}\tag{6.9}$$

Dividing by ν in (6.8) we obtain

$$\begin{aligned}\|\widetilde{G}-\widehat{G}\|_\infty &= \|\eta/\nu\|_\infty\\ &\le \|\eta\|_\infty\ \inf_{|z|=1}\ |\nu(z)|^{-1}\end{aligned}\tag{6.10}$$

We have to show that, for β_j sufficiently close to α_j and N sufficiently large, $|\nu|/\|\nu\|$ is uniformly bounded away from 0. As in (4.5) there exists $c>0$ such that

$$|x(z)| \ge c \tag{6.10}$$

for all $z \in \partial\mathbb{D}$ and all unit maximizing vectors x of H_G. Furthermore, if $\alpha_1, \ldots, \alpha_l$ are all the poles of G in $\mathbb{D}$ and

$$a(z) = \prod_{j=1}^{l} z - \alpha_j$$

then every maximizing vector x of H_G lies in $H^2 \ominus aH^2$. Thus, if ν is a maximizing vector of T and ν' is a maximizing vector H_G we have $\nu - \nu' \in H^2 \ominus z^N qaH^2$ and so, by Lemma 6,

$$\|\nu - \nu'\|_\infty \le \left\{ N + \sum_{j=1}^{m} \left(\frac{1 + |\beta_j|}{1 - |\beta_j|} \right)^{\frac{1}{2}} + \sum_{j=1}^{l} \left(\frac{1 + |\alpha_j|}{1 - |\alpha_j|} \right)^{\frac{1}{2}} \right\} \|\nu - \nu'\|$$

Hence, for β_j sufficiently close to $\alpha_j, 1 \le j \le m$, we have

$$\|\nu - \nu'\| \le (N + L) \, \|\nu - \nu'\| \tag{6.12}$$

where

$$L = 3 \sum_{j=1}^{l} \left(\frac{1 + |\alpha_j|}{1 - |\alpha_j|} \right)^{\frac{1}{2}}.$$

By hypothesis, $\|H_G\|$ is simple singular value of H_G. Hence, by Lemma 7, there is a constant $K > 0$ such that, if $0 < \epsilon < 1$ and ν is a $K\epsilon^2$-maximizing unit vector for H_G, then there is a unit maximizing vector ν' for H_G such that $\|\nu - \nu'\| < \epsilon$. Apply this observation to the ν above (which we now take to be a unit vector). By (6.7) ν is δ-maximizing where

$$\delta = 1 - \sqrt{1 - M^2} \; + \frac{4Cr^N}{\|H_G\|},$$

so that there exists a unit maximizing vector ν' for H_G such that

$$\|\nu - \nu'\| < \sqrt{\frac{\delta}{K}}$$

and so, by (6.12)

$$\|\nu - \nu'\|_\infty < \left\{ 1 - \sqrt{1 - M^2} \; + \frac{4Cr^N}{\|H_G\|} \right\}^{\frac{1}{2}} \frac{(N + L)}{\sqrt{K}}. \tag{6.13}$$

Choose $N_0 \in \mathbb{N}$ so large that $N \ge N_0$ implies

$$\frac{4Cr^N \; (N + L)}{\|H_G\| \; K} < \frac{c^2}{8}.$$

Corresponding to $N > N_0$ there are neighborhoods W_j^N of α_j, $1 \le j \le m$, such that $\beta_j \in W_j^N$ implies

$$(1 - \sqrt{1 - M^2}) \; \frac{(N + L)^2}{K} < \frac{c^2}{8},$$

M being given by (6.5). Then for $N > N_0$ and $\beta_j \in W_j^N, 1 \le j \le m$, (6.13) shows that

$$\|\nu - \nu'\|_\infty < \frac{c}{2},$$

ν' being a unit maximizing vector for H_G, and hence, in view of (6.11),

$$|\nu(z)| \ge \frac{c}{2}, \qquad z \in \partial \mathbb{D}.$$

Combining this with inequalities (6.9) and (6.10) we have: $N > N_0$ and $\beta_j \in W_j^N, 1 \le j \le m$, imply

$$\|\widetilde{G} - \widehat{G}\|_\infty \le A \sum_{j=1}^{m} |\alpha_j - \beta_j| + B r^{N/2}$$

where

$$A = \frac{4}{c} \sum_{j=1}^{k} \left(\frac{1 + |\rho_j|}{1 - |\rho_j|} \right)^{\frac{1}{2}} \left\{ \min_{1 \le j \le m} \; (1 - |\alpha_j|^2) \right\}^{-1} \|H_G\|$$

and

$$B = \frac{(32C\|H_G\|)^{\frac{1}{2}}}{c} \sum_{j=1}^{k} \left(\frac{1 + |\rho_j|}{1 - |\rho_j|} \right)^{\frac{1}{2}}.$$

We recall that in these formulae $\rho_1, \ldots, \rho_k$ are the zeros in $\mathbb{D}$ of $G - \widehat{G}, c$ is the infimum of the modulus of a unit maximizing vector on $\partial \mathbb{D}$ and C is a constant related to the rate of decay of the Fourier coefficients of G (cf. (6.1)).

REFERENCES

[AAK] V.M. Adamyan, D.Z. Arov and M.G. Krein, *Infinite Hankel matrices and generalized Caratheodory-Fejér and Riesz problems*, Functional Analysis and its Applications 2 (1968), 1–18 (English translation).

[C] D.N. Clark, *On the spectra of bounded Hermitian Hankel matrices*, Amer. J. Math., 90 (1968), 627–656.

[F] B. Francis, *An Introduction to H_∞ control*, Lecture Notes in Control and Information Sciences No. 88, Springer Verlag, Berlin (1986).

[G] J.B. Garnett, *Bounded Analytic Functions*, Academic Press, New York/London (1981).

[G1] K. Glover, *All optimal Hankelnorm approximations of linear multivariable systems and their L^∞ error bounds*, Int. J. Control 39 (1984) 1115–1193.

[GR] W.B. Gragg and L. Reichel, *On singular values of Hankel operators of finite rank*, Linear Algebra and Applications (to appear).

[H1] J.W. Helton, *Worst case analysis in the frequency domain: an H^∞ approach to control*, IEEE Trans. Auto Control (1985).

[H2] J.W. Helton, *Broadband gain equalization directly from data*, IEEE Trans. Circ. Syst. (1981).

[Ho] K. Hoffman, *Banach Spaces of Analytic Functions*, Prentice Hall, Englewood Cliffs, NJ (1962).

[HSY] J.W. Helton, P.G. Spain and N.J. Young, *Tracking poles and representing Hankel operators directly from data*, In preparation.

[PY] V. Pták and N.J. Young, *Functions of operators and the spectral radius*, Linear Algebra and its Applications 29 (1980), 357–392.

[S] D. Sarason, *Generalized Interpolation in H^∞*, Trans. AMS 127 (1967), 179–203.

[SB] L.M. Silverman and M. Bettayeb, *Optimal approximation of linear systems*, Proc. JACC (1980).

[T] L.N. Trefethen, *Near circularity of the error curve in complex Chebyshev approximation*, J. Approx. Theory 31 (1981) 344–367.

[Y1] N.J. Young, *An Introduction to Hilbert Space*, Cambridge University Press (1988).

[Y2] N.J. Young, *The rate of convergence of a matrix power series*, Linear Algebra and its Applications 35 (1981), 181–186.

[Y3] N.J. Young, *Singular value decomposition of an infinite Hankel matrix*, Linear Algebra and its Applications 50 (1983), 639–656.

NONLINEAR CONTROLLER DESIGN VIA APPROXIMATE NORMAL FORMS*

ARTHUR J. KRENER†

1. Introduction. Over the past several years a group of faculty and graduate students at UC Davis have been developing a set of tools for the design of controllers and observers for nonlinear systems. Our approach has been based on normal forms and approximate normal forms for nonlinear systems. When a nonlinear system admits a normal form the design of a controller or observer is greatly simplified and standard linear design tools can be employed. The people that have been involved in this program are Mont Hubbard, Sinan Karahan, Andrew Phelps, Yi Zhu Ruggero Frezza and myself. This work has been supported in part by AFOSR. In this paper I'll give an overview of our program.

2. Normal Forms. Following Kailath's terminology, [10], there are four normal forms for linear systems, i.e., controllable, observable, controller and observer form. The first two are relatively straightforward to obtain, provided the system is controllable or observable. However, the latter two are more useful in the design of stabilizing state feedback control laws and asymptotic state observers. If a linear system is both controllable and observable then it admits all four normal forms.

In [14] we discussed the nonlinear generalizations of the four linear normal forms. Unfortunately, even controllable and observable nonlinear systems do not admit all four nonlinear normal forms. A nonlinear system which admits controller normal form is sometimes said to be state feedback linearizable in the sense of Hunt–Su [8] and Jakubczyk–Respondek [9]. For a system in controller normal form, the design of a stabilizing state feedback control law is a straightforward task. However, most systems do not admit a controller normal form and even when one does, the transformation of a system into controller normal form involves solving a system of first order linear partial differential equations which can be quite difficult.

Similar remarks are even more appropriate for observer normal form. For a system in observer form, the design of an observer is a straightforward task. But very few systems admit such forms and the computation of observer normal form is, in general, extremely difficult.

For these reasons, we have introduced approximate versions of nonlinear controller and observer form [15, 16]. These may be thought of as finding systems nearby to the original which admit controller or observer form. The computation of such a system is relatively straightforward, and reduces to solving a set of linear equations. Unfortunately, these linear equations are not always solvable and they increase in size quite rapidly with the dimension of the system.

---

*Research supported in part by AFOSR–85–0267

†Institute of Theoretical Dynamics and Department of Mathematics, University of California, Davis, CA 95616

We start by introducing modified versions of controller and observer normal forms of the nonlinear system.

$$\dot{\xi} = f(\xi) + g(\xi)u \tag{2.1a}$$

$$y = h(\xi) \tag{2.1b}$$

$$\xi(0) \sim \xi^\circ = 0 \tag{2.1c}$$

around the nominal operating point ξ°, which for convenience we assume to be 0. We assume $f(0) = 0$ and $h(0) = 0$. If this is not the case, then in many important cases it can be made so by a possibly time varying change of state and output coordinates. As is usual, the state ξ is n dimensional, the control u is m dimensional and the output y is p dimensional. It is relatively straightforward generalization to consider systems where y depends directly on u, as in

$$y = h(\xi) + k(\xi)u, \tag{2.1d}$$

however to simplify the exposition we shall not do so.

We are interested in studying (2.1) under the pseudogroup of state coordinate transformations around $\xi^\circ = 0$. In [14] we studied arbitrary change of coordinates and attempted to bring the systems into normal form based on prime systems. Such normal forms are closely related to Brunovsky form and its dual. In this article we shall restrict our attention to changes of state coordinates $x = x(\xi)$ whose Jacobian at $\xi^\circ = 0$ is the identity

$$\frac{\partial x}{\partial \xi}(0) = I.$$

Such transformations have two virtues. The first is that they leave invariant the first order linear approximation to (2.1),

$$\dot{z} = Az + Bu + 0(z,u)^2 \tag{2.2a}$$

$$y = Cz + 0(z)^2 \tag{2.2b}$$

$$z(t) = \xi(t) + 0(z)^2 \tag{2.2c}$$

where

$$A = \frac{\partial f}{\partial \xi}(0) \tag{2.3a}$$

$$B = g(0) \tag{2.3b}$$

$$C = \frac{\partial h}{\partial \xi}(0) \tag{2.3c}$$

The second is that the nonlinear coordinates ξ and the normal form coordinates x agree to first order,

$$\xi = x + \phi(x) \tag{2.4a}$$

where

$$\phi(0) = 0, \quad \frac{\partial \phi}{\partial x}(0) = 0. \tag{2.4b}$$

Typically, the original coordinates in which the system is described have some natural meaning and the coordinates have different dimensions, e.g., distance, velocity, mass, etc. Property (2.4) means that at least to first order the normal form coordinates have the same dimensions and intuitive meanings as the natural coordinates.

The system (2.1) admits a modified controller form if there exists a change of state coordinates (2.4) which transforms (2.1) into

$$\dot{x} = Ax + Bu + B(\alpha(x) + \beta(x)u) \tag{2.5a}$$

$$y = Cx + \gamma(x) \tag{2.5b}$$

It follows from (2.4) that the nonlinear terms are quadratic or higher in (x, u), i.e.

$$\alpha(0) = 0, \quad \frac{\partial \alpha}{\partial x}(0) = 0 \tag{2.6a}$$

$$\beta(0) = 0, \tag{2.6b}$$

$$\gamma(0) = 0, \quad \frac{\partial \gamma}{\partial x}(0) = 0. \tag{2.6c}$$

We require that the $m \times m$ matrix $1 + \beta(x)$ be invertible for x of interest. These conditions (2.4) and (2.6) insure that A, B, C are given by (2.3). Hence the linear part of modified controller form of (2.1) is the same as the first order approximation (2.2) to (2.1).

The system (2.1) admits a modified observer form if there exists a change of state coordinates (2.4) which transforms (2.1) into

$$\dot{x} = Ax + Bu + \alpha(\overline{y}) + \beta(\overline{y})u, \tag{2.7a}$$

$$y = \overline{y} + \gamma(\overline{y}), \tag{2.7b}$$

$$\overline{y} = Cx. \tag{2.7c}$$

It follows from (2.1) that the nonlinear terms are quadratic or higher

3. Poincaré Linearization. Henri Poincaré considered the problem of transforming a nonlinear vector field into a linear field by a change of coordinates around a critical point. We briefly describe his theory, a fuller description can be found in Guckenheimer and Holmes [6] and Arnold [1].

We are given a single vector field

$$\dot{\xi} = f(\xi) \tag{3.1c}$$

$$f(0) = 0 \tag{3.1b}$$

with a critical point at $\xi^\circ = 0$. We are interested in finding a change of coordinates (2.4) which transforms (3.1) into a linear vector field,

$$\dot{x} = Ax \tag{3.2}$$

where A is given by (2.3a).

Poincaré noted that one could develop the change of coordinates term by term in homogeneous powers of x. At degree two we seek an n dimensional vector field $\phi^{(2)}(x)$ whose entries are homogeneous polynomials of degree 2 in x such that under the change of coordinates

$$\xi = x + \phi^{(2)}(x) \tag{3.3}$$

the differential equation (3.1) is transformed to

$$\dot{x} = Ax + O(x)^3 \tag{3.4}$$

whose $O(x)^3$ denotes cubic and higher terms in x. Superscripts in parentheses will be used to indicate that the function is homogeneous of the degree of the superscript in its arguments. If we expand $f(\xi)$ in homogeneous powers of ξ,

$$f(\xi) = A\xi + f^{(2)}(\xi) + f^{(3)}(x) + \cdots \tag{3.5}$$

then (3.1) is transformed into (3.4) iff $\phi^{(2)}$ (ξ) satisfies the so called homological equation

$$[Ax, \phi^{(2)}(x)] = f^{(2)}(x) \tag{3.6a}$$

where [,] is the Lie–Jacobi bracket

$$[Ax, \phi^{(2)}(x)] = \frac{\partial \phi^{(2)}}{\partial x}\ (x)\ Ax - A\phi^{(2)}(x) \tag{3.6b}$$

It is straightforward to verify that $[Ax,\ \cdot]$ is linear map from homogeneous vector fields of degree 2 into homogeneous vector fields of degree 2. Moreover the homogeneous n dimensional vector fields of degree 2 form a linear space of dimension $n^2(n+1)/2$. Hence (3.6a) is solvable for arbitrary $f^{(2)}$ iff zero is not an eigenvalue of the linear mapping defined by $[Ax,\ \cdot]$. Poincaré noted that the eigenvalues of this mapping are related to the eigenvalues of A in a simple fashion. To see why, suppose A is semisimple, i.e., there exists a basis $v^1, \ldots, v^n$ of eigenvectors of A

$$Av^i = \lambda_i\ v^i \tag{3.7a}$$

possibly over the complex numbers.

Let $w_1, \ldots, w_n$ be a cobasis of left eigenvectors of A,

$$w_i A = \lambda_i w_i \tag{3.7b}$$

Then the space of n vector fields homogeneous of degree 2 has as a basis

$$\phi^k_{ij}(x) = v^k (w_i x)\ (w_j x) \tag{3.8}$$

when $1 \leq i \leq j \leq n$ and $1 \leq k \leq n$. A straightforward calculation yields

$$[Ax, \phi_{ij}^k(x)] = (\lambda_i + \lambda_j - \lambda_k)\ \phi_{ij}^k(x).$$

Hence the eigenvalues of $[Ax,\ \cdot]$ on vector fields homogeneous of degree 2 are

$$\lambda_i + \lambda_j - \lambda_k \tag{3.9}$$

when $1 \leq i \leq n$ and $1 \leq k \leq n$.

Hence the homological equation (3.6a) is solvable if no expression of the form (3.9) is zero. Of course this is a sufficient but not necessary condition because a particular $f^{(2)}$ might well be the range of $[Ax,\ \cdot]$, e.g., $f^{(2)} = 0$.

If (3.6a) is solvable one can proceed to look for a transformation canceling the third degree terms in f,

$$\xi = x + \phi^{(3)}(x) \tag{3.10}$$

and $[Ax,\ \cdot]$ is linear mapping of these vector fields homogeneous of degree 3 into themselves. The eigenvalues of this mapping are

$$\lambda_i + \lambda_j + \lambda_k - \lambda_\ell \tag{3.12}$$

where $1 \leq i \leq j \leq k \leq n,\ 1 \leq \ell \leq n$.

Hence (3.11) is solvable for arbitrary $f^{(3)}$ iff none of (3.12) is zero. This generalizes to higher degree. If one of (3.9) or (3.12) or their generalization is zero then there is "resonance" and linearization is not always possible. We refer the reader to [1] and [6] for more details.

4. Approximate Controller Form. S. Karahan in his Ph.D. thesis [12] studied the application of Poincaré's method to finding controller forms and approximate controller forms. We give a brief description of his work.

One starts by expanding (2.1) into homogeneous powers of (ξ, u),

$$\dot{\xi} = A\xi + Bu + f^{(2)}(\xi) + g^{(1)}(\xi)u + \cdots \tag{4.1a}$$

$$y = C\xi + h^{(2)}(\xi) + \cdots \tag{4.1b}$$

One seeks a change of coordinates

$$\xi = x + \phi^{(2)}(x) \tag{4.2}$$

transforming (4.1) into approximate controller form

$$\dot{x} = Ax + Bu + B(\alpha^{(2)}(x) + \beta^{(1)}(x)u) + O(x,u)^3 \tag{4.3a}$$

$$y = Cx + \gamma^{(2)}(x) + O(x)^3 \tag{4.3b}$$

Following Poincaré, we see that this will happen iff

$$[Ax,\ \phi^{(2)}(x)] + B\ \alpha^{(2)}(x) = f^{(2)}(x) \tag{4.4a}$$

$$[Bu,\ \phi^{(2)}(x)] + B\beta^{(1)}(x)u = g^{(1)}(x)u \tag{4.4b}$$

where (4.4b) must hold for each constant u. We refer to these as the generalized homological equations. Like the homological equations, they are linear equations but they are generally not square. The space of unknown $\phi^{(2)}(x), \alpha^{(2)}(x)$ and $\beta^{(1)}(x)$ is $n^2(n+1)/2 + mn(n+1)/2 + m^2n$ dimensional. The constraint space of $f^{(2)}$ and $g^{(1)}$ is $n^2(n+1)/2 + n^2m$ space. These dimensions agree iff $n = 2m+1$. Generally the map $\phi^{(2)}, \alpha^{(2)}, \beta^{(1)} \longmapsto f^{(2)}, g^{(1)}$ is not of full rank so it is not always solvable even when $n = 2m+1$.

Karahan has analyzed this mapping using a basis and cobasis related to the controllability matrix $(B, AB, \ldots, A^{n-1}B$. We refer the reader to [12] for details.

Since the system (4.4a) is generally not solvable one is forced to seek approximate solutions. One way of doing this is to find $\tilde{f}^{(2)}$ and $\tilde{g}^{(1)}$ in the range of the mapping (4.2) which is closest in some least squares sense to the given $f^{(2)}$ and $g^{(1)}$. Moreover one would like to choose the smallest $\phi^{(2)}, \alpha^{(2)}$ and $\beta^{(1)}$ which maps into $\tilde{f}^{(2)}$ and $\tilde{g}^{(1)}$. Again we refer the reader to [12] for more details.

Before closing this section it should be mentioned how an approximate controller form (4.3) can be used to stabilize a nonlinear system (2.1) (or equivalently (4.1)) by state feedback. The standard approach is to approximate the nonlinear system to first order by (2.2), choose a stabilizing feedback law for (2.2), $u = Fz$, transform this back into original coordinates,

$$u = F\xi. \tag{4.5}$$

Expressed in homogeneous terms the closed loop dynamics is

$$\dot{\xi} = (A + BF)\xi + f^{(2)}(\xi) + g^{(1)}(\xi)F\xi + O(\xi)^3 \tag{4.6}$$

and hence the system is locally stable around $\xi^\circ = 0$. Of course, if it is too far from $\xi^\circ = 0$, the quadratic and higher terms may drive it unstable.

In the normal form approach, we typically will use the same stabilizing state feedback gain F but to apply it to the second order linearization (4.3) rather than the first order linearization (2.2). The resulting feedback is

$$u + \alpha^{(2)}(x) + \beta^{(1)}(x)u = Fx \tag{4.7}$$

which results in x coordinates the closed loop system

$$\dot{x} = (A + BF)x + O(x)^3 \tag{4.8}$$

Generally speaking, it is better to implement the feedback in the original ξ coordinates taking advantage of the fact that the inverse to (4.2) is

$$x = \xi - \phi^{(2)}(\xi) + O(\xi)^3 \tag{4.9}$$

Neglecting higher than quadratic terms we obtain from (4.7) the feedback

$$u = F\xi - (F\phi^{(2)}(x) + \alpha^{(2)}(\xi) + \beta^{(1)}(\xi)F\xi) + O(\xi)^3. \tag{4.10}$$

Note that to first degree the standard feedback (4.5) and the feedback (4.9) agree. However, the second degree terms of (4.10) cancel the second degree terms of (4.6) to obtain in x coordinates (4.8). One expects that (4.10) is asymptotically stabilizing over a larger neighborhood of ξ° then (4.5).

Of course one can also seek a higher degree approximate controller form. The dimensions of the homological and generalized homological equations grow exponentially in the degree of the approximation. Hence this approach may not be recommended. It might be more efficient and effective to find approximate controller forms of degree two around several operating points rather than an approximate controller form of degree three around a single point.

5. Approximate Observer Form. The work I'm about to describe is joint with Andrew Phelps. We seek a change of coordinates of the form (4.2) which transforms (4.1) into approximate observer form

$$\dot{x} = Ax + Bu + \alpha^{(2)}(\overline{y}) + \beta^{(1)}(\overline{y})u + O(x,u)^3 \tag{5.1a}$$

$$y = Cx + \gamma^{(2)}(\overline{y}) + O(x)^3 \tag{5.1b}$$

$$\overline{y} = Cx. \tag{5.1c}$$

As before this is possible iff we can solve another set of generalized homological equations

$$[Ax, \phi^{(2)}(x)] + \alpha^{(2)}(Cx) = f^{(2)}(x) \tag{5.2a}$$

$$[Bu, \phi^{(2)}(x)] + \beta^{(1)}(Cx)u = g^{(1)}(x)u \tag{5.2b}$$

$$\gamma^{(2)}(Cx) - C\phi^{(2)}(x) = h^{(2)}(x) \tag{5.2c}$$

As before (5.2b) must hold for each constant u.

These equations are linear mapping from the space of functions $\phi^{(2)}(x), \alpha^{(2)}(Cx), \gamma^{(2)}(Cx)$ to the space of functions $f^{(2)}(x), g^{(1)}(x), h^{(2)}(x)$. the dimension of the domain is $n^2(n+1)/2 + np(p+1)/2 + m\ 2p + p^2(p+1)/2$ and that of the range is $n^2(n+1)/2 + mn(n+1)/2 + pn(n+1)/2$. In general, these equations (5.2) are not solvable so as before one must seek a least squares solution. We shall report on that in more detail at another time.

If (2.1) (equivalently (4.1)) can be transformed to approximate observer form then it is easy to construct an observer. We choose H so that $A + HC$ is sufficiently stable. An approximation $\hat{x}(t)$ to $x(t)$ is defined to evolve according to

$$\begin{aligned}\dot{\hat{x}} &= (A + HC)\hat{x} + Bu - H(y - \gamma^{(2)}(y)) \\ &\quad + \alpha^{(2)}(y - \gamma^{(2)}(y)) + \beta^{(2)}(y - \gamma^{(2)}(y))u\end{aligned} \tag{5.3}$$

then the error $\tilde{x}(t) = x(t) - \tilde{x}(t)$ satisfies

$$\dot{\tilde{x}} = (A + HC)\tilde{x} + O(x, \hat{x}, u)^3. \tag{5.4}$$

Hence if the initial error is not too large and u is also not too large, we can expect $\tilde{x}(t) \to 0$ as $t \to \infty$.

Of course, it is preferable to implement the observer in natural coordinates so we transform (5.3) using $\widehat{\xi} = \hat{x} + \phi^{(2)}(\hat{x})$ to obtain

$$\begin{aligned}\dot{\widehat{\xi}} &= A\widehat{\xi} + Bu + H(\hat{y} - y) + f^{(2)}(\widehat{\xi}) + g^{(1)}(\widehat{\xi})u \\ &+ \alpha^{(2)}(y) - \alpha^{(2)}(\hat{y}) + (\beta^{(2)}(y) - \beta^{(2)}(\hat{y}))u \\ &+ H(\gamma^{(2)}(y) - \gamma^{(2)}(\hat{y})) + \frac{\partial \phi^{(2)}}{\partial \widehat{\xi}}(\widehat{\xi})H(\hat{y} - y) \\ &+ O(\xi, \widehat{\xi}, u)^3\end{aligned} \tag{5.5a}$$

$$\hat{y} = C\widehat{\xi} + h^{(2)}(\widehat{\xi}) \tag{5.5b}$$

$$\widehat{\xi}(0) = \xi^\circ = 0. \tag{5.5c}$$

Notice that the linear part of (5.5) is the observer for (2.1) one would obtain from the linear approximation (2.2), namely

$$\dot{\hat{z}} = A\hat{z} + Bu + H(\hat{y} - y) \tag{5.6a}$$

$$\hat{y} = C\hat{z} \tag{5.6b}$$

$$\hat{z}(0) = \xi^\circ = 0. \tag{5.6c}$$

The error $\tilde{z} = \xi - \hat{z}$ between (2.1) and (5.6) satisfies

$$\dot{\tilde{z}} = (A + HC)\tilde{z} + O(\xi, \widehat{\xi}, u)^2 \tag{5.7a}$$

while the error of the observer (5.5) expressed in $\tilde{x}$ coordinates satisfies (5.4). Hence one expects (5.5) to perform better as an observer for (2.1) over a larger operating range.

As with the state feedback (4.10), the second degree terms of the observer (5.5) are a correction to the standard linear observer for the quadratic nonlinearities of the original system. In implementations one would replace the state ξ in the state feedback control law (4.9) with the estimate $\widehat{\xi}$ from (5.5).

One can continue this process and look for a third degree change of coordinates which transforms the system into approximate observer form where the error terms are $O(\xi, u)^3$. One obtains in this fashion third order corrections to the state feedback (4.10) and observer (5.5). Viewed in this light, we see that the approximate normal form approach allows us to start with a standard linear design based on the linear approximation (2.2) and build in a succession of higher degree corrections to overcome the nonleararities of (4.1). Throughout we can keep the same feedback gain K and observer gain H, and these can be chosen by standard linear design techniques applied to the linear approximation (2.2).

6. Coprime Factorizations. This work is joint with Yi Zhu [19]. Suppose we have a system in controller normal form

$$\dot{x}_c = Ax_c + Bu + B(\alpha_c(x_c) + \beta_c(x_c)u) \tag{6.1a}$$

$$y = Cx_c + \gamma_c(x_c) \tag{6.1b}$$

$$x_c(0) = 0 \tag{6.1c}$$

where the c-subscripts indicate coordinates and functions associated to controller normal form. We view (6.1) as defining an input/output map

$$G : u(\cdot) \longmapsto y(\cdot) \tag{6.2a}$$

from functions $u(t)$ to $y(t)$ for $t \geq 0$.

We seek a right factorization of G

$$G = N \circ M^{-1}$$

where N and M are input/output maps

$$M : v(\cdot) \longmapsto u(\cdot) \tag{6.2b}$$

$$N : v(\cdot) \longmapsto y(\cdot), \tag{6.2c}$$

M is invertible and $\circ$ denotes composition. There is a large and growing literature on coprime factorization of both linear and nonlinear systems. A sampling is 2 − 5, 7, 10, 11, 13, 17, 18, 20 − 26]. In particular our approach follows [3, 4].

To describe the input/output maps M and N we shall use a state space realization. In particular we define M to be the input/output map of

$$\dot{\xi}_c = (A + BF)\xi_c + Bv \tag{6.3a}$$

$$\alpha_c(\xi_c) + (1 + \beta_c(\xi_c))u = F\xi_c + v \tag{6.3b}$$

$$\xi_c(0) = 0 \tag{6.3c}$$

where (6.3b) defines u as a function of ξ_c and v.

We consider the composition $N = G \circ M$, this is realized by the $2n$ dimensional system (6.1, 3) described in ξ_c, x_c coordinates. Let $e = x_c - \xi_c$ then

$$\begin{aligned}\dot{e} = Ae + B(&-F\xi_c - v + \alpha_c(x_c) \\ &+ (1 + \beta_c(x_c))(1 + \beta_c(\xi_c))^{-1}(F\xi_c + v - \alpha_c(\xi_c)))\end{aligned} \tag{6.4}$$

If $e(t) = 0$ then $\dot{e}(t) = 0$. Since $e(0) = 0$ we conclude that $e(t) = 0$ then $\dot{e}(t) = 0$ is unaffected by the input $v(t)$.

A controllable realization of N is

$$\dot{\zeta}_c = (A + BF)\zeta_c + Bv \tag{6.5a}$$

$$y = C\zeta_c + \gamma_c(\zeta_c) \tag{6.5b}$$

$$\zeta_c(0) = 0 \tag{6.5c}$$

Hence we conclude that $G = N \circ M^{-1}$ where N and M are realized by (6.5) and (6.3). Notice that M is invertible since $(1 + \beta_c)$ is invertible by assumption.

Notice also that if (A, B) is a controllable pair then we can choose F so that (6.3) and (6.5) are stable systems. Hence we have factored G over the ring of stable nonlinear systems. We are being deliberately vague about the precise definition of a stable nonlinear system. It is clear that (5.3, 5) are "stable" under any reasonable definition.

Of course, we are interested in *coprime* factorizations over the ring of stable nonlinear systems. Again we should not try to make this concept precise but following Hammer [7] and others we shall say that $G = N \circ M^{-1}$ is a coprime factorization if there exists $\widetilde{P}$, the input/output map of a stable system,

$$\widetilde{P} : \begin{pmatrix} u \\ y \end{pmatrix} \longmapsto w \tag{6.6a}$$

such that the composition

$$\widetilde{P} \circ \begin{pmatrix} M \\ N \end{pmatrix} : v \longmapsto \begin{pmatrix} u \\ y \end{pmatrix} \longmapsto w \tag{6.6b}$$

is the identity, $w = v$.

The input/output map of $\binom{M}{N}$ can be realized by an n dimensional system

$$\dot{\xi}_c = (A + BF)\xi_c + Bv \tag{6.7a}$$

$$\alpha_c(\xi_c) + (1 + \beta_c(\xi_c))u = F\xi_c + v \tag{6.7b}$$

$$y = C\xi_c + \gamma_c(\xi_c) \tag{6.7c}$$

$$\xi_c(0) = 0 \tag{6.7d}$$

A left inverse of (6.7) is

$$\dot{z}_c = Az_c + Bu + B(\alpha_c(z_c) + \beta_c(z_c)u) \tag{6.8a}$$

$$w = \alpha_c(z_c) + (1 + \beta_c(z_c))u - Fz_c \tag{6.8b}$$

$$z_c(0) = 0 \tag{6.8c}$$

If $e = \xi_c - z_c$ then

$$\dot{e} = Ae + B(\alpha_c(\xi_c) - \alpha_c(z_c) + (\beta_c(\xi_c) - \beta_c(z_c))u)$$

If $e(t) = 0$ the $\dot{e}(t) = 0$ and since $e(0) = 0$ it follows that $e(t) = 0$ for all $t \geq 0$. If $e(t) = \xi_c(t) - z_c(t) = 0$ then $w(t) = v(t)$ so (6.8) inverts (6.7).

However we do not know that (6.8) is stable. To insure the stability of (6.8), we must add to (6.8a) an extra term. This term must stabilize (6.8) and also must be zero when $\xi_c = z_c$ so that (6.8) remains a left inverse of (6.7). How do we find such a term?

Notice that the dynamics (6.8a) is the same as the dynamics of the original system (6.1a) and notice that the other output y of (6.7) does not appear in (6.8). Perhaps we can inject y into (6.8a) to stabilize it? This is more or less equivalent to asking whether output injection can be used to stabilize the original system (6.1). This is always possible for systems in observer form, hence we assume that there exists a change of coordinates

(6.9) $$x_c = x_0 + \phi(x_0)$$

satisfying (2.4b) transforming (6.1) into observer form

(6.10a) $$\dot{x}_0 = Ax_0 + Bu + \alpha_0(Cx_0) + \beta_0(Cx_0)u$$

(6.10b) $$y = Cx_0 + \gamma_0(Cx_0)$$

(6.10c) $$x_0(0) = 0$$

Suppose we consider a similar change of coordinates for (6.8)

(6.11) $$z_c = z_0 + \phi(z_0)$$

to obtain

(6.12a) $$\dot{z}_0 = Az_0 + Bu + \alpha_0(Cz_0) + \beta_0(Cz_0)u$$

(6.12b) $$w = \alpha_0(z_0 + \phi(z_0)) + (1 + \beta_c(z_0 + \phi(z_0)))u - F(z_0 + \phi(z_0))$$

(6.12c) $$z_0(0) = 0.$$

We add to (6.12a) the term

(6.13a) $$\alpha_0(\overline{y}) - \alpha_0(Cz_0) + (\beta_0(\overline{y}) - \beta_0(Cz_0))u + H(Cz_0 - \overline{y})$$

to obtain

(6.12aa) $$\dot{z}_0 = (A + HC)z_0 + Bu + \alpha_0(\overline{y}) + \beta_0(\overline{y})u - Hy$$

where $\overline{y}$ is a function of y of (6.7)c) defined by

(6.13b) $$y = \overline{y} + \gamma_0(\overline{y}) = C\xi_0 + \gamma_0(C\xi_0)$$

and ξ_0 is the state of (6.7) in observer coordinates

(6.13c) $$\xi_c = \xi_0 + \phi(\xi_0)$$

Notice that (6.13a) is zero whenever $\xi_0 = z_0$, hence the input/output map $\widetilde{P}$ of the (6.12aa, b, c) is stable.

In summary, we have shown that if a nonlinear system admits both controller and observer form than its input/output map G can be factored into the composition $N \circ M^{-1}$ of input/output maps of stable systems N and M. Moreover this composition is coprime in the sense that the input/output map $\binom{M}{N}$ has a left inverse $\widetilde{P}$ which is realized by a stable system.

We have not presented this as a theorem because we are reluctant at this point in time to give formal definitions of coprimeness and stability for nonlinear systems. However the above development is very analogous to the linear theory [3, 4]. See also Hammer [7]

Unfortunately the analogy is not so straightforward for left coprime factorizations. The theory of left coprime factorizations for nonlinear systems has some substantial differences with the linear theory.

We start with a system in observer form (6.10) realizing an input/output map G. We define another input output map

$$\widetilde{M} : \begin{pmatrix} u \\ y \end{pmatrix} \longmapsto w, \tag{6.14}$$

by

$$\dot{\xi}_0 = (A + HC)\xi_0 - H\overline{y} + \alpha_0(\overline{y}) + \beta_0(\overline{y})u \tag{6.15a}$$

where $\overline{y}$ is an invertible function of the input y defined by

$$y = \overline{y} + \gamma_0(\overline{y}) \tag{6.15b}$$

and the output is

$$w = -C\xi_0 + \overline{y} \tag{6.15c}$$

$$\xi_0(0) = 0 \tag{6.15d}$$

Consider the serial connection of (6.10) and (6.15), this is not a realization of the $\widetilde{M} \circ G$ but it is a realization of $\widetilde{N} = \widetilde{M} \circ \binom{I}{G}$. (This is the first important difference with the linear theory). If we define $\xi_0 = x_0 - \xi_0$ then $\widetilde{N}$ is realized by

$$\dot{\zeta}_0 = (A + HC)\zeta_0 + Bu \tag{6.16a}$$

$$w = C\zeta_0 \tag{6.16b}$$

$$\zeta_0(0) = 0 \tag{6.16c}$$

because in ξ_0, x_0 coordinates for (6.10, 15) only the ξ_0 coordinates are observable from the output w. We consider $\widetilde{N}, \widetilde{M}$ as a left factorization of G, although it is really a left factorization of $\binom{I}{G}$ in the sense that

$$\widetilde{M} \circ \begin{pmatrix} I \\ G \end{pmatrix} = \widetilde{N} \tag{6.17}$$

Notice that we cannot compose this on the left with $\widetilde{M}^{-1}$ since $\widetilde{M}$ is not invertible as a mapping from $\binom{u}{y}$ to w.

Perhaps the best way of viewing the situation is

$$\begin{pmatrix} I & 0 \\ 0 & \widetilde{M} \end{pmatrix} \circ \begin{pmatrix} I \\ G \end{pmatrix} = \begin{pmatrix} I \\ \widetilde{N} \end{pmatrix} \tag{6.18a}$$

or

$$\begin{pmatrix} I \\ G \end{pmatrix} = \begin{pmatrix} I & 0 \\ 0 & M \end{pmatrix}^{-1} \circ \begin{pmatrix} I \\ N \end{pmatrix} \tag{6.18b}$$

The matrix notation is somewhat misleading because $\widetilde{M}$ depends on both u and y.

In any case, if (C, A) is an observable pair then both (6.15) and (6.16) can be made stable by proper choice of H. In particular, the nonlinearities in (6.15a) are the memoryless functions of the inputs u and y hence (6.15) is BIBO stable.

Next we address the coprimeness of the above factorization.

We consider the input/output map

$$[-\widetilde{N}, \widetilde{M}] : \begin{pmatrix} u \\ y \end{pmatrix} \longmapsto w \tag{6.19a}$$

where again the matrix notation is somewhat misleading since both u and y are inputs to $\widetilde{M}$, i.e.,

$$w = -\widetilde{N}(u) + \widetilde{M}\begin{pmatrix} u \\ y \end{pmatrix}. \tag{6.19b}$$

This input/output map can be realized by an n dimensional system

$$\dot{\xi}_0 = (A + HC)\xi_0 + Bu + \alpha(\overline{y}) + \beta_0(\overline{y})u - H\overline{y} \tag{6.20a}$$

where $\overline{y}$ is an invertible function of the input y defined by

$$y = \overline{y} + \gamma_0(\overline{y}) \tag{6.20b}$$

and the output w is given by

$$w = -C\xi_0 + \overline{y} \tag{6.20c}$$

We wish to find a input/output map P realized by a stable system so that P is a right inverse of $[-\widetilde{N}, \widetilde{M}]$,

$$P : v \longmapsto \begin{pmatrix} u \\ y \end{pmatrix} \tag{6.21a}$$

$$[-\widetilde{N}, \widetilde{M}] \circ P : v \longmapsto w = v. \tag{6.21b}$$

We start by constructing an inverse for (6.20),

$$\dot{z}_0 = Az_0 - Hv + Bu + \alpha_0(\overline{y}) + \beta_0(\overline{y})u \tag{6.22a}$$

$$\overline{y} = Cz_0 + v \tag{6.22b}$$

$$y = \overline{y} + \gamma_0(\overline{y}) \tag{6.22c}$$

$$u = ? \tag{6.22d}$$

$$z_0(0) = 0 \tag{6.22e}$$

We leave unspecified for the moment the output u which also appears in the dynamics (6.22a). Notice that if $e = \xi_0 - z_0$ is the error between the states of (6.20) and (6.22) then $\dot{e} = 0$ whenever $e = 0$. Since $e(0) = 0$ we conclude that $e(t) = 0$ for all $t \geq 0$ and so by (6.20c) and (6.22b) we have $w(t) = v(t)$. In other words (6.22) is a right inverse of (6.20).

What about the stability of (6.22)? We would like to choose the output u in such a way that (6.22a) is stable in some sense. If we ignore the –Hv term of (6.22a) this looks like the original system is observer form. This is not exactly true because $\overline{y}$ is defined by (6.22b) with v present. Suppose the original system can be transformed into controller form (6.1) by a change of coordinates (6.9). If we apply a similar change of coordinates (6.11) to (6.22) we obtain

$$\begin{aligned}\dot{z}_c = {} & Az_c + Bu + B(\alpha_c(z_c) + \beta_c(z_c)u) - Hv \\ & - \frac{\partial \phi}{\partial z_0}(Hv) + (1 + \frac{\partial \phi}{\partial z_0})(\alpha_0(Cz_0 + v) - \alpha_0(Cz_0) \\ & + (\beta_0(Cz_0 + v) - \beta_0(Cz_0))u)\end{aligned} \tag{6.23a}$$

Suppose we choose an F such that $(A + BF)$ is stable and define u by

$$\alpha_c(z_c) + \beta_c(z_c)u = Fz_c. \tag{6.22dd}$$

When the input $v = 0$, (6.23a) becomes

$$\dot{z}_c = (A + BF)z_c. \tag{6.23b}$$

Unfortunately we cannot conclude that (6.23a) is BIBO stable since the input v is multiplied by a function of the state.

We conclude by noting that a "nonlinear Bezout identity" holds for the above. In other words beside $\widetilde{P}$ being a left inverse (6.6b) for $\binom{M}{N}$ and P a right inverse (6.21b) for $[-\widetilde{N}, \widetilde{M}]$, it is also true that

$$[-\widetilde{N}, \widetilde{M}] \circ \begin{pmatrix} M \\ N \end{pmatrix} : v \longmapsto w = 0 \tag{6.24a}$$

and

$$\widetilde{P} \circ P : v \longmapsto \begin{pmatrix} u \\ y \end{pmatrix} \longmapsto w = 0 \tag{6.24b}$$

In abuse of notation we summarize these equations by

$$\begin{pmatrix} \widetilde{P} \\ [-\widetilde{N}, \quad \widetilde{M}] \end{pmatrix} \circ \left(\begin{pmatrix} M \\ N \end{pmatrix} P \right) = \begin{pmatrix} 1 & 0 \\ 0 & I \end{pmatrix} \tag{6.25}$$

The verification of (6.24) is straightforward.

From the work of Doyle [3] Francis [4] and others the, the existence of a nonlinear Bezout identity suggests that it might be possible to develop a nonlinear version of Youla's Q parameterization of all stable and stabilizing controller of a linear system. This generalization would apply to those nonlinear systems which admit both controller and observer form. This class is very thin, but perhaps such a result could be extended approximately to those systems that approximately admit controller and observer form. Work in these areas is continuing.

7. Concluding Remarks. We have briefly described an approach to nonlinear compensator design based on nonlinear normal forms and approximately normal forms. This approach is being pursued by a group of researchers at U.C. Davis with support from AFOSR. The principal advantage of the normal forms approach is that to a large extent it reduces problems in nonlinear design to problems in linear design. We are developing software tools which utilize this approach as a compliment to existing linear design software so that these linear design packages can be used for nonlinear systems that admit at least approximate normal forms.

REFERENCES

[1] ARNOLD, V.I., *Geometrical Methods in the Theory of Ordinary Differential Equations*, SpringerVerlag, NY (1983).

[2] C. DESOER, R.W. LIU, J. MURRAY AND R. SAEKS, *Feedback system design: the fractional representation approach, IEEE Trans. Automat. Control, AC–25 (1980), pp. 399–412.*

[3] J. DOYLE, *Lecture Notes in Advances in Multivariable Control*, Office of Naval Research/Honeywell Workshop, Minneapolis, MN (1984).

[4] B. FRANCIS, *A Course in H_∞ Control Theory*, 88, Springer-Verlag (1987).

[5] B. FRANCIS AND J. DOYLE, *Linear Control Theory with an H_∞ Optimality Criterion*, SIAM J. Control and Optimization, 25, No. 4, July (1987), pp. 815–844.

[6] J. GUCKENHEIMER AND P. HOLMES, *Nonlinear Oscillations, Dynamical Systems and Bifurcation of Vector Fields*, Springer-Verlag, NY (1983).

[7] H. HAMMER, *Fraction representations of nonlinear systems: A simplified approach*, Int. J. of Control, 46, 2 (August 187), pp. 455–472.

[8] L.R. HUNT AND R.SU, *Linear equivalents of nonlinear time varying systems*, Proc. MTNS, Santa Monica (1981), pp. 119–123.

[9] B. JAKUBCZYK AND W. RESPONDEK, *On the linearization of control systems*, Bull. Acad. Polon. Sci., Ser. Math. Astron. Phys. 28 (1980), pp. 517–522.

[10] T. KAILATH, *Linear Systems*, Prentice Hall, Englewood Cliffs (1980).

[11] M. VIDYASAGAR, *Control System Synthesis: A Factorization Approach*, MIT Press (1985).

[12] S. KARAHAN, *Higher order linear approximation to nonlinear systems*, Ph.D., Mechanical Engineering, University of California, Davis (1988).

[13] P. KHARGONEKAR AND E. SONTAG, *On the relation between stable matrix factorizations and regulable realizations of linear systems over rings*, IEEE Trans. on Auto. Contr. 27 (1982), pp. 627–636.

[14] A. J. KRENER, *Normal forms for linear and nonlinear systems. In Differential Geometry, the Interface between Pure and Applied Mathematics*, M. Luksik, C. Martin and W. Shadwick, eds. Contemporary Mathematics 68, American Mathematical Society, Providence (1986), 157–189.

[15] A.J. KRENER, S. KARAHAN, M. HUBBARD AND R. FREZZA, *Higher order linear approximations to nonlinear control systems*, Proceedings, IEEE Conf. on Decision and Control, Los Angeles (1987), pp. 519–523.

[16] A. PHELPS AND A.J. KRENER, *Computation of observer normal form using MACSYMA*, In Nonlinear Dynamics and Control, C. Byrnes, C. Martin and R. Saeks, eds. North Holland (1988).

[17] E. SONTAG, *Smooth stabilization implies coprime factorization*, IEEE Trans. Auto. Control (to appear, 1989).

[18] E. SONTAG, *Nonlinear control via equilinearization*, Los Angeles (1987), pp. 1363–1367.

[19] YI ZHU, *Masters Thesis*, Applied Mathematics, University of California, Davis (1988).

[20] C.A. DESOER, C.A. LIN, *Nonlinear unity feedback systems and Q–parametrization*, Analysis and Optimization of Systems, A. Bensoussan and J.L. Lions ed., *Lecture Notes in Control and Information Sciences*, 62, Springer-Verlag (1984).

[21] C.A. DESOER, M.G. KABULI, *Nonlinear plants, factorizations and stable feedback systems*, Proceedings of 26th IEEE Conference on Decision and Control, Los Angeles (December 1987), pp. 155–156.

[22] C.A. DESOER, M.G. KABULI, *Stabilization and robustness of nonlinear unity–feedback system: factorization approach*, International Journal of Control, 47, 4 (1988), pp. 1133–1148.

[23] C.A. DESOER, M.G. KABULI, *Right Factorization of a Class of Time–Varying Nonlinear Systems*, IEEE Transactions on Automatic Control, 33, 8 (August 1988), pp. 755–757.

[24] J. HAMMER, *Non–linear systems, stabilization, and coprimeness*, International Journal of Control, 42, 1 (1985), pp. 1–20.

[25] C.N. NETT, C.A. JACOBSON, M.J. BALAS, *A connection between state–space and doubly coprime fractional representations*, IEEE Transaction on Automatic Control, 29, 9 (September 1984), pp. 831–832.

[26] D.C. YOULA, H.A. JABR AND J.J. BONGIORNO, JR., *Modern Wiener–Hopf design of optimal controllers, Part II: The multivariable case*, IEEE Transactions on Automatic Control, 21 (June 1976), pp. 319–338.

FEEDBACK WITH DELAYS: STABILIZATION OF LINEAR TIME-DELAY AND TWO-DIMENSIONAL SYSTEMS*

E. BRUCE LEE† AND WU-SHENG LU‡

Abstract. Provided is an introduction to the techniques for stabilization of systems having a time delay or a two-dimensional structure. Examples are used to tie the theoretical results to engineering techniques for designing feedback controllers which involve simple delay schemes.

1. Introduction. The task of comparing controllers seems to be without end. There is always some new direction to take as we rummage through the set of all controllers seeking a better one. Our goal here is to share some directions when robust stabilization is the stated objective. We shall describe a search for controllers which provide stabilization of certain infinite dimensional systems as represented by linear differential-difference equations or by linear two-dimensional difference equations. We shall also describe how simple versions of delay feedback can be used when finite dimensional systems are subject to stabilization (see examples below).

Our search begins by noting some different forms that stabilization can take based on the various concepts of stability: homogeneous Cauchy initial value problem having solutions which all approach the null solution (asymptotic stability); bounded inputs giving rise to bounded outputs (BIBO stability); or a given property being shared by nearby systems (structural stability).

Even a finer division of directions will be provided because of the increase in possibilities that arise for the infinite dimensional models.

In Section II stabilization of time delay type systems (represented by differential-difference equations or transfer functions involving ratios of exponential polynomials) will be described, while Section III describes the situation for stabilization of two-dimensional systems (as represented by difference equations in two independent variables or transfer functions involving rational functions in two complex variables).

In general the results are stated in the mathematical format of theorem with proof, but on occasion results are related directly to the development of an idea. The general notation is from the text "*Foundations of Optimal Control Theory*" by Lee and Markus (New edition; Krieger Pub. Co. 1986). Stabilizability by state feedback for finite dimensional linear type systems is introduced there starting on page 96. The general approach which will be used is dependent on results from modern operator theory and exploits in particular spectral results for linear transformations as found in the text "*Linear Operator Theory in Engineering and Science*" by Naylor and Sell (Holt, Rinehart and Winston, 1971).

Before presenting the theoretical results we wish to relate the theory to the applications through several examples.

---

*This research was supported by the National Science Foundation under Grant Numbers DMS-8607687 and DMS-87 22402 and the Airforce Office of Scientific under Grant No. 860088.

†Department of Electrical Engineering, University of Minnesota, Minneapolis, MN 55455

‡Department of Electrical and Computer Engineering, University of Victoria, Victoria, Canada

The examples 1 and 2 below show how simple delay feedback leads to robust stabilization even for nondelay type systems and shows some of the peculiarities of the pure time-delay type systems including some ties to the quadratic optimal control theory.

We shall assume a feedback configuration (block diagram) as shown below in Figure 1 where U represents controlled "inputs" to the system, W represents other inputs, Y represents sensed outputs and Z represents other outputs. The system to be controlled is represented by its transfer function $G(s)$ and the controller (feedback) to be designed is represented by its transfer function $K(s)$.

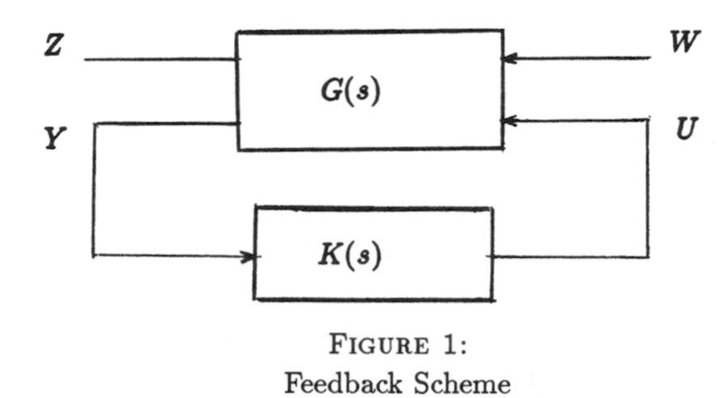

FIGURE 1:
Feedback Scheme

To illustrate the use of delayed feedback the stabilizing controller will be sought in the form $U = KY$ with $K = K(s) = k_0 + k_1e^{-hs} + k_2e^{-2hs} + \cdots + k_le^{-lhs}$; in Laplace transform notation.

The controller design task is then to select the constants $k_0, k_1, \ldots, k_l$ and the delay duration $h > 0$ to get stability and possibly optimal performance. Note such controllers are stable and very easy to implement (see summary of work of Chyung [11] for some indication of their other properties). Also controllability means that generalized pole placement is possible (Result of Morse [45] which will be introduced in the next section). The third example will admit a more complicated delay feedback scheme. Example 4 relates to a study of the optimality system and characteristic roots of the system concluding with plots of root locations of exponential polynomials as a function of the delay parameter.

Example 1. Consider a system with transfer function (U to Y)

$$G(s) = \frac{1}{s(s+1)(s+2)} = \frac{1}{s^3+3s^2+2s} = \frac{b}{s^3+a_3s^2+a_2s+a_1} \quad .$$

It is easy to find parameters k_0, k_1, k_2 and $h > 0$ such that $K(s)$ is stabilizing in the feedback loop of Figure 1. For this example take $K(s) = (10 - 5\mu + 3\mu^2)$ with $\mu = e^{-hs}$. Note that then we have a delay parameter $h > 0$ yet to be selected and that robustness can be checked with respect to the system parameters a_1, a_2, a_3, and b which have nominal values of $a_3 = 3, a_2 = 2, a_1 = 0$, and $b = 1$.

Figure 2 is a Nyquist plot indicating (with Figure 3) that for h in the range $0.49 \le h \le 1.12$ the above feedback controller with transfer function $K(s) = (10 - 5\mu + 3\mu^2)$ for $\mu = e^{-hs}$ is stabilizing for the nominal system (no encirclement of $-1, 0)$).

Also figure 3 indicates that for $h = 0.8$ the stability margin is almost the best. Selection of $h = 0.8$ then fully specifies the controller. Figure 4 shows a typical response of the closed loop feedback system (step input), while figures 5, 6 and 7 show Nyquist plots for parameter variations near the nominal as a check on structural stability. It can be concluded that the stability achieved with the delay feedback is indeed robust with respect to parameter variations near the nominal.

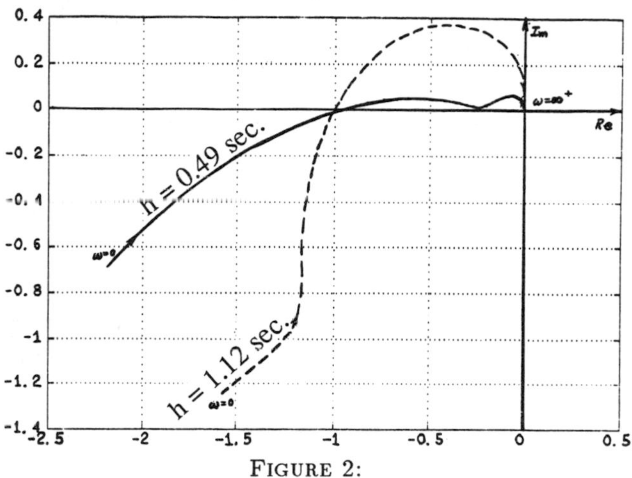

FIGURE 2:
Marginal Stability $h = 0.49$ to 1.12 sec.

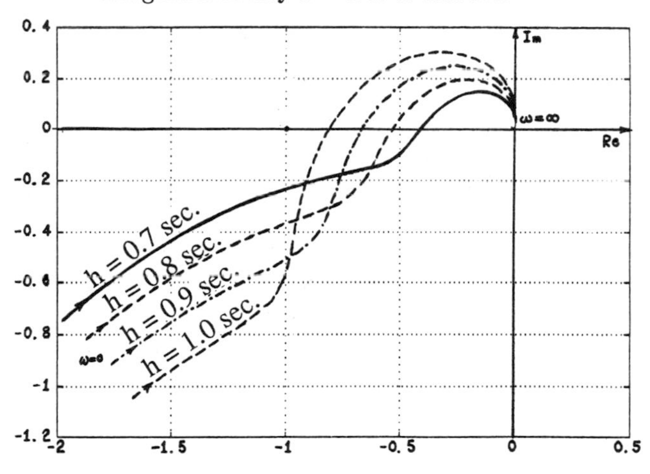

FIGURE 3:
Sensitivity of Nyquist Plot due to Delay Timings

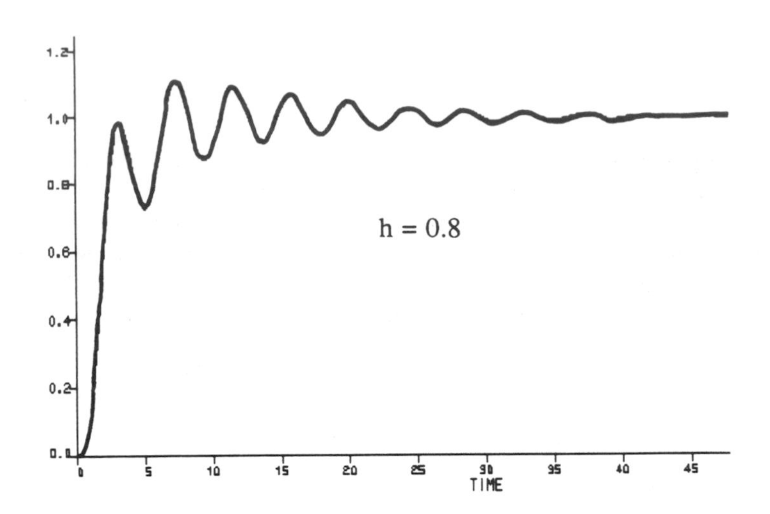

FIGURE 4:
Step response

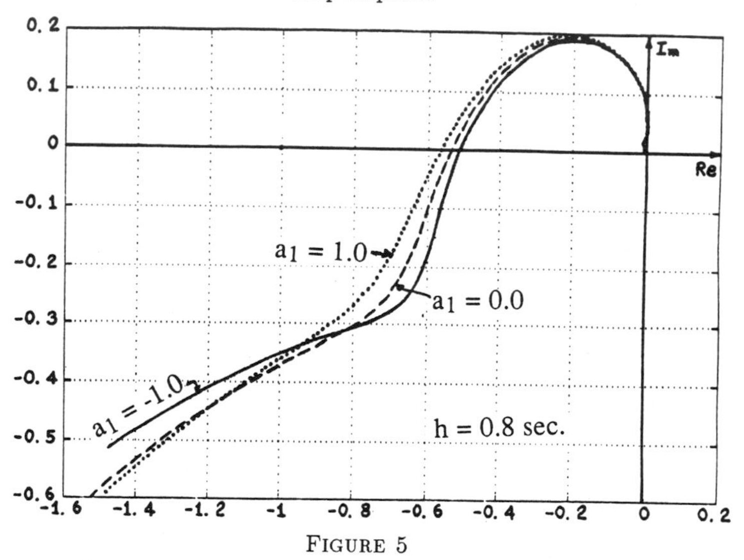

FIGURE 5
Parameter variation stability plot for $a_1 = -1$ to $+1$

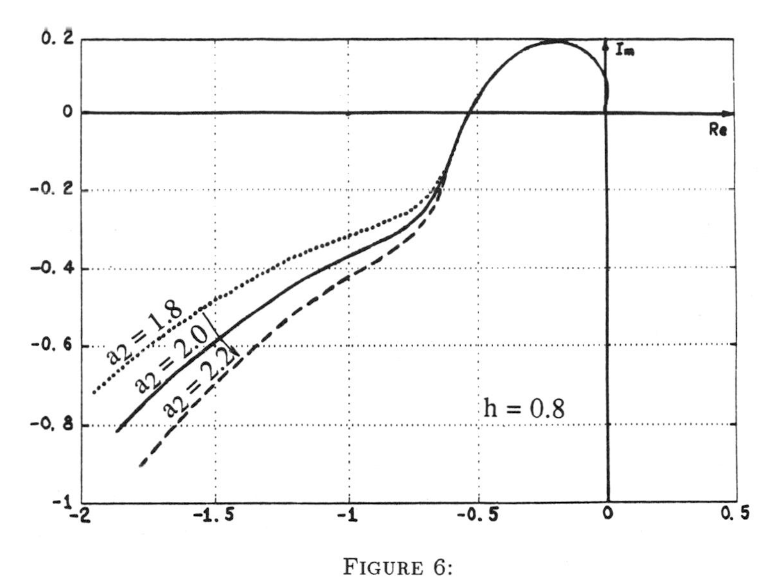

FIGURE 6:
Parameter variation stability plots for $a_2 = 1.8$ to 2.2

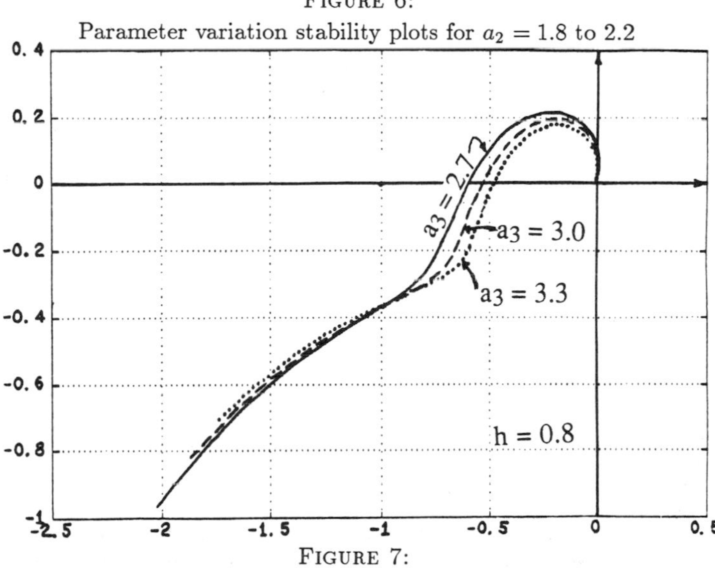

FIGURE 7:
Parameter variation stability plots for $a_3 = 2.7$ to 3.3

Example 2. Consider a system with transfer function $G(s) = \dfrac{1}{s^2+3}$. In addition to illustrating that feedback with delays can lead to stabilization in feedback scheme of figure 1 we shall show some connection to quadratic optimal control theory, which is also illustrated in more detail by the next example. State type equations will be needed for the quadratic theory so we now introduce a minimal realization of the system with transfer function $\frac{1}{s^2+3}$ by state equations:

$$\begin{aligned} \dot{x}_1 &= x_2 \\ \dot{x}_2 &= -3x_1 + u(t)\ , \end{aligned} \qquad \left[\begin{array}{c} A = \begin{bmatrix} 0 & 1 \\ -3 & 0 \end{bmatrix}, \quad B = \begin{bmatrix} 0 \\ 1 \end{bmatrix} \\ \\ C = [\ 1 \quad 0\ ], \quad D = [\ 0\ ] \end{array} \right]$$

with output $\quad y(t) = x_1$.

Note first that such a system is open loop unstable (characteristic equation is $\lambda^2 + 3 = 0$) and the simple feedback scheme $u(t) = ky(t) = kx_1(t)$ does not lead to desirable stability properties for any real parameter value k.

However, the simple delay feedback scheme $u(t) = k_0 y(t) + k_1 y(t-h) = k_0 x_1(t) + k_1 x_1(t-h)$ does lead to desirable stability properties for many different values of the parameters k_0, k_1, and $h > 0$ (characteristic equation is $\lambda^2 + 3 - k_0 - k_1 e^{-h\lambda} = 0$).

Take $k_0 = -1$, $k_1 = -1$ and $h = 2$ to get characteristic equation $\lambda^2 + 4 + e^{-2\lambda} = 0$.

Using the argument principle [4] it is easy to show that the function $f(\lambda) = \lambda^2 + 4 + e^{-2\lambda}$ has no zero in the (closed) right half (complex) plane (RHP). A sketch of the image under $f(\lambda)$ of the D-shaped curve which encircles the RHP is shown as figure 8; clearly indicating the lack of zeros in the RHP.

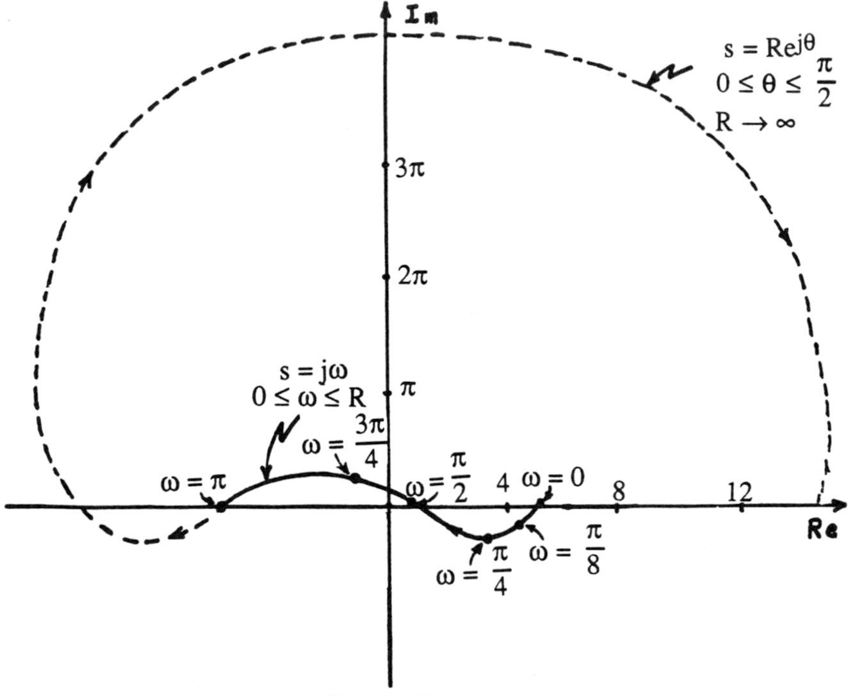

FIGURE 8:
Sketch of $f(s) = s^2 + e^{-2s} + 4$ with $s \epsilon D^+$.

Now we ask the question of whether such a delay feedback controller $u(t) = -x_1(t) - x_1(t-2)$ is optimal in perhaps the quadratic sense.

In figure 1, if we consider as two other output variables (Z-variables):

$$\xi_1(t) = 2x_1(t) + 2x_1(t-2)$$
$$\xi_2(t) = x_2(t) + x_2(t-2)$$

and the quadratic cost criterion

$$\int_0^\infty [\xi_1^2(t) - \xi_2^2(t) + u^2(t)]dt$$

then a generalized optimality system is satisfied.

To see this introduce the delay operator (right shift) notation $zf(t) = f(t-h)$ as in the next section; so

$$\xi_1(t) = (2+2z)x_1(t) = C_1(z)x(t) = [2+2z \quad 0] \begin{bmatrix} x_1(t) \\ x_2(t) \end{bmatrix}$$

and $$\xi_2(t) = (1+z)x_2(t) = C_2(z)x(t) = [0 \quad 1+z] \begin{bmatrix} x_1(t) \\ x_2(t) \end{bmatrix}.$$

Then the optimality system is [28]

$$\begin{bmatrix} \dot{x} \\ \dot{p} \end{bmatrix} = \left[\begin{array}{c:c} A & -BB^* \\ \hdashline -C_1^*C_1 + C_2^*C_2 & -A^* \end{array}\right] \begin{bmatrix} x \\ p \end{bmatrix}$$

where $*$ is defined by $C^*(z) = C^T\left(\frac{1}{z}\right)$.

Compute the characteristic function of the optimality system as

$$\det \left[\begin{array}{cc:cc} \lambda & -1 & 0 & 0 \\ 3 & \lambda & 0 & 1 \\ \hdashline 8+4z+4/z & 0 & \lambda & -3 \\ 0 & -\left(2+z+\frac{1}{z}\right) & 1 & \lambda \end{array}\right]$$

$$= (\lambda^2 + 4 + z)(\lambda^2 + 4 + 1/z), \text{ where } z = e^{-2\lambda}.$$

Note $$-C_1^*C_1 + C_2^*C_2 = -\begin{bmatrix} 2+2/z \\ 0 \end{bmatrix} [2+2z \quad 0] + \begin{bmatrix} 0 \\ 1+1/z \end{bmatrix} [0 \quad 1+z]$$

$$= \begin{bmatrix} 2+2/z & 0 \\ 0 & 1+1/z \end{bmatrix} \begin{bmatrix} -1 & 0 \\ 0 & 1 \end{bmatrix} \begin{bmatrix} 2+2z & 0 \\ 0 & 1+z \end{bmatrix}$$

$\underline{\underline{\Delta}} - C^*QC$; so the Q matrix in the quadratic criterion is not positive definite or even nonnegative and thus an optimal controller need not exist. Existence usually depends on being able to solve an algebraic Riccati equation [7]. The appropriate Generalized Algebraic Riccati Equation (GARE) is provided in reference [31]. This is given by considering state feedback $u(t) = K(z)x(t)$ with GARE:

$$[P \quad -I]\begin{bmatrix} A & -BB^* \\ -C^*QC & -A^* \end{bmatrix}\begin{bmatrix} I \\ P \end{bmatrix} = 0$$

and seeking a symmetric solution $P = P^*$. Then $K = -B^*P|_{z=e^{-hs}}$ will specify the feedback controller.

In our example take $P(z) = \begin{bmatrix} 0 & 1+1/z \\ 1+z & 0 \end{bmatrix}$; clearly $P = P^*$ and P satisfies GARE for the above C^*QC. For take $K = -B^*P = -[0 \quad 1]\begin{bmatrix} 0 & 1+1/z \\ 1+z & 0 \end{bmatrix} = -[1+z \quad 0]$ and compute according to GARE that $C^*QC = -A^*P - PA + PBB^*P$

$$= -\begin{bmatrix} 0 & -3 \\ 1 & 0 \end{bmatrix}\begin{bmatrix} 0 & 1+1/z \\ 1+z & 0 \end{bmatrix} - \begin{bmatrix} 0 & 1+1/z \\ 1+z & 0 \end{bmatrix}\begin{bmatrix} 0 & 1 \\ -3 & 0 \end{bmatrix}$$
$$+ \begin{bmatrix} 0 & 1+1/z \\ 1+z & 0 \end{bmatrix}\begin{bmatrix} 0 & 0 \\ 0 & 1 \end{bmatrix}\begin{bmatrix} 0 & 1+1/z \\ 1+z & 0 \end{bmatrix}$$
$$= \begin{bmatrix} 8+4z+4/z & 0 \\ 0 & -(2+z+1/z) \end{bmatrix}, \quad \text{as given above.}$$

Example 3. In this example we illustrate the infinite dimensional nature of the delay type systems (state space is composed of segments of time functions) and that the state space can be dependent on the delays involved in the quadratic cost functional.

This example also shows that a system can be stabilized by a feedback controller with the resulting spectrum being finite, but the given pole configuration is not the result of minimizing a quadratic functional of the form

$$J(u) = \int_0^\infty \langle Qx(t), x(t)\rangle + \langle u(t), u(t)\rangle dt$$

with Q having real number entries.

However, it may be the result of minimizing a generalized quadratic function of the form

$$J(u) = \int_0^\infty \|\Sigma P_i x(t-h_i)\|^2 + \|u(t)\|^2 \; dt$$

with P_i real constant matrices and h_i certain delays.

As the example, (From the Ph.D. thesis of Alena Fernandez, Univ. of Minnesota 1983) consider a differential-difference equation as the model for the controlled system

$$\begin{aligned}\dot{x}_1(t) &= x_2(t-a)\\ \dot{x}_2(t) &= x_1(t-a) + x_2(t) + u(t) \quad , \quad a > 0 \ .\end{aligned}$$

Or in delay operator notation as in the previous example:

$$\dot{x}(t) = \begin{bmatrix} 0 & z \\ z & 1 \end{bmatrix} x(t) + \begin{bmatrix} 0 \\ 1 \end{bmatrix} u(t), \text{ so } A(z) = \begin{bmatrix} 0 & z \\ z & 1 \end{bmatrix}, \ b(z) = \begin{bmatrix} 0 \\ 1 \end{bmatrix}$$

and z is right shift (delay) operator i.e. $zx(t) = x(t-a)$, with $a > 0$ the delay duration.

Take $-u(\cdot) = (Kx)(t) = 2x_1(t) + x_1(t-2a) + 2\int_{-a}^{0} x_2(t+\theta)d\theta$

$$+ \int_{-3a}^{-2a} x_2(t+\theta)d\theta + 4x_2(t)$$

as the feedback controller of figure 1, which stabilizes the system and the poles are located at $\lambda = -1$ and $\lambda = -2$, assuming $y(t) = x(t)$ is the system output.

To check this note the closed loop system is modeled as

$$\begin{aligned}\dot{x}_1(t) &= x_2(t-a)\\ \dot{x}_2(t) &= x_1(t-a) - 3x_2(t) - 2x_1(t) - x_1(t-2a)\\ &\quad - 2\int_{-a}^{0} x_2(t+\theta)d\theta - \int_{-3a}^{-2a} x_2(t+\theta)d\theta \ .\end{aligned}$$

Therefore

$$\begin{aligned}\ddot{x}_2(t) &= \dot{x}_1(t-a) - 3\dot{x}_2(t) - 2\dot{x}_1(t) - \dot{x}_1(t-2a)\\ &\quad - 2\int_{-a}^{0} \dot{x}_2(t+\theta)d\theta - \int_{-3a}^{-2a} \dot{x}_2(t+\theta)d\theta\\ &= x_2(t-2a) - 3\dot{x}_2(t) - 2x_2(t-a) - x_2(t-3a)\\ &\quad - 2x_2(t) + 2x_2(t-a) - x_2(t-2a) + x_2(t-3a)\\ &= -3\dot{x}_2(t) - 2x_2(t) \ .\end{aligned}$$

So $\lambda^2 + 3\lambda + 2 = 0$ is the characteristic equation and the closed loop poles are at $\lambda = -2$, $\lambda = -1$.

The optimality system associated with such a differential-difference equation model can be given as

$$\begin{bmatrix} \dot{x} \\ \dot{p} \end{bmatrix} = \begin{bmatrix} A(z) & -bb^T \\ -Q & -A^T\left(\frac{1}{z}\right) \end{bmatrix} \begin{bmatrix} x \\ p \end{bmatrix}$$

where Q is the "weight" for x in the cost functional i.e.

$$J(u) = \int_0^\infty (\langle Qx, x\rangle(t) + \|u(t)\|^2)dt$$

if $Q \epsilon R^{n\times n}$ and

$$J(u) = \int_0^\infty (\|(C(z)x)(t)\|^2 + \|u(t)\|^2)dt$$

if $$Q = Q(z) = C^T\left(\frac{1}{z}\right) C(z).$$

This optimality system for delay type systems follows from work of Lee on generalized quadratic cost functionals [28].

Proceeding in a formal fashion the characteristic function of the optimality system is (at $z = e^{-as}$)

$$\Delta(s,z) = \det \left[\begin{array}{cc|cc} s & -z & 0 & 0 \\ -z & s-1 & 0 & 1 \\ \hline q_{00} & q_{01} & s & 1/z \\ q_{10} & q_{11} & 1/z & s+1 \end{array}\right]$$

$$= s \begin{vmatrix} s-1 & 0 & 1 \\ q_{01} & s & 1/z \\ q_{11} & 1/z & s+1 \end{vmatrix} + z \begin{vmatrix} -z & 0 & 1 \\ q_{00} & s & 1/z \\ q_{10} & 1/z & s+1 \end{vmatrix}$$

$$= s^4 - s^2(z^2 + 1 + \frac{1}{z^2} + q_{11}) - s(zq_{10} - \frac{1}{z}\, q_{01} - \frac{1}{z^2} + z^2) + q_{00} + 1.$$

By inspection of the coefficients of s^2 and s it can be seen that they cannot be real numbers if q_{11}, q_{10}, and q_{01} are real numbers (i.e. the quadratic term cannot be obtained as a continuous operator from M_2 to M_2 with M_2 as described in the next section).

But the matrix $Q(z)$ given by

$$Q(z) = C^T\left(\frac{1}{z}\right)C(z) \quad \text{with} \quad C(z) = A_0 + A_1 z + A_2 z^2$$

where $$A_0 = \begin{bmatrix} \sqrt{2} & 0 \\ 0 & -1 \end{bmatrix}, \; A_1 = \begin{bmatrix} 0 & \sqrt{2} \\ 1 & 0 \end{bmatrix}, \; A_2 = \begin{bmatrix} 0 & 0 \\ 0 & 1 \end{bmatrix}$$

gives $\Delta(s,z) = (s^2-4)(s^2-1)$.

For calculate $C^T\left(\frac{1}{z}\right)C(z) = \begin{bmatrix} \sqrt{2} & 1/z \\ \sqrt{2}/z & -1+1/z^2 \end{bmatrix}\begin{bmatrix} \sqrt{2} & \sqrt{2}\ z \\ z & -1+z^2 \end{bmatrix}$
$= \begin{bmatrix} 3 & 3z-1/z \\ 3/z-z & 4-1/z^2-z^2 \end{bmatrix} = \begin{bmatrix} q_{00} & q_{01} \\ q_{10} & q_{11} \end{bmatrix}$, and substitute these values for q_{00}, q_{01}, q_{10} and q_{11} into the previous expression for $\Delta(s,z)$.

Also, note that the state feedback involves one delay interval from the delay in the original Differential-Difference equation model and two delay intervals from the two delays in the cost criterion.

Example 4. The last example related stabilization to an optimality system. Here we relate some of the properties of such optimality systems (for systems with commensurate delays) such as root locations of their characteristic exponential polynomial. Consider then the matrix polynomial $A(z) = \sum_{i=0}^{N} A_i z^i$, $(A_i \epsilon R^{n\times n})$, and the exponential polynomial

$$\det\begin{bmatrix} \lambda I - A(z) & H \\ Q & \lambda I + A^T\left(\frac{1}{z}\right) \end{bmatrix}_{z=e^{-h\lambda}} = \det \hat{\Delta}(\lambda)$$

Such exponential polynomials are described in the book of Bellman and Cooke [4]. It is known from work of Manitius and Tran [42] that:

(1) $\det \widehat{\Delta}(\lambda) = \det \widehat{\Delta}(-\lambda)$;

(2) $\det \widehat{\Delta}(\lambda) = \lambda^{2n} + \theta_1(z)\lambda^{2n-1} + \cdots + \theta_{2n}(z)$ where $\theta_i(z) = \sum_{k=-iN}^{iN} \theta_{i,k} z^k$ and $\theta_{i,k}$ are real coefficients easily calculated by Faddeev's algorithm in recursive form: and

(3) Approximates of roots of $\det \widehat{\Delta}(\lambda)$ can be found using Kuhn saturated triangles.

Using these results one can study the root locations of optimality systems.

Figures 9, 10, and 11 are typical plots of roots of such exponential polynomials using a Kuhn saturated triangle based algorithm [6].

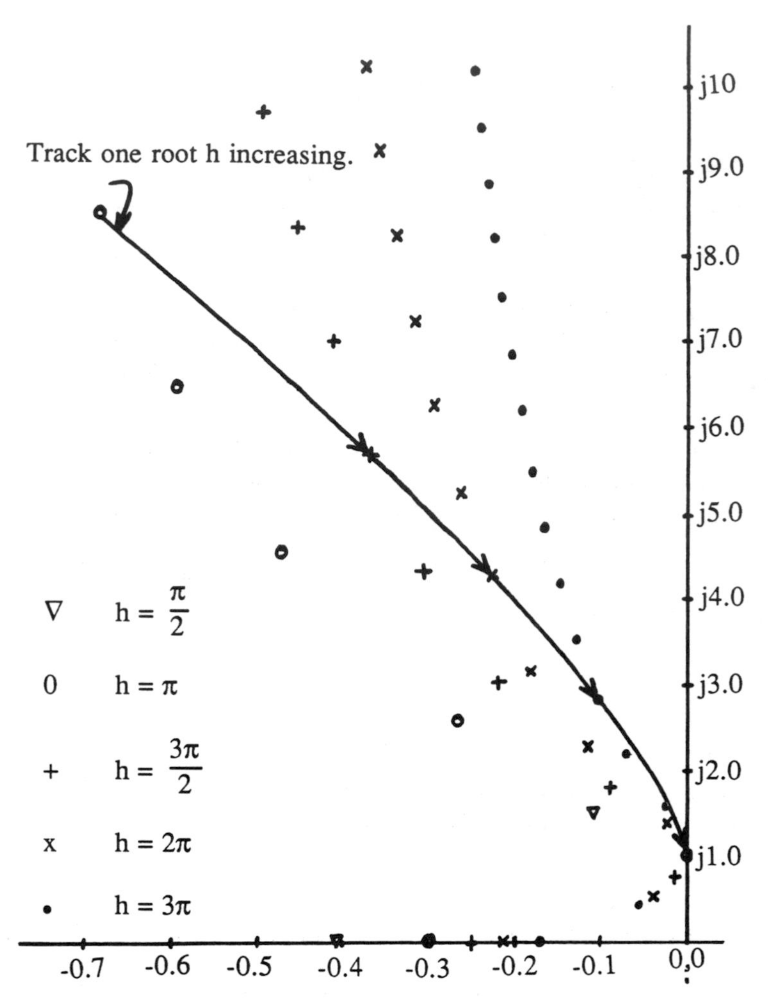

FIGURE 9: Root Chains for Mixed Delay h
$\Delta(s, e^{-hs}) = s^2 + s + se^{-hs} + 1 = 0$

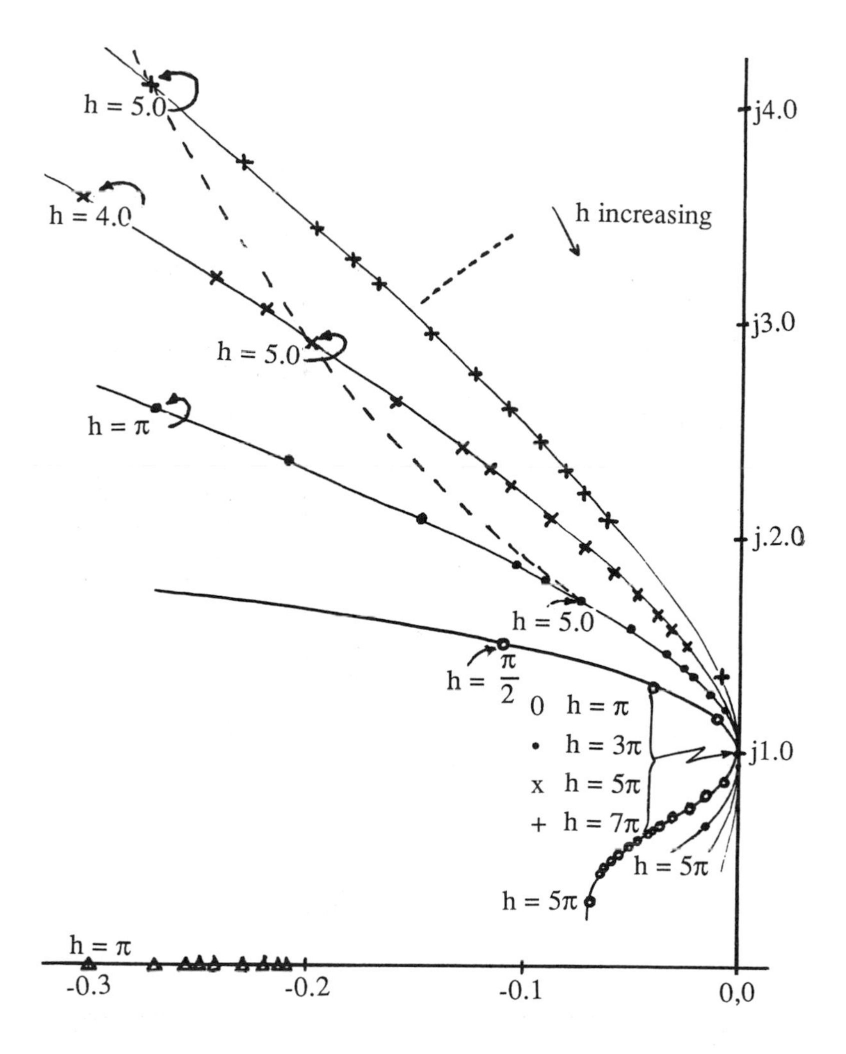

FIGURE 10: Loci for Increasing Delay h

$$\Delta(s, e^{-hs}) = s^2 + s + se^{-hs} + 1 = 0$$

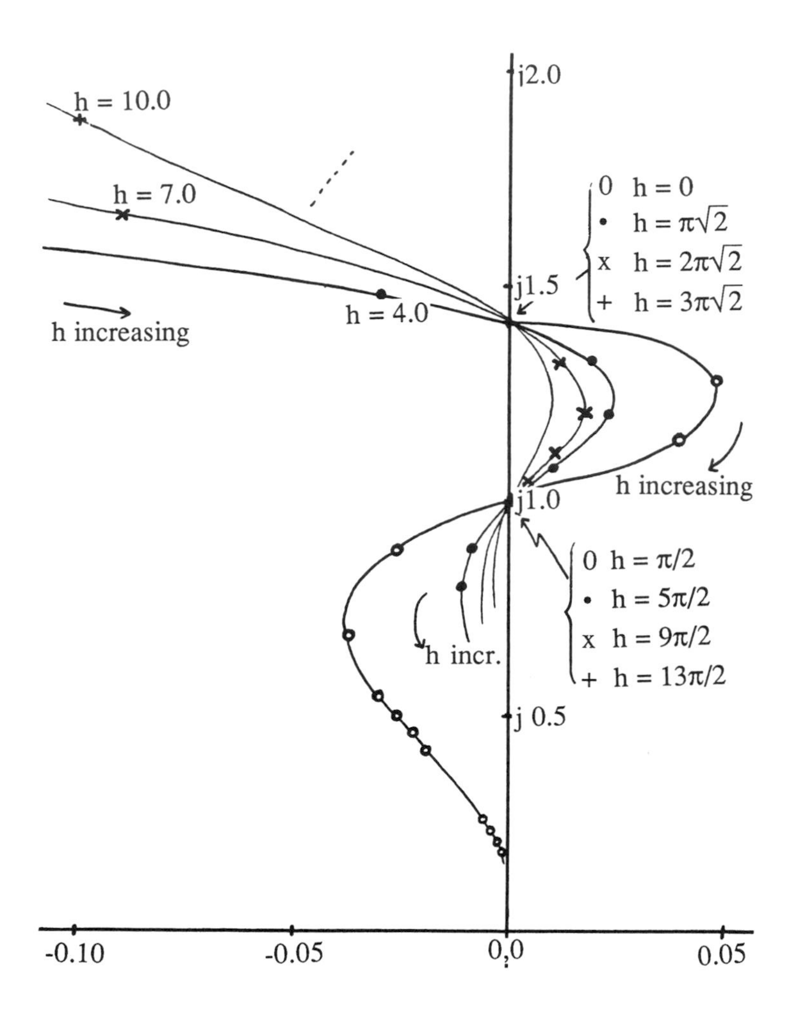

FIGURE 11: Loci for Increasing Delay h
$\Delta(s, e^{-hs}) = s^2 + s + se^{-hs} + 1 = 0$

2. Stabilization of Time-Delay Systems. In this section we shall describe recent results for the stabilization of time delay type systems; interlacing new results with our previous results and related results of others. This is just a brief introduction to the theory of such stabilization.

2.1 Delay Systems in M_2-space.

Let R denote the field of real numbers and $R[z]$ the ring of polynomials in the indeterminate z with coefficients in R.

Consider linear multivariable systems with commensurate delays as modeled by the linear differential-difference equations (Cauchy initial value problem in vector-matrix notation involves finding $x(t)$ to the right of the initial data given the input function $u(t)$)

$$\dot{x}(t) = A(z)x(t) + B(z)u(t), \tag{1a}$$

$$y(t) = C(z)x(t) \tag{1b}$$

with initial data $x(t) = \phi(t)$, $u(t) = \psi(t)$, $t\epsilon[-lh, 0]$ where $x(t)\epsilon R^n$, $u(t)\epsilon R^m$, $y(t)\epsilon R^p$, $A(z)\epsilon R^{n\times n}[z]$, $B(z)\epsilon R^{n\times m}[z]$, $C(z)\epsilon R^{p\times n}[z]$, $l = \max(\deg A(z), \deg B(z))$, and z stands for the delay (right shift) operator such that $zx(t) = x(t-h)$ with delay duration $h > 0$. An equivalent characterization of (1a) can be obtained by introducing a product space (Hilbert space M_2) to describe it in an abstract setting. For the sake of simplicity assume in (1a) that $A(z) = A_0 + A_1 z$, and $B(z) = B$ where A_0, A_1 and B are constant matrices. Then (1a) can be rewritten as an abstract Cauchy problem [41]:

$$\dot{\tilde{x}} = \widetilde{A}\tilde{x} + \widetilde{B}u, \ \tilde{x}(0) = (\phi(0), \ \phi) \ . \tag{2}$$

Here $\tilde{x} \in M^2([-h, 0], R^n) = R^n \times L_2([-h, 0], \ R^n)$ i.e. $\tilde{x} = (x^0, x^1)$ with $x^0 \in R^n$, $x^1 \epsilon L_2([-h, 0], R^n)$, equipped with the norm

$$\|\tilde{x}\|_{M_2} = (\|x^0\|^2_{R^n} + \|x^1\|^2_{L_2})^{1/2}$$

and the inner product

$$(\tilde{x}, z)_{M_2} = x^{0T}z^0 + \int_{-h}^{0} x^{1T}(\theta)z^1(\theta)d\theta \ .$$

The operators $\widetilde{A}$ and $\widetilde{B}$ in (2) are defined by

$$\widetilde{A}\tilde{x} = \{A_0x^0 + A_1x^1(-h), \frac{d}{d\theta}\, x^1) \tag{3}$$

with domain

$$D(\widetilde{A}) = \{\tilde{x}\epsilon M_2 |\ x^1\epsilon W_2^{(1)}([-h, 0], R^n), x^1(0) = x^0\}$$

where $W_2^{(1)}([-h, 0], R^n)$ is a Sobolev space, and

$$\widetilde{B}u = (Bu, \ 0).$$

The operator $\widetilde{A}$ defined by (3) is the infinitesimal generator of the C_0-semigroup $S(t)$ on M_2 characterized by

$$S(t)(\phi(0), \phi) = (x(t), x(t+\theta)) \qquad t \geq 0,\ \theta \epsilon [-h, 0]\ .$$

Further notice that the resolvent operator $R(\lambda, \widetilde{A})$ is compact and consequently, the spectrum of $\widetilde{A}$ is a point-spectrum, which is given by

$$\sigma(\widetilde{A}) = \{s | \det \Delta(s, e^{-sh}) = 0\} \tag{4}$$

where

$$\Delta(s, e^{-sh}) = sI - A(e^{-sh}) \tag{5}$$

i.e. $\sigma(\widetilde{A})$ consists of the eigenvalues of (1).

Let $s_j \epsilon \sigma(\widetilde{A})$ with multiplicity k_j and denote the generalized eigenspace of $\widetilde{A}$ corresponding to s_j by $Ej = \mathrm{Ker}(s_j I - \widetilde{A})^{k_j}$. The projection Π_{sj} of $\tilde{x}$ onto Ej is given by [41] [2]

$$\Pi_{sj}\tilde{x}(t) = \frac{1}{2\pi i} \int_{\Gamma_j} R(s, \widetilde{A})\tilde{x}(s) ds = \Phi_{sj}\xi(t)$$

where Γ_j is a rectifiable, simple closed curve containing only the eigenvalue s_j and $\xi(t)$ is an d_j dimensional vector of coefficients, satisfying the following ordinary differential equation

$$\dot{\xi}(t) = A_{sj}\ \xi(t) + B_{sj}\ u(t) \tag{6}$$

where A_{sj} is a $d_j \times d_j$ matrix whose only eigenvalue is s_j, and B_{sj} is a $d_j \times m$ matrix such that $\Phi_{sj} = \Pi_{sj}\widetilde{B}$.

2.2 Spectral Controllability and Finite Spectrum Assignability. The spectral mode associated with s_j is said to be controllable if the pair (A_{sj}, B_{sj}) in (6) is controllable. The system represented by (2), and thus (1a), is called spectrally controllable if all its spectral modes are controllable. The authors of [49] and [5] have shown that the controllability of (A_{sj}, B_{sj}) is equivalent to rank $[\Delta(s_j, e^{-s_j h}), B] = n$. Hence the system represented by (2) is spectrally controllable if and only if rank $[\Delta(s, e^{-sh}), B] = n$ for all $s \in C$, where C is the field of complex numbers. Similarly one can prove that the system represented by (1a) is spectrally controllable if and only if

$$\mathrm{rank}\ [\Delta(s, e^{-sh}),\ B(e^{-sh})] = n \quad \text{for all} \quad s \in C\ . \tag{7}$$

Another concept relevant to spectral controllability is finite spectrum assignability associated with the retarded delay system. The system represented by (1a) is said to be finite spectrum assignable if there exists a feedback composed of current and past 'states' and finite integrals over past 'states' and inputs such that the n coefficients of the characteristic function det $\Delta_f(s, e^{-sh})$ of the feedback system ($\Delta_f(s, e^{-sh})$ = closed loop characteristic function) can be assigned arbitrarily. In particular, one may for a finite spectrum assignable system find a feedback compensator such that the spectrum of the closed-loop system is a set of finite points in the s-plane; thus the dynamical behavior of the compensated system is the same as that of a finite- dimensional system as shown in examples above. One of the most interesting results, among many, that relate spectral controllability to finite spectrum assignability is stated as the following theorem.

THEOREM 1. *A controlled system as represented by differential-difference equations (1a) is finite spectrum assignable if and only if it is spectrally controllable.*

Proofs of the theorem can be found in [40] for the case of control delays, in [63] for the single-input case and in [62] for the multi-input case. Clearly such a result makes it worthwhile to find easy-to-check tests for spectral controllability. In this regard a result of [59] is worth noting. To state the result define the pair $(A(z), B(z))$ associated with system equation 1a to be $R(z)$-controllable if rank $[A(z)|\ B(z)] = n$ for all but finitely many $z \in C$, where $[A(z)|\ B(z)] = [B(z)\ A(z)B(z) \ldots A^{n-1}(z)B(z)]$.

THEOREM 2. *The system represented by (1a) is spectrally controllable if and only if (i) system represented by (1a) is $R(z)$–controllable and (ii) rank $[\Delta(s, e^{-sh}),$ $B(e^{-sh})] = n$ for $s \epsilon S_0 \cap \sigma(\widetilde{A})$, where $\sigma(\widetilde{A})$ is given by (4) and $S_0 = \{s|\ \text{rank}[A(e^{-s})|$ $B(e^{-sh})] < n\}$.*

In [59] the proof was given for the case of constant B but it also works for $B = B(e^{-sh}) \epsilon R^{n \times n}[e^{-sh}]$. Note that the $R(z)$-controllability is easy to check (it is a generic property even for the case $m = 1$) while the set S_0 is always finite so that condition (ii) requires finitely many steps to verify.

2.3 Other Algebraic Stabilizability Conditions. It is known [4] that the null solution of the homogeneous $(u(t) = 0)$ differential-difference equation system (1) is asymptotically stable if and only if the corresponding characteristic equation (5) satisfies the condition

$$\det \Delta(s, e^{-s}) \neq 0 \qquad \text{for} \qquad \text{Res} \geq 0.$$

The (null solution of) system represented by (1a) is said to be ***asymptotically stable independent of delay*** (i.o.d.) if

$$\det \Delta(s, e,^{-sh}) \neq 0 \qquad \text{for} \qquad Res \geq 0 \quad \text{and all} \quad h \geq 0 \quad .$$

A system as represented by equations (1a) is said to be ***stabilizable*** if there exists a feedback control law $u(t) = Kx(t) + v(t)$ where the entries of K might be constants, polynomials in $z = e^{-sh}$, causal rational in s and z, or other physically implementable feedback, such that the closed-loop null solution of equation system

$\dot{x} = (A + KB)x + Bv$ is asymptotically stable. If for a set of preassigned poles (or coefficients) a feedback controller can be designed such that the roots (or the coefficients) of the characteristic equation of the resulting closed-loop system coincide with the preasigned set, then the system is said to be ***pole assignable*** (or coefficient assignable).

The relation of coefficient assignability as well as pole assignability to stabilizability is obvious. For instance pole assignability implies stabilizability (i.o.d.) since one can obtain a closed-loop characteristic equation that is free of e^{-sh}. Another easy-to-make observation is that coefficient assignability implies pole assignability but the converse is not true in general.

By Theorem 1 a feedback controller that uses current and past states and finite integrals over past states and inputs exists under the spectral controllability hypothesis in order to stabilize the system. In this regard one may pose algebraic conditions more restrictive than spectral controllability on the system under consideration to simplify the structure of the feedback controller while preserving coefficient or pole assignability. On the other hand one may pose conditions milder than spectral controllability under which the system can still be stabilized if one searches for the stabilizer in a broader class. Let us start discussing these issues with a well-known result due to Morse [45].

THEOREM 3. *For each set of polynomials $\{\beta_i, 1 \leq i \leq n\}$ in $R[z]$ there exists $K \in R^{m \times m}[z]$ such that*

$$\det[sI - (A + BK)] = \prod_{i=1}^{n}(s - \beta_i)$$

if and only if $\big(A(z), B(z)\big)$ is $R[z]$-controllable, i.e.

$$\text{rank}[A(z)|\ B(z)] = n \text{ for all } z \in C. \tag{8}$$

The proof of the theorem given in [45] is constructive. Another constructive but matrix based proof was given in [32]. By Morse's theorem pole assignability using polynomial type of feedback $u(t) = K(z)x(t)$ is equivalent to the system's $R[z]$-controllability. If $m = 1$, it is apparent that $R[z]$-controllability also implies coefficient assignability. However, for the multi-input case the question of coefficient assignability is still open [58].

It has been known for some time [30, page 100] that for the finite-dimensional case controllability is a generic property. More precisely fixing n and m and mapping each constant pair (A, B) to a point in the Euclidean space R^N with $N = n(n+m)$ in a natural manner, then the property of controllability over C holds on a set $\Omega \subset R^N$ containing an open and dense subset. As far as the genericity of $R[z]$-controllability for systems that can be modeled as (1a) is concerned, let us consider all matrix pairs $\big(A(z), B(z)\big)$ over $R[z]$ such that the maximal degrees of the entries $A(z)$ and $B(z)$ are bounded by given integers ν and μ, respectively ($\deg A(z) \leq \nu$ and $\deg \beta(z) \leq \mu$). The following result [29] indicates that $R[z]$-controllability is a generic property whenever there are at least two independent controls involved in a delay system.

THEOREM 4. *Assume that $\nu > 0$ and, $\mu > 0$ (non-trivial polynomial case). $R[z]$-controllability is generic in the space of parameters if and only if $m \geq 2$.*

The results of [29] cover the noncommensurate case showing that $R[z_1, \ldots, z_k]$-controllability is generic in the space of parameters if and only if $m \geq k+1$.

By the result of Theorem 4 $R[z]$-controllability is nongeneric in the case of one control variable. Indeed with $m = 1$ (8) is equivalent to $\det[A(z)|B(z)] =$ nonzero constant, which doesn't happen very often. One may therefore seek other feedback controllers to stabilize the system in case it fails to be $R[z]$-controllable. For example the following result proved in [25] indicates that a polynomial feedback might be found to stabilize the system even if it is not $R[z]$-controllable but satisfies a weaker controllability condition.

THEOREM 5. *The system represented by (1a) is stabilizable (i.o.d.) by polynomial feedback if*

$$\operatorname{rank}[A(z)|B(z)] = n \qquad \text{for} \qquad |z| \leq 1 \tag{9}$$

COROLLARY 1. *(1a) is stabilizable (i.o.d.) by polynomial feedback if*

$$\operatorname{rank}[sI - A(z), B(z)] = n \qquad \text{for} \qquad s \in C \quad \text{and} \quad |z| \leq 1.$$

To prove this theorem a classic result from analytic function theory [25] (see also [58]) was used which allows one to approximate a rational function analytic in $|z| \leq 1$ by a polynomial. Further Kamen et al [26] show that a finite-dimensional compensator can be constructed to stabilize a retarded system with noncommensurate delays if some controllability and observability conditions are met as stated in the theorem below.

THEOREM 6. *Consider a system as represented by (1a) with $z = (z_1, z_2, \ldots, z_k) = (e^{-sh_1}, e^{-sh_2}, \ldots, e^{-sh_k})$. The system is stabilizable by a finite-dimensional feedback controller realized as $H(sI - F)^{-1}G + J$ over R if and only if*

$$\operatorname{rank}[sI - A(z),\ B(z)] = n, \quad \text{and} \quad \operatorname{rank}\begin{bmatrix} sI - A(z) \\ C(z) \end{bmatrix} = n \tag{10}$$

for all $s \in \overline{H}$ and $z = (e^{-sh_1}, e^{-sh_2}, \ldots, e^{-sh_k})$ where $\overline{H} = \{s \in C | \operatorname{Re} s \geq 0\}$.

The theorem was established through the use of a result in [27], which relates that there is a feedback controller $H(z)(sI - F(z))^{-1}G(s) + J(z)$ over $R[z]$ that stabilizes (1a) i.o.d., if and only if (10) holds for all $(s, z) \epsilon H \times \Delta^k$ where Δ^k denotes the k-fold Cartesian product of the closed unit disk $\Delta = \{z \in C\ |z| \leq 1\}$. In conjunction with a variation of a result of Mergelyan [44]; being a claim that if we let $\widetilde{H} = H \cup \{\infty\}$ and $\Gamma_\gamma(\widetilde{H}) =$ the algebra of functions holomorphic in the open right-half plane H and continuous on the boundary of $\widetilde{H}$, and which have real coefficients in Taylor expansions about real points, and letting $\psi : \widetilde{H} \to \Delta$ be any conformal equivalence, then any $f \epsilon \Gamma_\gamma(\widetilde{H})$ may be uniformly approximated on $\widetilde{H}$ by polynomials of the form $a_n\psi(s)^n + \cdots + a_1\psi(s) + a_0$, where $a_n, \ldots, a_0 \in R$. Namely for any $f \in \Gamma_\gamma(\widetilde{H})$ and any $\epsilon > 0$, there is a polynomial $\Pi_\epsilon(s) = a_n\psi(s)^n + \cdots + a_1\psi(s) + a_0$ such that

$$\sup_{s\in\widetilde{H}} |f(s) - \Pi_\epsilon(s)| < \epsilon \quad .$$

One way of obtaining such a polynomial $\Pi_\epsilon(s)$ is to use the Fourier series of $g(z) = f[\psi^{-1}(z)]$. Assume $g(z)\epsilon\Gamma_\gamma(\Delta)=$ the algebra of functions holomorphic on Δ and continuous on the unit circle T, and which have real coefficients, and

$$g(e^{j\omega}) \sim \sum_0^\infty c_n e^{jn\omega} \quad , \quad \omega \in [0, 2\pi].$$

For each integer $N \geq 0$ define

$$S_N(z) = \sum_{n\leq N} c_n z^n \quad , \quad z \in \overline{\Delta} \, ,$$

then the Nth Cesáro sum define by

$$P_N = \frac{1}{N+1} \sum_0^N S_i(z)$$

converges uniformly to $g(z)$ on $\overline{\Delta}$. Thus approximating a function in $\Gamma_\gamma(\widetilde{H})$ reduces to computing Fourier coefficients.

Obviously $R(z)$-controllability is weaker than the first condition in (10). An interesting question is what can one do with polynomial state feedback for a single-input linear system with commensurate delays that is assumed to be $R(z)$-controllable in terms of stabilizability or coefficient assignability? A result reported in [33] gives the answer. Let us consider a single-input $R(z)$-controllable pair $\big(A(z), b(z)\big)$ and denote

$$[A|b]^{-1} = \begin{bmatrix} h_1 \\ h_2 \\ \vdots \\ h_n \end{bmatrix} \quad \text{and} \quad T = \begin{bmatrix} h_n \\ h_n A \\ \vdots \\ h_n A^{n-1} \end{bmatrix}$$

Then $(\widehat{A}, \hat{b}) = (TAT^{-1}, Tb)$ is a canonical pair over $R[z]$ given by

$$\widehat{A} = \begin{bmatrix} 0 & \vdots & & \\ 0 & \vdots & I_{n-1} & \\ & \vdots & & \\ 0 & \vdots & & \\ \cdots & \cdots & \cdots & \cdots \\ -a_1 & -a_2 & \ldots & -a_n \end{bmatrix} , \quad \hat{b} = \begin{bmatrix} 0 \\ 0 \\ \vdots \\ 0 \\ 1 \end{bmatrix}$$

where $a_i \in R[z]$, $1 \leq i \leq n$, are determined by $\det[sI - A(z)] = s^n + \sum_{i=1}^n a_i s^{i-1}$. Notice that $T \in R^{n\times n}(z), \widehat{A} \in R^{n\times n}[z]$ and the characteristic polynomial of the closed-loop system when applying a polynomial state feedback $u(t) = k(z)x(t)$ is $\det[sI - (A(z) + b(z)k(z))] = \det[sI - (\widehat{A} + \hat{b}\hat{k})] = s^n + \sum_{i=1}^{n}(a_i - \hat{k}_i)s^{i-1}$ where $\hat{k}(z) = k(z)T^{-1} = [\hat{k}_1(z), \ldots, \hat{k}_n(z)]$. Now let $T = D^{-1}(z)N(z)$ be an irreducible matrix fraction description (MFD). By [41, lemma 6.6–1]it is seen that $k(z) = \hat{k}(z)T = \hat{k}D^{-1}(z)N(z)$ will be an $1 \times n$ polynomial vector if and only if there exists $f(z) \in R^{1\times n}[z]$ such that $\hat{k}(z) = f(z)D(z)$. Denote the i-th row vector by $d_i(z)$ and $\mathfrak{D} = \text{span}_{R[z]}\{d_i(z), 1 \leq i \leq n\}$, we observe that $\mathfrak{D}$ is a submodule in $R^n[z]$ and $k(z) = \hat{k}(z)T$ will belong to $R^{1\times n}[z]$ if and only if $\hat{k}(z) \in \mathfrak{D}$. In other words we have the following result (stated as a theorem).

THEOREM 7. *Given a single-input $R(z)$-controllable delay system as represented by the pair $(A(z), b(z))$, the coefficients of* $\det[sI - (A + bk)]$ *can be assigned to be $\alpha(z) \in R^{1\times n}[z]$ by a polynomial state feedback $u(t) = k(z)x(t)$ if and only if $a(z) - \alpha(z) \in \mathfrak{D}$, where $a(z) = [a_1(z) \ldots a_n(z)]$ and $\mathfrak{D}$ is a submodule (in $R^n[z]$) spanned by $d_i(z)$ $(1 \leq i \leq n)$, and the desired feedback gain is given by $k(z) = (a(z) - \alpha(z))T$.*

REMARK. Denote $T = P(z) + H(z)$ with $P(z) \in R^{n\times n}[z]$ and $H(z) \in R^{n\times n}(z)$ strictly proper. Assume $H(z) = D_1^{-1}(z)N_1(z)$ is an irreducible MFD. It then is easy to see that the above theorem also holds if $\mathfrak{D}$ is replace by $\mathfrak{D}_1$, where $\mathfrak{D}_1$ is the submodule generated by the rows of $D_1(z)$.

Recently Tadmor [61] has introduced the idea of trajectory stabilizing controls for the delay type systems. Specifically he considers linear delay type systems modeled by functional differential equation:

$$\dot{x}(t) = \int_0^h d\alpha(\theta)x(t-\theta) + \int_0^h d_\beta(\theta)u(t-\theta) \tag{11}$$

The matrix valued measures $\alpha(\theta)$ and $\beta(\theta)$ determine the dependence on past x values and past u values in the evolution of the system. $h > 0$ is the delay duration, and $x \in R^n$, while $u \in R^m$.

The results are of interest when it is possible to improve the decay rate, but the improvement cannot be sustained by ordinary stabilizing controllers, which are usually asymptotically proportional to $x(t)$.

The results are based on the idea of γ-stabilizable systems introduced by Olbrot [48], where systems are excluded if the stabilizing control must grow exponentially.

γ-trajectory stabilizing means given any initial (Cauchy) data for (11) there exists a control $u(\cdot)$ and a real number δ such that the functions $t \to e^{-\gamma t}\, x(t)$ and $t \to e^{-\delta t}\, u(t)$ are of class $L_1(0, \infty)$. (In applications both γ and δ will be negative.) The Olbrot concept is then *γ-state stabilizing*, occuring where $\delta = \gamma$.

THEOREM 8. (Olbrot) *(11) is γ-state stabilizable if and only if*

$$\operatorname{rank}[\Delta(s), \quad B(s)] = n \quad \forall \; Res \geq \gamma$$

where $$\Delta(s) = sI - \int_0^h e^{-\theta s} \, d\alpha(\theta)$$

and $$B(s) = \int_0^h e^{-\theta s} \, d\beta(\theta) \; .$$

THEOREM 9. (Tadmor) *The system* (11) *is γ-trajectory stabilizable if and only if there is an $m \times m$ nonsingular, rational matrix valued function $R(s)$, with properties*

(i) $R(s)$ has form

$$R(s) = \left(\frac{P_1}{s-\lambda_1} + Q_1\right)\left(\frac{P_2}{s-\lambda_2} + Q_2\right)\cdots\left(\frac{P_l}{s-\lambda_l} + Q_l\right)$$

where each matrix Pj is an orthogonal projection, $Q_l = I - P_l$ and each λ_j is an eigenvalue of homogeneous part of (11) *with $Re\lambda j \geq \gamma$.*

(ii) The function $\overline{B}(s) = B(s)R(s)$ is the Laplace transform of a finite measure β supported within $[0, h]$.

(iii) $\operatorname{rank}[\Delta(s), \overline{B}(s)] = n$ *for all $Res \geq \gamma$.*

We should like to also briefly mention a functional analytic approach to stabilize a linear system with distributed delays recently reported by Fiagbedzi and Pearson [14]. The main feature of their approach is the use of a transformation that converts the delay system to a high order delay-free system whose spectrum coincides with the unstable spectrum of the delay system. The assumption of spectral stabilizability then permits the determination of a feedback compensator that stabilizes the delay free system and the delay system as well.

As a final topic in this subsection we relate some recent developments in the quadratic optimal control theory for time delay type systems. In particular Gibson [17] has introduced approximations, as have Banks, Rosen and Ito [3], into the theory of feedback control for the linear hereditary systems. Also, the quadratic theory has been further generalized to handle control action delays and delays in the cost criterion [13] [35]. Such optimal feedback controllers can lead to stability of the overall system.

2.4 Neutral Delay Systems. A single-input neutral system with finitely many noncommensurate delays may be modeled by differential-difference equations

$$D(z)\dot{x}(t) = A(z)x(t) + b(z)u(t) \tag{12}$$

where $z = (z_1 \ldots, z_k)$ with delay operators z_i, i.e $z_i x(t) = x(t - h_i)$, $h_i > 0$, h_i's are noncommensurable, $A(z) \in R^{n \times n}[z]$ and $D(z) \in R^{n \times n}[z]$ and $b(z) \in R^{n \times 1}[z]$. The characteristic equation of (12) is then given by

$$\Delta(s, e^{-sh}) = \det[sD(e^{-sh}) - A(e^{-sh})]$$

where $e^{-sh} = (e^{-sh_1} \dots e^{sh_R})$ and all its zeroes from the spectrum of (12). It has been shown [4] that the null solution of the homogeneous counterpart of (12) is asymptotically stable if the spectrum of (12) is contained in the open left-half s-plane C^-. Throughout this subsection it is assumed that there exists $\delta > 0$ such that $\det D(z) \neq 0$ for all $z \in X_\delta = \{z = (z_1, \dots, z_k) | \, |z_i| \leq 1 + \delta, \; 1 \leq i \leq k\}$. Such a $D(z)$ is called a formally stable operator. Now writing $\Delta(s, e^{-sh})$ as

$$\Delta(s, e^{-sh}) = \det D(e^{-sh}) \cdot \det[sI - F(e^{-sh})]$$

where $F(e^{-sh}) = D^{-1}(e^{-sh})A(e^{-sh})$, it then follows that the spectrum of Δ is contained in C^- if and only if *so* is spectrum of F. This observation motivated us to consider an associated system as represented by

$$\dot{x}(t) = F(z)x(t) + g(z)u(t) \tag{13}$$

where $F(z) = D^{-1}(z)A(z)$ and $g(z) = D^{-1}(z)b(z)$.

Differing from the retarded case, it is known [4] [55] that there maybe infinitely many unstable eigenvalues for a neutral delay system and consequently the stabilization of an unstable neutral system is even more difficult than that of an unstable retarded sysm. Under the formal D-stability assumption, we would like to handle stabilization through the associated system in a purely algebraic way. To be more precise, using the notation utilized in [8], let $S_\delta = \{p(z) \in R(z) | p(z) \neq 0 \text{ for } z \in X_\delta\}$ and $R_\delta = S_\delta^{-1} R[z] \underline{\underline{\Delta}} \{q(z)/p(z) \; | q(z) \in R[z], p(z) \in S_\delta\}$. Clearly the localization R_δ forms a ring under usual addition and multiplication. An element $r = q/p \in R_\delta$ is an unit of R_δ whenever $q \in S_\delta$. Thus $D(z) \epsilon R^{n \times n}[z]$ is formally stable if $\det D(z) \in S_\delta$ for some $\delta > 0$. Obviously formal stability of $D(z)$ implies that $D^{-1}(z) \in R_\delta^{n \times n}$. The set $R_\delta^{n \times n}$ forms a noncommutative ring under usual addition and multiplication. An element $U(z) \in R_\delta^{n \times n}$ is an unit whenever $\det U$ is an unit in R_δ. Since we have assumed $D(z)$ in (11) is formally stable, $F(z)$ and $g(z)$ in (13) belong to $R_\delta^{n \times n}$ and $R_\delta^{n \times 1}$ respectively, hence (13) is called the R_δ-associated system. (13) is said to be R_δ-controllable if $[F|g]$ is an unit in $R_\delta^{n \times n}$. Thus (F, g) is R_δ-controllable if and only if $\det[F|g] \in S_\delta$. From the above observation, a feedback compensator in $R_\delta^{1 \times n}$ which stabilizes the R_δ-associated system (13) will also yield a stable closed-loop system associated with (12).

Now let

$$\det[sI - F(z)] = s^n + f_1(z)s^{n-1} + \dots + f_n(z), \quad f_i \in R_\delta, \quad 1 \leq i \leq n$$

and

$$f(z) = [f_1(z) \dots f_n(z)],$$

for any desired coefficient vector

$$\tilde{f}(z) = [\tilde{f}_1(z) \dots \tilde{f}_n(z)].$$

The state feedback controller

$$u(t) = -k(z)x(t) + \nu(t) \tag{14}$$

with

$$k(z) = (\tilde{f} - f)\Gamma^T(F|g)^{-1} \tag{15}$$

results in a closed-loop system whose characteristic polynomial $\det[sI - F(z) + g(z)k(z)]$ is associated with the desired coefficient vector $\tilde{f}(z)$ where Γ is the inverse of the lower triangular Toeplitz matrix

$$\begin{bmatrix} 1 & & & \\ f_1 & 1 & & \\ \cdots & \cdots & \cdots & \cdots \\ f_{n-1} & f_{n-2} & \ldots & 1 \end{bmatrix}.$$

Expression (15) is known as the generalized Bass-Gura formula [8]. R_δ-associated system (13) is said to be arbitrarily coefficient-assignable by state feedback if for any $\tilde{f} \in R_{\delta^{1\times n}}$ there exists a feedback (14) with $k(z) \in R_\delta^{1\times n}$ such that the resulting closed-loop system has coefficient vector $\tilde{f}$ associated with its characteristic polynomial. By (15), (13) is arbitrarily coefficient-assignable by a state feedback if and only if (F, g) is an R_δ-controllable pair.

Now write $k(z)$ in (15) as

$$(\tilde{f} - f)\Gamma^T[F|g]^{-1} = \begin{bmatrix} \frac{n_1(z)}{w_1(z)} & \cdots & \frac{n_n(z)}{w_n(z)} \end{bmatrix} \quad \text{with} \quad n_i(z),\ w_i(z) \in R[z]$$

and define $w(z)$= the least common multiple of $w_i(z), 1 \leq i \leq n$, with $w(0) = 1$. Then $(\tilde{f} - f)\Gamma^T[F|g]^{-1} = w^{-1}(z)N(z)$ for some $N(z) \in r^{1\times n}[z]$. Clearly $k(z)$ in (15) belongs to $R_\delta^{1\times n}$ if and only if $w(z) \in S_\delta$. In such case the control law (14) can be implemented in a feedback-feedforward scheme as

$$u(t) = -N(z)x(t) + \big(1 - w(z)\big)u(t) + w(z)\nu(t)$$

where the term $\big(1 - w(z)\big)u(t)$ involves only data $u(t-h)$, $u(t-2h)$, etc.

Similar to the retarded delay case, a parametrization of the attainable set of coefficient vectors $\tilde{f}(z)$ can be obtained for the system (11) with $k = 1$ (i.e. the commensurate delay case). To see this let $\Gamma^T[F|g]^{-1} = D^{-1}(z)N(z)$ be an irreducible factorization of $\Gamma^T[F|g]^{-1} \in R^{n\times n}(z)$, where $D(z)$ and $N(z)$ are $n \times n$ polynomial matrices. Then

$$k(z) = (\tilde{f} - f)D^{-1}(z)N(z)$$

Note that $k(z) \in R_\delta^{1\times n}$ if and only if there exists a polynomial $w(z) \in S_\delta$ such that $w(z)k(z) \in R^{1\times n}[z]$, i.e.

$$w(z)(\tilde{f} - f)D^{-1}(z)N(z) \in R^{1\times n}[z]. \tag{16}$$

Further observe that (16) holds if and only if $w(z)(\tilde{f} - f) = h(z)D(z)$ i.e.

$$\tilde{f} = f + w^{-1}(z)h(z)D(z)$$

for some $h(z) \in R^{1\times n}[z]$. Denoting the i-th row of $D(z)$ by $d_i(z)$ $(1 \leq i \leq n)$ and noting that $w^{-1}h \in R_d^{1\times n}$, we have [39] the following result.

THEOREM 10. *Given a single-input neutral system as represented by equation (11) with commensurate delays and with formally stable D-operator, the coefficient vector associated with* $\det[sI - F + gk]$ *can be assigned to be* $f(z)$ *by the feedback controller (14) with* $k(z) \in R_\delta^{1\times n}$ *if an only if* $\tilde{f} - f \in D_\delta$ *where* D_δ *is the submodule (in* $R_\delta^{1\times n}$*) spanned by* $\{d_i(z),\ 1 \le i \le n\}$.

2.5 Robust Stabilization: A Factorization Approach. The following notation will be used throughout this subsection:

I :	Identity matrix of appropriate dimension.
$R[z][s]$:	Set of all polynomials in s with coefficients in $R[z]$,
H :	Set of all polynomials $a(s,z) \in R[z][s]$ which are monic w.r.t.s., and $a(s, e^{-sh}) \neq 0 \quad \mathrm{Re}s \ge 0$ and all $h \ge 0$,
$R_p[z](s)$:	Set of all $b(s,z)/a(s,z)$ where $a(s,z)$, $b(s,z)$ are in $R[z][s]$, $a(s,z)$ monic w.r.t.s; and $\deg_s b(s,z) \le \deg_s a(s,z)$;
S :	Set of all $b(s,z)/a(s,z) \in R_p[z](s)$ with $a(s,z) \in H$;
$S^{n\times n}$:	Set of all $n \times n$ matrices whose entries are all in S;
Y :	Set of all elements $b/a \in S$ such that $a/b \in R_p[z](s)$;
J :	Set of all $b(s,z)/a(s,z)$ such that both $b(s,z)/a(s,z))$ and $a(s,z)/b(s,z)$ are in S; and
$\overline{\sigma}(M)$:	The largest singular value of matrix M.

Note that S forms a commutative ring with unit element under usual addition and multiplication, and any $p(s,z) \in R_p[z](s)$ can be written as $n(s,z)/d(s,z)$ for some $n(s,z)$ and $d(s,z)$ in S.

For $p(s,z) \in S$, define

$$\|p\| = \sup_{\mathrm{Re}s \ge 0} |p(s, e^{-sh})| = \sup_{\omega \in R} |p(j\omega, e^{-j\omega h})| \tag{17}$$

Clearly $\|\cdot\|$ is a norm mapping S to R^+, the set of all nonnegative real numbers. With $\|\cdot\|$ in (17), S becomes a normed algebra over R. Further for $P \in S^{n\times m}$ define

$$\|P\| = \sup_{\mathrm{Re}s \ge 0} \overline{\sigma}\big(P(s, e^{-sh})\big) = \sup_{\omega \in R} \overline{\sigma}\big(P(j\omega, e^{-j\omega h})\big) \tag{18}$$

then $\|\cdot\|$ in (18) is a norm mapping $S^{n\times m}$ to R^+.

When $n = m$, $S^{n\times n}$ is a ring. A square matrix $U \in S^{n\times n}$ is said to be S-unimodular if U is an unit in $S^{n\times n}$, i.e., there exists $V \in S^{n\times n}$ such that $UV = I$. Clearly $U \in S^{n\times n}$ is S-unimodular iff $\det U$ is an unit in S, i.e. $\det U \in J$.

We conclude that if $P \in S^{n\times n}$ satisfies $\|P\| < 1$ then $I + P$ is S-unimodular. In particular, if $p \in S$ satisfies $\|p\| < 1$ then $1 + p \in J$.

When $n > 1$, the computation of $\|P\|$ is not easy. The following lemma gives an easy-to-check sufficient condition for $I + P$ being S-unimodular.

LEMMA 1. *$I + P$ with $P \in S^{n\times n}$ is S-unimodular if $\|g\| < 1$ where $g = \det(I + P) - 1$.*

Proof. Since $\det(I + P) = 1 + g$ and $\|g\| < 1$, $\det(I + P) \in J$. Therefore $I + P$ is S-unimodular.

□

REMARK. Either $\|P\| < 1$ or $\|g\| < 1$ are sufficient conditions for $I + P$ being S-unimodular, and each of them does not imply the other unless $n = 1$.

The pair $(N, D) \in S^{n\times m} \times S^{m\times m}$ is a right-coprime fractional representation (*r.c.f.r.*) of $P \in R_p^{n\times m}[z](s)$ if $\det D \in Y$ and $P = ND^{-1}$ represents a right-coprime factorization ($r, c.f.$) of P i.e. there exists $U \in S^{m\times n}$ and $V \epsilon S^{m\times m}$ such that $UN + VD = I$. The pair $(\widetilde{D}, \widetilde{N}) \in S^{n\times n} \times S^{n\times m}$ is a left-coprime fractional representation (*l.c.f.r.*) of $P \in R_p^{n\times m}[z](s)$ if $\det \widetilde{D} \in Y$ and $P = \widetilde{D}^{-1}\widetilde{N}$ represents a left-coprime factorization (*l.c.f.*) of P i.e. there exists $\widetilde{U} \in S^{m\times n}$ and $\widetilde{V} \in S^{n\times n}$ such that $\widetilde{N}\widetilde{U} + \widetilde{D}\widetilde{V} = I$. Now consider the feedback compensation scheme as depicted in Fig. 12 where $P(s, z) \in R_p^{m\times m}[z](s)$ represents a linear retarded time-delay system to be controlled and $C(s, z) \in R^{m\times n}[z](s)$ is the 'compensator' to be designed such that the overall closed-loop system, denoted by (P, C), is internally stable (S-stable).

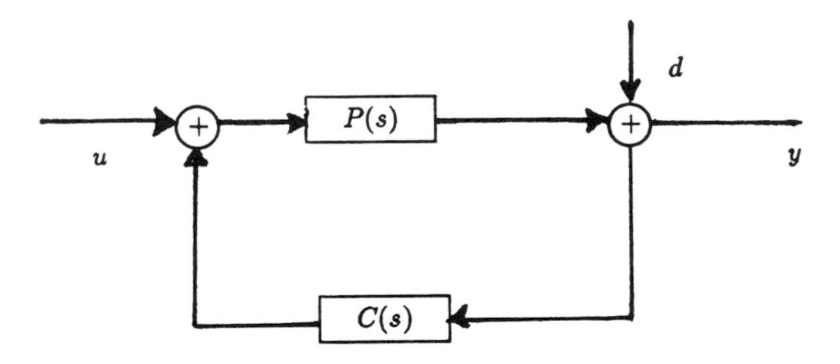

FIGURE 12

It is straightforward to show that letting (N_p, D_p) and $(\widetilde{D}_p, \widetilde{N}_p)$ be any *r.c.f.* and *l.c.f.* of P, (i.e. there exists $U, V, \widetilde{U}$ and $\widetilde{V}$ such that $UN_p + VD_p = I$ and $\widetilde{N}_p\widetilde{U} + \widetilde{D}_p\widetilde{V} = I$). Then all compensators in $R^{m\times n}[z](s)$ that stabilize P are given by $(P) = \{(V - RN_p)^{-1}\ (U + R\widetilde{D}_p) : R \in S^{m\times n},\ \det(V - R\widetilde{N}_p) \neq 0\} = \{(\widetilde{U} + N_pT)(\widetilde{V} - N_pT) - 1 :\ T \in S^{m\times n}, \det(\widetilde{V} - N_pT) \neq 0\}$. For the scalar case, let $p(s, z) = n(s, z)/d(s, z)$ be any coprime factorization, i.e. there exist $u(s, z)$ and $\nu(s, z)$ in S such that $un + \nu d = 1$, then all the stabilizers (having the above special form) of $p(s, z)$ are given by

$$(p) = \left\{\frac{u + rd}{v - rn}\ : r \in S,\quad v - rn \neq 0\right\} \tag{19}$$

As far as the coprime factorization of the system transfer function is concerned, one may follow [47] in conjunction with [46] and [25] to conclude that $p(s,z)$ admits a coprime factorization if

$$\det R \neq 0 \qquad \text{for} \qquad |z| \leq 1 \tag{20}$$

where $R = [b\ AB \dots A^{n-1}b]$,

$$A = \begin{bmatrix} -a_{n-1}(z) & 1 & 0 & \dots & 0 \\ -a_{n-2}(z) & 0 & 1 & \dots & 0 \\ \dots & \dots & \dots & \dots & \dots \\ -a_1(z) & 0 & 0 & \dots & 1 \\ -a_0(z) & 0 & 0 & \dots & 0 \end{bmatrix}, \quad b = \begin{bmatrix} b_{n-1}(z) \\ b_{n-2}(z) \\ \vdots \\ b_0(z) \end{bmatrix}$$

and a_i's and b_i's are given by

$$p(s,z) = \frac{b_{n-1}(z)s^{n-1} + \dots b_0(z)}{s^n + a_{n-1}(z)s^{n-1} + \dots + a_0(z)} \quad . \tag{21}$$

It follows from [47] that if condition (20) is satisfied then the compensator $c = u/v$ can be constructed by solving for a state feedback stabilization scheme and then designing a full-order state observer. For the multivariable case, sufficient condition similar to (20) for the existence of a *r.c.f.* or a *l.c.f.* of P can be stated.

With the above preliminaries we are now in a position to discuss the robustness of the compensators. The results reported here might be regarded as a slight improvement of the results in [46, ch.3]. Assume (P,C) is S-stable, P admits a *r.c.f.*$(D_p, N_p), C$ admits an *l.c.f.*, (D_c, N_c), and

$$D_cD_p + N_cN_p = I \tag{22}$$

Further let $P' = N_p' D_p'^{-1}$ be right factorization of P' and define

$$\Delta = D_cD_p' + N_cN_p',$$

thus

$$[D_c \;\; N_c \begin{bmatrix} D_p' - D_p \\ N_p' - N_p \end{bmatrix} = \Delta - I, \tag{23}$$

and

$$(\Delta^{-1}N_c)N_p' + (\Delta^{-1}D_c)D_p' = I \tag{24}$$

Let (N_p, D_p, γ) be a family of systems represented by transfer functions $P' = N_p' D_p'^{-1}$ satisfying

$$\left\| \begin{bmatrix} D_p' \\ N_p' \end{bmatrix} - \begin{bmatrix} D_p \\ N_p \end{bmatrix} \right\| < \gamma \qquad \text{for some} \quad \gamma > 0.$$

Clearly if

$$P' \in (N_p, D_p, \gamma_0) \quad \text{where} \quad \gamma_0 = \frac{1}{\|[D_c \;\; N_c]\|} \tag{25}$$

(23) then implies $\|I - \Delta\| < 1$ and, therefore, Δ is S-unimodular. In this case (24) means that *every* P' in (N_p, D_p, γ_0) admits a *r.c.f.* and C stabilizes each system of (N_p', D_p', γ_0).

Instead of (25), one can also pose slightly different conditions on the fractional perturbation to guarantee the S-stability of (P', C). For instance, if

$$\|D_p' - D_p\| \leq \alpha \|D_c\|^{-1} \quad \text{and} \quad \|N_p' - N_p\| \leq \beta \|N_c\|^{-1} \tag{26}$$

with $\alpha + \beta < 1$, then (23) implies $\|I - \Delta\| < 1$ and, therefore, C stabilizes (N_p, D_p, γ_0). Further define $g_p = \det(I - \Delta) - 1$, it follows from

$$\|g_p\| < 1 \tag{27}$$

then, by lemma 1, Δ is S-unimodular, and (P', C) is S-stable.

We summarize the above discussion as follows.

THEOREM 11. *Assume $C(s,z)$ internally stabilizes the nominal system represented by transfer function $P(s,z)$ where P and C admit a r.c.f. and an l.c.f. respectively. Then $C(s.z)$ also stabilizes the true system represented by the transfer function $P'(s,z)$ if P' admits a r.c.f. and satisfies one of the conditions (25), (26), or (27).*

Notice that if the nominal system transfer function also admits an *l.c.f.*$(\widetilde{D}_p, \widetilde{N}_p)$, then the stabilizer $C = D_c^{-1} N_c$ can be replaced by any member of $(P) = \{(D_c - R\widetilde{N}_p)^{-1}(N_c + R\widetilde{D}_p) : R \in S^{m \times n}, \ \det(D_c - R\widetilde{N}_p) \neq 0\}$. Therefore, one can make condition (25) less conservative by searching an appropriate $R_0 \in S^{m \times n}$ such that $\det(D_c - R_0\widetilde{N}_p) \neq 0$ and

$$\begin{aligned} \|[D_c - R_0\widetilde{N}_p \quad N_c + R_0\widetilde{D}_p]\| &= \inf_{\substack{R \in S^{m\times n} \\ \det(D_c - R\widetilde{N}_p) \neq 0}} \|[D_c - R\widetilde{N}_p \quad N_c + R\widetilde{D}_p]\| \\ &= \inf_{\substack{R \in S^{m\times n} \\ \det(D_c - R\widetilde{N}_p) \neq 0}} \|X - RW\| \end{aligned} \tag{28}$$

where $X = [D_c \; N_c]$ and $W = [\widetilde{N}_p - \widetilde{D}_p]$.

(28) represents a challenging optimization task. For the finite-dimensional case and its relation to the optimal Hankel-norm approximation, see [24]. For the scalar delay case, a similar task is considered [15] [16] [65].

It is known [4] [21] that for any fixed $\sigma_0 \in R$, the poles of a retarded delay system whose real parts are greater than or equal to σ_0 consist of a finite set. Based on this fact, a multivariable retarded delay system represented by transfer function $P(s,z) \in R_p^{n \times m}[z](s)$ admits an additive decomposition:

$$P(s,z) = P_1(s,z) + P_2(s) \tag{29}$$

where $P_1(s,z)$ is the sum of the principal parts of the Laurent expansions of $P(s,z)$ at its poles in the closed right-half plan (RHP), and $P_2(s) = P(s,z) - P_1(s,z)$is a finite-dimensional, causal rational transfer function matrix [46]. Assume $P_1(s,z) = C(z)(sI - A(z))^{-1}(B(z)$ with $A(z) \in R^{N\times N}[z]$, $B(z) \in R^{N\times m}[z], C(z) \in R^{n\times N}[z]$ and assume that there exists $K(z) \in R^{m\times N}[z]$ and $L(z) \in R^{N\times n}[z]$ such that $[sI - (A - BK)]^{-1}$ and $[sI - (A - LC)]^{-1}$ belong to $S^{N\times N}$. Define

$$\begin{aligned} N &= (C - P_2K)(sI - A + BK)^{-1}B + P_2, & D &= I - K(sI - A + BK)^{-1}B \\ U &= K(sI - A + LC)^{-1}L, & V &= I + K(sI - A + LC)^{-1}(B - LP_2) \\ \widetilde{D} &= I - C(sI - A + LC)^{-1}L, & \widetilde{N} &= C(sI - A + LC)^{-1}(B - LP_2) + P_2, \\ \widetilde{V} &= I + (C - P_2K)(sI - A + BK)^{-1}L, & \widetilde{U} &= K(sI - A + BK)^{-1}L, \end{aligned}$$

then [46] all eight matrices defined above have entries in $S; (N, D)$ is a *r.c.f.r.* of $P(s,z)$ and $(\widetilde{D}, \widetilde{N})$ is an *l.c.f.r.* of $P(s,z)$; and

$$\begin{bmatrix} V & U \\ -\widetilde{N} & \widetilde{D} \end{bmatrix} \begin{bmatrix} D & -\widetilde{U} \\ N & \widetilde{V} \end{bmatrix} = \begin{bmatrix} I & 0 \\ 0 & I \end{bmatrix}$$

Therefore using the additive decomposition of a given delay system transfer function matrix P we can also construct all compensators that stabilize $P : (P) = \{V - R\widetilde{N})^{-1}(U + R\widetilde{D}) : R \in S^{m\times n}, \det(V - R\widetilde{N}) \neq 0\} = \{(U + DT)(\widetilde{V} - NT)^{-1} : T \in S^{m\times n}, \det(\widetilde{V} - NT) \neq 0\}$.

Through the use of the results of Glover [19] [18] and Nett [46. Ch. 6], recently Curtain and Glover [10] have shown how to use a finite dimensional controller to stabilize an infinite-dimensional system with transfer function P subject to an additive perturbation Δ. To explain their results in a time-delay system setting, we assume (i) the retarded delay system transfer function $P(s,z) \in R_p^{n\times m}[z](s)$ has no poles on the imaginary axis, and $\Delta(s,z) \in R_p^{n\times m}[z](s)$; (ii) P and $P + \Delta$ have an equal number of poles in the closed RHP; and (iii)

$$\|W_1 \Delta W_2\| < \epsilon \tag{30}$$

where W_1, W_2, W_1^{-1}, and W_2^{-1} are stable rational functions. Next decompose W_1PW_2 as

$$W_1PW_2 = \widehat{P}_1(s,z) + \widehat{P}_2(s) \tag{31}$$

where $\widehat{P}_1(s,z) \in S^{n\times m}$ and $\widehat{P}_2(s)$ is such that $\widehat{P}_2(\infty) = 0$ and $\widehat{P}_2^* = [P_2(-s)]^*$ is a stable rational transfer function matrix, i.e. $\widehat{P}_1(s,z)$ is a stable infinite-dimensional retarded delay system while $\widehat{P}_2(s)$ is completely unstable (all poles in open right-half plane) but finite-dimensional system. The system denoted by $(P, W_1, W_2, K, \epsilon)$ is said to be robustly stable if $(P + \Delta, K)$ is internally stable for all perturbation $\Delta \in R_p^{n\times m}[z](s)$ such that $(P, W_1, W_2, K, \epsilon)$ is robustly stable.

It is observed [10] that $(P, W_1, W_2, K, \epsilon)$ is robustly stable if and only if

$$K = W_2 K_2 (I + \widehat{P}_1 K_2)^{-1} W_1 \tag{32}$$

for some $K_2 \in R^{m \times n}$ such that $(\widehat{P}_2, I, I, K_2, \epsilon)$ is robustly stable. This means that the robust stabilizability of the system only depends on the robust stabilizability of the finite-dimensional unstable component of the system [19] [10]. The above observation in conjunction with the finite-dimensional result [[19], Th. 4.1] leads to the following necessary and sufficient condition for robustly stabilizing a retarded delay system.

THEOREM 12 [10]. *Under the assumptions (i)–(iii), the system as represented by (P, W_1, W_2, ϵ) is robustly stabilizable if and only if*

$$\sigma_{\min}(\widehat{P}_2^*) \geq \epsilon. \tag{33}$$

By (32) it is seen that the robust controller K represents a system that is in general infinite dimensional. To find a finite-dimensional one, note that RH_∞ the set of all stable rational functions is dense in S. This fact was proved by Nett [46] using Walsh's theorem [see 43. p. 98]. Hence for any given $\delta > 0$, there exists $\widetilde{P}_1(s) \in RH_\infty$ such that $\|\widetilde{P}_1(s,h) - \widetilde{P}_1(s)\| \leq \delta$. Let $\widetilde{P} = W_1^{-1}(\widetilde{P}_1 + \widetilde{P}_2)W_2^{-1}$ then $\widetilde{P}$ is finite-dimensional, and $W_1 \widetilde{P} W_2 = \widetilde{P}_1 + \widetilde{P}_2$ with $\widetilde{P}_1$ and $\widetilde{P}_2 \in RH_\infty$ and $\|W_1(P - \widetilde{P})W_2\| \leq \delta$. Now if we take $\delta < \epsilon \leq \sigma_{\min}(P_2^*)$ and let K_2 be a rational compensator such that $(\widetilde{P}_2, I, I, K_2, \epsilon)$ is robustly stable and such that $\det(I + \widetilde{P}_1 K_1)(\infty)$ is bounded away from zero then if follows from [19] that the *controller* represents a finite dimensional system

$$K = W_2 K_2 (I + \widetilde{P}_1 K_2)^{-1} W_1$$

and is such that $(P, W_1, W_2, K, \epsilon - \delta)$ is robustly stable [10].

3. Stabilization of Two-Dimensional (2-D) Systems. In this section we shall describe recent results for stabilization of two dimensional type systems; interlacing new results with our previous results and results of others. These results follow from the above delay theory since for example stability independent of delay is essentially a two variable criterion.

3.1 2-D Systems and Roesser's State-Space Model. Consider 2-D systems as characterized by the rational function of two complex variables

$$g(z_1, z_2) = \frac{b(z_1, z_2)}{a(z_1, z_2)} = \frac{\sum_{i=0}^{p} \sum_{j=0}^{q} b_{ij} z_1^i z_2^j}{\sum_{i=0}^{p} \sum_{j=0}^{q} a_{ij} z_1^i z_2^j}, \qquad a_{oo} = 1 \quad . \tag{34}$$

Next write $g(z_1, z_2)$ as the formal power series

$$g(z_1, z_2) = \sum_{l=0}^{\infty} \sum_{k=0}^{\infty} h_{lk} z_1^l z_2^k$$

The 2-D sequence $\{h_{lk}\}$ can be used to relate an input sequence $\{u(i,j)\}$ to an output sequence $\{y(i,j)\}$ through the 2-D convolution sum

$$y(i,j) = \sum_{l=0}^{\infty} \sum_{k=0}^{\infty} h_{lk} u(i-l, j-k) \,. \tag{35}$$

The sequence $\{h_{lk}\}$ can be considered to be the "impulse response" of the system since if follows from (35) that $\{h_{ij}\}$ is the response to the sequence $\{\delta(i,j)\}$ where $\delta(0,0) = 1$ and $\delta(i,j) \equiv 0$ for $(i,j) \neq (0,0)$.

The system as represented by the convolution sum (35) is said to be bounded-input bounded-output (BIBO) stable if for any given $M > 0$ there exists an $N > 0$ such that $|u(i,j)| < M$ for all (i,j) implies that $|y(i,j)| < N$ for all (i,j). By (35) it is easy to see that (34) represents a BIBO stable system if and only if $\{h_{lk}\} \epsilon l^1$, i.e.

$$\sum_{l=0}^{\infty} \sum_{k=0}^{\infty} |h_{lk}| < \infty \,. \tag{36}$$

More feasible stability criteria may be described in terms of the location of the zeros of the denominator polynomial $a(z_1, z_2)$ of the transfer function (34). Differing from the 1–D case, however, the zeros of the numerator $b(z_1, z_2)$ play also a significant role in the stability analysis of 2–D systems [20]. For the sake of simplicity we assume throughout this section that the system has no nonessential singularities of the second kind on the unit bicircle $T^2 = \{(z_1, z_2) |\ |z_1| = 1,\ |z_2| = 1\}$, i.e. there are no points $(z_1, z_2) \in T^2$ such that $a(z_1, z_2) = b(z_1, z_2) = 0$. It has been shown in [57] that BIBO stability obtains if and only if $a(z_1, z_2) \neq 0$ for $(z_1, z_2) \in U^2 = \{(z_1, z_2) |\ |z_1| \leq 1,\ |z_2| \leq 1\}$. Quite a few variations of this criterion are available in the literature, see e.g., [22] [12] [60].

One may also realize the 2-D system as represented by (34) in a local state-space setting and do stability analysis in that setting. For example, (34) can be realized in Roesser's local state-space from [53] in terms of a four-tuple (A, b, c, d) involving the difference equations

(37a)

$$\begin{bmatrix} x^h(i+1,j) \\ x^v(i,j+1) \end{bmatrix} = \begin{bmatrix} A_1 & A_2 \\ A_3 & A_4 \end{bmatrix} \begin{bmatrix} x^h(i,j) \\ x^v(i,j) \end{bmatrix} + \begin{bmatrix} b_1 \\ b_2 \end{bmatrix} u(i,j) \triangleq Ax + bu$$

(37b)

$$y(i,j) = [C_1 \quad C_2] \begin{bmatrix} x^h(i,j) \\ x^v(i,j) \end{bmatrix} + du(i,j) \triangleq cx + du$$

by considering z_1 be a left-shift operation and z_2 to be an up shift operation on a two-dimensional array of data using column row indices. Further by introducing initial data along the boundaries for the array of x data one can consider the initial value problem in the two-dimensional setting, that is let

$$x^h(0,j) \equiv \xi(j),\ x^v(i,0) \equiv \eta(i) \quad \text{for } i \geq 0 \text{ and } j > 0. \tag{37c}$$

Here $x^h(i,j) \in C^n$ and $x^v(i,j) \in C^m$ form the 'local state' at (i,j) and A, b, c, d are constant, complex matrices with appropriate dimensions.

In general one has $n \geq p$ and $n \geq q$. If $n = p$ and $m = q$, (37) is called the absolutely minimal realization of $g(z_1, z_2)$ and $a(z_1, z_2)$ becomes

$$a(z_1, z_2) = \det \begin{bmatrix} I - z_1 A_1 & -z_1 A_2 \\ -z_2 A_3 & I - z_2 A_4 \end{bmatrix}. \tag{38}$$

Define $A^{i,j}$ recursively as

$$A^{i,j} = A^{1,0} A^{i-1,j} + A^{0,1} A^{i,j-1} \qquad \text{for} \qquad (i,j) > (0,0)$$
$$A^{i,j} = 0 \quad \text{for} \quad i < 0 \quad \text{or} \quad j < 0$$

with

$$A^{0,0} = I, A^{1,0} = \begin{bmatrix} A_1 & A_2 \\ 0 & 0 \end{bmatrix}, \text{ and } A^{0,1} = \begin{bmatrix} 0 & 0 \\ A_3 & A_4 \end{bmatrix}.$$

It can then be shown [53] that the solution of the initial value problem can be written as

$$\begin{aligned} y(i,j) = c\Bigg(& \sum_{r=0}^{i} A^{i-r,j} \begin{bmatrix} 0 \\ \eta(r) \end{bmatrix} + \sum_{r=0}^{j} A^{i,j-r} \begin{bmatrix} \xi(r) \\ 0 \end{bmatrix} \\ &+ \sum_{(0,0)\leq(l,k)<(i,j)} c(A^{i-l-1,j-k} \begin{bmatrix} b_1 \\ 0 \end{bmatrix} + A^{i-l,j-k-1} \begin{bmatrix} 0 \\ b_2 \end{bmatrix}\Bigg) u(l,k) \\ &+ du(i,j) \text{ all } (i,j) \geq 0. \end{aligned} \tag{39}$$

By (35) and (39) in conjunction with the causality of the system ($a_{oo} = 1$) one concludes that

$$h_{lk} = c\left(A^{l-1,k} \begin{bmatrix} b_1 \\ 0 \end{bmatrix} + A^{l,k-1} \begin{bmatrix} 0 \\ b_2 \end{bmatrix}\right) .$$

It also follows from (39) that in this state–space setting the system when represented in local state form (37) can be said to be BIBO stable if a bounded input sequence $\{u,(i,j)\}$ along with zero initial conditions (i.e. $\xi(i) = \eta(i) \equiv 0$ for all $i \geq 0$) produces a bounded output sequence $\{y(i,j)\}$.

Writing $a(z_1, z_2)$ of (38) as

$$\begin{aligned} a(z_1, z_2) &= \det(I - z_1 A_1) \det\{I - z_2[A_4 + A_3(z_1^{-1} I - A_1)^{-1} A_2]\} \\ &= \det(I - z_2 A_4) \det\{I - z_1[A_1 + A_2(z_2^{-1} I - A_4)^{-1} A_3]\} \end{aligned}$$

the state-space versions of the stability criteria [22] [12] can be stated as [37] the following Lemma. Note a square matrix A of real (or complex numbers) is said to be stable if $|zI - A| \neq 0$ for $|z| \geq 1$.

LEMMA 2. *The following statements are equivalent:*

1. (37) *is BIBO stable;*
2. (i) A_1 *is stable, and*
   (ii) $A_4 + A_3(zI - A_1)^{-1}A_2$ *with* $|z| = 1$ *is stable;*
3. (i) A_4 *is stable, and*
   (ii) $A_1 + A_2(zI - A_4)^{-1}A_3$ *with* $|z| = 1$ *is stable;*
4. (i) A_1 *is stable;*
   (ii) A_1 *has no eigenvalues on* T, *and*
   (iii) $A_4 + A_3(zI - A_1)^{-1}A_2$ *with* $|z| = 1$ *has no eigenvalues on the unit circle;*
5. (i) A_1 *is stable,*
   (ii) A_4 *has no eigenvalues on* T, *and*
   (iii) $A_1 + A_2(zI - A_4)^{-1}A_3$ *with* $|z| = 1$ *has no eigenvalues on the unit circle.*

Here $T = \{z|\ |z| = 1\}$.

A possible way of verifying 2(ii) and 3(ii) of Lemma 2 is to use a Lyapunov approach [38] as stated below.

LEMMA 3. *A matrix* $F(z)$ *with a complex parameter* $z \in T$ *is stable if and only if for any given positive-definite Hermitian matrix* $W(z)$ *with* $z \in T$, *there exists a unique positive-definite Hermitian matrix* $G(z)$ *such that*

$$G(z) - F^*(z)G(z)F(z) = W(z)$$

The Lyapunov theorem reported in [51] [36], described below, gives a sufficient but not necessary BIBO stability condition [1].

LEMMA 4. *BIBO stability obtains for system represented in form (37) if there exists a positive-definite matrix* $G = G_1 \oplus G_2$ *such that* $G - A^TGA$ *is positive-definite where* $\oplus$ *denotes the direct sum.*

Another stability concept is asymptotic stability of solutions of a 2-D system (internal stability) In the Roesser's state-space setting, it is defined as follows [23]. A system represented in the form (37) is said to be asymptotically stable if for zero input and finite $\sup_j \|\xi(j)\|$ and $\sup_i \|\eta(i)\|$, $\sup_{i,j} \|x(i,j)\|$ is finite and $\lim_{\substack{i\to\infty \\ j\to\infty}}(x(i,j) = 0$, where $\|\cdot\|$ denotes the Euclidean norm of a vector, and $x(i,j) = \begin{bmatrix} x^h(i,j) \\ x^v(i,j) \end{bmatrix}$. Obviously if both input and boundary condition are zero, then (37) has only zero (null) solution $\{x(i,j) \equiv 0\}$. Thus an asymptotically stable system is such a system that the solution of the homogeneous equation (corresponding to $u(i,j) \equiv 0$) under finite boundary-disturbance will converge to the null solution as $i \to \infty$ and $j \to \infty$. The following theorem gives a necessary and sufficient condition for asymptotic stability [23].

THEOREM 13. *The null solution of the homogeneous difference equation system (37) is asymptotically stable if and only if $a(z_1, z_2) \neq 0$ for $(z_1, z_2)\epsilon U^2$, where $a(z_1, z_2)$ is defined by (38).*

REMARKS. 1. Assume that (37) has no nonessential singularities of the second kind on T^2. Then BIBO stability of (37) leads to the asymptotic stability of the same system and vice versa, even though they are two different stability concepts.

2. Denote the set of all 1-D sequences that are square-summable by l_1^2 and the set of all 2-D sequences that are square-summable by l_2^2 then the formula (39) that relates both input and boundary conditions to the output can be written as

$$y(i,j) = T_1(\xi, \eta) + T_2(u)$$

where T_1 represents an operator acting on two 1-D sequences to yield a 2-D sequence and T_2 is an operator mapping one 2-D sequence to another. It is easy to show that T_1 and T_2 are such that

$$l_1^2 \times l_1^2 \xrightarrow{T_1} l_2^2$$

$$l_2^2 \xrightarrow{T_2} l_2^2$$

if the system is BIBO stable. In other words, if the spectrum of A is in U^2 then an l_2^2 input sequence along with l_1^2 boundary conditions guarantees an l_2^2 solution of (37).

3.2 Stabilization: State-Space Approach. Having looked at the stability concepts in detail, we are now in a position to consider the stabilization question when the system represented by (37) is unstable in the BIBO sense. The first stabilization scheme we are going to discuss is the *constant* state feedback control, that is, one may ask whether a feedback controller,

$$u(i,j) = -k \begin{bmatrix} x^h(i,j) \\ \\ x^v(i,j) \end{bmatrix} + \nu(i,j) ,$$

with $k \in C^{1\times(n+m)}$ and an external input ν exists such that the resulting closed-loop system

$$\begin{bmatrix} x^h(i+1,j) \\ \\ x^v(i,j+1) \end{bmatrix} = (A - bk) \begin{bmatrix} x^h(i,j) \\ \\ x^v(i,j) \end{bmatrix} + b\nu(i,j)$$

is BIBO stable, in which case we say that the system is stabilizable. The following theorem is an immediate consequence of Lemma 4.

THEOREM 14. *A system represented by (37) is stabilizable by state feedback $u = -kx + \nu$ if there exists positive-definite matrices W and $G = G_1 \oplus G_2$ such that the Riccati equation*

$$\widehat{A}^* K + K^* \widehat{A} - K^* D K + Q = 0$$

with $\widehat{A} = B^*GA$, $D = B^*GB$, and $Q = G - A^*GA - W$ *has a solution* K.

For detailed discussion of the solvability of the above Riccati equation, see [38].

It turns out [38] [50] that stabilizing a discrete 2-D system by a constant state feedback is, in general, very difficult. Through the use of the past local state (delayed feedback) in the feedback controller, one may stabilize a 2-D system much more easily as is seen from the following result [34].

THEOREM 15. *A system represented by (37) is stabilizable by* $u(i,j) = kx(i,j) + \nu$ *with* $k = [k_1, k_2(z)]$, $k_1 \in C^{1\times n}, k_2(z) \in C^{1\times m}[z]$ *if* (A_1, b_1) *is a controllable pair and* $(f(z), g(z))$ *defined by*

$$f(z) = A_4 + z^n(A_3 + b_2k_1)adj[z^{-1}I - (A_1 + b_1k_1)]A_2$$

and

$$g(z) = b_2 + z^n(A_3 + b_2k_1)adj[z^{-1}I - (A_1 + b_1k_1)]b_1$$

is controllable over $C[z]$, *where* k_1 *is chosen such that*

$$\det[z^{-1}I - (A_1 + b_1k_1)] = z^{-n}$$

and $u = k_2(z)x + \nu$ *is a polynomial feedback controller that stabilizes* $(f(z), g(z))$.

3.3 Stabilization Using Frequency Domain Approaches. It was shown [64] that using causal, dynamic output feedback of form $y = \dfrac{d(z_1,z_2)}{c(z_1,z_2)}u$ it is impossible to do arbitrary coefficient assignment. A Wolovich type compensation scheme is considered by Chiasson and Lee [9]. The compensator has transfer function of the form

$$u = \left(\frac{l(w_1,w_2)}{\delta(w_1,w_2)}u + \frac{m(w_1,w_2)}{\delta(w_1,w_2)}y\right) + \nu$$

where $w_1 = z_1^{-1}, w_2 = z_2^{-1}$, l, m, δ are polynomials with δ stable, and ν denotes the external signal. This compensator leads to the transfer function from $\nu(w_1,w_2)$ to $y(w_1,w_2)$ as

$$\frac{y(w_1,w_2)}{\nu(w_1,w_2)} = \frac{b(w_1,w_2)\delta(w_1,w_2)}{d(w_1,w_2)} \tag{40}$$

where

$$d = (l + \delta)a + bm \tag{41}$$

If one could choose $d = \alpha\delta$ with the desired denominator polynomial a then the closed-loop transfer function (40) becomes

$$\frac{y}{\nu} = \frac{b}{\alpha}$$

Thus the design problem is reduced to finding 2-D polynomials l, m and δ such that δ is stable, l/δ and m/d are causal and such that

$$(\alpha - a)\delta = la + mb \tag{42}$$

Assume that

$$a(w_1, w_2) = \sum_{i=0}^{q} a_i(w_1)w_2^i = \sum_{j=0}^{p} \tilde{a}_j(w_2)w_1^j \text{ and } b(w_1, w_2) = \sum_{i=0}^{q} b_i(w_1)w_2^i = \sum_{j=0}^{p} \tilde{b}_j(w_2)w_1^j$$

and denote the greatest common divisor of a_i and b_i by (a_i, b_i). The following theorem [9] characterizes the structure of δ with which (42) is solvable.

THEOREM 16. *Equation (42) has a solution with a causal compensator whose characteristic polynomial is $\delta(w_1, w_2)$ if*

$$\delta = a + \gamma b \tag{43}$$

for some $\gamma \in R$. Conversely, if $(a_q(w_1), b_q(w_1)) = 1$ or $(\tilde{a}_p(w_2), \tilde{b}_p(w_2)) = 1$, then, generically, a solution to (42) exists only if (43) holds for some $\gamma \in R$.

Similar to the 1-D case, a coprime fractional representation scheme can also be developed for 2-D systems [52]. Let $R[z_1, z_2]$ be the set of all 2-D polynomials with real coefficients and let $R(z_1, z_2)$ be the set of all 2-D rational functions with real coefficients, and let S be the subset of $R(z_1, z_2)$ whose denominator polynomials have no zeros inside the prescribed unstable region Ω on the (z_1, z_2) plane. The following result with a constructive proof can be found in [52].

THEOREM 17. *For a given 2-D plant $p(z_1, z_2) = f(z_1, z_2)/g(z_1, z_2) \in R(z_1, z_2)$ with $f, g \in R[z_1, z_2]$ factor-coprime there exist a compensator $c(z_1, z_2) \in R(z_1, z_2)$ such that the closed-loop system shown in Figure 1 (section 2) is internally stable if and only if there exist $u(z_1, z_2)$ and $v(z_1, z_2)$ in S such that $uf + vg = 1$. If such u and v exist, then all stabilizer of the plant can be characterized as $c(z_1, z_2) = (u - tg)/(v + tf)$ where t varies in S with $v + tf \neq 0$. Furthermore, if the unstable region is complex-conjugate symmetric, i.e. $(z_1, z_2) \in \Omega$ implies that $(z_1, \bar{z}_2), (\bar{z}_1, z_2)$, and $(\bar{z}_1, \bar{z}_2)$ are also in Ω, then a necessary and sufficient condition for the existence of a solution u, v in S to the Bezout identity $uf + vg = 1$ is that f and g do not have common zeros in Ω.*

Finally we mention that some work on asymptotic tracking has been recently done for $2 - D$ systems by Sebek [56].

Acknowledgement. The authors acknowledge with thanks the assistance of Dr. Stephen D. Brierley and Dr. Charles Liu in providing the computer simulation results of figures 2–11.

REFERENCES

[1] B.D.O. ANDERSON, P. AGATHOKLIS, E.I. JURY AND M. MANSOUR, *Stability and matrix Lyapunov equation for discrete 2-D systems*, IEEE Trans. Circuits Syst., CAS-33 (1986), 261–267.

[2] H.T. BANKS AND A. MANITIUS, *Projection series for functional differential equations with applications to optimal control problems*, J. Differential Equations, 18 (1975) 296–332.

[3] H.T. BANKS, I.G. ROSEN, AND K. ITO, *A spline based technique for computing Riccati operators and feedback controls in regulator problems for delay equations*, SIAM J. on Scientific and Statistical computing, 5 (1984).

[4] R.E. BELLMAN AND K.L. COOKE, *Differential-Difference Equations*, New York: Academic (1963).

[5] R.P.M. BHAT AND N. KOIVO, *Modal characterizations of controllability and observability in time-delay systems*, IEEE Trans. Automat. Contr., AC-21 (1976) 292–293.

[6] S. BRIERLEY, *Stability and stabilization of generalized linear systems*, Ph.D. Thesis, University of Minnesota (1984).

[7] R. BROCKETT, *Finite dimensional linear systems*, John Wiley & Sons, Inc., New York (1970).

[8] C.I. BYRNES, M.W. SPONG, AND T.J. TARN, *A several complex variables approach to feedback stabilization of linear neutral delay-differential systems*, Math. Syst. Theory, 17 (1984) 97–133.

[9] J.N. CHIASSON AND E.B. LEE, *Control of multipass processes using 2-D system theory*, Proc. 24th CDC (1985), 1734–1737.

[10] R.F. CURTAIN AND K. GLOVER, *Robust stabilization of infinite-dimensional systems by finite dimensional controllers*, Syst. Contr. Lett., 7 (1986), 41 47.

[11] D. CHYUNG, *Application of time delays in output feedback control systems*, Proc. Johns Hopkins Conf. on Information Science & Systems (1982), 41–42.

[12] R.A. DECARLO, J. MURRAY, AND R. SAEKS, *Multivariable Nyquist Theory*, In Contro., 25, (1977), 657–675.

[13] M.C. DELFOUR, *Linear optimal Control of systems with state and control variable delays*, Automatica 20 (1984) 69–77.

[14] Y.A. FIAGBEDZI AND A.E. PEARSON, *A multi-state reduction technique for feedback stabilizing time-lag systems*, submitted for publication.

[15] D.S. FLAMM, *Control of delay systems for sensitivity*, Ph.D. Thesis, Department EECS, MIT (1986).

[16] C. FOIAS, A. TANNENBAUM, AND G. ZAMES, *Weighted sensitivity minimization for delay systems*, IEEE Trans. Automat. Contr. AC–31 (1986), 763–766.

[17] S. GIBSON, *Linear quadratic optimal control of hereditary differential systems; infinite dimensional Riccati equations and numerical approximations*, SIAM J. Contr. and Optim., 21 (1983), 95–139.

[18] K. GLOVER, *All optimal Hankel-norm approximations of linear multivariable systems and their L^∞-error bounds*, Int. J. Contrl. vol. 39 (1984) 1115–1193.

[19] K. GLOVER, *Robust stabilization of multivariable linear systems: Relations to approximation*, Int. J. Contr. 43 (1986) 741–766.

[20] D. GOODMAN, *Some stability properties of two-dimensional linear shift-invariant digital filter*, IEEE Trans. Circuits Syst., CAS-24 (1977) 201–208.

[21] J.K. HALE, *Functional Differential Equations*, Springer-Verlag, New York (1982).

[22] T.S. HUANG, *Stability of two-dimensional recursive filters*, IEEE Trans. Audio Electroacoust., Au-20 (1972) 158–163.

[23] T. KACZOREK, *Two-Dimensional Linear Systems*, Lecture Notes in Control & Information Sciences, 68, Springer-Verlag, Heidelberg (1985).

[24] T. KAITATH, *Linear Systems*, Prentice Hall, Englewood Cliffs, NJ: (1980).

[25] E.W. KAMEN, *Linear systems with commensurate time delays: Stability and stabilization independent of delay*, IEEE Trans. Automat. Contr. AC-27 (1982) 367–375.

[26] E.W. KAMEN, P.P. KHARGONEKAR, AND A. TANNENBAUM, *Stabilization of time-delay systems using finite-dimensional compensators*, IEEE Trans. Automat. Contr. AC-30 (1985) 75–78.

[27] E.W. KAMEN, P.P. KHARGONEKAR, AND A. TANNENBAUM, *Pointwise stability and feedback control of linear systems with noncommensurate time delays*, Acta Appl. Math., 2 (1984) 159–184.

[28] E.B. LEE, *Generalized Quadratic optimal controllers for linear hereditary systems*, IEEE Trans. Automat. Contr. AC-25 (1980), 528–531.

[29] E.B. LEE AND A.W. OLBROT, *On reachability over polynomial rings and a related genericity problem*, Int. J. Syst. Sci., 13 (1982), 109–113.

[30] E.B. LEE AND L. MARKUS, *Foundations of Optimal Control Theory*, John Wiley & Sons Inc. New York (1967).

[31] E.B. LEE, S.H. ŻAK AND S.D. BRIERLEY, *Stabilization of generalized linear systems via the algebraic Riccati equation*, Int. J. Contr. 39 (1984) 1025–1041.

[32] E.B. LEE AND S.H. ŻAK, *On spectrum placement for linear time invariant delays systems*, IEEE Trans. Automat. Contr. AC-27 (1982), 446–449.

[33] E.B. LEE AND W.-S. LU, *Coefficient assignability for linear systems with delays*, IEEE Trans Automat. Contr., AC-29 (1984), 1048–1052.

[34] E.B. LEE AND W.-S. LU, *Stabilization of two-dimensional systems*, IEEE Trans. Automat. Contr. AC-30 (1985), 409–411.

[35] E.B. LEE AND Y.C. YOU, *Optimal Syntheses for infinite-dimensional linear delayed output systems: A semicausality approach*, Appl. Math and Optim. 19 (1989), 113–136.

[36] J.H. LODGE AND M.M. FAHMY, *Stability and overflow oscillations in 2-D state-space digital filters*, IEEE Acoust., Speech, Signal Processing, ASSP-29 (1981) 1161–1171.

[37] W.-S. LU AND E.B. LEE, *Stability analysis for two-dimensional systems*, IEEE Trans. Circuits Syst., CAS-30 (1983), 455–461.

[38] W.-S. LU AND E.B. LEE, *Stability analysis for two-dimensional systems via a Lyapunov approach*, IEEE Trans. Circuits Syst., CAS-32 (1985), 61–68.

[39] W.-S. LU, E.B. LEE, AND S.H. . ŻAK, *On the stabilization of linear neutral delay-differential systems*, IEEE Trans. Automat. Contr., AC-31 (1986), 65–67.

[40] A. MANITIUS AND A.W. OLBROT, *Finite spectrum assignment for systems with delays*, IEEE Trans. Autot. Contr., AC–24 (1979), 541–553.

[41] A. MANITIUS AND R. TRIGGIANI, *Function space controllability of linear retarded systems: A derivation from abstract operator conditions*, SIAM J. Contr. Opt., 16 (1978), 699–645.

[42] A. MANITIUS AND H. TRAN, *Computation of Closed-loop eigenvalues associated with the optimal regulator problem for functional differential equations*, IEEE Trans. on Automat. Contr. AC-30 (1985), 1245–1248.

[43] A.I. MARKUSHEVICH, *Theory of function of a Complex Variable*, III, Prentice Hall (1967).

[44] S.N. MERGELYAN, *Uniform approximations of functions of a complex variable*, Uspehi Math. Nauk. (N.S.), American Math. Sco., 7 (1952), 31–122.

[45] A.S. MORSE, *Ring model for delay-differential systems*, Automatica 12 (1976), 529–531.

[46] C.N. NETT, *The fractional representation approach to robust linear feedback design: A self-contained exposition*, M.S. Thesis, Department of ECSE, RPI, Troy, New York (1984).

[47] C.N. NETT, C.A. JACOBSON, M.J. BALAS, *A connection between state-space and doubly coprime fractional representations*, IEEE Trans. Automat. Contr. AC-29 (1984), 831–832.

[48] A. OLBROT, *Stabilizability, detectability and spectrum assignment for linear autonomous systems with general time delays*, IEEE Trans on Automat Contr., AC-23 (1978), 887–890.

[49] L. PANDOLFI, *On feedback stabilization of functional differential equations*, Boll. Unione Math. Italiana, 12 (1975), 626–635.

[50] P.N. PARASKEVOPOULOS, *Eigenvalue assignment at linear multivariable 2-D systems*, Proc. IEE, 126 (1979), 1204–1208.

[51] M.S. PIEKARSKI, *Algebraic characterization of matrices whose multivariable characteristic polynomial is Hurwitzian*, Proc. Int. Symp. Operator Theory, Lubbock, TX (Aug., 1977), 121 126.

[52] V.R. RAMAN AND R.-W. LIU, *A constructive algorithm for the complete set of compensators for two-dimensional feedback system design*, IEEE Trans. Automat. Contr. AC-31 (1986) 166-170.

[53] R.P. ROESSER, *A discrete state-space model for linear image processing*, IEEE Trans Automat. Contr. AC-20 (1975), 1–10.

[54] M.G. SAFANOV AND M.S. VERMA, *L^∞-optimization and Hankel approximation*, IEEE Trans. Automat. Contr. AC–30 (1985), 279–280.

[55] D. SALAMON, *On Control and observation of neutral systems*, Ph.D. Thesis, University of Bremen, West Germany (1982).

[56] M. ŠEBEK, *Asymptotic tracking in 2–D and delay–differential systems*, Automatica (IFAC), 24 (1988), 711–713.

[57] J.L. SHANKS, S. TREITEL, AND J.H. JUSTIC, *Stability and synthesis of two-dimensional recursive filters*, IEEE Trans Audio Electroacoust., AU-20 (1972), 115–128.

[58] E.D. SONTAG, *Linear systems over commutative rings: A survey*, Ricerche di Automatica, 7 (1976), 1–34.

[59] M.W. SPONG AND T.J. TARN, *On the spectral controllability of delay-differential equations*, IEEE Trans Automat. Contr, AC-26, 527–528 (1981).

[60] M.G. STRINTZIS, *Test of stability of multidimensional filters*, IEEE Trans. Circuits Syst., CAS-24 (1977), 432–437.

[61] G. TADMOR, *Trajectory stabilizing controls in hereditary linear systems*, SIAM J. Control. and Optim., 26 (1988) 138–154.

[62] K. WATANABE, *Finite spectrum assignment and observer for multivariable systems with commensurated delays*, IEEE Trans. Automat. Contr. AC-31, 543–550 (1986).

[63] K. WATANABE, M. ITO, AND M. KENEKO, *Finite spectrum assignment problem of systems with multiple commensurate delays in states and control*, Int. J. Contr., 39 (1984), 1073–1082.

[64] S.Y. ZHANG, *On the coefficient assignment of 2-D systems*, IEEE Trans. Automat. Contr. AC–30 (1985), 1248–1251.

[65] K. ZHOU AND P.P. KHARGONEKAR, *On the weighted sensitivity minimization problem for delay systems, Syst. and Contr. Lett., 8 (1987), 307–312.*

CONTROL STRUCTURE SELECTION: ISSUES AND A NEW METHODOLOGY

MANFRED MORARI* AND JAY H. LEE*

Abstract. A unified approach to the problem of control structure selection for centralized and decentralized controllers is presented. The approach is based on the structured singular value (μ) theory and hence incorporates many important practical issues (e.g. model uncertainty, restrictions on the controller structure). Using a high-purity distillation column as an example, it is demonstrated that the proposed control structure selection method can be successfully applied to a realistic system yielding a control structure and a controller with robust performance.

Introduction. When designing a control system, an engineer is faced with the problem of control structure selection before beginning the actual design of the controller. "Control structure selection" refers to the decisions regarding whether or not a specific available actuator will be influenced by a specific sensor through the controller. These decisions are of paramount importance since they determine some of the inherent properties of the system (*e.g.*, nonminimum-phase behavior, actuator constraints, sensitivity to model uncertainty) limiting the achievable performance of the closed-loop system. The control structure decisions are concerned mainly with the following two issues:

1. Will the controller be centralized or decentralized (*i.e.*, restricted to diagonal/block diagonal structure)?
2. Among the available actuators and sensors, which will be used by the control system, and, in the case of a decentralized controller, how will they be paired?

The choice between a centralized and a decentralized controller is governed by practical considerations. Decentralized controllers offer the advantages of hardware simplicity, fewer tuning parameters and more easily achieved sensor/actuator failure tolerance. However, due to the restrictions on the freedom of the controller structure, the achievable performance of a decentralized controller is often worse than (or at best equal to) that of the best centralized controller. The problem of sensor/actuator selection arises when physical constraints or design considerations limit the number of sensors/actuators. This problem as well as the sensor/actuator pairing problem for a decentralized controller, are combinatorial in nature and can be quite formidable for a large-scale system: An efficient screening method is clearly required.

In this paper, we present a unified approach to the problem of control structure selection. We will consider the following two problems:

1. Actuator/sensor selection for a centralized controller.
2. Actuator/sensor selection and input-output pairing for decentralized controllers.

*Chemical Engineering 206–41, California Institute of Technology, Pasadena, California 91125

Using a high-purity distillation column as an example, we demonstrate that the proposed methods can be successfully applied to a realistic system to yield a control structure and a controller satisfying the performance specifications, in the face of many practical issues (*e.g.*, model uncertainties, restrictions on the controller structure). Throughout this paper, we assume that all plants are described by linear, time-invariant, *stable* transfer function matrices.

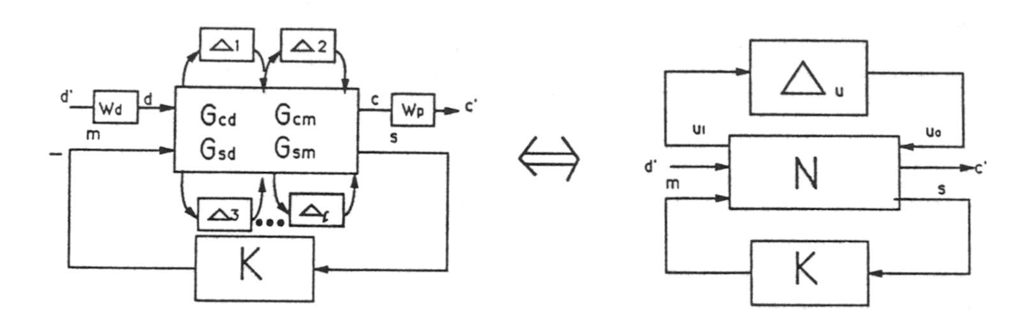

$$\Delta_u = \left\{ \Delta_u : \Delta_u = \begin{bmatrix} \Delta_1 & & \\ & \ddots & \\ & & \Delta_\ell \end{bmatrix} ; \overline{\sigma}(\Delta_i(j\omega)) \leq 1 \quad \forall\omega, \; i = 1, \ldots, \ell; \right.$$

$$\left. \mathcal{F}(\Delta u, N) \in \mathcal{R}_s\{\dim(C') + \dim(s)\} \times \{\dim(d') + \dim(m)\} \right\}$$

$$\mathcal{F}(\Delta u, N) = N_{22} + N_{21}\Delta u(I - N_{22}\Delta u)^{-1}N_{12}$$

$$\mathcal{R}_s^{m_i \times n_i}(s) = \text{Ring of } m_i \times n_i \text{ stable rational transfer function matrices}$$

Figure 1: General Robust Performance Problem

2. Problem Formulation. Fig. 1 presents the general block diagram of a feedback control system. d' and c' denote the weighted disturbance vector and the weighted controlled variable vector, respectively. The uncertainties of the system are often modelled as induced-2-norm bounded perturbations (Δ_i's) to the nominal frequency response model. They are normalized through frequency-dependent weights such that $\overline{\sigma}(\Delta_i(j\omega)) \leq 1 \quad \forall\omega \; \forall i$. It is assumed that all perturbed systems have the same RHP poles (hence, stable in this study) as the nominal system.

Two popular performance measures are the H_2- and H_∞-norms of the closed-loop transfer function matrix. The H_2- and H_∞-norms are defined as follows:

$$\|G\|_2 = \left(\frac{1}{2\pi}\int_{-\infty}^{\infty} \text{trace}\,[G(j\omega)^* G(j\omega)]\, d\omega\right)^{\frac{1}{2}}$$

$$\|G\|_\infty = \sup_\omega \bar{\sigma}(G(j\omega)) \tag{1}$$

where $\bar{\sigma}$ denotes the maximum singular value. The H_2-norm measures the weighted integral square error of the output for a fixed set of inputs. The H_∞-norm measures the worst-possible weighted integral square of the output for a class of norm-bounded inputs. We will adopt the H_∞-norm as the performance measure throughout this paper.

DEFINITION 1. ROBUST PERFORMANCE. *The closed loop system shown in Fig. 1 is said to achieve "robust performance" if the closed-loop system is nominally stable (i.e., stable with $\Delta_u = 0$) and*

$$\max_{\Delta_u \in \mathbf{\Delta}_u} \|F_{c'd'}(K, \Delta_u)\|_\infty < 1 \tag{2}$$

where $F_{c'd'}(K, \Delta_u)$ is the closed-loop transfer function matrix from d' to c'.

Remark: In the time-domain, robust performance guarantees that $\|c'(t)\|_2 < 1$ for any plant within the set defined by the uncertainty description and any exogenous signal $d'(t)$ such that $\|d'(t)\|_2 \leq 1$.

DEFINITION 2. STRUCTURED SINGULAR VALUE (μ). *Let $M \in \mathcal{C}^{n\times m}$ and define the set Δ as follows:*

$$\Delta = \left\{ \tilde{\Delta} : \tilde{\Delta} = \begin{bmatrix} \Delta_1 & & \\ & \ddots & \\ & & \Delta_k \end{bmatrix} ; \Delta_i \in \mathcal{C}^{m_i \times n_i}; \sum_{i=1}^{k} m_i = m, \sum_{i=1}^{k} n_i = n \right\} \tag{3}$$

Then $\mu_\Delta(M)$ (μ of M with respect to the uncertainty structure Δ) is defined as

$$\mu_\Delta(M) = \begin{cases} \left[\min_{\tilde{\Delta}} \left\{\bar{\sigma}(\tilde{\Delta}) : det(I + M\tilde{\Delta}) = 0,\ \tilde{\Delta} \in \Delta\right\}\right]^{-1} \\ 0 \text{ if } \exists \text{ no } \tilde{\Delta} \in \Delta \text{ such that } det(I + M\tilde{\Delta}) = 0 \end{cases} \tag{4}$$

THEOREM 1. *(referring to Figure 1),*

$$\max_{\Delta_u \in \mathbf{\Delta}_u} \|F_{c'd'}(K, \Delta_u)\|_\infty < 1 \tag{5}$$

if and only if

$$\sup_\omega \mu_{\begin{bmatrix} \Delta_u^* & \\ & \Delta_p \end{bmatrix}} (N_{11}(j\omega) - N_{12}K(I - N_{22}K)^{-1}N_{21}(j\omega)) < 1 \tag{6}$$

where

$$\Delta_u^* = \left\{ \Delta_u^* : \Delta_u^* = \begin{bmatrix} \Delta_1 & & \\ & \ddots & \\ & & \Delta_\ell \end{bmatrix} ; \Delta_i \in \mathcal{C}^{m_i \times n_i}; \sum_{i=1}^{\ell} m_i = dim\{u_o\}, \sum_{i=1}^{\ell} n_i = dim\{u_i\} \right\}$$

(7)

$$\Delta_p = \left\{ \Delta : \Delta \in \mathcal{C}^{dim\{d'\} \times dim\{c'\}} \right\}$$

Several approaches are available toward the problem of control structure selection. The most immediate and straightforward approach is to define the objective as minimizing the following quantity over all available control structure candidates:

(8) $$\min_K \sup_\omega \mu_{\begin{bmatrix} \Delta_u^* & \\ & \Delta_p \end{bmatrix}} (N_{11}(j\omega) - N_{12}K(I - N_{22}K)^{-1}N_{21}(j\omega))$$

where K is a rational transfer function matrix which may be restricted to a diagonal/block diagonal structure if the controller is to be decentralized. In other words, we synthesize K achieving the minimum μ for each control structure candidate, then compare the μ's and select the control structure and the corresponding controller achieving the lowest μ. There are, however, significant theoretical and practical drawbacks to this formulation: The minimization problem expressed through Eq.8 is computationally formidable. The algorithm available currently (*i.e.*, μ-synthesis) is unreliable and requires large CPU-time. In addition, it does not allow restriction of the controller to a decentralized structure. In view of the combinatorial nature of the problem, the approach is clearly not feasible for large-scale problems.

Therefore, we adopt another approach and redefine the objective as follows:

Find *a* control structure for which

(9) $$\min_K \sup_\omega \mu_{\begin{bmatrix} \Delta_u^* & \\ & \Delta_p \end{bmatrix}} (N_{11}(j\omega) - N_{12}K(I - N_{22})^{-1}N_{21}(j\omega)) < 1$$

Since μ is not a direct measure of robust performance, but instead is a measure of how much the magnitudes of the uncertainties and the performance weights can (should) be scaled simultaneously while preserving (to achieve) robust performance, the objective (9) seems more fitting to the practical problem than the minimization of μ. Of course, there may exist more than one control structure satisfying (9). Then, we have the freedom of tightening the performance specifications in order to find the structure which yields the most desirable performance.

In order for the approach to be of practical use, however, we need a way to establish the *existence* of a controller K meeting the condition

(10) $$\sup_\omega \mu_{\begin{bmatrix} \Delta_u^* & \\ & \Delta_p \end{bmatrix}} (N_{11}(j\omega) - N_{12}K(I - N_{22}K)^{-1}N_{21}(j\omega)) < 1$$

for a particular control structure, without having to synthesize the μ-optimal controller. Necessary *and* sufficient conditions for the existence of a controller (either centralized or decentralized) achieving robust performance, are not available at this time. Hence our approach is to develop "tight" necessary conditions and "tight" sufficient conditions. The necessary conditions can be used as screening tools to eliminate candidates for which no controller exists which achieves robust performance. Sufficient conditions should help the engineer locate a structure for which a controller with robust performance can be easily designed. In other words, with sufficient conditions, we would like to reach a compromise between μ-optimality and complexity of controller design.

In the next three sections, we state a series of necessary conditions and sufficient conditions for the existence of robustly performing centralized and decentralized controllers.

3. General Methodology. In this section, we develop a general control structure selection method which is designed to be both computationally simple and for which the subsequent design of a robustly performing controller is straightforward. We base our criteria on the robust performance norm bounds on the transfer function matrices that are relevant to the sensitivity and the robustness of the closed-loop system. Skogestad & Morari[4] showed a means for transforming the necessary and sufficient condition for robust performance in the μ-terminology into sufficient conditions in terms of norm bounds on desired transfer function matrices. The procedure for deriving the norm bound on a particular transfer function matrix L for robust performance can be summarized as follows:

1. Fig. 2(a) represents the general robust performance problem (see Theorem 1). The first step is to parametrize the controller K in terms of the transfer function matrix L on which the norm bound is desired, as shown in Fig. 2(b).
2. After combining the matrices N and J into R (Fig. 2(c)), the matrix L is treated as a real scalar parameter c_L times an uncertainty block Δ_L as shown in Fig. 2(d).
3. The norm bound on L guaranteeing robust performance is calculated via Theorem 2.

THEOREM 2. *Let $M \in \mathcal{C}^{n\times m}$ be written as*

$$M = R_{11} + R_{12}L(I - R_{22}L)^{-1}R_{21} \tag{11}$$

where

$$R_{11} \in \mathcal{C}^{n\times m}, R_{12} \in \mathcal{C}^{n\times p}, R_{21}, \in \mathcal{C}^{k\times m}, R_{22} \in \mathcal{C}^{k\times p} \text{ and } L \in \mathcal{C}^{p\times k} \tag{12}$$

Define

$$f(c_L) = \mu_{\begin{bmatrix} \Delta & \\ & \Delta_L \end{bmatrix}} \begin{bmatrix} R_{11} & R_{12} \\ c_L R_{21} & c_L R_{22} \end{bmatrix} \tag{13}$$

where

$$\Delta = \left\{ \Delta : \Delta = \begin{bmatrix} \Delta & & \\ & \ddots & \\ & & \Delta_\ell \end{bmatrix}; \sum_{i=1}^{\ell} m_i = m \quad \sum_{i=1}^{\ell} n_i = n, \quad \Delta_i \in \mathcal{C}^{m_i\times n_i} \right\}$$

(14)
$$\Delta_L = \{\Delta : \Delta \in \mathcal{C}^{p\times k}\}$$
$$c_L \in \Re^+$$

Assume

$$\mu_\Delta(R_{11}) < 1 \quad \textit{and} \quad \textit{det}(I - R_{22}L) \neq 0 \tag{15}$$

then

$$\mu_\Delta(M) < 1 \tag{16}$$

if

$$\bar{\sigma}(L) < c_L^* \tag{17}$$

where

$$c_L^* = \min_{\tilde{c}_L} \{\tilde{c}_L : f(\tilde{c}_L) = 1, \tilde{c}_L > 0\} \tag{18}$$

Theorem 2 implies that choosing the scaling factor c_L such that $f(c_L(\omega)) = 1$ $\forall\omega$ provides the frequency-by-frequency bound of $\bar{\sigma}(L)$ guaranteeing robust performance (assuming K is chosen such that the closed-loop system is nominally stable).

Remarks:

1. $f(c_L)$ is a non-decreasing function of c_L. Hence

$$f(c_L) < 1 \text{ for } 0 \leq c_L < c_L^* \tag{19}$$

   Thus c_L^* can be easily found through a simple search-procedure (*e.g.*, bisection method).

2. There may exist many sets of J, L parametrizing K. The norm bounds on different L's can be combined over different frequency ranges.
3. Condition (17) is only sufficient for robust performance as robust performance must be guaranteed for every L (as opposed to a particular L), which satisfies $0 \leq \bar{\sigma}(L)(\omega) < c_L^*(\omega)$. The requirement $\mu_\Delta(R_{11}) < 1$ requires $f(c_L) < 1$ for $c_L = 0$.
4. c_L is the tightest norm bound of L in the sense that, for each $c_L \geq c_L^*$, there exists at least one L such that

$$\bar{\sigma}(L) = c_L \text{ and } f(c_L) \geq 1 \tag{20}$$

5. Meeting the bounds does *not* guarantee that the closed-loop system is nominally stable.

We base our control structure selection criteria upon these norm bounds: Specifically, one should select the control structure yielding the norm bounds which are easiest to meet. In order to provide a meaningful basis for selection the norm bounds should be evaluated for those operators with direct relevance to the sensitivity and the robustness of the closed-loop system.

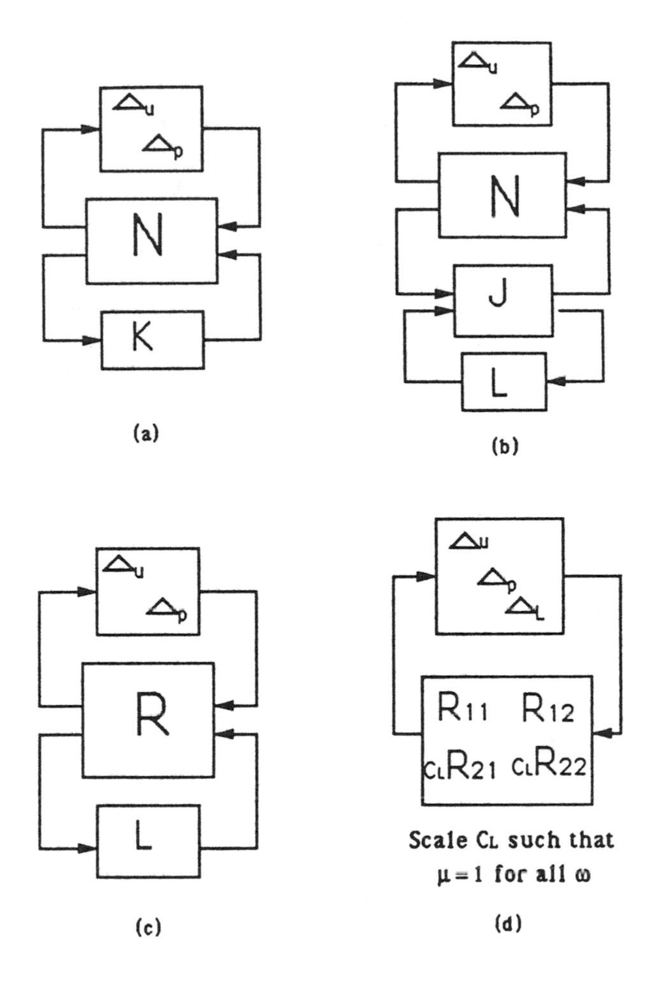

Figure 2: Derivation of Norm Bound on L for Robust Performance

The obvious drawback of the method is that the bounds are only *sufficient* for robust performance. Hence the bounds cannot be used (or should be used only with extreme caution) to eliminate any candidate. In addition, the conservativeness may be so severe that none of the available control structures yields norm bounds which can be met, even though control structures with robustly performing controllers exist. As will be shown in the next section, the conservativeness can often be reduced dramatically by restricting the controller to a specific form.

This "norm-bound method" will be the basis for the further developments in the next sections. It will be shown that, together with some necessary conditions for robust performance developed under a specific uncertainty assumption, it forms an effective method for finding a control structure and a controller leading to robust performance.

4. Actuator/Sensor Selection for a Centralized Controller. In this section, we consider the problem of control structure selection for a centralized controller: The actuators and sensors are to be selected among the available candidates. We adopt the particular uncertainty description shown in Fig. 3. The uncertainty blocks Δ_I and Δ_O are best viewed as the uncertainties on the actuators and sensors respectively; however, they may also account for other structured uncertainties occurring within the plant. They may occur with either a full or a block diagonal structure depending on the system. Although we cannot claim that the uncertainty description is adequate for all cases, there is little doubt that these are important types of uncertainties which are part of every physical system. Next, we look at the robust performance problem of the closed-loop system in Fig. 3.

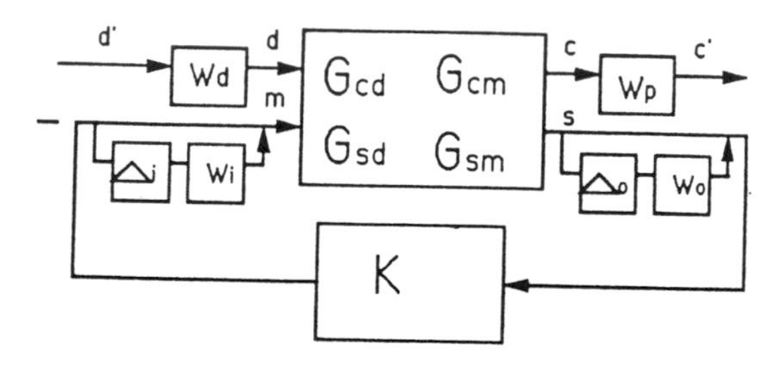

Figure 3: Uncertainties on the Actuator and Sensors

4.1 Robust Performance Condition. For the closed-loop system in Fig.3, the robust performance condition can be stated as follows:

$$\text{Robust Performance} \Leftrightarrow (1) \sup_{\omega} \mu_{\bar{\Delta}}(M(j\omega)) < 1$$

$$\text{where } \bar{\Delta} = \begin{bmatrix} \Delta_I & & \\ & \Delta_O & \\ & & \Delta_p \end{bmatrix} \tag{21}$$

(2)The mapping of Nyquist D-contour under $\det(I + G_{sm}K(j\omega))$ does not encircle the origin.

where

$$M(s) = \begin{bmatrix} KG_{sm}(I+KG_{sm})^{-1}W_I & K(I+G_{sm}K)^{-1}W_O & K(I+G_{sm}K)^{-1}G_{sd}W_d \\ G_{sm}(I+KG_{sm})^{-1}W_I & G_{sm}K(I+G_{sm}K)^{-1}W_O & (I+G_{sm}K)^{-1}G_{sd}W_d \\ W_pG_{cm}(I+KG_{sm})^{-1}W_I & W_pG_{cm}K(I+G_{sm}K)^{-1}W_O & W_p[G_{cd}-G_{cm}K(I+G_{sm}K)^{-1}G_{sd}]W_d \end{bmatrix} \quad (22)$$

4.2 Robust Performance for a Controller With Integral Action. Controllers with integral action are used almost always in the chemical process industries for good low-frequency performance. When the controlled variables (also called "primary variables) are measured and the controller includes integral action, then offset-free performance (*i.e.*, asymptotically vanishing error for asymptotically constant external inputs) is achieved even in the presence of modelling error. However, for systems where only "secondary variables" are measured, as we will show next, integral action is not always desirable because the offset-free response of the secondary variables can lead to offsets in the primary variables. Then, the selection of the secondary measurements becomes crucial to ensure that the integral control of the secondary variables implies the control of the primary variables within the specification.

To aid the selection process we develop a necessary and sufficient condition for robust performance *at the steady state*, for controllers with integral action. Because the condition pertains only to the steady state, it is a *necessary* condition for robust performance. Thus it can be a valuable screening tool to eliminate undesirable control structure candidates. $T(0) = G_{sm}K(I+G_{sm}K)^{-1}(0) = I$ and $S(0) = (I+G_{sm}K)^{-1}(0) = 0$ when K contains integrator(s), and (22) becomes

$$M(0) = \begin{bmatrix} W_I(0) & G_{sm}^{-1}W_O(0) & G_{sm}^{-1}G_{sd}W_d(0) \\ 0 & W_O(0) & 0 \\ 0 & W_pG_{cm}G_{sm}^{-1}W_O(0) & W_p[G_{cd}-G_{cm}G_{sm}^{-1}G_{sd}]W_d(0) \end{bmatrix} \quad (23)$$

THEOREM 3. *Suppose that W_p and W_d do not contain any pole at the origin. Then*

$$\mu_{\tilde{\Delta}}(M(0)) < 1$$

if and only if

$$\mu_{\Delta_I}(W_I(0)) < 1, \mu_{\Delta_O}(W_O(0)) < 1 \text{ and } \bar{\sigma}[W_p(G_{cd}-G_{cm}G_{sm}^{-1}G_{sd})W_d(0)] < 1 \quad (24)$$

Remarks:

1. The condition (24) is necessary and sufficient for robust performance at the *steady-state*, hence it is a *necessary* condition for robust performance.
2. The condition (24) is *necessary* for existence of a bound c_S^* of form (17) (*i.e.*, $\bar{\sigma}(S(0)) < c_S^*(0)$), which guarantees robust performance (condition

$\bar{\sigma}(S(0)) < c_S^*(0)$ implies that the closed-loop system must achieve robust performance at $\omega = 0$ with $S(0) = 0$).

3. The condition $\bar{\sigma}[W_p(G_{cd} - G_{cm}G_{sm}^{-1}G_{sd})W_d(0)] < 1$ is always satisfied if the measured variables (s) include the controlled variables (c) and W_p and W_d do not contain any pole at the origin.
4. For cases where W_p and/or W_d contain one or more poles at the origin, an equivalent condition to (24) can be obtained by evaluating $\lim_{s\to 0} \mu_{\tilde{\Delta}}(M(s))$.

The two conditions, $\mu_{\Delta_I}(W_I) < 1$ and $\mu_{\Delta_O}(W_O) < 1$, are necessary for robust stability and sensor/actuator selection do not usually affect them. Hence, the control structure should be chosen such that $\bar{\sigma}[W_p(G_{cd} - G_{cm}G_{sm}^{-1}G_{sd})W_d(0)] < 1$.

4.3 Robust Performance Norm Bounds on Sensitivity and Complementary Sensitivity Functions. We apply the procedure described earlier to obtain norm bounds on the sensitivity (S) and complementary sensitivity (T) matrices for robust performance. One can easily parametrize K in terms of T as shown in Fig. 4 and obtain the diagrams shown in Fig. 5(a). For the sensitivity function matrix, one obtains an equivalent diagram by using the fact that $S = I - T$ (see Fig. 5(b)).

The motivation behind choosing the sensitivity and complementary sensitivity functions is that, often in practice, the design and tuning of the controller is based entirely on the performance of the secondary variables (*i.e.*, sensitivity and complementary sensitivity functions). It is desirable to select measurements so as to allow such a simple design procedure by providing the engineer with the norm bounds on the sensitivity and complementary sensitivity function matrices for robust performance of the primary variable. Using the method presented in Sec. 2, one can easily derive $c_S^*(\omega)$ and $c_T^*(\omega)$ such that robust performance of the primary variables is guaranteed if either $\bar{\sigma}(S(j\omega)) < c_S^*(\omega)$ or $\bar{\sigma}(T(j\omega)) < c_T^*(\omega)$ at each ω for all ω.

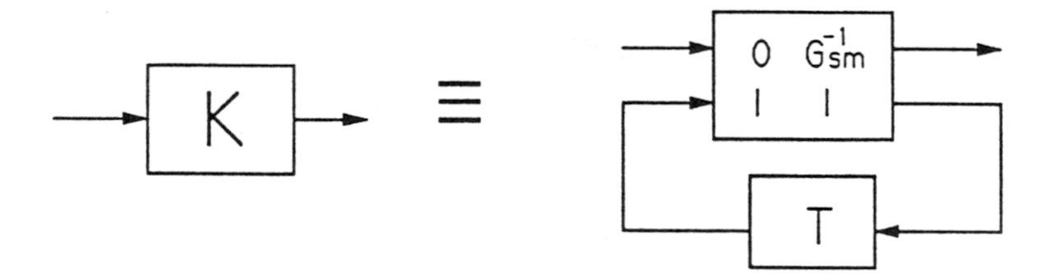

Figure 4: Parametrization of K in terms of T.

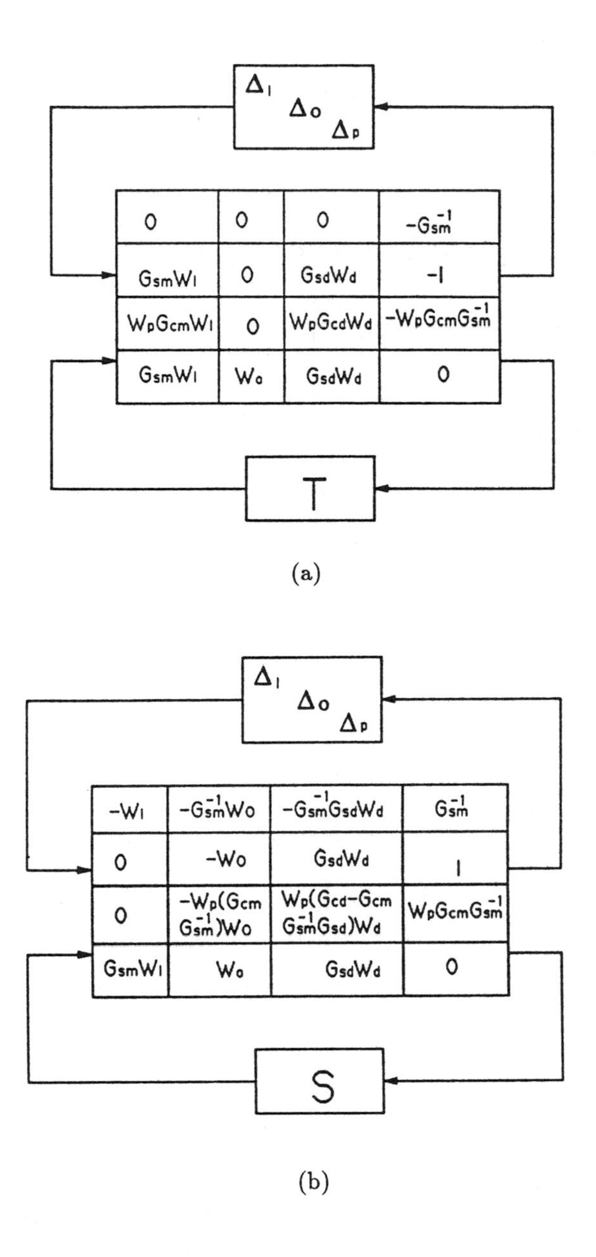

Figure 5: Derivation of norm bounds of T and S.

It is important to notice that the derived norm bounds are only sufficient for robust performance. Taking Δ_T and Δ_S as full blocks, the bounds must guarantee robust performance for every $K \in \mathcal{K}_a$ where

$$(25)\quad \mathcal{K}_a = \left\{\tilde{K} : \bar{\sigma}[T(\tilde{K})(j\omega)] < c_T^*(\omega) \text{ or } \bar{\sigma}[S(\tilde{K})(j\omega)] < c_S^*(\omega) \text{ at each } \omega,\ \forall\omega\right\}$$

Noting that $\mathcal{K}_b \subset \mathcal{K}_a$ where

$$\mathcal{K}_b = \left\{\tilde{K} : T(\tilde{K}) = \frac{g(s)}{1+g(s)}I,\quad S(\tilde{K}) = \frac{1}{1+g(s)}I \text{ and } \left|\frac{g(j\omega)}{1+g(j\omega)}\right| < c_T^*(\omega),\right.$$
$$\left.\left|\frac{1}{1+g(j\omega)}\right| < c_S^*(\omega) \text{ at each } \omega, \text{ for all } \omega,\quad g(s) \in \mathcal{R}_T(s)\right\}$$

(26)

$\mathcal{R}_T(s) =$ Set of all rational transfer functions $g(s)$ such that $\dfrac{g(s)}{1+g(s)}$ is stable,

we can reduce the conservativeness of the bound if we restrict K to the set $\mathcal{K}_{\text{inverse}}$, and consequently, Δ_T and Δ_S to δ_T and δ_S where

$$(28)\qquad \mathcal{K}_{\text{inverse}} = \left\{\tilde{K} : \tilde{K} = g(s)G_{sm}^{-1}(s),\quad g(s) \in \mathcal{R}_T(s)\right\}$$

$$(29)\qquad \delta_T = \{\delta I : \quad \delta \in \mathcal{C}\}$$

$$(30)\qquad \delta_S = \{\delta I : \quad \delta \in \mathcal{C}\}.$$

Here we assumed that G_{sm} has a stable inverse. Extension to the cases where G_{sm} is nonminimum phase is straightforward, and the interested readers are referred to Morari and Zafiriou[3].

4.4 Numerical Example: High-Purity Distillation. Experience indicates that, for high-purity distillation columns (shown in Fig. 6), the placement of the temperature sensors (T_a and T_b) is of vital importance for the achievable closed-loop performance. The control problem is shown in Fig. 7.

- *Controlled Variables.* The composition of the bottom and distillate products (x_B and y_D).
- *Manipulated Variables.* The reflux flowrate (L) and boilup (V).
- *Measured Variables.* Two tray temperatures (T_a and T_b). Because of the column symmetry with respect to the feed tray, it is assumed that the temperature sensors are to be placed symmetrically.
- *Disturbances.* Feed flowrate (F), feed composition (z_F), and uncompensated pressure variation (P). P can be thought of as a measurement error; it affects only the measured variables and not the controlled variables as a first-order approximation.
- *Model.* A linearized state-space model with 41 states, based on material balances[5].

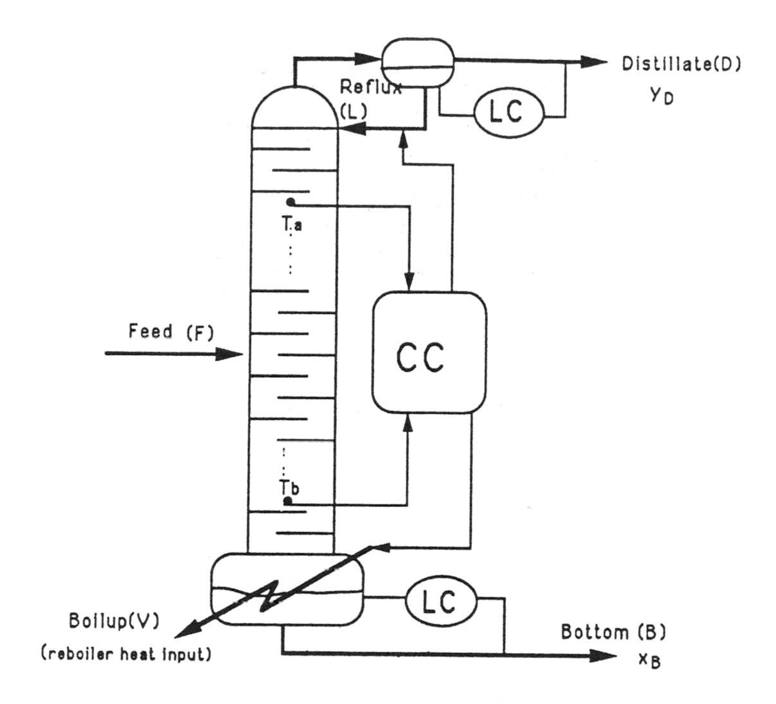

Number of stages: 40
Constant relative volatility: 1.5
Feed stage: 21
Feed flowrate: 1.0 (kmol/min)
Feed composition: 0.5
Feed condition: saturated liquid
Distillate flowrate: 0.5 (kmol/min)
Distillate composition: 0.99
Bottom flowrate: 0.5(kmol/min)
Bottom composition: .01
Reflux ratio (reflux/distillate): 5.464

Figure 6: High-Purity Distillation Column

- *Uncertainties.* The uncertainties on the manipulated variables are considered. This type of uncertainty has been shown to be most important for high-purity distillation columns[5]. W_I and Δ_I are chosen as follows (see Fig. 8(a)):

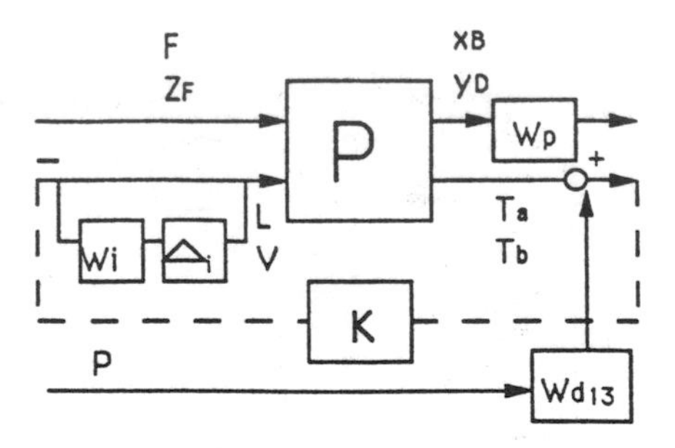

Figure 7: Distillation Column Control Problem

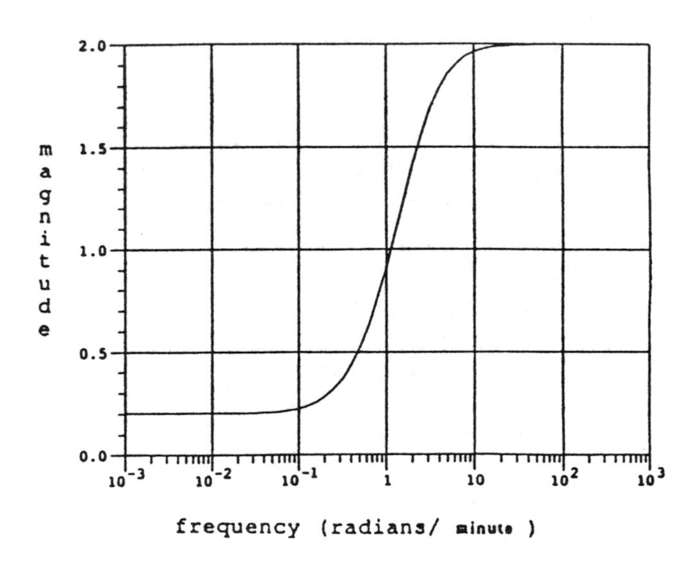

(a) $|W_I|$

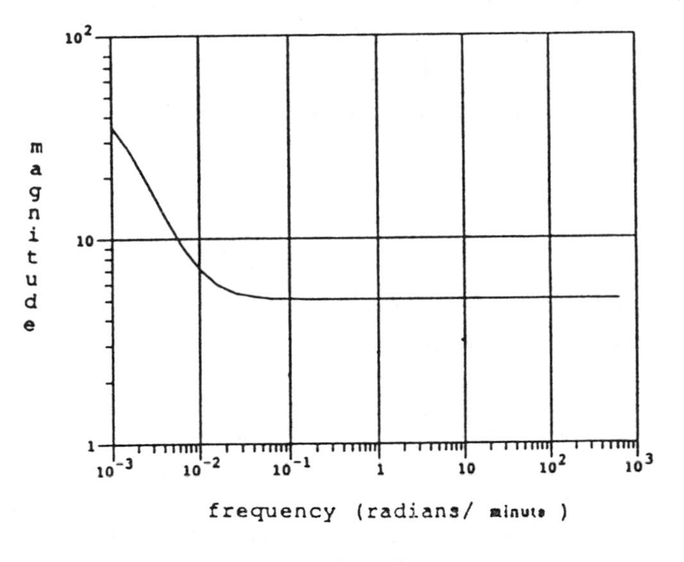

(b) $|W_p|$

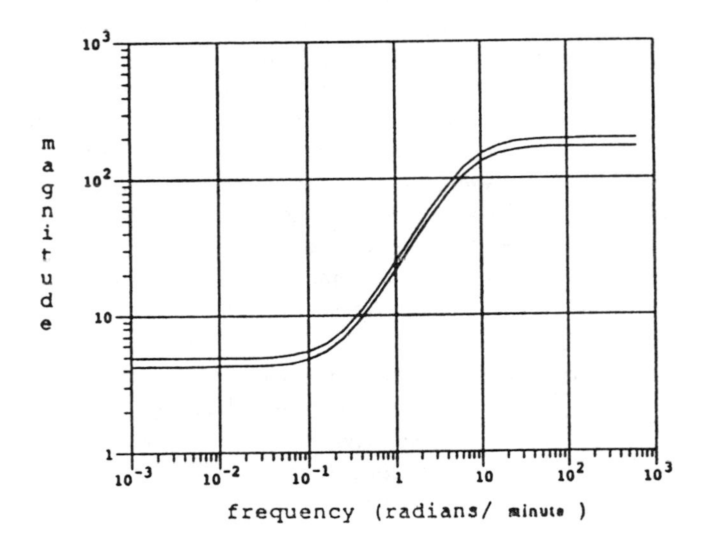

(c) $|W_{d_{13}}|$

Figure 8: Uncertainty, Disturbance, and Performance Weights.

$$W_I = 0.2\frac{5s+1}{0.5s+1}I \qquad \Delta_I = \begin{bmatrix} \Delta_1 & \\ & \Delta_2 \end{bmatrix}$$

- *Performance and Disturbance Weights.* The performance weight W_p and the disturbance weight W_d are chosen as follows (see Fig. 8),

$$W_p = 0.5\frac{100s+1}{10s+0.01}I$$
$$W_d = \left[I \quad 0.04\frac{5s+1}{0.125s+1}\begin{bmatrix} \left(\frac{dT}{dP}\right)_{T=T_a} \\ \left(\frac{dT}{dP}\right)_{T=T_b} \end{bmatrix} \right] \tag{32}$$

Note that this choice is equivalent to

$$W_p = I$$
$$W_d = 0.5\frac{100s+1}{10s+0.01}\left[I \quad 0.04\frac{5s+1}{0.125s+1}\begin{bmatrix} \left(\frac{dT}{dP}\right)_{T=T_a} \\ \left(\frac{dT}{dP}\right)_{T=T_b} \end{bmatrix} \right] \tag{33}$$

- *Steady-State Performance.* The necessary condition for robust performance at steady-state with an integral controller (Theorem 3) is applied first. The plot of $\bar{\sigma}[W_p \left\{ G_{cd} - G_{cm}G_{sm}^{-1}G_{sd} \right\} W_d(0)]$ vs. measurement location is shown in Fig. 9. The measurement set T_7/T_{35} is clearly the best choice; in fact, it is the only measurement set meeting the robust performance condition at steady-state.

  This result can be interpreted physically. There exists a direct static, linear (when the linearized model is used) relationship between the temperature of the reboiler (T_1) and the composition of the bottom product (x_B) on the one hand and the temperature of the condenser (T_{41}) and the composition of the distillate product (y_D) on the other. These relationships become less direct when the temperatures are measured at the trays (*e.g.*, T_{17}/T_{25}) far away from the condenser and the reboiler, and hence the integral control of these temperatures does not result in the offset-free response of the compositions. However, the temperatures measured at (or close to) the condenser and the reboiler have poor signal-to-noise ratio. Thus, placement of the temperature sensors involves compromise between these two factors. The measurement set T_7/T_{35} is apparently the best compromise between the "inferential error" and the sensitivity to the measurement error.

Next we must check that, for this particular measurement set, we can design a controller such that robust performance is achieved. This can easily be done by deriving the robust performance bounds on $|\frac{g(j\omega)}{1+g(j\omega)}|$ and $|\frac{1}{1+g(j\omega)}|$ (*i.e.*, the norm of sensitivity and complementary sensitivity function matrices) for the controller $K = g(s)G_{sm}^{-1}(s)$. To clarify the effects of the different sensor placements, we will look at the following three measurement sets:

$$s_1 = \begin{pmatrix} T_1 \\ T_{41} \end{pmatrix}, \quad s_2 = \begin{pmatrix} T_7 \\ T_{35} \end{pmatrix}, \quad s_3 = \begin{pmatrix} T_{17} \\ T_{25} \end{pmatrix}$$

- *Robust Performance Bounds on* $|\frac{g(j\omega)}{1+g(j\omega)}|$ *and* $|\frac{1}{1+g(j\omega)}|$. The bounds for all three measurements are shown in Fig. 10. As expected, the effect of the measurement error dominates for s_1 and the "inferential error" dominates for s_3. Only s_2 yields bounds which can be met. The following complementary sensitivity function matrix meets at least one of the respective bounds at every frequency (see Fig. 11):

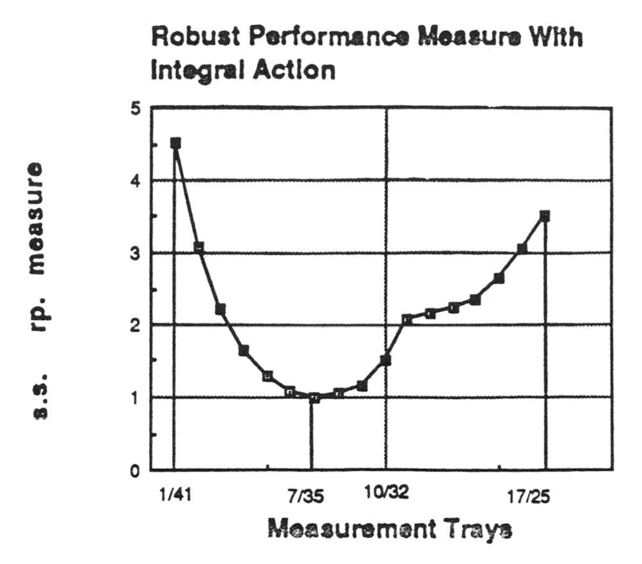

Figure 9: Steady State Robust Performance Measure With Integral Action

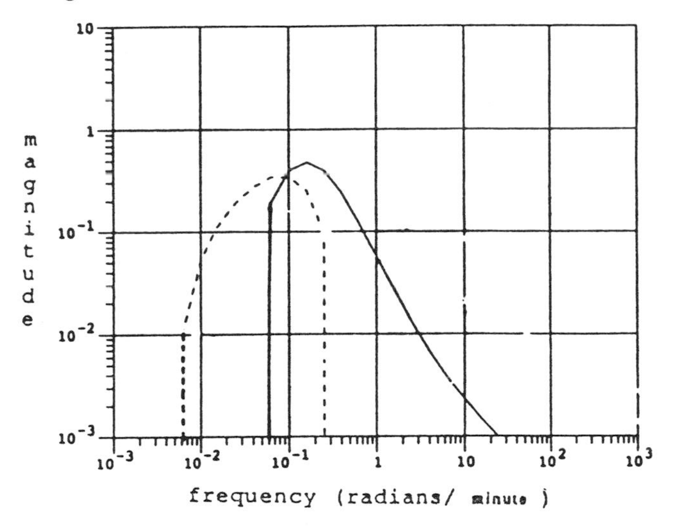

(a) s_1

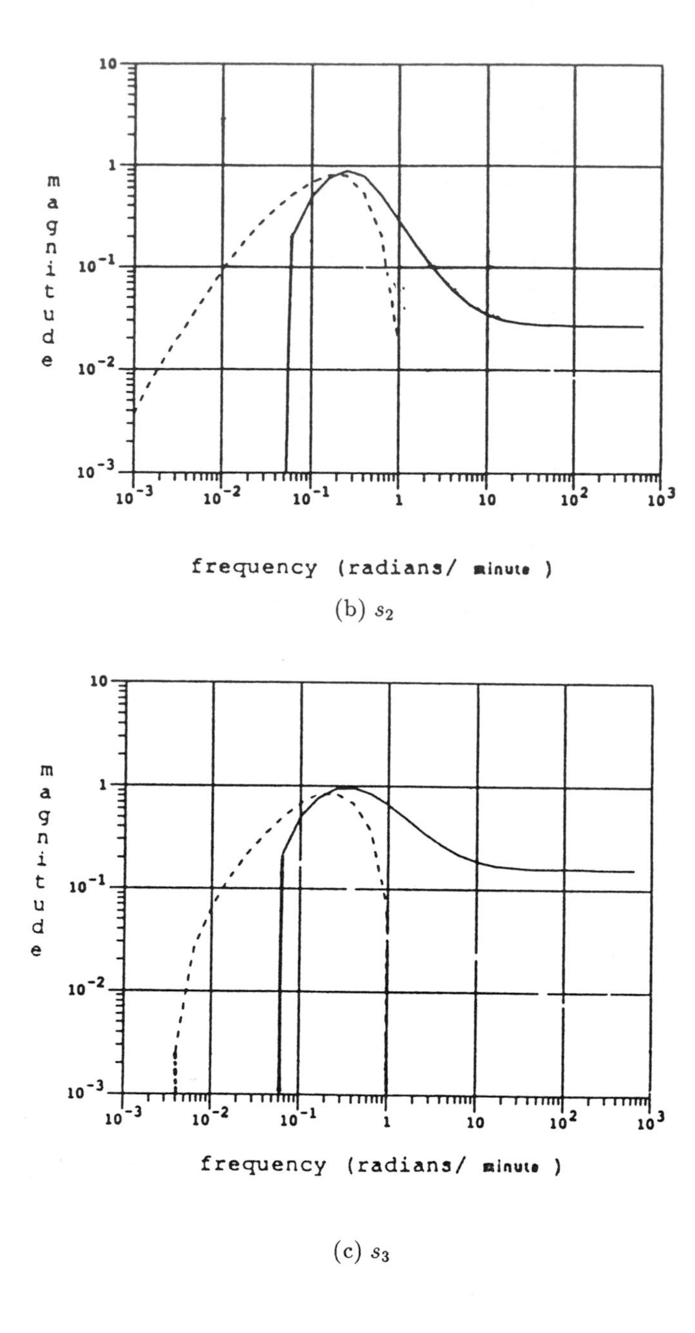

(b) s_2

(c) s_3

Figure 10: Robust Performance Bounds of $|\frac{g}{1+g}|$ and $|\frac{1}{1+g}|$.

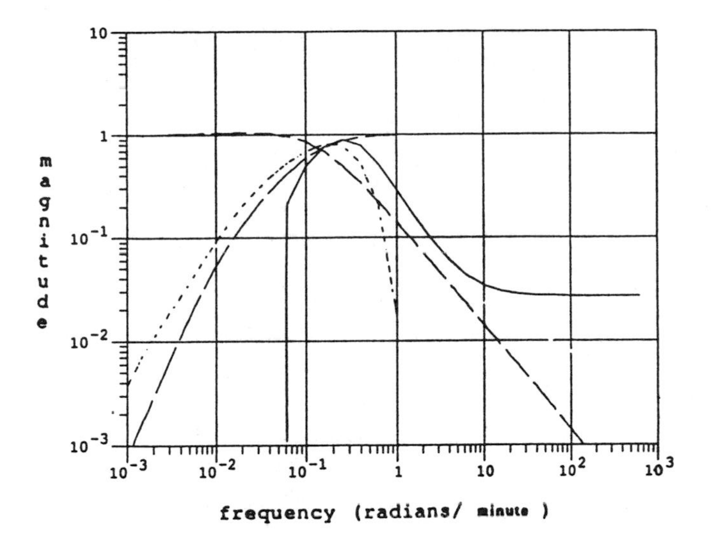

Figure 11: Robust Performance for s_2 with $\frac{g(s)}{1+g(s)} = \frac{107.5s+1}{750s^2+107.5s+1}$.

$$T = \frac{g(s)}{1+g(s)} I = \frac{107.5s+1}{750s^2+107.5s+1} I \tag{35}$$

The μ-plots for robust performance (Fig. 12) show that robust performance is achieved only for the measurement set s_2.

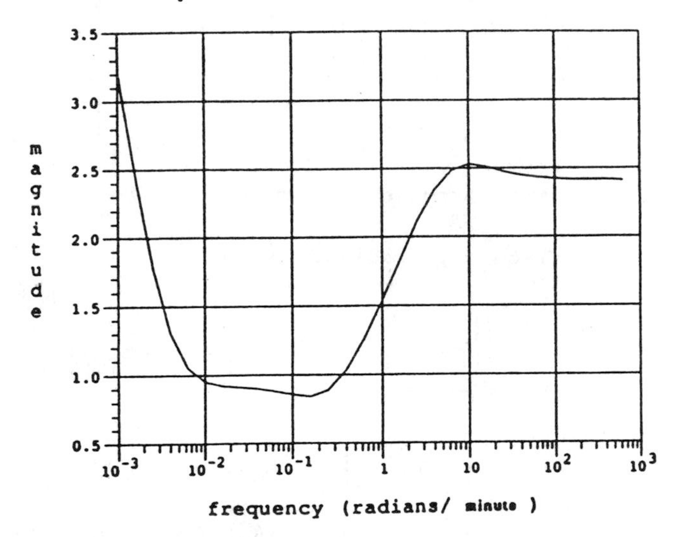

(a) s_1

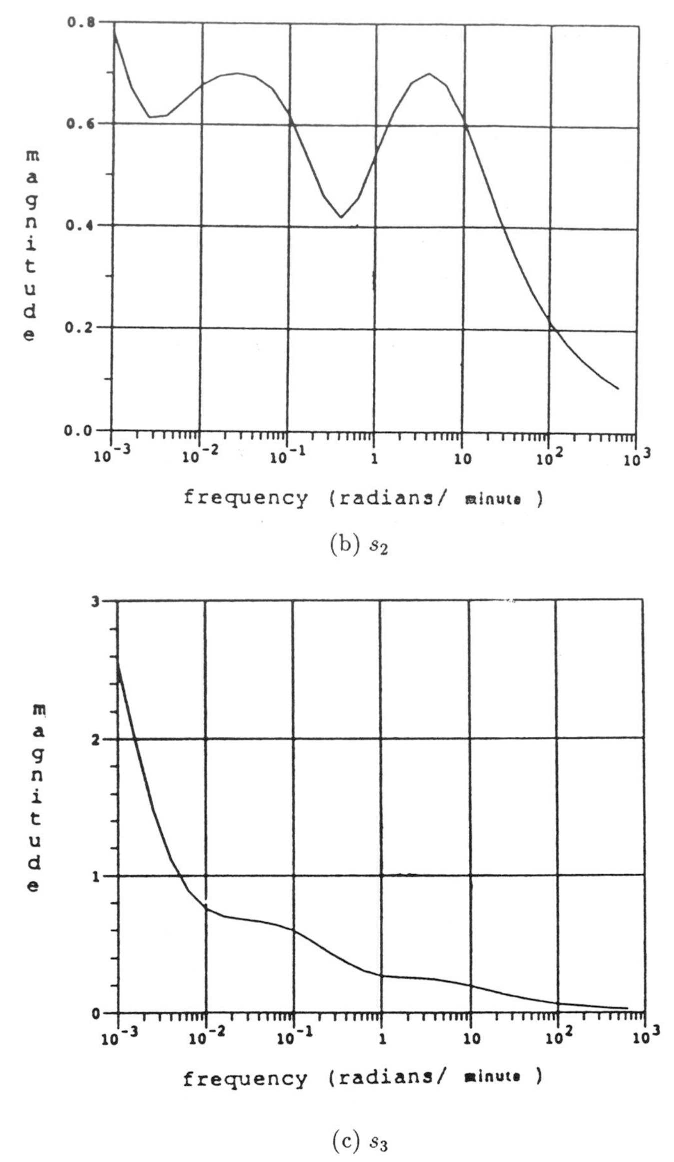

(b) s_2

(c) s_3

Figure 12: Structured Singular Values (μ) for Robust Performance With $\frac{g(s)}{1+g(s)} = \frac{107.5s+1}{750s^2+107.5s+1}$.

5. Actuator/Sensor Selection and Pairing for a Decentralized Controller.

Decentralized control means that the controller K in Fig. 1 is constrained to a diagonal/block diagonal structure. This controller structure constraint makes the problem of decentralized control structure selection much more difficult since no synthesis procedures for H_2/H_∞-optimal decentralized controllers are available. The problem consists of two parts: the actuator/sensor selection and the pairing of the selected actuators and sensors. To simplify the notation, we assume that the controller is to be completely decentralized (*i.e.*, diagonal) for the subsequent development in the section.

The approach we take is analogous to that taken previously for the case of the centralized controller. We develop some necessary conditions for stability, failure tolerance, and robust performance with an integral controller. These necessary conditions are based on steady-state information and can be effective in screening control structure candiates. Next, on the basis of an optimally scaled "small gain theorem," we develop a set of sufficient conditions for nominal stability of the closed-loop system with a decentralized controller. These frequency-domain conditions (called "μ-interaction measures") guarantee nominal stability of the closed-loop system and can also be effective in rating the degree of undesirable interactions among the individual loops for each control structure candidate. Finally, we adopt the previously discussed norm-bound approach to develop the sufficient conditions for the existence of K with robust performance, for further selection and design. This approach has many desirable features such as independent loop-design and incorporation of failure tolerance. When combined, these conditions form an effective control structure selection method for cases where the controller is restricted to a decentralized structure.

5.1. Stability and Failure Tolerance for a Controller with Integral Action.

Consider the block diagram in Fig. 13.

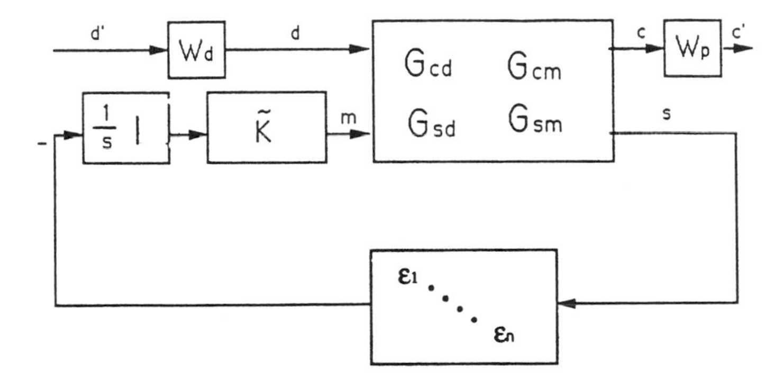

Figure 13: Definition of Decentralized Integral Controllability.

DEFINITION 3. DECENTRALIZED INTEGRAL CONTROLLABILITY (DIC). $G(s) =$

$\begin{bmatrix} G_{cd} & G_{cm} \\ G_{sd} & G_{sm} \end{bmatrix}$ *is Decentralized Integral Controllable (DIC) if there exists a controller* $\tilde{K} = \begin{bmatrix} k_1(s) & & \\ & \ddots & \\ & & k_n(s) \end{bmatrix}$ *such that the closed-loop system is stable for all* ϵ_i *satisfying* $0 \le \epsilon_i \le 1$ *with the corresponding* i^{th} *integrator removed when* $\epsilon_i = 0$.

This definition implies that any subset of the loops can be detuned or taken out of service while maintaining stability.

Unfortunately, no necessary *and* sufficient condition for DIC is available at this time. However, some necessary conditions for DIC have been derived which can serve as screening tools.

THEOREM 4. *A system* $G = \begin{bmatrix} G_{cd} & G_{cm} \\ G_{sd} & G_{sm} \end{bmatrix}$ *with* $G_{sm} = \begin{bmatrix} g_{11} & \cdots & g_{1n} \\ \vdots & \ddots & \\ g_{n1} & & g_{nn} \end{bmatrix}$ *is DIC only if all of the following conditions are met:*

1. $\mathrm{Re}\{\lambda_i(G_{sm}(0)\tilde{G}_{sm}^{-1}(0))\} \ge 0, \;\; \forall i$ where

$$\tilde{G}_{sm} = \begin{bmatrix} g_{11} & & \\ & \ddots & \\ & & g_{nn} \end{bmatrix}$$

2. $\mathrm{Re}\{\lambda_i(G_{sm}^{+}(0))\} \ge 0, \;\; \forall i$ where

$$G_{sm}^{+}(0) = G_{sm}(0) \cdot \begin{bmatrix} sgn(g_{11}) & & \\ & \ddots & \\ & & sgn(g_{nn}) \end{bmatrix}$$

3. $\gamma_{ii} \ge 0, \;\; \forall i$ where

$$\begin{bmatrix} \gamma_{11} & \cdots & \gamma_{1n} \\ & \ddots & \\ \gamma_{n1} & \cdots & \gamma_{nn} \end{bmatrix} = G_{sm}(0) \times (G_{sm}^{-1}(0))^T \tag{38}$$

Here $\times$ denotes the Schur or element-by-element product.

Proof. See Morari and Zafiriou [3]

5.2. Robust Performance for a Controller with Integral Action. In addition to the conditions for DIC, the following steady-state robust performance condition for a controller with integral action (Theorem 3), must be satisfied:

$$\begin{gathered} \mu_{\Delta_I}[W_I(0)] < 1, \quad \mu_{\Delta_O}[W_O(0)] < 1 \text{ and} \\ \bar{\sigma}[W_p(G_{cd} - G_{cm}G_{sm}^{-1}G_{sd})W_d(0)] < 1 \end{gathered} \tag{39}$$

This condition and the necessary conditions for DIC, are effective screening tools for eliminating control structure candidates.

5.3 Norm Bounds for Nominal Stability (μ-Interaction Measures). In this section, we develop a set of sufficient conditions for nominal stability of the closed-loop system (with a decentralized controller), "μ-interaction measures," which are also useful as control structure selection tools.

Based on a scaled version of the small gain theorem, we can state the following sufficient conditions for *nominal* stability[2]:

THEOREM 5. *Let*

$$\tilde{\Delta} = \begin{bmatrix} \Delta_1 & & \\ & \ddots & \\ & & \Delta_n \end{bmatrix}$$

$$\begin{aligned} L_T(s) &= (G_{sm} - \tilde{G}_{sm})\tilde{G}_{sm}^{-1}(s) \\ \tilde{T}(s) &= \tilde{G}_{sm}\tilde{K}(I + \tilde{G}_{sm}\tilde{K})^{-1} \\ L_S(s) &= (G_{sm} - \tilde{G}_{sm})G_{sm}^{-1}(s) \\ \tilde{S}(s) &= (I + \tilde{G}_{sm}\tilde{K})^{-1} \end{aligned} \tag{40}$$

Suppose $\tilde{T}(s)$ and $\tilde{S}(s)$ are stable.
Then the closed-loop system in Fig. 14 is stable if

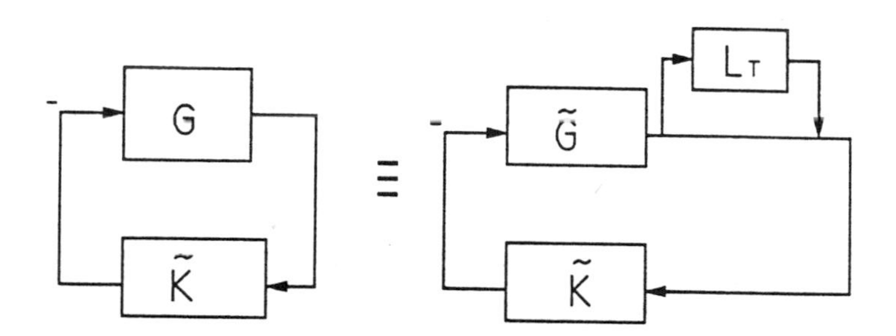

Figure 14: μ-Interaction Measures

$$\mu_{\tilde{\Delta}}(L_T(j\omega)\tilde{T}(j\omega)) < 1 \qquad \forall\omega$$

or if

$$\mu_{\tilde{\Delta}}(\tilde{S}(j\omega)L_S(j\omega)) < 1 \qquad \forall\omega$$

Remarks:

1. The conditions (41) and (42) cannot be combined over different frequency ranges.
2. If $K = g(s)\tilde{G}_{sm}^{-1}$ (assuming $\tilde{G}_{sm}^{-1}$ is stable), the conditions (41) and (42) become
$$\left|\frac{g(j\omega)}{1+g(j\omega)}\right| < \frac{1}{\mu_{\tilde{\Delta}}(L_T(j\omega))} \text{ and } \left|\frac{1}{1+g(j\omega)}\right| < \frac{1}{\mu_{\tilde{\Delta}}(L_S(j\omega))}$$
respectively.
3. $\mu_{\tilde{\Delta}}(L_T(j\omega))$ and $\mu_{\tilde{\Delta}}(L_S(j\omega))$ (or $\rho(L_T(j\omega))$ and $\rho(L_S(j\omega))$ if $\tilde{T}(s)$ and $\tilde{S}(s)$ are restricted to scalar-times-identity matrices) are the so called "μ-interaction measures" [2] indicating the degree of interaction among the loops.

5.4 Norm Bounds for Robust Performance. The norm bound method developed in Sec. 2 can easily be extended to the case of decentralized control. By parametrizing $\tilde{K}(s)$ in terms of $\tilde{T}(s)$ as shown in Fig. 15, one can obtain the bound on $\bar{\sigma}(\tilde{T}(j\omega))$ for robust performance. Using the fact that $\tilde{S} = I - \tilde{T}$, the bound on $\bar{\sigma}(\tilde{S}(j\omega))$ for robust performance can also be obtained. These two bounds can be combined over different frequency ranges.

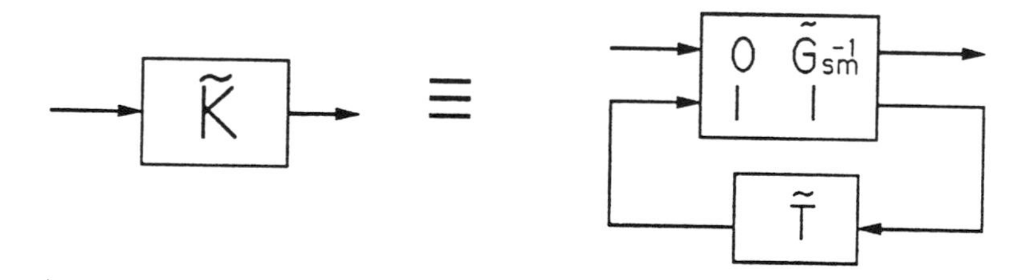

15: Parametrization of $\tilde{K}$ in terms of $\tilde{T}$.

One can reduce the conservativeness of the bounds by restricting the controller to a specific form. For example, suppose that $\tilde{K}$ is parametrized as $\tilde{K} = g(s)\tilde{W}(s)$ where $\tilde{W}(s)$ is known. Then one can obtain the bounds on $|\frac{g(i\omega)}{1+g(j\omega)}|$ and $|\frac{1}{1+g(j\omega)}|$ by parametrizing $\tilde{K}$ as shown in Fig. 15 with $\tilde{W}(s)$ and $\frac{g}{1+g}I$ replacing $\tilde{G}_{sm}^{-1}$ and $\tilde{T}$ respectively. The resulting bounds will be much less conservative as they involve only *scalar* transfer function. The "optimal choice" for $\tilde{W}(s)$ is not clear. For G_{sm} with little interaction (*i.e.*, the off-diagonal elements are negligible), the appropriate choice for $\tilde{W}(s)$ may be $\tilde{G}_{sm}^{-1}(s)$, assuming $\tilde{G}_{sm}^{-1}(s)$ is stable.

Numerical Example

See Skogestad and Morari[4] for a numerical example.

6. Conclusion. This paper presents a unified approach to the problem of control structure selection for centralized and decentralized controllers. With the objective of robust performance, a series of necessary conditions and sufficient conditions for the existence of a controller achieving robust performance were developed. Necessary conditions serve as screening tools to eliminate candidate structures. Sufficient conditions (norm bounds on various transfer function matrices) are intended to help the engineer locate quickly a control structure for which a robustly performing controller can easily be synthesized.

For decentralized controllers, the problem is much more difficult because no quantitative measure of the achievable performance for a particular decentralized control structure (even without model uncertainty) is available. This measure of achievable performance, along with the solution for the H_2- or H_∞-optimal decentralized controller would be extremely beneficial. In addition, development of more efficient screening tools (*i.e.*, tighter necessary conditions) and design tools (*i.e.*, tighter sufficient conditions) appears a promising approach toward a better method for control structure selection and decentralized control system design.

Acknowledgement: Support from the National Science Foundation and the Petroleum Research Fund administered by the American Chemical Society is gratefully acknowledged.

REFERENCES

[1] DOYLE, J. C., *Analysis of Feedback Systems with Structured Uncertainties*, Inst. Electrical Engineers Proc., Pt. D, 129, D(6) (1982), pp. 242–247.

[2] GROSDIDIER, P. AND M. MORARI, *Interaction Measures for Systems Under Decentralized Control Automatica*.

[3] MORARI, M. AND E. ZAFIRIOU, *Robust Process Control*, Prentice Hall, Englewood Cliffs, NJ, 1989.

[4] SKOGESTAD, S. AND M. MORARI, *Some New Properties of the Structured Singular Value*, IEEE Trans. Autom. Control (December, 1988).

[5] SKOGESTAD, S., M. MORARI AND J. DOYLE, *Robust Control of Ill-Conditioned Plants: High-Purity Distillation*, IEEE Trans. Autom. Control (December, 1988).

[6] SKOGESTAD, S. AND M. MORARI, *Robust Performance of Decentralized Control Systems by Independent Designs*, Automatica, in press.

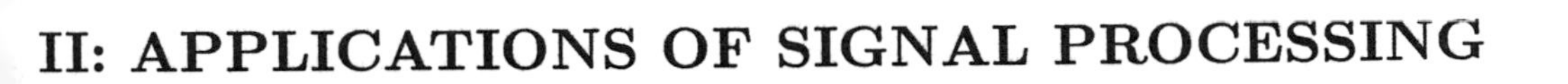

II: APPLICATIONS OF SIGNAL PROCESSING

TOMOGRAPHY IN RADAR

MARVIN BERNFELD*

Abstract. The following discussion introduces an approach for mapping radar backscatter intensity as a function of range and Doppler. The approach is a new application of Johann Radon's idea for reconstructing two-dimensional distributions based on line-integral projections. The best known previous application is in clinical radiology where it is called Computerized Tomography. The basis, by which such reconstructions in the range-Doppler domain can be provided, consists of the ridge-like ambiguity function which is characteristic of linear FM (CHIRP) pulse compression; electronically changing the FM rate is analogous to the mechanical steering normally employed to provide different perspectives of a distribution. The approach is promising because it offers several advantages compared to the alternative involving coherent pulse Doppler radar.

Introduction. The idea of reconstructing a two-dimensional distribution from corresponding line-integral projections, which are available for all aspect angles distributed over 180 degrees, was introduced in 1917 by Johann Radon. These mathematical results have since inspired developments in diverse fields. The best known is in clinical radiology where it is referred to as Computerized Tomography. The value to clinical radiology is an ability to routinely produce transaxial x-ray images (tomo in Greek means slice) involving certain biological cross-sections that are normally inaccessible without destructive surgery.

Radio astronomy, geophysical probing, and non-destructive testing by industry are some other fields where this idea has also been successfully exploited (1–4). Moreover, the principles are not unfamiliar to radar engineers. The images produced by (spot light) synthetic aperture radar (SAR) have been perceived as reconstructions based on line-integral projections (5). Microwave imaging is another radar example where it has been recognized that similar projections are utilized in imaging rotating objects (6).

Johann Radon's mathematical method for reconstructing two-dimensional distributions was recently adopted yet another time (7–9). This time the application is to reconstruct the distribution of radar backscatter intensity as a function of range delay and Doppler shift. The basis for this new application is that the line-integral projections in the range-Doppler domain can be obtained with a radar system incorporating linear FM (CHIRP) pulse compression in which the FM rate can be changed from pulse to pulse.

The basic theory supporting this new application of tomographic mathematical methods is provided in the following discussion. The emphasis is on explaining how multiple projections of a range-Doppler distribution can be made available by measurements involving CHIRP pulse compression signals (7). In conjunction with this discussion, a perspective is also provided regarding the alternative method involving coherent pulse doppler radar. In addition, an introduction to the radar ambiguity function is provided so that its role in the reconstruction of range-Doppler distributions via the alternative approaches can be understood.

*Raytheon Co., 430 Boston Post Road, Wayland, MA 01778

Conventional Radar Imaging. The conceptual system for conventional mapping of radar echo intensity as a function of range and Doppler consists of a matrix of active correlators with (say) rows and columns representing range and Doppler, respectively. A generic row in this matrix is shown in Figure 1.

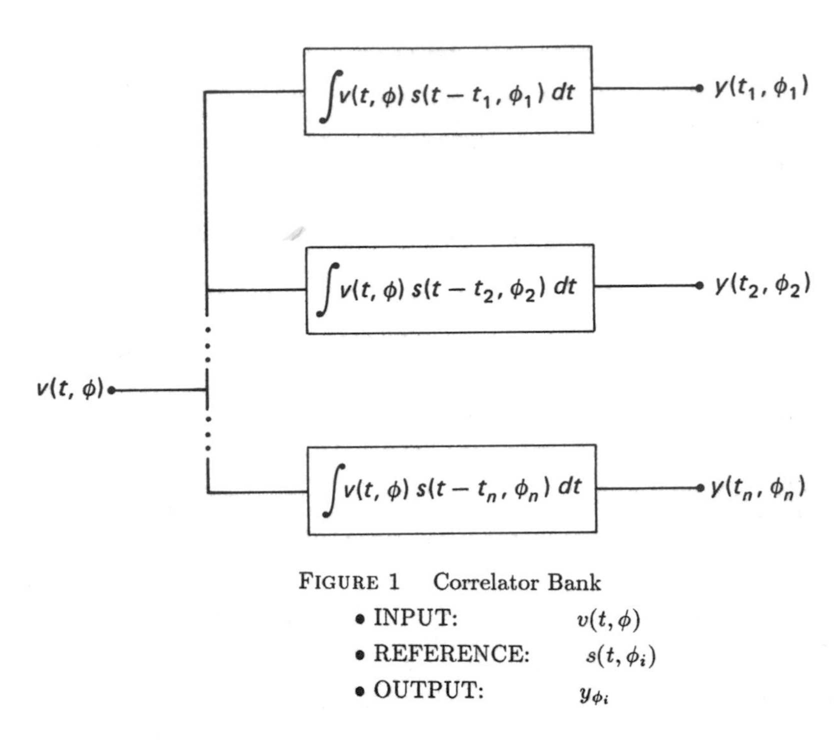

FIGURE 1 Correlator Bank

- INPUT: $v(t,\phi)$
- REFERENCE: $s(t,\phi_i)$
- OUTPUT: y_{ϕ_i}

Each row of the matrix contains correlation references - replicas of the transmitted signal - that are mutually synchronized for a certain range displacement and individually displaced to progressively different offset frequencies relative to the carrier of the transmitted signal. In place of the active correlators, an alternative arrangement instead would be to employ a matrix of range-gated matched filters with similar spectral displacements across the columns.

An exact reconstruction of range-Doppler backscatter distributions can never be provided via the preceding system. This limitation affects all imaging systems and is caused by point spreading, the phenomena in nature pronounced by Lord Rayleigh's resolution criteria which quantifies our inability to see things as they really are. Point spreading in range-Doppler space is represented by the radar ambiguity function, which is a measure of the difference between a waveform and its time displaced and frequency shifted replica.

The Ambiguity Function. The Ambiguity Function was introduced into radar in a monograph on radar theory by P.M. Woodward (10). It is mathematically defined by the product of complex conjugate functions: $\chi(\tau,\phi)\chi^*(\tau,\phi)$, where $\chi(\tau,\phi)$ denotes the complex response of a matched filter as a function of range delay

(τ) and Doppler shift (ϕ). Alternative expressions can be given for $\chi(\tau, \phi)$ that are interchangeable through Parseval's theorem. Thus, by utilizing $u(t)$ and $U(f)$ to denote the Gabor-complex representation of a radar waveform and the corresponding Fourier transform, respectively, Parseval's theorem yields:

$$\chi(\tau, \phi) = \exp\{j\pi\phi t\} \int_{-\infty}^{\infty} u(t - \tau/2)\; u(t + \tau/2)\; \exp\{-j2\pi\phi t\} dt$$

$$= \exp\{j\pi\phi t\} \int_{-\infty}^{\infty} U(f - \phi/2)\; U(f + \phi/2)\; \exp\{-j2\pi f \tau\} df$$

The magnitude $|\chi(\tau, \phi)|$ can be observed at the output of a bank of matched filters which are individually tuned to a different center frequency.

Until now, the importance of the radar ambiguity function was derived because of its role in selecting the most suitable transmitter waveform for different radar applications. This importance is expanded here, wherein it now becomes a key part in the following approach for mapping backscatter intensity as a function range and Doppler based on tomographic mathematical methods.

A general perspective of the ambiguity function is provided in figure 2, including certain properties that are frequency exploited during waveform investigations. The 4th property possesses particular significance in the application of tomographic methods for reconstructions in range-Doppler space. It states that the ambiguity function is sheared parallel to the Doppler axis, as a function of range delay, when quadratic phase modulation (i.e., linear FM) is attached to the original waveform. In case the original waveform is a linear FM pulse, to start, the effect is to pseudo rotate its ridge-like ambiguity function (figure 3). Figures 3 and 4 provide several examples of $|\chi(\tau, \phi)|$, for analog and discrete phase modulated pulses. Figure 4 also exhibits the elusive "thumbstack" ambiguity function which is often the theoretical target in waveform design investigations. The so called maximum length-shift register, discrete phase modulations tend to yield an approximation of this ambiguity function which gets better with time-bandwidth product.

FIGURE 2
Radar Ambiguity Function

Let the radar transmission signal $s(t)$ be represented by the Gabor-analytic complex notation, where

$$s(t) = \text{REAL}\{\psi(t)\}$$
$$\psi(t) = u(t)\ \exp\{j2\pi f_0 t\}$$
$$u(t) = a(t)\ \exp\{j\theta(t)\}$$

The radar ambiguity function is equal to $|\chi(\tau, \phi|^2$ where

$$\chi(\tau,\phi) = \exp\{j\pi\phi t\} \int_{-\infty}^{\infty} u(t-\tau/2)\ u(t+\tau/2)\ \exp\{-j2\pi\phi t\}dt$$
$$= \exp\{j\pi\phi t\} \int_{-\infty}^{\infty} U(f-\phi/2)\ U(f+\phi/2)\ \exp\{-j2\pi f\tau\}df$$

INTERPRETATIONS:

1. A measure of the difference between a signal and its time-frequency shifted self.
2. A matched filter response for different Dopplers (ϕ).
3. A point spread function in range-Doppler space.

UTILITY:

A mathematical tool for selecting waveforms to employ in given radar applications.

PROPERTIES:

1. $|\chi(\tau,\phi)| \leq |\chi(0,0)|$
2. $\int_{-\infty}^{\infty}\int_{-\infty}^{\infty} |\chi(\tau,\phi)|^2\ d\tau d\phi = |\chi(0,0)|$
3. $\int_{-\infty}^{\infty} |\chi(\tau,\phi)|^2\ d\tau = \int_{-\infty} |\chi(\tau,0)|^2 \exp\{j2\pi\phi\tau\}d\tau$
4. If $u(t)$ yields $\chi(\tau,\phi)$, then:

   $$u(at)\ \exp\{jkt^2\} \text{ yields } (1/|a|)\ \chi(a\tau, 1/a(\phi + k\tau/\pi))$$

5. The ambiguity function and the Wigner distribution are a transform pair.

(a) CW pulse

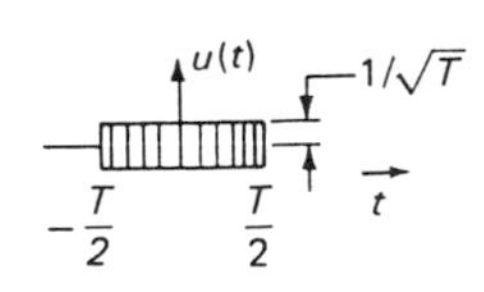

$$u(t) = \frac{1}{\sqrt{T}} \left(-\frac{T}{2} \leq t \leq \frac{T}{2}\right)$$
$$= 0 \quad \text{elsewhere}$$

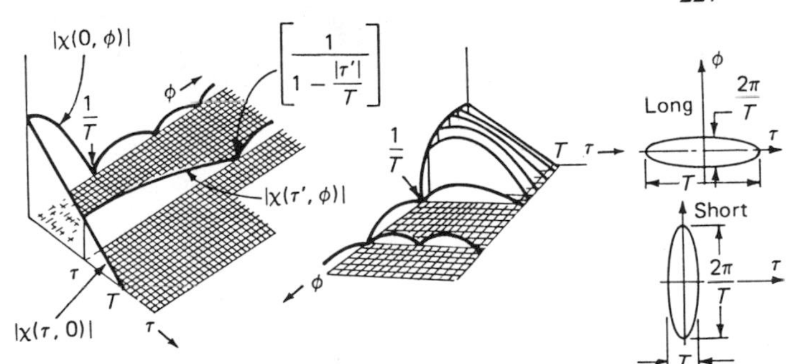

(b) Linear FM

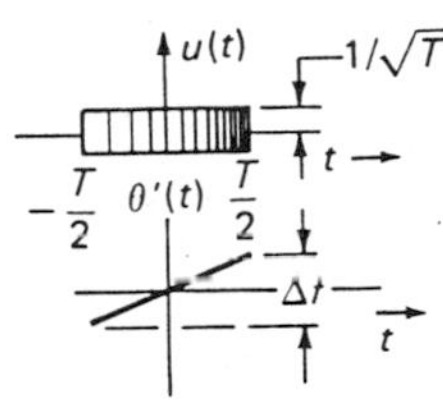

$$u(t) = \sqrt{\frac{1}{T}} \exp\,[jbt^2] \left(-\frac{T}{2} \leq t \leq \frac{T}{2}\right)$$
$$= 0 \quad \text{elsewhere}$$

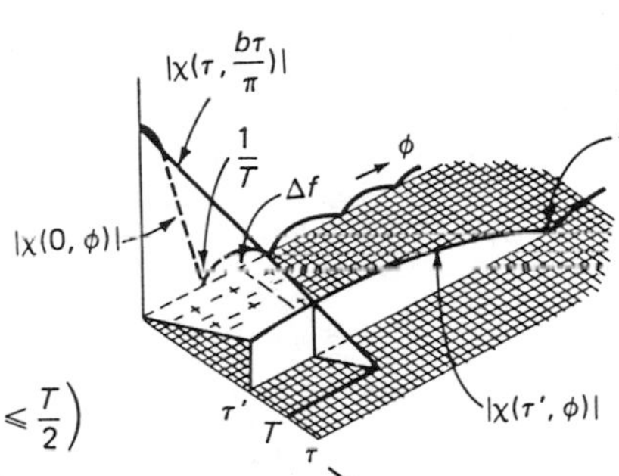

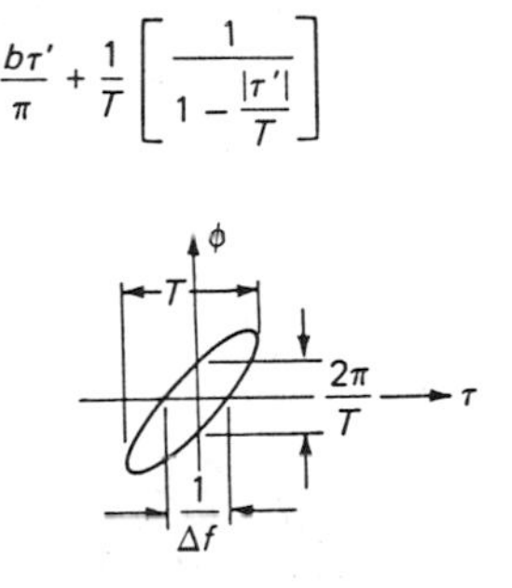

(c) Quadratic FM

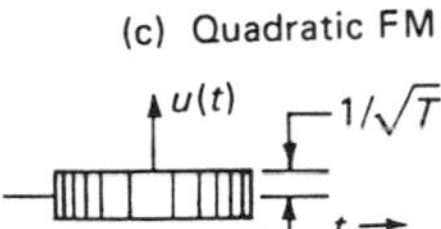

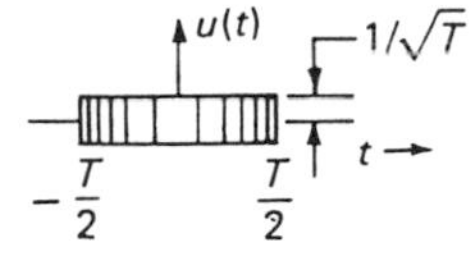

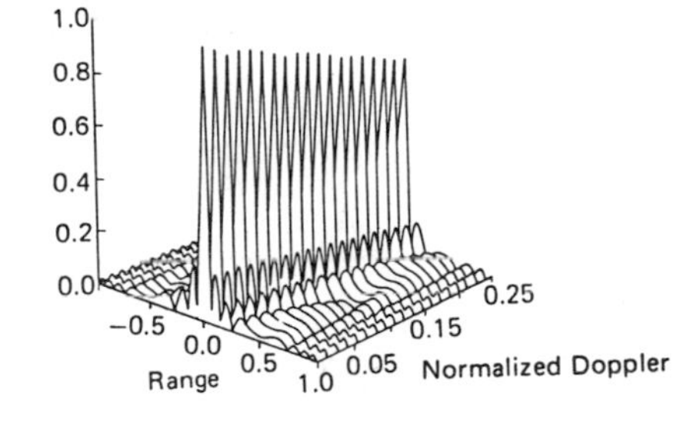

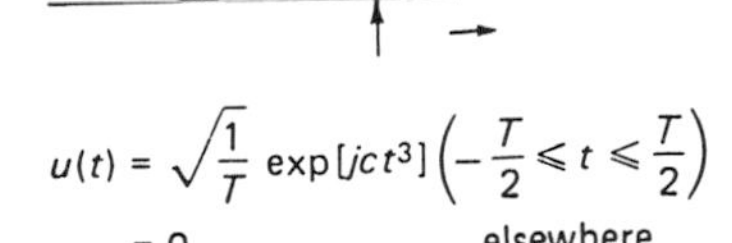

$$u(t) = \sqrt{\frac{1}{T}} \exp[jct^3] \left(-\frac{T}{2} \leq t \leq \frac{T}{2}\right)$$
$$= 0 \quad \text{elsewhere}$$

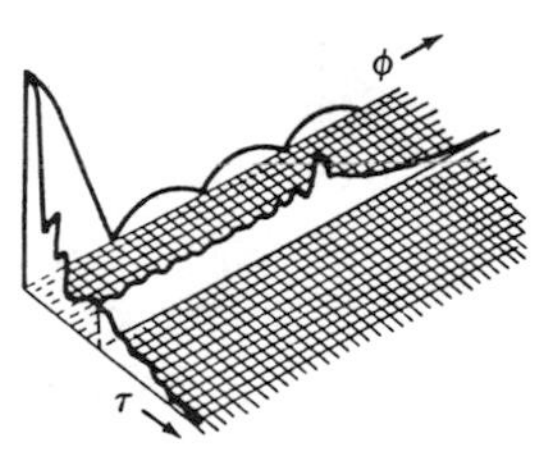

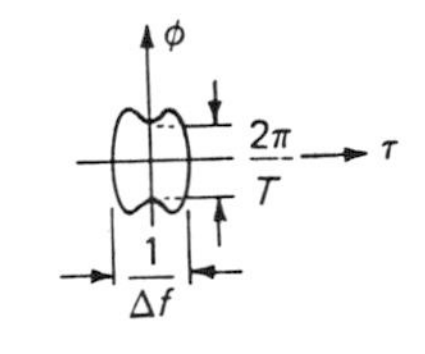

FIGURE 3 Ambiguity function: analog wave form examples
(a) CW Pulse
(b) Linear FM
(c) Quadratic FM

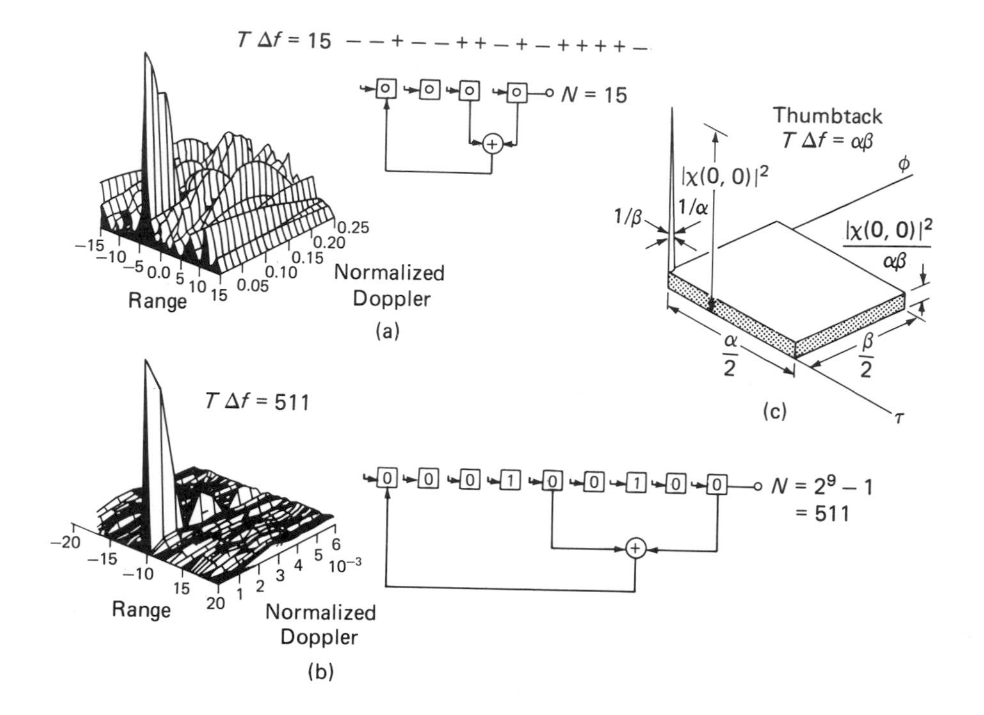

FIGURE 4 Ambiguity Functions: Maximum Length-Shift Register Discrete Phase Examples

(a) $T\Delta f = 15$

(b) $T\Delta f = 511$

(c) "Thumbtack" Ambiguity Function

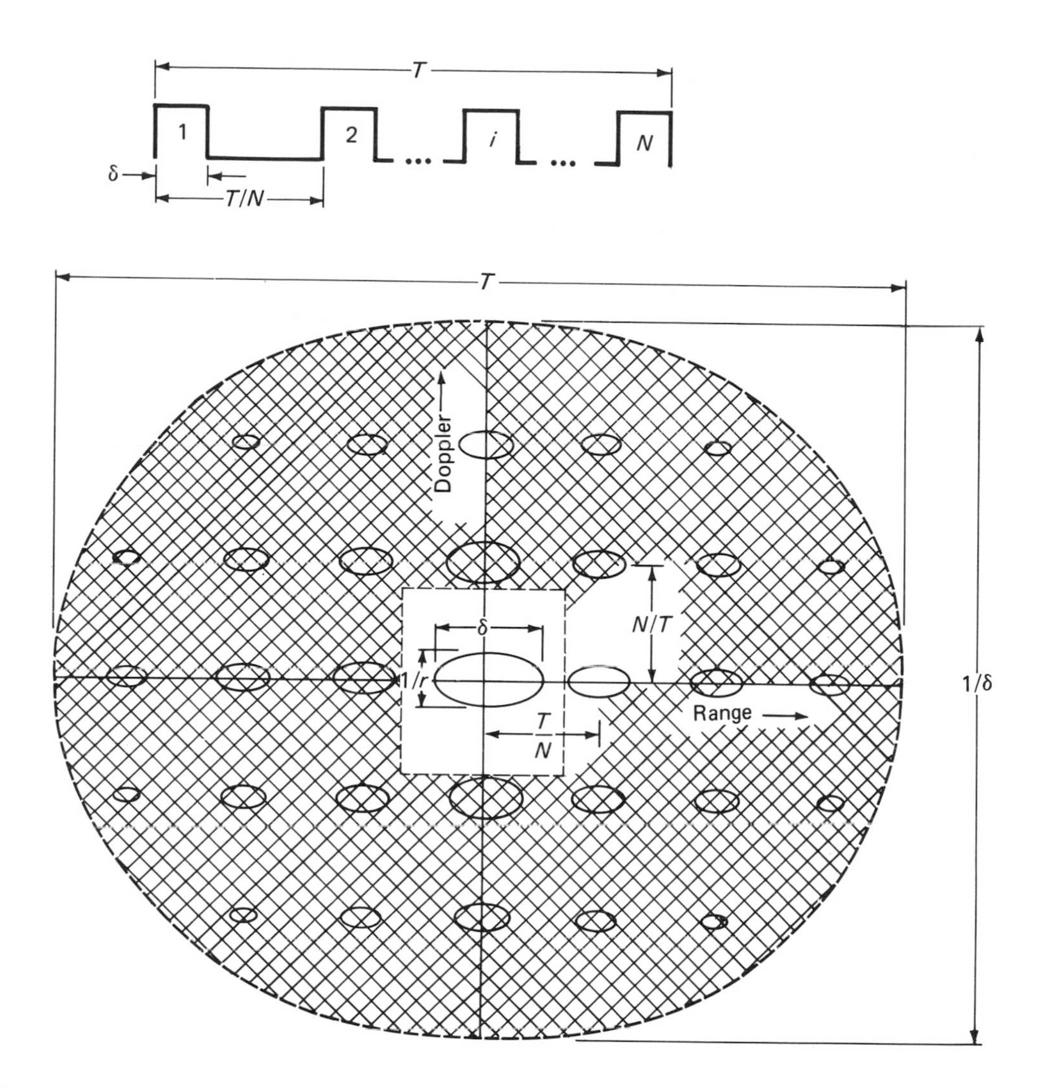

FIGURE 5 Pulse Train Ambiguity Function
T - Pulse Train Duration
δ - Pulse Width
T/N - Pulse Period

The radar waveform that has been utilized previously for mapping the backscatter intensity as a function of range and Doppler has a train of phase-coherent pulses. The point spreading produced by this waveform is indicated in figure 5 by the array of peak-ambiguity function areas that is displayed. The dimensions of these areas are as indicated determined by pulse train parameters. These areas reduce to points and the ambiguity function becomes an infinite "bed of nails" when the pulse train is transformed, idealistically, into an infinite train consisting of impulses (or pulses with infinite bandwith, affected via internal phase modulation).

Pulse Doppler Radar. Topographically, the receiving system that is employed for pulse train processing, to map range-Doppler distributions, consists of multiple range-gating channels in which gated pulse train signals feed to a bank of adjacent frequency filters. A particular implementation of this receiving system is indicated in figure 6. In this system, the pulse train processing is preceded by linear FM (CHIRP) pulse compression; the bandwith of the pulses is normally enhanced via phase modulation, to obtain a better approximation of the ideal "bed of nails" ambiguity function without sacrificing detection performance. The linear FM is employed in deference to other waveform modulations because its relative insensitivity to moderate Doppler shifts allows a single pulse compression filter instead of a bank of such filters. In the particular system that is shown in figure 6, the analog output obtained from a single pulse compression filter is converted, via analog-to-digital circuits, to a digital representation that is corner turned (i.e., column stored, then, read row-by-row) and, subsequently, transformed by fast Fourier transform (FFT). The conversion from analog to digital representation effectively range gates the analog signal, and the FFT processor is equivalent to a Doppler filter bank.

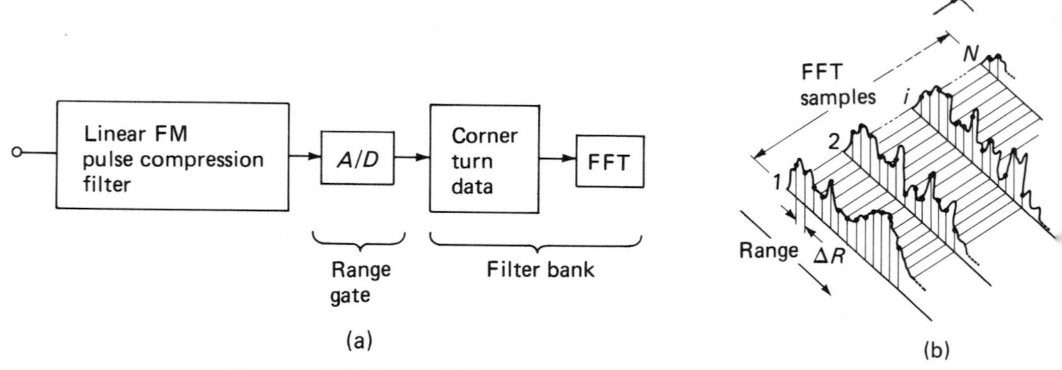

FIGURE 6 Coherent Pulse Doppler Radar Signal Processing Configuration (a) and Input Signal (b)

The system incorporating the signal processing shown in figure 6 fits the general description of coherent pulse Doppler radars. The radars in this class can be employed for other purposes beside mapping backscatter intensity as a function of range and Doppler; moving target detection is one of the applications. Unfortunately, the coherent signal processing will provide access to only the central region of the ambiguity domain, whose boundaries are established by the pulse repetition frequency (PRF) in one dimension and by the period between pulses in the other dimension. Any backscatter distributions outside these boundaries are folded back onto the central region. The fold-back distributions can not be distinguished from the true distribution of backscatter energy and is commonly termed "aliasing".

The ideal central region is sufficiently large so that no backscatter extends past these boundaries. But simultaneous expansion of the central region, to enlarge the area of range and Doppler coverage, is barred because of reciprocal coupling between the PRF and the pulse period (figure 7). As a result, backscatter images in the range-Doppler domain are frequently degraded because of alising.

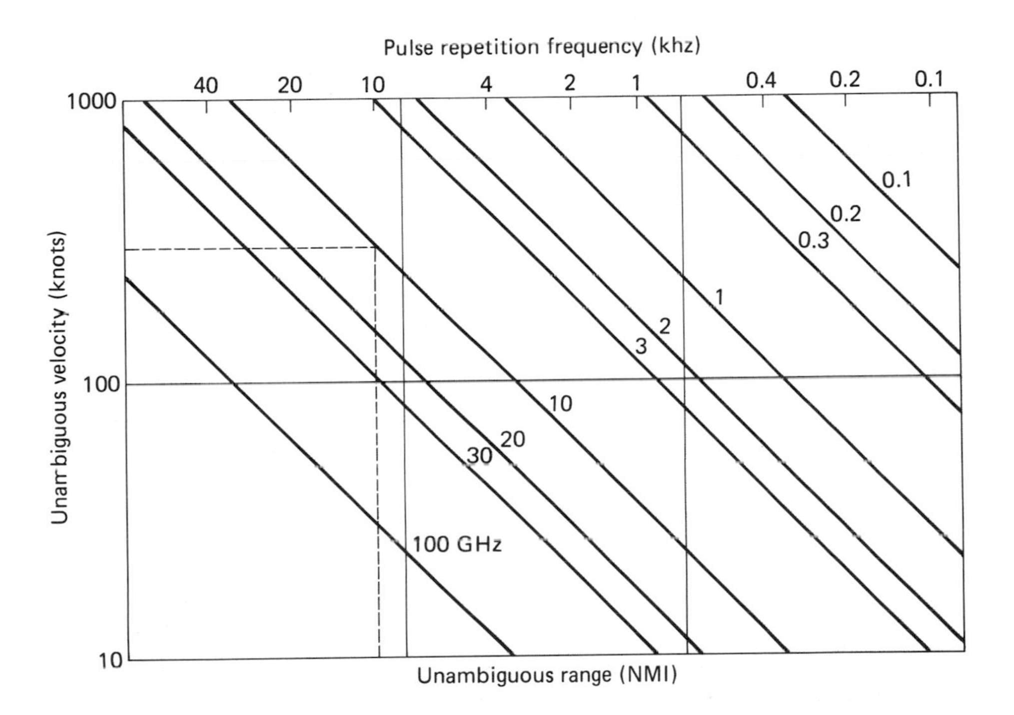

FIGURE 7 Parametric plots of unambiguous velocity (knots) versus unambiguous range (NM) for various carrier frequencies. For example, with Δ carrier of 10 Gigahertz, the unambiguous velocities will be less than 300 knots while the unambiguous ranges will be less than 7 nautical miles

The likelihood for degradations such as above becoming severe in future applications of coherent pulse Doppler radars will increase when demands for greater resolution in range are met. With wider bandwidths to satisfy these demans, the carrier frequency must also be scaled upward. This step is necessary to prevent distorted waveform features because of insufficient bandwidth-to-frequency ratio. As a result, diminishing Doppler velocities will yield Doppler shifts that exceed the boundaries of the central region of the ambiguity function. Hence, degradation due to fold-back will be more likely to occur in the future when mapping realistic backscatter distributions without any theoretical restriction that will limit their expense in range and Doppler. This is an important characteristic of the new mapping approach.

Tomography in Radar. To affect the reconstruction of a two-dimensional distribution, assuming the corresponding line-integral projections are available, any one of three established approaches can be followed. These consist of (1), the Inverse Radon Transform, (2), an application of the Projection-Slice Theorem, and, (3), algebraic reconstruction techniques.

The application of one of these techniques, for mapping radar backscatter intensity in range delay and Doppler frequency shift, is made possible by the following analogy with respect to line integral projections.

In the general description for radar matched filtering, the response for backscattered input signals can be represented in complex notation by the following two-dimensional integral with respect to range delay and Doppler shift.

$$\gamma(t) = \int_{-\infty}^{\infty} \int_{-\infty}^{\infty} C(\tau, \phi)\chi(\tau - t, \phi) \exp\{-j2\pi f_0(\tau - t)\}\ d\tau d\phi$$

where $\chi(\tau, \phi)$ is the matched filter response that was previously defined, and, the $C(\tau, \phi)$ denotes the combined reflection coefficient as a function of range delay and Doppler shift.

For an intuitive interpretation of the matched filter input signal see figure 8. Assuming that the backscattering is randomly located with respect to range and velocity, the combined reflection coefficient is also random. If the latter is everywhere statistically independent as function of range delay and Doppler shift, the expectation $E\{|C(\tau, \phi)|^2\}$ can be interpreted as the joint probability density for the distribution of backscattering as a function of range delay and Doppler shift. The joint density is normally also time dependent, but, in the majority of cases, the variations are relatively narrowband compared observation times so the time dependency is not a factor that must be addressed at this time.

FIGURE 8 Perspectives of the Backscatter Model

(a) Point Scatterer Model
(b) Combined backscatter-signal samples for a dwell of N pulses
(c) Composite complex reflection coefficient in range-Doppler space

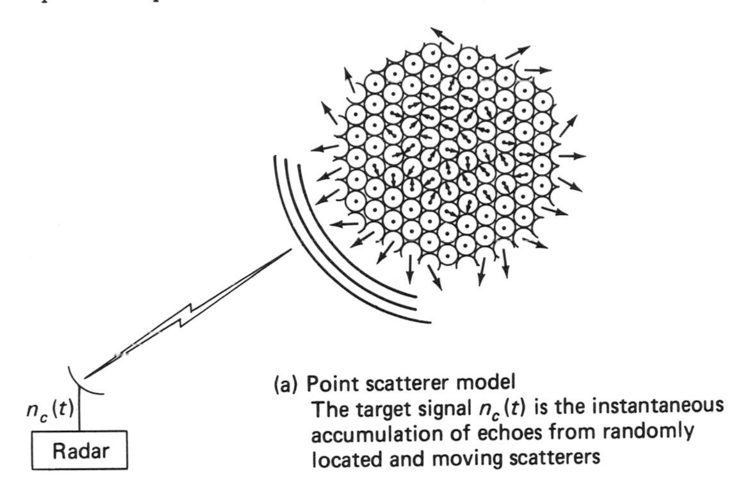

(a) Point scatterer model
The target signal $n_c(t)$ is the instantaneous accumulation of echoes from randomly located and moving scatterers

(b) $$n_c(t) = \sum_{i=1}^{n} \sqrt{b_i S_{R_i}(t)}$$

where:

b_i = scattering power (cross-section)

$S_{R_i}(t)$ is characterized by

θ_i = reflection phase angle

ϕ_i = frequency displacement

τ_i = time displacement

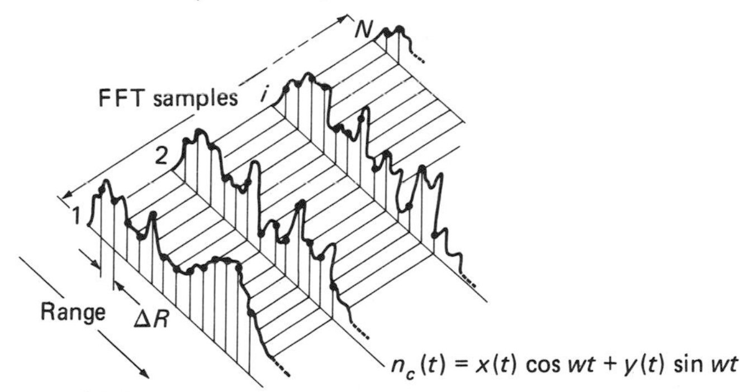

(c) Target can also be described as a two-dimensional random process comprised of composite complex reflection coefficients

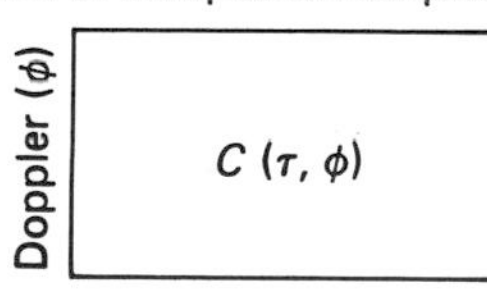

Time displacement (τ)

It is possible to derive the following result based on the preceding description for matched filter input that is distributed in range and Doppler (figure 9).

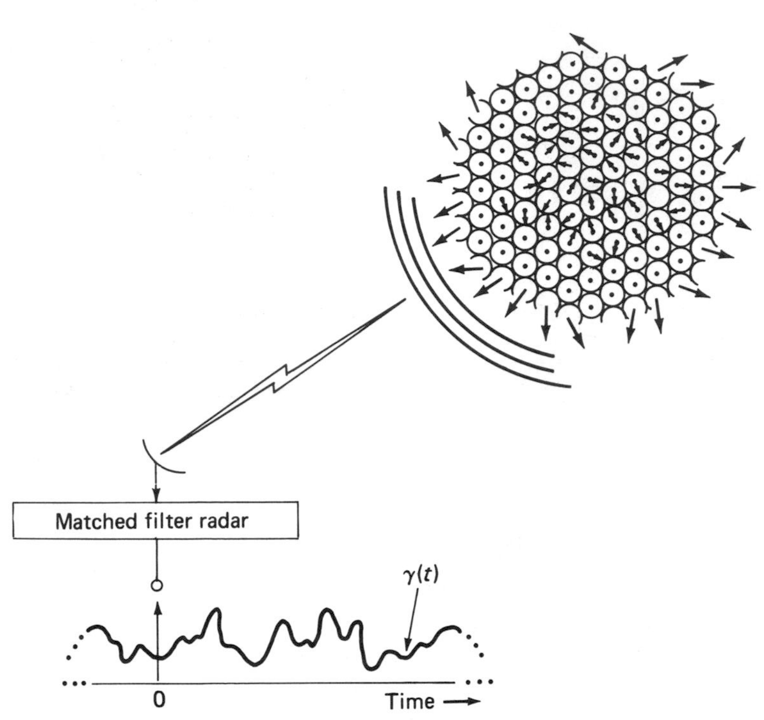

•M.F. response:

$$\gamma(t) = \int_{-\infty}^{\infty}\int_{\infty}^{\infty} C(\tau, \phi)\ \chi(\tau - t, \phi)\ \exp\{-j\, 2\pi\, f_0(\tau - t)\}\ d\tau/d\phi$$

•Instantaneous power:

$$E\{\gamma(t)\ \gamma^*(t)\} = \int_{-\infty}^{\infty}\int_{\infty}^{\infty} E\{|C(\tau, \phi|^2\} \cdot |\chi(\tau - t, \phi)|^2\ d\tau/d\phi$$

where:

$C(\tau, \phi)$ = composite-complex-random reflection coefficient

$|\chi(\tau, \phi)|^2$ = radar ambiguity function

FIGURE 9 Matched Filter Radar Output Signal

$$E\{\gamma(t)\gamma^*(t)\} = \int_{-\infty}^{\infty}\int_{-\infty}^{\infty} E\{|C(\tau, \phi)|^2\}|\chi(\tau - 5t, \phi)|^2\ d\tau d\phi$$

It is observed that this result is proportional to instantaneous power. Specifically, it is equal to a convolution, involving the average input backscattering power and the ambiguity function, where the integration is performed paratrically as a function of time, over the entire range-Doppler domain. Since linear FM (CHIRP) pulses characteristically possess ridge-like ambiguity functions (figure 10), when this waveform is employed, the regular two-dimensional convolution is reduced to a continuous precession of approximate line integrations where the slope of the integration path equals the FM rate. The matched filter output, in this case, is analogous to the line-integral projections employed by tomography (figure 11). By electronically changing the linear FM rate before individual pulse transmissions, a different perspective of the range-Doppler space is obtained. The different pulse compression responses that are obtained. The different pulse compression responses that are obtained. The different pulse compression responses that are obtained, as a result, can be combined as in tomography.

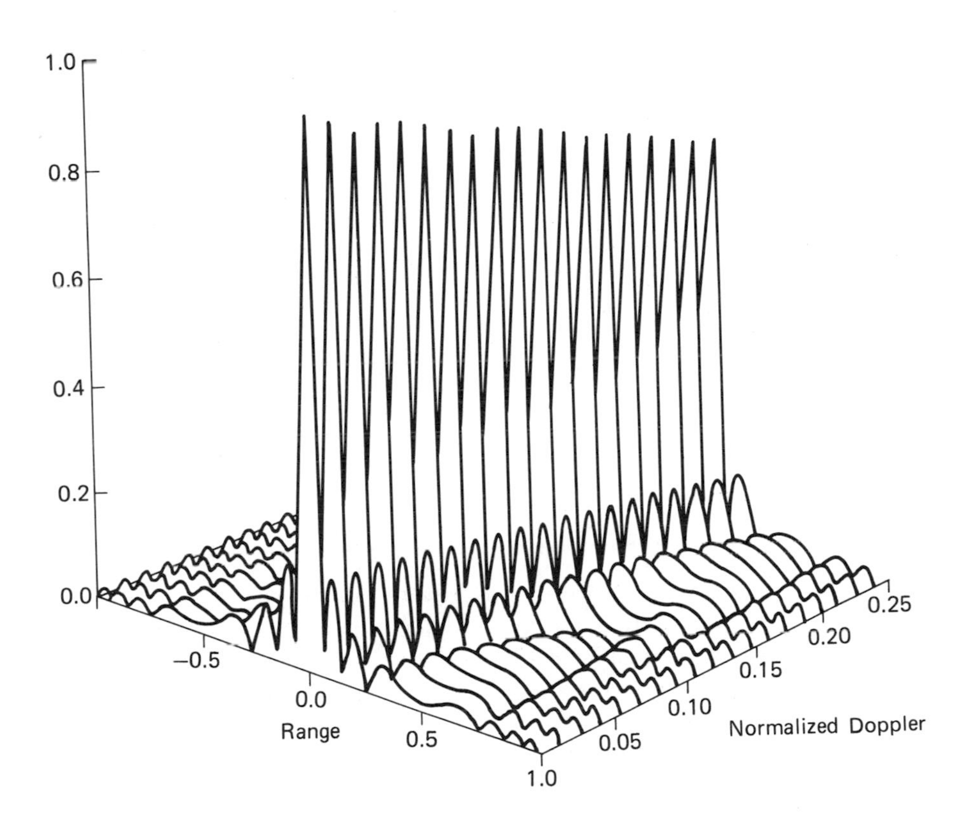

FIGURE 10 Ridge-like ambiguity function corresponding to linear FM (CHIRP) pulse (semi $\tau - \phi$ plane illustration)

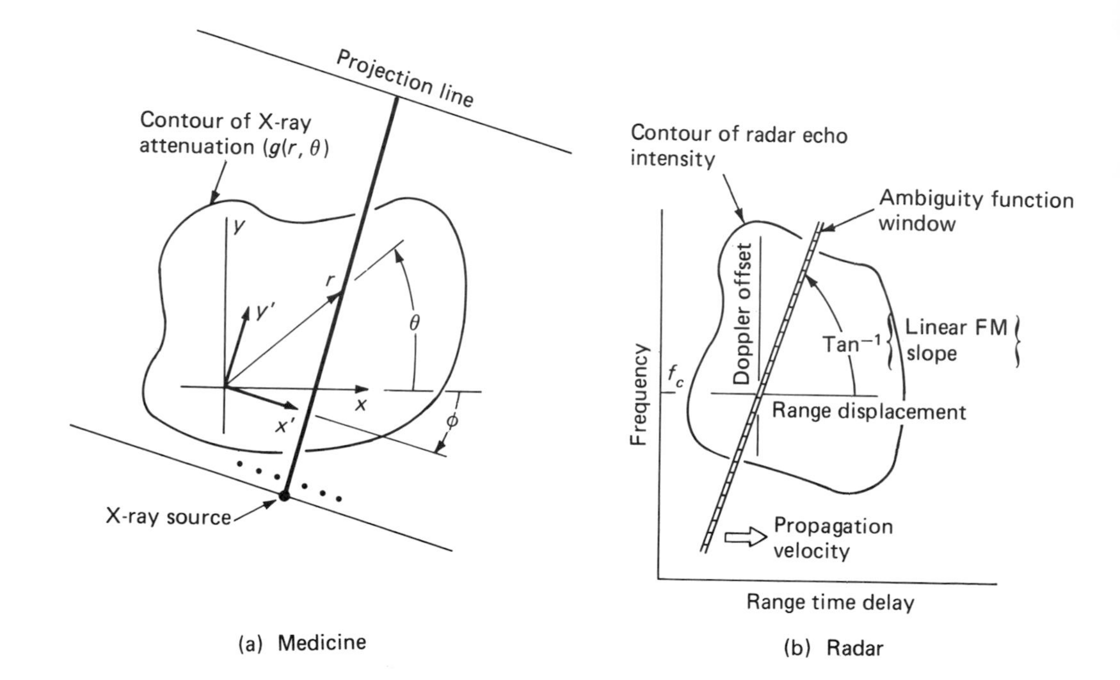

FIGURE 11 Medicine-radar analogy. In medicine, the integration path is steered as well as transported across the object by means of mechanical procedures. In radar, the integration path is steered electronically by changing the linear FM rate while it is transported across the object via electromagnetic propagation.

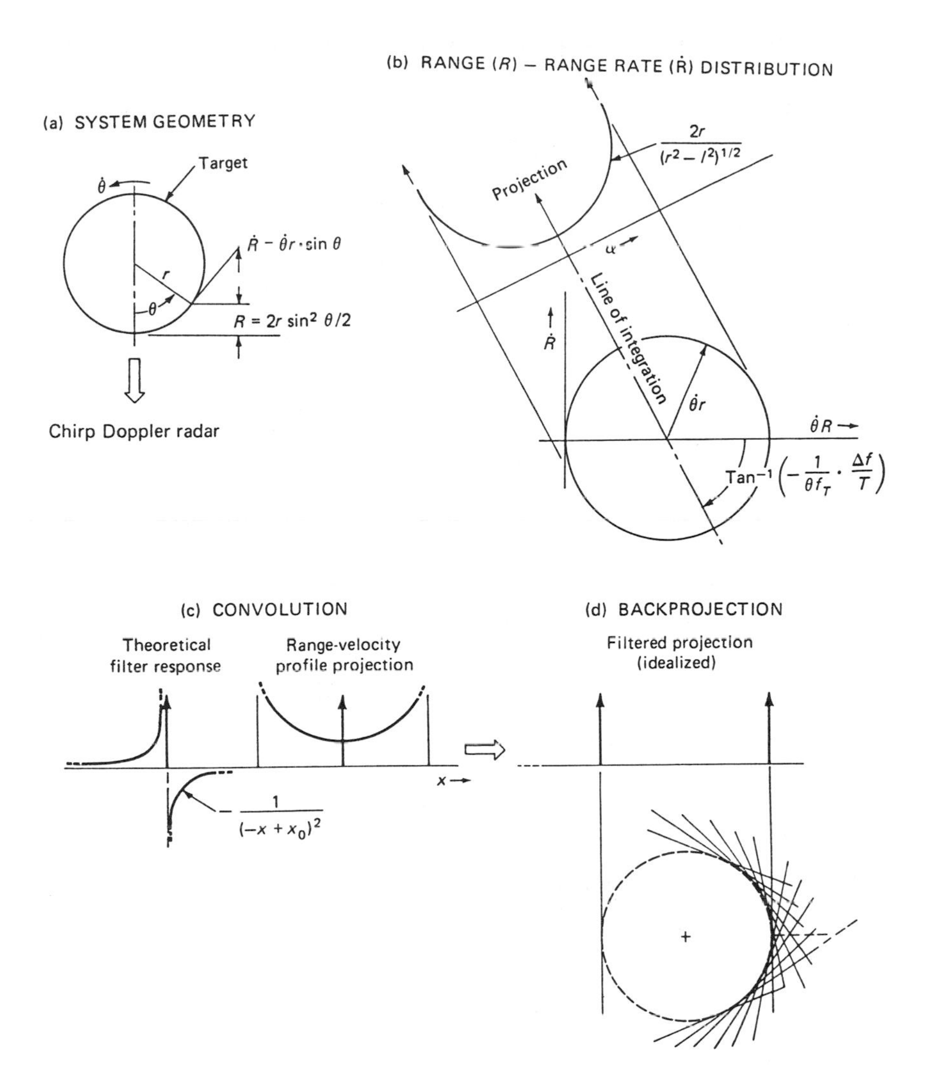

FIGURE 12 Rotating-ring target and states of reconstruction with respect to $R - \dot{R}$ distribution

To reconstruct the distribution of backscatter intensity as a function of range and Doppler, any of the proven reconstruction algorithms from tomography can be employed to combine the matched filter responses that are obtained with different linear FM slopes. The stages of reconstruction, in one example, are illustrated in figure 12. The target model in this example is a rotating ring whose radius and rotation rate are denoted by r and o, respectively. The differential radar range (R) to any point on the right as well as the range rate at this point are stated in figure 12a. It is assumed that the absolute radar range to this target is many times greater than the target radius. Moreover, it is assumed that the backscattering around the ring is uniform. For the sake of simplicity, any theoretical ramifications arising from the latter assumption have been ignored.

The corresponding backscatter distribution as a function of range and range rate can be derived by solving the parametric equations that are given for these variables. The solution is also ring when range is scaled by the rotation rate (figure 12b). In practice the distribution would be an ellipse since access to range scaling is unavailable. Side stepping this issue and also assuming idealized line integrals, the projections of the ring distribution will be proportional to $2r/(r^2 - l^2)^{1/2}$, for all aspect angles, where (referring to figure 12b) l denotes the distance from the center line through the distribution (12). The perspective represented by individual projections is determined by the linear FM rate $\Delta f/T$, where Δf and T denote the swept bandwidth and time duration of the pulse, respectively.

In the final stage of this example, the backscatter distribution is reconstructed via Inverse Radon Transform computations. This involves a procedure known as "filtered backprojection" in which the multiple projections are linearly combined as indicated in figure 12c; there are alternative methods for combining the projections as noted above. In the illustration (figure 12c), the filtering is represented by an idealized impulse response and the output is idealized by representing this with an impulse pair. A partial reconstruction of the range-range rate distribution is illustrated in figure 12c.

Conclusion. The preceding discussion has described a new approach for mapping radar backscatter intensity as a function of range and Doppler. The approach is promising because it offers certain advantages compared to the alternative involving coherent pulse Doppler radar. These advantages are (1), PRF and pulse-to-pulse phase stability are not requirements, (2), the pulse transmissions do not have to be time-synchronized or scheduled in any special order, and, (3), no theoretical restrictions have been found which limit areas that can be mapped in the range-Doppler domain. Moreover, the radar ambiguity function is exploited in contrast with its normal role in radar as a point spreading characteristic.

Future Research. The issues for future research include:

1. The soft line-integrals.
2. Incomplete angular projection data.
3. The time dependence of backscatterer distributions.
4. The statistical reduction of pulse compression output signals to Radon transforms.

REFERENCES

[1] R.N. Bracewell, *Strip Integration in Radio Astronomy*, Aust. J. Phys., 9 (1956), 198–217.

[2] K.A. Dines and R.J. Lytle, *Computerized Geophysical Tomography*, Pro. IEEE, 67 (1979), 1065–1073.

[3] R.A. Crowther, D.J. DeRosier, and, A. Klug, *The reconstruction of a three dimensional structure from projections and its applications to electron microscopy*, Proc. Roy. Soc. London, A 317 (1970), 319–340.

[4] P. Reimers and J. Goebbels, *New Possibilities of Nondestructive Evaluation by X-RAY Computed Tomography*, *Materials Evaluation*, 41 (1983), 732–737.

[5] D.C. Munson, Jr., J.D. O'Brien, and, W.K. Jenkins, *A Tomographic Formulation of Spotlight-Mode Synthetic Aperature Radar*, Proc. IEEE, 71 (1983), 917–925.

[6] D.L. Mensa, S. Halvey, and, G. Wade, *Coherent Doppler Tomography for Microwave Imaging*, Proc. IEEE, 71 (1983), 254–261.

[7] M. Bernfeld, *CHIRP Doppler Radar*, Proc. IEEE, 72 (1984), 540–541.

[8] E. Feig and A. Grünbaum, *Tomographic Methods in Range-Doppler Radar*, Inverse Problems, 2 (1986).

[9] D.L. Snyder, H.L. Whitehouse, J.T. Wohlschlaeger, R.C. Lewis, *A New Approach to Radar/Sonar Imaging*, SPIE, 696 (1986).

[10] P.M. Woodward, *Probability and Information theory With Applications to Radar*, Pergmon Press (1953).

[11] C.E. Cook, M. Bernfeld, *Radar Signals, An Introduction to the Theory and Application*, Academic Press (1967).

[12] A. Papoulis, *Optical Systems, Singularity Functions, Complex Hankel Transforms*, J. Opt. Soc. America, 57, No. 2 (1967), 207–213.

LOCAL AND GLOBAL TOMOGRAPHY*

A. FARIDANI,† F. KEINERT,‡ F. NATTERER,†
E.L. RITMAN,# AND K.T. SMITH$

Abstract. The formulas for global tomography (the reconstruction of the x-ray attenuation coefficient f) and local tomography (the reconstruction of Λf) are reviewed. Physical and numerical imperatives influencing the numerical implementation of the formulas are discussed. Comparisons of local and global tomography are made, and illustrated in a number of example. The table of contents is as follows:

1. Definitions and formulas. The divergent and parallel beam x-ray transforms of a function f on R^n are defined by

$$D_a f(\theta) = \int_0^\infty f(a + t\theta)dt, \qquad \theta \in S^{n-1} \tag{1.1}$$

$$P_\theta f(x) = \int_{-\infty}^\infty f(x + t\theta)dt, \qquad x \in \theta^\perp \tag{1.2}$$

The x-ray attenuation coefficient f is assumed to be square integrable with bounded support. $D_a f(\theta)$ is the total attenuation along the ray with origin a, the x-ray source, and direction θ. It is assumed that the x-ray sources lie on a sphere A of radius R surrounding the support of f. $P_\theta f(x)$ is the attenuation along the line through x (a point on the "film" $\theta^\perp$) with direction θ.

The operator Λ, which is the basis of inversion formulas, is defined by

$$\Lambda u = (n-1)C(n,1)\sum_{j=1}^{n}(x_j/|x|^{n-1}) * \partial u/\partial x_j, \quad (\Lambda u)\hat{\ }(\xi) = |\xi|\hat{u}(\xi).$$

$$\text{with} \qquad C(n,\alpha) = \frac{\Gamma((n-\alpha)/2)}{2^\alpha \pi^{n/2}\Gamma(\alpha/2)}.$$

The convolution is a Cauchy principal value. The $\hat{\ }$ is the Fourier transform. $\mathcal{H}^s$ (below) is the Sobolev space of order s.

The general approximate global inversion formula, proved in [8], is

*This research has been supported by the Alexander von Humboldt Foundation, the Institute for Mathematics and its Applications, the National Science Foundation under grant DMS87-2716, and the National Institutes of Health under grant HL04664

†Mathematics Department, University of Münster
‡Mathematics Department, Iowa State University
#Biodynamics Research Institute, Mayo Clinic
$Mathematics Department, Oregon State University

THEOREM 1.3. *If* $e \in \mathcal{H}^{1/2}$, $(1+|x|)^{1-n}e \in L^1$, *and* $|\xi|^{-1}\hat{e} \in L^1_{loc}$, *then*

$$e * f(x) = \int_A \int_{S^{n-1}} (D_a f(\theta) + D_a f(-\theta))|\langle a, \theta\rangle| k(E_\theta(x-a)) d\theta da \tag{1.4}$$

$$\text{with} \qquad k = (1/4R)C(n,1)\Lambda P_\theta e. \tag{1.5}$$

E_θ is the orthogonal projection on the subspace $\theta^\perp$. In general k is a function of $\theta \in S^{n-1}$ and $x \in \theta^\perp$. The dependence on θ is suppressed because e is normally independent of θ.

In ordinary (= global) tomography the "point spread function" e is an approximate δ-function, and $e * f$ is an approximation to f. Such formulas were introduced by A.V. Lakshminarayanan [2] with the point spread function $e(x) = (1/2)J_1(|x|)/|x|$ in dimension $n = 2$.

In local tomography, introduced in [8,9], e is replaced by Λe, in which case, since $\Lambda^2 = -\Delta$, Theorem 1.3 becomes

THEOREM 1.6. *If* $e \in \mathcal{H}^{3/2}$ *and* $(1+|x|)^{1-n}\Lambda e \in L^1$, *then*

$$\Lambda e * f(x) = \int_A \int_{S^{n-1}} (D_a f(\theta) + D_a f(-\theta))|\langle a, \theta\rangle| K(E_\theta(x-a)) d\theta da \tag{1.7}$$

$$\text{with} \qquad K = -(1/4R)C(n,1)\Delta P_\theta e. \qquad (\Delta = \text{Laplacian}) \tag{1.8}$$

Formally, $\Lambda e * f = e * \Lambda f$. If e is an approximate δ-function, $\Lambda e * f$ becomes an approximation to Λf, which is very different from f quantitatively but has the same singularities, and presents a surprisingly similar gray level image.

Ordinary tomography is global because the kernel k in (1.5) can never have bounded support: $\hat{k}(\xi) \sim |\xi|\hat{e}(\xi)$ with $\hat{e}(0) \neq 0$. If the point spread function e has bounded support, it normally has very small support, and the kernel K in (1.8) has equally small support. Therefore the Lambda tomography of Theorem 1.6 is local. The value of $\Lambda e * f(x)$ is obtained from attenuation measurements along rays passing very close to x (within a mm. or so on a medical scale). Moreover, in practice the inner integral in (1.7) requires only $3-8$ multiplications, as opposed to the several hundred required in (1.4). Note that the integrals for global and local tomography are identical except in respect to the kernels.

2. Dimension 2. In most current scanners the x-ray detectors (which correspond to the directions θ) lie on an arc centered at the x-ray source and are numbered in the clockwise direction along this arc. For each source a it is therefore natural to index the points θ with the clockwise angle ϕ from θ to $-a$. If $a = R(\cos\alpha, \sin\alpha)$, then $\theta = -(\cos(\alpha-\phi), \sin(\alpha-\phi))$, and we write $D_\alpha f(\phi)$ for $D_a f(\theta)$. Since the source circle A surrounds the support of f, $D_a f(\theta) = 0$ if $\langle a, \theta\rangle > 0$; i.e. $D_\alpha f(\phi) = 0$ if $|\phi| > \pi/2$.

For fixed x and a let $x - a = -|x-a|(\cos(\alpha-\gamma), \sin(\alpha-\gamma))$. Then

$$|\langle a, \theta\rangle| = R\cos\phi, \qquad \text{and } E_\theta(x-a) = -|x-a|\sin(\gamma-\phi)$$

on the line $\theta^\perp$. If, as will be assumed, e is radial so that k is independent of θ, formulas (1.4) and (1.7) become

$$(2.1) \qquad e * f(x) = \int_0^{2\pi}\int_{-\pi/2}^{\pi/2} D_\alpha f(\phi)\cos\phi\, k(|x-a|\sin(\gamma-\phi))d\phi d\alpha$$

with $k = (1/4\pi)\Lambda P_\theta e$, and

$$(2.2) \qquad \Lambda e * f(x) = R\int_0^{2\pi}\int_{-\pi/2}^{\pi/2} D_\alpha f(\phi)\cos\phi\, K(|x-a|\sin(\gamma-\phi))d\phi d\alpha$$

with $K = -(1/4\pi)\Lambda P_\theta e$

These formulas require evaluation of the inner integral at each point x of the two dimensional support of f. At least in the case of global tomography, this involves too much computation, and an approximation is made. A δ-function in R^n is homogeneous of degree $-n$. As an approximate δ-function in R^1, $P_\theta e$ is approximately homogeneous of degree -1, so $\Lambda P_\theta e$ and $\Delta P_\theta e$ are approximately homogeneous of degrees -2 and -3, respectively. The approximate formulas become

$$(2.3) \qquad e * f(x) \sim R\int_0^{2\pi} |x-a|^{-2}\int_{-\pi/2}^{\pi/2} D_\alpha f(\phi)\cos\phi\; k(\sin(\gamma-\phi))d\phi d\alpha$$

$$(2.4) \qquad \Lambda e * f(x) \sim R\int_0^{2\pi} |x-a|^{-3}\int_{-\pi/2}^{\pi/2} D_\alpha f(\phi)\cos\phi\; K(\sin(\gamma-\phi))d\phi d\alpha$$

but with different kernels k and K, as described in the next section.

The inner integrals still depend on x (via γ). In practice,

$$(2.5) \qquad F_\alpha(\gamma) = \int_{-\pi/2}^{\pi/2} D_\alpha f(\phi)\cos\phi\, k(\sin(\gamma-\phi))d\phi$$

is computed for sufficiently many values of γ, and the others are obtained by interpolation.

This ingenious approximation, due to Lakshminarayanan [2], works surprisingly well if the diameter of the support of f is not large relative to the distance from the source circle to the support of f, i.e. if the relative variation of $|x-a|$ is not large. In the case of local tomography where the inner integral entails only $3-8$ multiplications, it may be possible to skip the approximation, but this has not been tested yet.

3. Point spread functions and kernels. The usual way to get point spread functions and kernels is to fix an initial function e_1 with $\hat{e}_1$ bounded and continuous at 0, and with $\hat{e}_1(0) = (2\pi)^{-n/2}$ (or integral $e_1 = 1$ if e_1 is integrable). If

$$e_r(x) = r^{-n} e_1(x/r), \tag{3.1}$$

as $r \to 0$, $e_r * f \to f$ in various senses, depending upon additional properties of e_1 and f. For small r, e_r is an approximate delta function. If k_1 is the global tomography kernel corresponding to e_1, then

$$k_r(y) = r^{-n} k_1(y/r) \tag{3.2}$$

is the kernel corresponding to e_r. If K_1 is the local tomography kernel corresponding to e_1, then

$$K_r(y) = r^{-n-1} K_1(y/r) \tag{3.3}$$

is the kernel corresponding to e_r.

The choice of r depends primarily on the required resolution, the presence of high contrast features, and the noise — with secondary adjustments for the x-ray detector spacing (i.e. the sampling of $D_a f$). For any point spread function e, let $\rho(e)$, the radius of e, be the smallest number ρ such that for some fixed small ϵ (usually $\sim .01$)

$$\int_{|x|\le\rho} e(x)dx \ge 1 - \epsilon. \tag{3.4}$$

Since $\rho(e_r) = r\rho(e_1)$, (3.1) provides a point spread function e_r of any desired radius. Too large a radius entails a loss of resolution; too small a radius amplifies noise and creates ringing and streaking artifacts. If the required resolution ρ_0 is defined to be half the diameter of the smallest details that must be seen, it usually is satisfactory to take r so that $\rho(e_r)$ is comparable to ρ_0. (The reconstruction $e * f$ is usually displayed as a gray level picture with 256 gray levels. Variations in $e * f$ are visible only if the magnitude is at least 1/512. If a jump of .35% on a circle of radius ρ_0 is to be seen, as commonly is necessary, and f is the characteristic function of the circle, then $e * f(x) > 1/(.0035 \times 512)$ must hold on most of the circle. With e_1 from (3.7) below, $e = e_r$, and r so that $\rho(e) = \rho_0$, this holds on a circle of radius about $.80\rho_0$.)

However, the approximations in (2.3) and (2.4) play a role. By (3.2), $k_r(|x-a|\sin\phi) = |x-a|^{-2} k_{r/|x-a|}(\sin\phi)$, so the approximations entail the replacement of k_r by some k_s with s fairly near $r/|x-a|$ for all sources a and all x in the reconstruction. When the latter lie in a circle with center 0 and radius $R_0 << R$, as they normally do, it is satisfactory to take $s = r/(R+R_0)$, with r fixed so that $\rho(e_r) \sim \rho_0$. For local tomography the prescription is the same.

Three typical point spread functions are as follows.

$$\hat{e}_1(\xi) = 1/2\pi, \text{ for } |\xi| < 1 \qquad \text{(Ram-Lak [5])} \tag{3.5}$$

$$\hat{e}_1(\xi) = (1/2\pi)((2/|\xi|)\sin(|\xi|/2))^3 \qquad \text{(Shepp-Logan [6])} \tag{3.6}$$

$$e_1(x) = ((2m+3)/2\pi)(1-|x|^2)^{m+1/2}, \text{ for } |x| < 1 \quad \text{([7])} \tag{3.7}$$

The corresponding global tomography kernels are given by

(3.5′)
$$k_1(y) = (1/4\pi^2)((1/y)\sin y - (2/y^2)\sin^2(y/2))$$

(3.6′)
$$k_1(y) = 1/(\pi^2(1-4y^2)), y \text{ integer, linear between integers}$$

(3.7′)
$$k_1(y) = C_m[H(y)\log\frac{|y+1|}{|y-1|} + \sum_{n=1}^{2m+1}\frac{(-1)^n}{n\, n!}H^n(y)((y+1)^n - (y-1)^n)]$$

$$\text{with } C_m = \frac{\Gamma(m+(5/2))}{2\pi^{5/2}m!}, \quad H(y) = y(1-y^2)^m.$$

The Ram-Lak and Shepp-Logan point spread functions do not have bounded support, and cannot be used for local tomography. The local tomography kernel for the point spread function in (3.7) is (with C_m as above)

$$K_1(y) = \pi C_m(1-y^2)^{m-1}(1-(2m+1)y^2) \tag{3.8}$$

Normally $6 \leq m \leq 15$. With $m = 6$, the point spread radius is

$$\rho(e_1) = .678 \quad (e_1 \text{ from } (3.7)). \tag{3.9}$$

In the Shepp-Logan case

$$\rho(e_1) = 1.37 \quad (e_1 \text{ from } (3.6)). \tag{3.10}$$

The Ram-Lak point spread radius has not been checked, since the considerations below do not apply to oscillatory kernels.

From the nature of Λ it is to be expected that if either the point spread function or kernel is given by a simple formula, the other will not. The simple point spread function in (3.7) leads to the kernel in (3.7'), which, fortunately has quickly convergent series expansions.

$$k_1(y) = \begin{cases} (1/4\pi^2)\sum_{j=0}^{\infty} A(m,-j)y^{2j} & \text{for } |y| < 1 \\ -(1/4\pi^2)\sum_{j=1}^{\infty} A(m,j)y^{-2j} & \text{for } |y| > 1 \end{cases} \tag{3.11}$$

with $A(m,j) = 1\cdot 3\cdots(2m+3)/(1+2j)\cdot(3+2j)\cdots(2m+1+2j)$.

The simple Shepp-Logan kernel comes from the point spread function

$$e_1(x) = 2\pi k(|x|) - 4\int_0^{|x|} k'(t)\arcsin(t/|x|)dt. \tag{3.12}$$

Since k' is constant between integers the integral is easy to calculate but the resulting expression is complicated and hard to analyse, as is the integral itself. For example we do not know the order of magnitude of the Shepp-Logan point spread function for large $|x|$, nor whether it oscillates indefinitely.

4. Numerical implementation of (2.3) and (2.4). Let the x-ray sources be situated at the points

$$a_i = R(\cos\alpha_i, \sin\alpha_i), \quad \alpha_i = 2\pi(i-1)/p, \quad i = 1, \dots, p$$

and the detectors at the angles

$$\phi_\ell = \pi\ell/2q, \quad \ell = -q, \dots, 0, \dots, q.$$

The first step in implementing formulas (2.3) and (2.4) is to compute the convolutions $F_{\alpha_i}(\phi_j)$ in (2.5), i.e. to compute

$$F_{\alpha_i}(\phi_j) = \int_{-\pi/2}^{\pi/2} D_{\alpha_i} f(\phi) \cos\phi \; k(\sin(\phi_j - \phi)) d\phi, \tag{4.1}$$

where k is a tomography kernel, either global or local. These are approximated by the discrete convolutions

$$\overline{F}_i(j) = (\pi/2q) \sum_{\ell=-q}^{q} D_{\alpha_i} f(\phi_\ell) \cos\phi_\ell \; \overline{k}(j-\ell) \tag{4.2}$$

where $\overline{k}(j)$ is an approximation to $k(\sin\phi_j)$.

The second step is to compute, for each x in the (discretized) support of f, the outer, or back projection, integral

$$R \int_0^{2\pi} |x - a_i|^{-m} F_\alpha(\gamma) d\alpha \tag{4.3}$$

where $m = 2$ or 3 according as the tomography is global or local. This integral is approximated by the Riemann sum

$$(2\pi R/p) \sum_{i=1}^{p} |x - a_i|^{-m} \overline{F}_i(\gamma_i) \tag{4.4}$$

where $\gamma_i = \arcsin(\langle x - a_i, a_i^\perp \rangle / R|x - a_i|)$, and $\overline{F}_i(\gamma_i)$ is obtained by linear interpolation: if $\gamma_i = (1-u)\phi_j + u\phi_{j+1}$, $0 \le u \le 1$, then

$$\overline{F}_i(\gamma_i) = (1-u)\overline{F}_i(j) + u\overline{F}_i(j+1). \tag{4.5}$$

The interpolation error is studied in [3,4].

The Shepp-Logan kernel in (3.6′) and the kernel in (3.7′) have a sharp maximum at 0 and single symmetric pair of sharp minima. In order that the discrete convolution (4.2) be a good approximation to the true convolution (4.1) it is advisable that the maximum and the minima occur at detectors, i.e. at points where $D_a f$ is sampled. If the detectors are fixed (as in an already existing scanner), the point spread radius must be adjusted so as to bring the minimum to a detector position.

The numbers r and s are determined initially by the requirements that $\rho(e_r) \sim \rho_0$ and $s = r/(R+R_0)$, then adjusted slightly so that the minimum of $k_s(\sin\phi)$ falls on a detector position. Normally the kernel minimum is placed at the first detector position. In the absence of noise this provides the best resolution attainable with the fixed detectors. In the presence of noise it may provide very poor resolution. (See [8].) In general, ρ_0, therefore r and s, should be determined on the basis of the problem imperatives, then s should be adjusted to bring the kernel minimum to the nearest detector position.

The role of the condition that the kernel minimum occur at a detector is clarified in [4]. It need not be satisfied for all kernels. With the Ram-Lak kernel the principal minimum falls to the right of the first detector if the detector spacing fulfills the Nyquist condition.

Any tomography kernel k has mean value 0, since the Fourier transform vanishes at 0. In parallel beam tomography experiments it was found necessary to preserve this condition by choosing the discrete approximation $\overline{k}$ in (4.2) to have mean value 0. For the kernel in (3.7′) this was achieved in [7] by choosing an integration process making (4.2) exact for piecewise linear $D_a f$. With the Shepp-Logan kernel the mean value is 0 when the kernel minimum falls on a detector position. With the Ram-Lak kernel the mean value is 0 if the detector spacing satisfies the Nyquist condition. It is proved in [1] that if the discrete kernel K has mean value $\mu \neq 0$, then the local parallel beam reconstruction contains an additive artifact with Fourier transform proportional to $\mu/|\xi|$ for small $|\xi|$, i.e. a cup artifact. It is also shown experimentally that similar artifacts appear in the divergent beam case.

5. Scanner parameters. Suppose that f has support in the circle of radius R_0, and that $e * f(x)$ is to be reconstructed with resolution ρ_0 in this circle at the points of an $n \times n$ matrix.

If the kernel minimum is placed at the first detector, detector spacing is determined as follows. Set $r = \rho_0/\rho(e_1)$ and $s = r/(R+R_0)$. If the minimum of k_1 occurs at y_1, then the minimum of $k_s(\sin\phi)$ occurs at $\arcsin(sy_1)$, and this is the detector spacing.

The number of x-ray sources is determined heuristically through the formula (see [8])

$$\int_{A^+} (D_a f(\theta) + D_a f(-\theta)) |\langle a, \theta\rangle| k_r(E_\theta x - E_\theta a) da = R \int_{\theta^\perp} P_\theta f(y) k_r(E_\theta x - y) dy$$

where A^+ is the semi-circle $\langle a, \theta\rangle > 0$, and the left side is the source integral in (1.4). This formula suggests that the number of sources in the semi-circle should be the number of y's needed for the right hand integral. Since the minimum of k_r occurs at ry_1, this calls for $p = 4R_0/ry_1$ sources around the full circle.

In low contrast situations (e.g. the .35% envisioned in section 3) noise plays a significant role. For the visibility of a .35% contrast circle of radius .8 ρ_0 (see section 3) that circle must contain about 4 pixels of the $n \times n$ matrix on which $e * f$ is displayed. Since each pixel represents a circle of radius R_0/n, if 4 such circles are to lie in a circle of radius $.8\rho_0$, then $n \geq 2R_0/.8\rho_0$.

Calculations like these cannot be taken too seriously, but only as starting points for experimentation. However, the resulting parameters agree pretty well with parameters in use. A standard current medical scanner, the General Electric 9800, aims at resolution $\rho_0 = 1.25$ mm. with .35% contrast. The source radius $R = 630$ mm. The object radius $R_0 = 257$ mm. The scanner has 984 sources around the source circle and 742 detectors in an angle of .8552. The matrix size is 512×512. For comparison, the calculated parameters with $\rho(e_r) = \rho_0$ are:

	Kernel in (3.7′), $m = 6$	Shepp-Logan kernel	local kernel
$\rho(e_1)=$	.678	1.37	.678
$r =$	1.8437	.9124	1.8437
$s =$	.0020	.0010	.0020
min k_r at	1.0177	.9142	.8831
min k_s at	.00115	.00103	.001
matrix	514× 514		514 × 514
# detectors	745	829	856
# sources	1010	1124	1160

The resolution of small features is very sensitive to the contrast. According to scanner specifications, if the contrast drops from .35% to .32%, ρ_0 increases from 1.25 mm. to 2 mm. For high contrast features it decreases to .375 mm.

Since the local kernel K_1 is 0 for $|y| > 1$, and its minimum is at $\sqrt{3/13}$, local numerical convolutions in (4.2) require 3,5, or 7 multiplications per point, according as the minimum is at the first, second, or third detector.

These parameter calculations are for $m = 6$ in (3.7). In practice a suitable detector for the kernel minimum is determined, then m (real) chosen so that the sum of the kernel values is small (and positive to counter the intrinsic cupping of Λ). When this is not possible, r is changed slightly, or a small constant is added to the kernel.

6. Remarks. Apart from the obvious numerical advantage coning from replacing several hundred multiplications by $3 - 8$, the main advantages of local tomography appear in the reconstruction of relatively small interior regions. Some of them are as follows.

(1) In medical examinations the x-ray dose is reduced by coning the x-ray beam.

(2) Better spatial resolution is obtained with the same number of detectors, or the same resolution with fewer detectors.

(3) Parts of an object too big to be x-rayed can be reconstructed.

(4) When the region of interest has substantially different density from neighboring regions (e.g. the heart or spine, surrounded by lungs) the x-ray exposure can be taylored to the region of interest.

(5) Interfaces are sharper.

Some of these advantages are illustrated in the examples below.

The principal disadvantage is that local tomography does not give a quantitative picture of the x-ray attenuation coefficient. Also, at high contrast interfaces the local reconstruction has a positive-negative swing that appears as a halo in a gray scale picture. This problem is more cosmetic than real, but the halo can hide small variations next to the interface.

The formulas and most of the advantages of local tomography were recognized about 1979. The first tests, involving a parallel beam algorithm applied to narrowly coned fan beam data, were done in 1984. One result is shown in [8]. Parallel beam formulas with a different approximation to Δ were found independently by Vainberg and Faingois [9]. The examples below come from fan beam algorithms applied to data from the DSR scanner at the Mayo Clinic and a Siemens scanner at the University Hospital in Münster.

7. Examples. The examples below show comparisons of local and global tomography applied to the same objects. Those from Siemens data are shown on 256 $\times$ 256 matrices; those from Mayo Clinic data are on 128 $\times$ 128 matrices. There are 256 gray levels with black = minimum density, white = maximum. Details mentioned in the discussion are all clearly visible on Polaroid photos of a high resolution CRT, but some may not reproduce well enough to be visible in the printed article.

1. Resolution phantom. The phantom is used by the Siemens Co. to test scanner resolution. The Hounsfield numbers of the components are: exterior: -1000, rim: 120, main interior: 0, dense block at right: 910: two upper blocks: -55, two lower blocks: 135, eight small holes in the two central blocks: 0. Approximate x-ray attenuation coefficients are obtained by adding 1000 and multiplying by .000195.

The global reconstruction was done on the Siemens scanner. The image is small because the phantom filled a small part of the scanner reconstruction circle. In the local reconstructions most of the exterior of the phantom was ignored. At present the dimensions of the small holes are not known to the authors.

Both intuition and the calculations in section 5 suggest that for a given resolution more sources and detectors are needed in local tomography than in global. Conversely, with the sources and detectors of the Siemens scanner fixed for global tomography, some problems might be expected in local tomography. In this case, with the kernel minimum at the standard 1st detector (the picture with $r = .003$), there are four quite visible streaks coming from the long edges of the dense block, and many faint ones in the upper left of the phantom. Increasing r to .004 (i.e. decreasing the resolution) removes all of the faint streaks and most of the four stronger ones, while the smallest hole with the smallest contrast remains visible, at least on the screen and on the photographs of the screen. As described in section 6, there is an overshoot at the boundary of the dense block and at the rim, the latter seen mainly in the fact that the rim with Hounsfield number 120 appears whiter than the blocks with Hounsfield number 135. There is also some overshoot at the boundaries of those blocks. At sharp boundaries between sharp density contrasts white appears whiter and black appears blacker. This phenomenon is characteristic of Λf, not an artifact. It is much less pronounced in reconstructions of living objects where the boundaries are not so sharp.

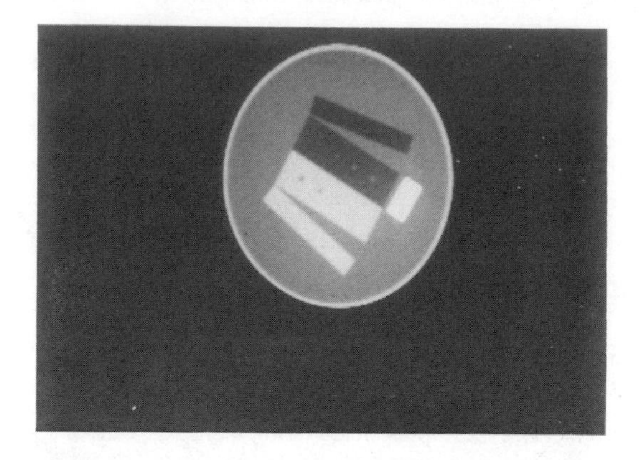

Resolution phantom, global reconstruction

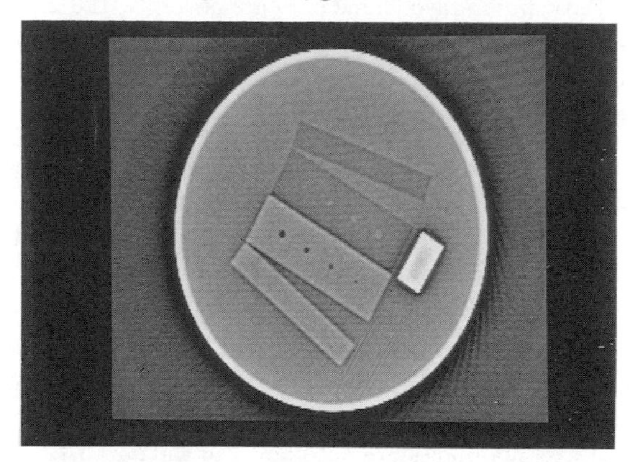

Resolution phantom, local reconstruction, $r = .003$

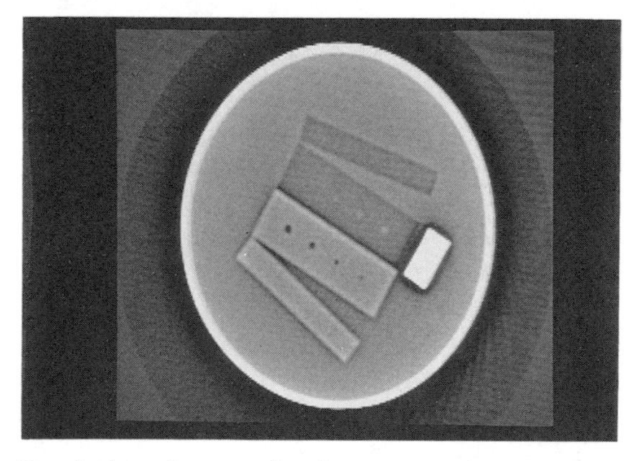

Resolution phantom, local reconstruction, $r = .004$

2. Abdomen. Again, the global reconstruction, was done on the Siemens scanner. Only 360 x-ray sources were used, so in the local reconstruction r was increased to .006, reducing somewhat the resolution. In all cases (like this one) where only part of the object is shown in the local reconstruction, that reconstruction was done with the x-ray beam coned to the part shown. In this case, and in the case of the head, it was coned by discarding the data from rays not meeting that part. In the heart it was coned with lead shields during the x-ray procedure.

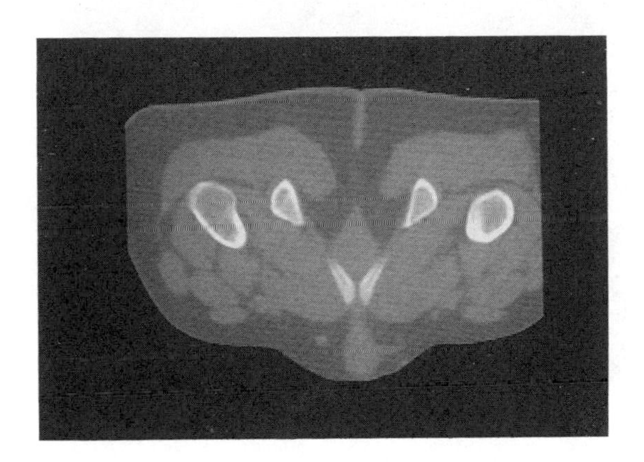

Abdomen, global reconstruction

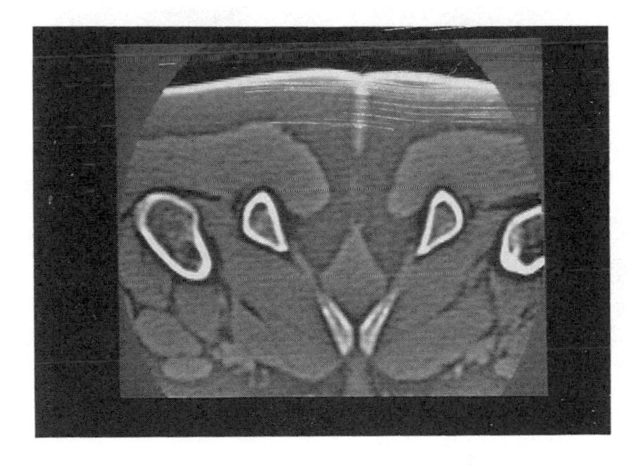

Abdomen, local reconstruction

3. Head. The scan was done on a Siemens scanner, but the global reconstruction was done at the University of Münster. The two pictures represent the same reconstruction displayed with different gray level windows to show different features. The light picture shows details in the dark areas; the dark picture shows details in the light areas. The picture of the local reconstruction ($r = .003$ again) seems to show both equally well.

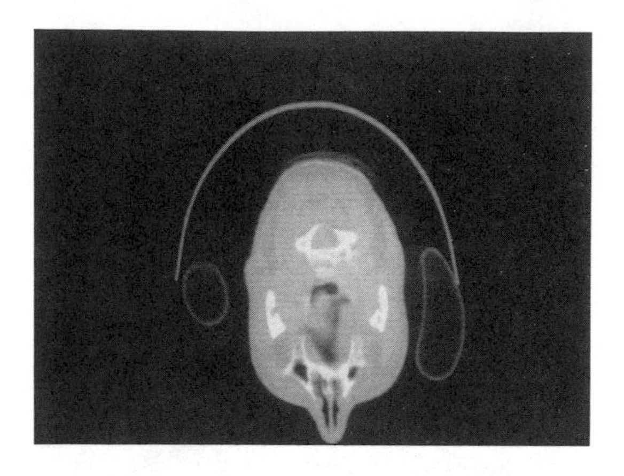

Head, global reconstruction

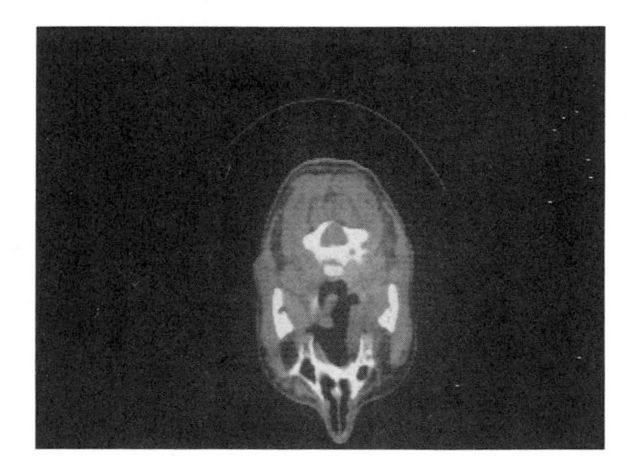

Head, global reconstruction

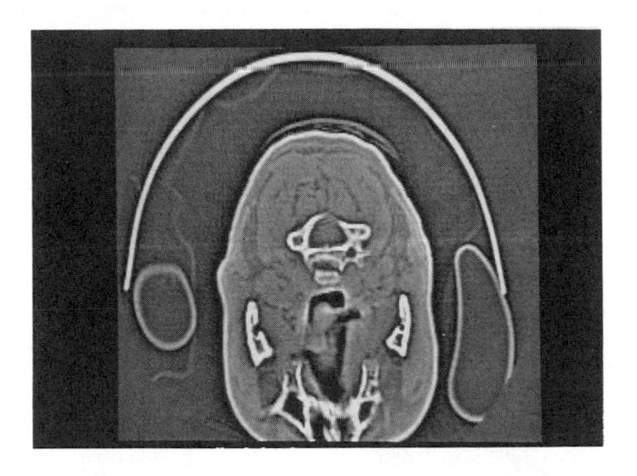

Head, local reconstruction

4. Coronary arteries. The pictures of the heart and coronary arteries come from scans of a realistic chest phantom made by the Humanoid Company and used at the Mayo Clinic. The arteries are filled with dilute contrast medium (about 40 mg. I/ml.) simulating a low level intravenous contrast injection. Three dimensional reconstructions were made from 120 cross sections about 1 mm. thick. The dye filled heart chambers were erased from the reconstructions, then the pictures were made from brightest point projections on a plane.

The scans were done by the Mayo Clinic DSR, a very fast beating heart scanner. The DSR employs fourteen x-ray tubes and records the attenuation on fourteen 2 dimensional fluorescent screens. One 3 dimensional scan takes .127 sec. Usually six scans, gated to the heart beat, are used in order to improve the photon statistics and to ensure that the contrast medium is present throughout the arteries. The attenuation recorded on the 30×30 cm. fluorescent screen is transferred to a 1 × 1 cm. CCD chip by a lens. During the experiment shown the DSR was in a process of modification from older TV cameras to ones with CCD chips, and only one of the latter had been installed. Experience has shown that better results are obtained from all fourteen cameras.

Because of the high speed and the 2 dimensional recording, the DSR provides noiser data and less dynamic range than conventional 1 dimensional crystal or xenon gas detectors. Also, the compression of the 30×30 cm. fluorescent screen to a 1×1 cm. CCD chip introduces geometric distortions. On the other hand, 2 dimensional recording appears to be necessary in 3 dimensional reconstructions of moving objects like the beating heart. Local tomography avoids the dynamic range problem (by confining the region reconstructed to the region of interest), and the geometric distortion problem (by reducing the compression). Moreover, when the region of interest is relatively small, (e.g. the heart vs. the full chest) local tomography allows cheaper and more accurate $3D$ scanner designs than the current fluorescent screen-lens-CCD chip combination.

The numerical implementation of the local reconstruction algorithm within the DSR reconstruction software is due to Dr. P.J. Thomas of the Biodynamics Research Unit at the Mayo Clinic. Because of the noisy DSR data, the kernel minimum was placed at the third detector.

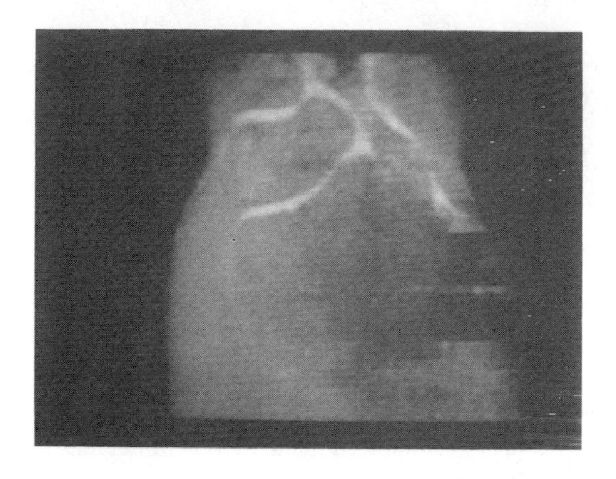

Coronary arteries, global reconstruction

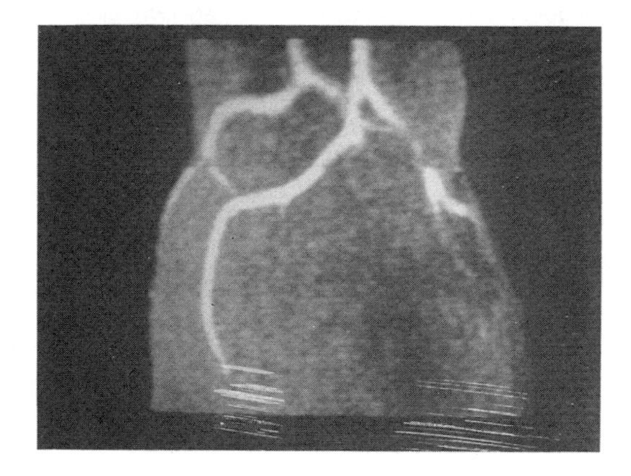

Coronary arteries, local reconstruction

REFERENCES

[1] A. FARIDANI AND K.T. SMITH, *Cup effects in tomography*, In preparation.

[2] A.V. LAKSHMINARAYANAN, *Reconstruction from divergent x-ray data*, SUNY Tech. Rpt., 92, Comp. Sci. Dept., Buffalo, NY (1975).

[3] F. NATTERER, *The Mathematics of Computerized Tomography*, J. Wiley & Sons, New York, 1986.

[4] F. NATTERER AND A. FARIDANI, *Basic algorithms in tomography*. These proceedings.

[5] N. RAMACHANDRAN AND A.V. LAKSHMINARAYANAN, *Three dimensional reconstruction from radiographs and electron micrographs: application of convolutions instead of Fourier transforms*, PNAS USA (1971), pp. 2236–2240.

[6] L.A. SHEPP AND B.F. LOGAN, *The Fourier reconstruction of a head section*, IEEE Trans. on Nuclear Sci. (1974), pp. 21–43.

[7] K.T. SMITH, *Reconstruction formulas in computed tomography*, Proc. Symp. Appl. Math., AMS (1983), pp. 7–23.

[8] K.T. SMITH AND F. KEINERT, *Mathematical foundations of computed tomography*, Applied Optics (1985), pp. 3950–3957.

[9] E.I. VAINBERG AND M.L. FAINGOIS, *Increasing the spatial resolution in computerized tomography*, in *Problems in Tomographic Reconstruction*, pp. 28–35. A.S. Alekseev, M.M. Lavrent'ev, and G.N. Preobrazhensky, ed. Siberian branch of the Academy of Sciences of the USSR, Novosibirsk, 1985.

THE PHASE PROBLEM OF X-RAY CRYSTALLOGRAPHY

HERBERT HAUPTMAN*

Abstract. The electron density function $\rho(r)$ in a crystal determines its diffraction pattern, that is, both the magnitudes and phases of its x-ray diffraction maxima, and conversely. If, however, as is always the case, only magnitudes are available from the diffraction experiment, then the density function $\rho(r)$ cannot be recovered. If one invokes prior structural knowledge, usually that the crystal is composed of discrete atoms of known atomic numbers, then the observed magnitudes are, in general, sufficient to determine the positions of the atoms, that is the crystal structure.

The intensities of a sufficient number of X-ray diffraction maxima determine the structure of a crystal. The available intensities usually exceed the number of parameters needed to describe the structure. From these intensities a set of numbers $E_{\mathbf{H}}$ can be derived, one corresponding to each intensity. However, the elucidation of the crystal structure also requires a knowledge of the complex numbers $E_{\mathbf{H}} = |E_{\mathbf{H}}| \exp{(i\phi_{\mathbf{H}})}$, the normalized structure factors, of which only the magnitudes $|E_{\mathbf{H}}|$ can be determined from experiment. Thus, a "phase" $\phi_{\mathbf{H}}$, unobtainable from the diffraction experiment, must be assigned to each $|E_{\mathbf{H}}|$, and the problem of determining the phases when only the magnitudes $|E_{\mathbf{H}}|$ are known is called "the phase problem." Owing to the known atomicity of crystal structures and the redundancy of observed magnitudes $|E_{\mathbf{H}}|$, the phase problem is solvable in principle.

CONTENTS

1. The Experimental Background. The fundamental experiment on which the science of X-ray crystallography is based was performed in 1912 by Friedrich and Knipping at the instigation of Max von Laue.

*Medical Foundation of Buffalo, Inc. Research Institute, 73 High Street, Buffalo, New York 14203–1196

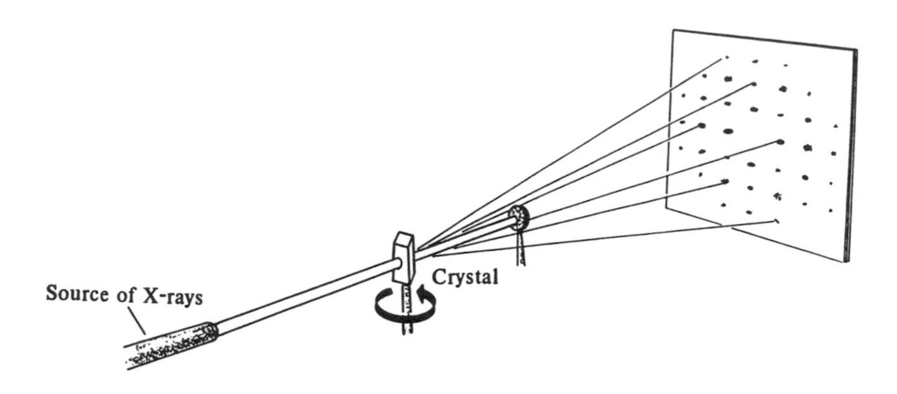

Figure 1. The X-ray diffraction experiment.

As shown in figure 1, when a beam of monochromatic X-rays is caused to stike a crystal, the incident ray is scattered in many different, discrete directions. The nature of the resulting so-called diffraction pattern, which is to say, the directions, intensities and phases of the scattered X-rays, is completely determined by the crystal structure, that is, the lattice parameters and the kinds and positions of the atoms in the crystal. By relating the crystal structure to its diffraction pattern this experiment was correctly seen to provide the key to the determination of crystal structures, and therefore, of molecular structures as well. At the present time, about 75 years after the celebrated experiment of Friedrich, Knipping, and von Laue, some 50,000 molecular structures have been determined by the technique of X-ray diffraction. This vast body of experimental knowledge constitutes the raw data on which the science of modern structural chemistry, with its myriad consequences, is based. It is for this reason that the 1912 experiment which demonstrated the scattering of X-rays by crystals must be regarded as one of the landmark experiments of the twentieth century.

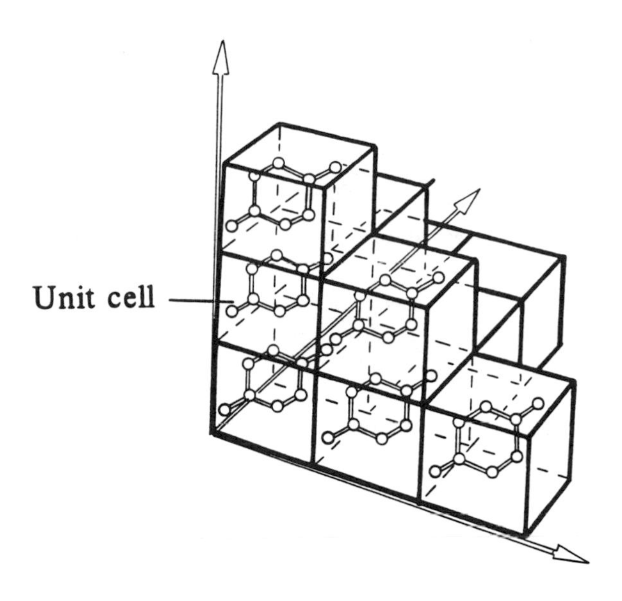

Figure 2. A triply periodic function of position defines a crystal structure.

2. The Mathematical Background. One imagines, as in Figure 2, that Euclidean space is partitioned by three families of parallel planes, consecutive planes in each family being equally spaced. In this way all of space is decomposed into congruent parallelepipeds, completely filling space, with no gaps between the fundamental parallelepipeds. One then distributes a finite set of points in each parallelepiped, the same number in each and identically arranged, so that each parallelepiped, together with its contents, is indistinguishable from every other one. If we imagine each point to be an atom, then there is associated with each atom an electron density function, and the superposition of these defines an electron density function $\rho(r)$ throughout all of space. Owing to its geometric construction, it is clear that $\rho(r)$ is a triply periodic function of the position vector r. From the mathematical point of view therefore, a crystal may be regarded as a non-negative triply periodic function, the density function $\rho(r)$, or as a triply periodic array of atoms. The crystal structure then is simply the density function $\rho(r)$ or, alternatively, the position vectors of the atoms in the fundamental parallelepiped, usually called the unit cell of the crystal. If the collection of atoms in the unit cell is regarded as a molecule, then knowledge of the crystal structure implies knowledge of the molecular structure as well.

3. The Phase Problem. Since the electron density function $\rho(r)$ is triply periodic it admits of a Fourier Series representation as follows:

$$\rho(r) = \frac{1}{V} \sum_{H} F_H \exp(-2\pi i H \cdot r) \tag{1}$$

where the summation is taken over all ordered triples of integers, here abbreviated by the so-called reciprocal lattice vector H, and the Fourier coefficient F_H is defined by

(2) $$F_H = \int_V \rho(r) \ \exp(2\pi i H \cdot r) dV$$

in which the triple integral is extended over the volume V of the unit cell.

Clearly the Fourier coefficient F_H, usually called the structure factor, is in general a complex number which may be written in polar form:

(3) $$F_H = |F_H| \exp(i\phi_H)$$

where $|F_h|$ is the magnitude of the complex number F_H and ϕ_H is said to be the phase of F_H. If one substitutes from Eq. (3) into Eq. (1), one obtains

(4) $$\rho(r) = \frac{1}{V} \sum_H |F_H| \exp\ i(\phi_H - 2\pi i H \cdot r),$$

the Fourier series representation of the density function $\rho(r)$, which is seen to depend on the magnitudes $|F_H|$ and the phases ϕ_H of the (complex) structure factors F_H.

The relationship between the diffraction pattern (Fig. 1) and the crystal structure stems from the fact that the magnitude $|F_H|$ is readily derived from the intensity of the X-ray scattered in the direction labeled by the reciprocal lattice vector H. The phase ϕ_H of this scattered ray, which is also needed if one is to calculate $\rho(r)$ from Eq. (4), is however lost in the diffraction experiment. It is this inability to measure the phases ϕ_H which gives rise to the phase problem of X-ray crystallography.

For almost 40 years after the scattering experiment of 1912 the crystallographic community argued that by calculating $\rho(r)$ from Eq. (4), using the experimentally determined values for the magnitudes $|F_H|$, but arbitrary values for the unobservable phases ϕ_H, one would obtain many electron density functions $\rho(r)$, all consistent with the observed magnitudes $|F_H|$. In short, observed magnitudes $|F_H|$ by themselves were deemed to be insufficient to determine a unique electron density function and therefore a unique crystal structure. Hence, it was believed that the phase problem - to deduce the values of the unobserved phases ϕ_H from the measured intensities of the scattered X-rays – was unsolvable, even in principle.

However, this argument is flawed because if arbitrary values for the phases ϕ_H are used in Eq. (4) one obtains, in general, electron density functions not in accord with real structures. For example, the electron density function $\rho(r)$ must be non-negative everywhere, whereas arbitrary values for the phases in Eq. (4) will yield density functions $\rho(r)$ which, in general, will be negative somewhere. Clearly then, such phases will be inadmissible. Although the non-negativity of $\rho(r)$ imposes severe restrictions on the possible values which the phases may have, it turns out that, except in the simplest of cases, the nonnegativity of $\rho(r)$ is not sufficiently restrictive to determine unique values for the phases. A stronger restriction is needed; but this too is available. The required condition is atomicity. Because molecules consist of atoms it turns out that not only is the electron density function non-negative everywhere, but it also has the property that it takes on large positive values at the positions of the atoms and drops down to relatively small values in the regions between the atoms. It remains to exploit this property of $\rho(r)$, and to this end one refers to Eq. (2).

The integrand of Eq. (2) makes its major contribution to the value of the integral at position vectors r corresponding to the atomic positions and makes a negligible contribution to the value of the integral at points between the atoms. Replacing the real crystal with the continuous electron density function $\rho(r)$ by an ideal crystal consisting of N non-vibrating point atoms per unit cell, the integral of Eq. (2) is replaced by a sum and the structure factor F_H is replaced by the normalized structure factor E_H defined by

$$E_H = |E_H| \exp(i\phi_H) = \frac{1}{\sigma_2^{1/2}} \sum_{j=1}^{N} Z_j \exp(2\pi i H \cdot r_j) \tag{5}$$

where N is the number of atoms in the unit cell, r_j is the position vector and Z_j is the atomic number of the atom labeled j and

$$\sigma_n = \sum_{j=1}^{N} Z_j^n \ , \ n = 2, 3, 4, \ldots \tag{6}$$

It now turns out that the magnitudes $|E_H|$ of the normalized structure factors E_H can again be calculated, at least approximately, from the observed intensities of scattered X-rays, but the phases ϕ_H are unobservable. Now however, since Eq. (5) captures the atomicity property of real structures and since all one now requires are the position vectors r_j, rather than the much more complicated electron density function $\rho(r)$, it turns out that the observed magnitudes $|E_H|$ are more than sufficient for the job. This is most readily seen by equating the magnitudes of the two sides of Eq. (5) in order to eliminate the unknown phases ϕ_H, in this way obtaining the system of equations

$$|E_H| = \frac{1}{\sigma_2^{1/2}} \left| \sum_{j=1}^{N} Z_j \exp(2\pi i H \cdot r_j) \right|, \tag{7}$$

one equation corresponding to each reciprocal lattice vector H for which the intensity of the corresponding scattered X-ray has been observed. The unknowns in Eq. (7) are simply the N atomic position vectors r_j each of which is defined by its three components. Hence, the number of unknowns is 3N. However, for structures of moderate complexity, say those having fewer than some 100 non-hydrogen atoms per unit cell, the number of measurable diffraction intensities usually exceeds the number of unknowns by a factor of an order of magnitude at least. Hence, for such structures the number of equations (7) is far more than needed to determine the unknown position vectors r_j. Once the atomic position vectors r_j have been determined, substitution into Eq. (5) yields not only the known magnitudes $|E_H|$ but the unknown phases ϕ_H as well. Thus, not only is the phase problem not an unsolvable one, even in principle, as had been generally believed for so many years, but it is actually a greatly overdetermined one. In short, the missing phase information is to be found in the measured intensities. It remains to show how this phase information is to be extracted.

The system (5) implies the existence of relationships among the normalized structure factors E_H since the (relatively few) unknown position vectors r_j may, at least in principle, be eliminated. By the term "direct methods" is meant that class of methods which exploits relationships among the normalized structure factors in order to go directly from the observed magnitudes $|E|$ to the needed phases ϕ. How this is done by probabilistic techniques is described in the sequel.

4. The Structure Invariants. Eq. (5) leads directly to

$$\begin{aligned}\left\langle E_H \exp(-2\pi i H \cdot r)\right\rangle_H &= \frac{1}{\sigma_2^{1/2}}\left\langle \sum_{j=1}^{N} Z_j \exp[2\pi i H \cdot (r_j - r)]\right\rangle_H \\ &= \frac{Z_j}{\sigma_2^{1/2}} \qquad \text{if } r = r_j \\ &= 0 \qquad \text{if } r \neq r_j \end{aligned} \tag{8}$$

Eq. (8) implies that the normalized structure factors $E_{\mathbf{H}}$ determine the crystal structure. However (5) does not imply that, conversely, the crystal structure determines the values of the normalized structure factors $E_{\mathbf{H}}$ since the position vectors $\mathbf{r}_j$ depend not only on the structure but on the choice of origin as well. It turns out nevertheless that the magnitudes $|E_{\mathbf{H}}|$ of the normalized structure factors are in fact uniquely determined by the crystal structure and are independent of the choice of origin, but that the values of the phases $\phi_{\mathbf{H}}$ depend also on the choice of origin. Although the values of the individual phases depend on the structure and the choice of origin, there exist certain linear combinations of the phases, the so-called structure invariants, whose values are determined by the structure alone and are independent of the choice of origin.

If the origin of coordinates is shifted to a new point having position vector $\mathbf{r}_0$ with respect to the old origin, then, from the definition, (5), of $E_{\mathbf{H}}$, it follows readily that the phase $\phi_{\mathbf{H}}$ of the normalized structure factor $E_{\mathbf{H}}$ with respect to the old origin is replaced by the new phase $\phi'_{\mathbf{H}}$ with respect to the new origin given by

$$\phi'_{\mathbf{H}} = \phi_{\mathbf{H}} - 2\pi \mathbf{H} \cdot \mathbf{r}_0. \tag{9}$$

Eq. (9) implies that the linear combination of three phases,

$$\psi_3 = \phi_{\mathbf{H}} + \phi_{\mathbf{K}} + \phi_{\mathbf{L}}, \tag{10}$$

is a structure invariant (triplet) provided that

$$\mathbf{H} + \mathbf{K} + \mathbf{L} = 0; \tag{11}$$

the linear combination of four phases,

$$\psi_4 = \phi_{\mathbf{H}} + \phi_{\mathbf{K}} + \phi_{\mathbf{L}} + \phi_{\mathbf{M}} \tag{12}$$

is a structure invariant (quartet) provided that

$$\mathbf{H} + \mathbf{K} + \mathbf{L} + \mathbf{M} = 0; \tag{13}$$

etc.

Since the values of the individual phases depend not only on the structure but on the choice of origin, it follows that magnitudes $|E|$ alone cannot determine unique values of the individual phases. It is clear that magnitudes $|E|$ alone determine only the values of the structure invariants (and not even uniquely at that, because of the enantiomorph problem, as clarified below) and only then, after suitable specification of the origin (and enantiomorph when necessary), may the individual phases be determined.

It should be noted finally that the theory of the structure invariants leads directly to recipes for origin specification called for by the techniques of direct methods. For example, when no crystallographic element of symmetry is present (space group P1) the rule states simply that the values of any three phases

$$\phi_{h_1 k_1 l_1}, \phi_{h_2 k_2 l_2}, \phi_{h_3 k_3 l_3},$$

where the determinant

$$\begin{vmatrix} h_1 k_1 l_1 \\ h_2 k_2 l_2 \\ h_3 k_3 l_3 \end{vmatrix} = \pm 1,$$

are to be specified arbitrarily, thus fixing the origin uniquely.

5. The Fundamental Principle of Direct Methods. It is known that the values of a sufficiently extensive set of cosine invariants (the cosines of the structure invariants) lead unambiguously to the values of the individual phases *(1)*. Magnitudes $|E|$ are capable of yielding estimates of the cosine invariants only or, equivalently, the magnitudes of the structure invariants; the signs of the structure invariants are ambiguous because the two enantiomorphous structures (related to each other by reflection through a point) which are permitted by the observed magnitudes $|E|$ correspond to two values of each structure invariant differing only in sign. However, once the enantiomorph has been selected by specifying arbitrarily the sign of a particular enantiomorph sensitive structure invariant (i.e. one different from 0 or π), then the magnitudes $|E|$ determine both signs and magnitudes of the structure invariants consistent with the chosen enantiomorph. Thus, for fixed enantiomorph, the observed magnitudes $|E|$ determine unique values for the structure invariants; the latter, in turn, as certain well defined linear combinations of the phases, lead to unique values of the individual phases. In short, the structure invariants serve to link the observed magnitudes $|E|$ with the desired phases ϕ (the fundamental principle of direct methods). It is this property of the structure invariants which accounts for their importance and which justifies the stress placed on them here.

6. The Neighborhood Principle. It has long been known that, for fixed enantiomorph, the value of any structure invariant ψ is, in general, uniquely determined by the magnitudes $|E|$ of the normalized structure factors. Recently it has become clear that, for fixed enantiomorph, there corresponds to ψ one or more small sets of magnitudes $|E|$, the neighborhoods of ψ, on which, in favorable cases, the value of ψ most sensitively depends; that is to say that, in favorable cases, ψ is primarily determined by the values of $|E|$ in any of its neighborhoods and is relatively independent of the values of the great bulk of remaining magnitudes. The conditional probability distribution of ψ, assuming as known the magnitudes $|E|$ in any of its neighborhoods, yields an estimate for ψ which is particularly good in the favorable case that the variance of the distribution happens to be small *(2,3)* (the neighborhood principle).

The first neighborhood of the triplet ψ_3, (Eq. 10), consists of the three magnitudes

$$|E_{\mathbf{H}}|,\ |E_{\mathbf{K}}|,\ |E_{\mathbf{L}}|. \tag{14}$$

The first neighborhood of the quartet ψ_4, (Eq. 12), consists of the four magnitudes

$$|E_{\mathbf{H}}|,\ |E_{\mathbf{K}}|,\ |E_{\mathbf{L}}|,\ E_{\mathbf{M}}|. \tag{15}$$

The second neighborhood of the quartet consists of the four magnitudes (15) plus the three additional magnitudes

$$|E_{H+K}|,\ |E_{K+L}|,\ |E_{L+H}|, \tag{16}$$

i.e. seven magnitudes $|E|$ in all (3).

The neighborhoods of all the structure invariants are now known.

7. The Solution Strategy *(4)*. One starts with the system of equation (5). By equating real and imaginary parts of Eq. (5) one obtains two equations for each reciprocal lattice vector **H**. The magnitude $|E_{\mathbf{H}}|$ and the atomic numbers Z_j are presumed to be known. The unknowns are the atomic position vectors r_j and the phases $\phi_{\mathbf{H}}$. Owing to the redundancy of the system (5), one naturally invokes probabilistic techniques in order to eliminate the unknown position vectors r_j, and in this way to obtain relationships among the unknown phases $\phi_{\mathbf{H}}$, dependent on the known magnitudes $|E|$, having probabilistic validity.

Choose a finite number of reciprocal lattice vectors $H, K, \ldots$ in such a way that the linear combination of phases

$$(17) \qquad \psi = \phi_{\mathbf{H}} + \phi_{\mathbf{K}} + \cdots$$

is a structure invariant whose value we wish to estimate. Choose satellite reciprocal lattice vectors $H', K', \ldots$ in such a way that the collection of magnitudes

$$(18) \qquad |E_{\mathbf{H}}|,\ |E_{\mathbf{K}}|, \ldots;\ |E_{\mathbf{H}'}|,\ |E_{\mathbf{K}'}|, \ldots$$

constitutes a neighborhood of ψ. The atomic position vectors r_j are taken to be the primitive random variables which are assumed to be uniformly and independently distributed. Then the magnitudes $|E_{\mathbf{H}}|, |E_{\mathbf{K}}|, \ldots;\ |E_{\mathbf{H}'}|,\ |E_{\mathbf{K}'}|, \ldots;$ and phases $\phi_{\mathbf{H}},\ \phi_{\mathbf{K}}, \ldots;\ \phi_{\mathbf{H}'}, \phi_{\mathbf{K}'}, \ldots$ of the complex normalized structure factors $E_{\mathbf{H}}, E_{\mathbf{K}}, \ldots; E_{\mathbf{H}'}, E_{\mathbf{K}'}, \ldots$, as functions, (5), of the position vectors r_j, are themselves random variables, and their joint probability distribution P may be obtained by techniques which are now standard. From the distribution P one derives the conditional joint probability distribution

$$(19) \qquad P(\Phi_{\mathbf{H}},\ \Phi_{\mathbf{K}}, \ldots \mid |E_{\mathbf{H}}|,\ |E_{\mathbf{K}}|, \ldots; |E_{\mathbf{H}'}|,\ |E_{\mathbf{K}'}|, \ldots),$$

of the phases $\phi_{\mathbf{H}}, \phi_{\mathbf{K}}, \ldots$, given the magnitudes $|E_{\mathbf{H}}|,\ |E_{\mathbf{K}}|, \ldots;\ |E_{\mathbf{H}'}|,\ |E_{\mathbf{K}'}|, \ldots$, by fixing the known magnitudes, integrating with respect to the unknown phases $\Phi_{\mathbf{H}'},\ \Phi_{\mathbf{K}'}, \ldots$ from 0 to 2π, and multiplying by a suitable normalizing parameter. The distribution (19), in turn, then leads directly to the conditional probability distribution

$$(20) \qquad P(\Psi|\ |E_{\mathbf{H}}|,\ |E_{\mathbf{K}}|, \ldots;\ |E_{\mathbf{H}'}|,\ |E_{\mathbf{K}'}|, \ldots)$$

of the structure invariant ψ, assuming as known the magnitudes (18) constituting a neighborhood of ψ. Finally the distribution (20) yields an estimate for ψ (e.g. the mode) which is particularly good in the favorable case that the variance of the distribution (20) happens to be small.

8. Estimating the Triplet. Let the three reciprocal lattice vectors H, K, and L satisfy (11). Refer to the set (14) for the first neighborhood of the triplet ψ_3 and to the previous paragraph for the probabilistic background.

Suppose that R_1, R_2, and R_3 are three specified non-negative numbers. Denote by

$$P_{1/3} = P(\Psi|R_1, R_2, R_3)$$

the conditional probability distribution of the triplet ψ_3, given the three magnitudes in its first neighborhood:

$$(21) \qquad |E_{\mathbf{H}}| = R_1,\ |E_{\mathbf{K}}| = R_2,\ |E_{\mathbf{L}}| = R_3.$$

Then, carrying out the program described earlier, one finds *(5)*

$$P_{1/3} = P(\Psi|R_1, R_2, R_3) \approx \frac{1}{K} \exp\,(A \cos \Psi) \tag{22}$$

where

$$A = \frac{2\sigma_3}{\sigma_2^{3/2}} R_1 R_2 R_3, \tag{23}$$

K is a normalizing constant not needed for the present purpose, and σ_n is defined by Eq. (6). Since $A > 0, P_{1/3}$ has a unique maximum at $\Psi = 0$; and it is clear that the larger the value of A the smaller is the variance of the distribution. See Figure 3, where $A = 2.316$ and Figure 4, where $A = 0.731$.

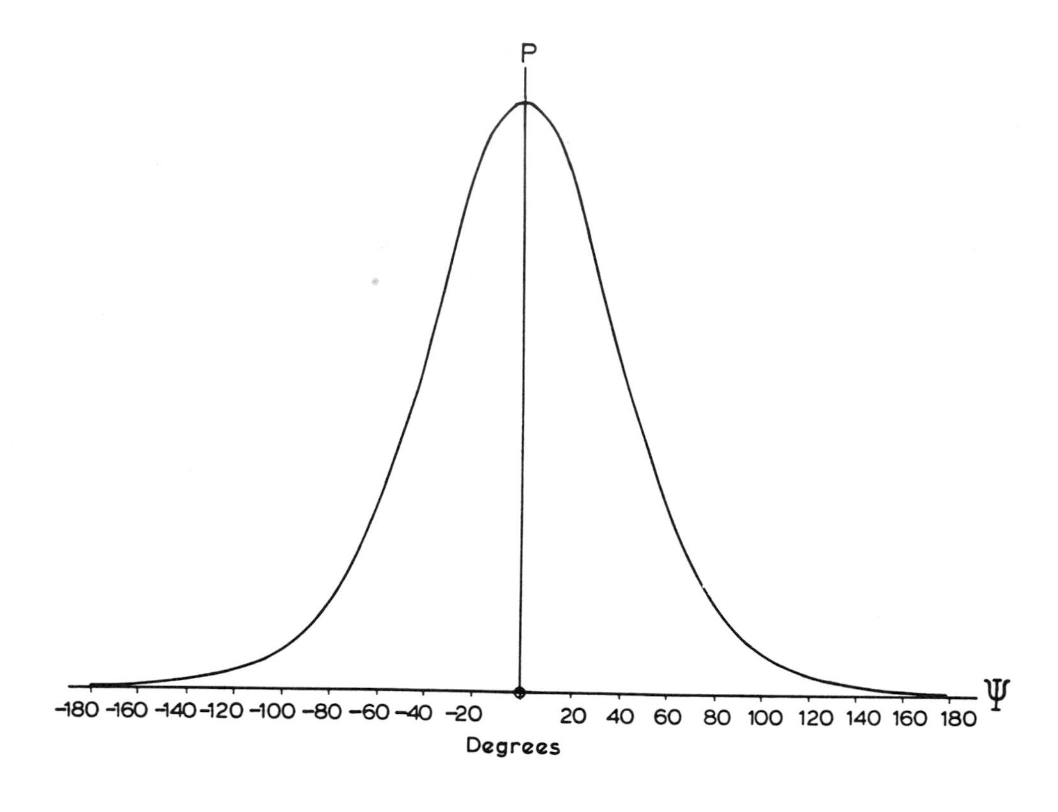

Figure 3. The distribution $P_{1/3}$, Eq. (22), for $A = 2.316$.

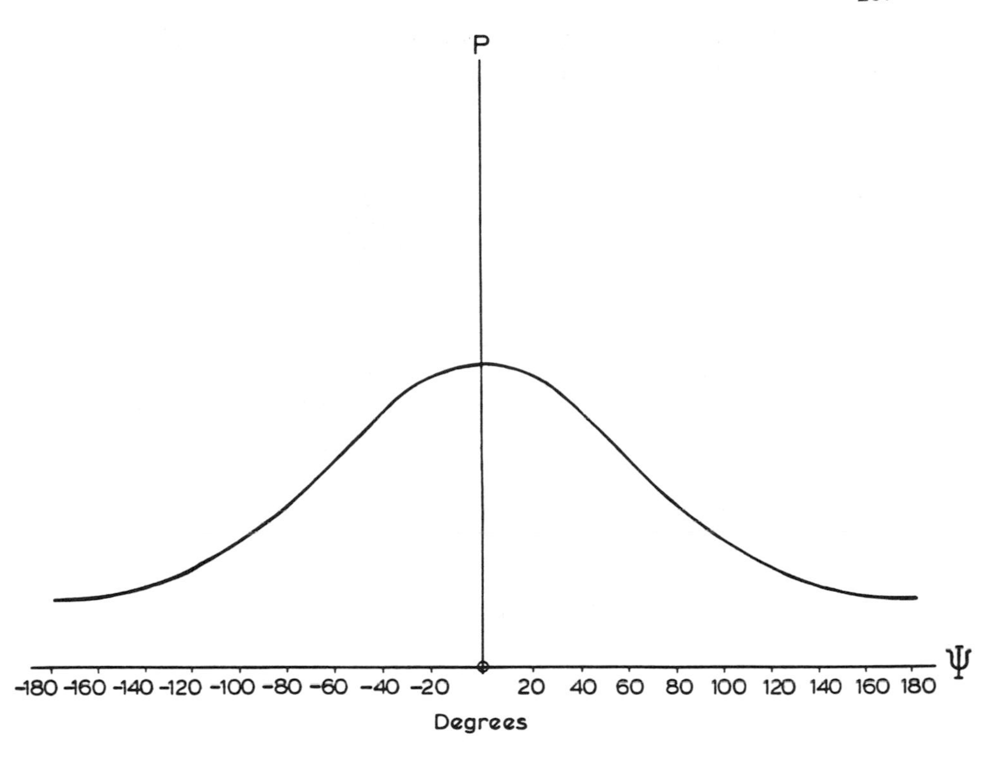

Figure 4. The distribution $P_{1/3}$, Eq. (22), for $A = 0.731$.

Hence in the favorable case that A is large, say, for example, $A > 3$, the distribution leads to a reliable estimate of the structure invariant ψ_3, zero in this case:

$$\psi_3 \approx 0 \quad \text{if } A \text{ is large.} \tag{24}$$

Furthermore, the larger the value of A, the more likely is the probabilistic statement (24). It is remarkable how useful this relationship has proven to be in the applications; and yet (24) is severely limited because it is capable of yielding only the zero estimate for ψ_3, and only those estimates are reliable for which A is large, the favorable cases. Note that the previously specified numbers R_1, R_2, R_3, identified with observed magnitudes $|E|$ by (21), as well as the atomic numbers Z_j, presumed to be known, appear via (23) as parameters of the distribution (22).

It should be mentioned in passing that a distribution closely related to (22) leads directly to the so-called tangent formula (6) which is universally used by direct methods practitioners:

$$\tan\ \phi_{\mathbf{h}} = \frac{\left\langle |E_{\mathbf{K}} E_{\mathbf{h}-\mathbf{K}}| \sin(\phi_{\mathbf{K}} + \phi_{\mathbf{h}-\mathbf{K}}) \right\rangle_{\mathbf{K}}}{\left\langle |E_{\mathbf{K}} E_{\mathbf{h}-\mathbf{K}}| \cos(\phi_{\mathbf{K}} + \phi_{\mathbf{h}-\mathbf{K}}) \right\rangle_{\mathbf{K}}}, \tag{25}$$

in which $\mathbf{h}$ is a fixed reciprocal lattice vector, the averages are taken over the same set of reciprocal lattice vectors $\mathbf{K}$, usually restricted to those vectors $\mathbf{K}$ for which $|E_{\mathbf{K}}|$ and $|E_{\mathbf{h}-\mathbf{K}}|$ are both large, and the sign of sin $\phi_{\mathbf{h}}(\cos\phi_{\mathbf{h}})$ is the same as the sign of the numerator (denominator) on the right hand side. The tangent formula is usually used to refine and extend a basis set of phases, presumed to be known. Because it employs the zero estimate of the triplet ψ_3, together with a measure (A) of variance, to link observed magnitudes $|E|$ with desired phases ϕ, the tangent formula serves as the simplest illustration of the fundamental principle of direct methods.

9. Estimating the Quartet. Only the seven magnitude, second neighborhood, estimate is given here. Proceed as in §8. Suppose that H, K, L and M are four reciprocal lattice vectors which satisfy Eq. (13). Refer to (15) and (16) for the seven mgnitudes $|E|$ which constitute the second neighborhood of the quartet ψ_4 (Eq. 12). Suppose that $R_1, R_2, R_3, R_4, R_{12}, R_{23}$, and R_{31} are seven non-negative numbers. Denote by

(26) $$P_{1/7} = P(\Psi|R_1, R_2, R_3, R_4; R_{12}, R_{23}, R_{31})$$

the conditional probability distribution of the quartet ψ_4, given the seven magnitudes in its second neighborhood:

(27) $$|E_{\mathbf{H}}| = R_1,\ |E_{\mathbf{K}}| = R_2,\ |E_{\mathbf{L}}| = R_3,\ |E_{\mathbf{M}}| = R_4;$$

(28) $$|E_{\mathbf{H}+\mathbf{K}}| = R_{12},\ |E_{\mathbf{K}+\mathbf{L}}| = R_{23},\ |E_{\mathbf{L}+\mathbf{H}}| = R_{31}.$$

The explicit form for $P_{1/7}$ has been found *(2, 3, 7)* but is too long to be given explicitly here. Instead, Figures 5 through 7 show the distribution (26) for typical values of the seven parameters Eqs. (27) and (28). Since the magnitudes $|E|$ have been obtained from a real structure with $N = 29$, comparison with the true value of the quartet is also possible. As already emphasized, the distribution (22) always has a unique maximum at $\Psi = 0$. The distribution (26), on the other hand, may have a maximum at $\Psi = 0$, or π, or any value between these extremes, as shown by Figures 5–7. Roughly speaking, the maximum of Eq. (26) occurs at 0 or π according as the three parameters R_{12}, R_{23}, R_{31} are all large or all small, respectively. Thus, in the special case that

(29) $$R_{12} \approx R_{23} \approx R_{31} \approx 0,$$

the distribution (26) reduces simply to

(30) $$P_{1/7} \approx \frac{1}{L}\ exp(-2B' \cos\Psi),$$

where

(31) $$B' = \frac{1}{\sigma_2^3}\ (3\sigma_3^2 - \sigma_2\sigma_4) R_1 R_2 R_3 R_4,$$

and L is a normalizing parameter not relevant here, which has a unique maximum at $\Psi = \pi$ (Fig. 7).

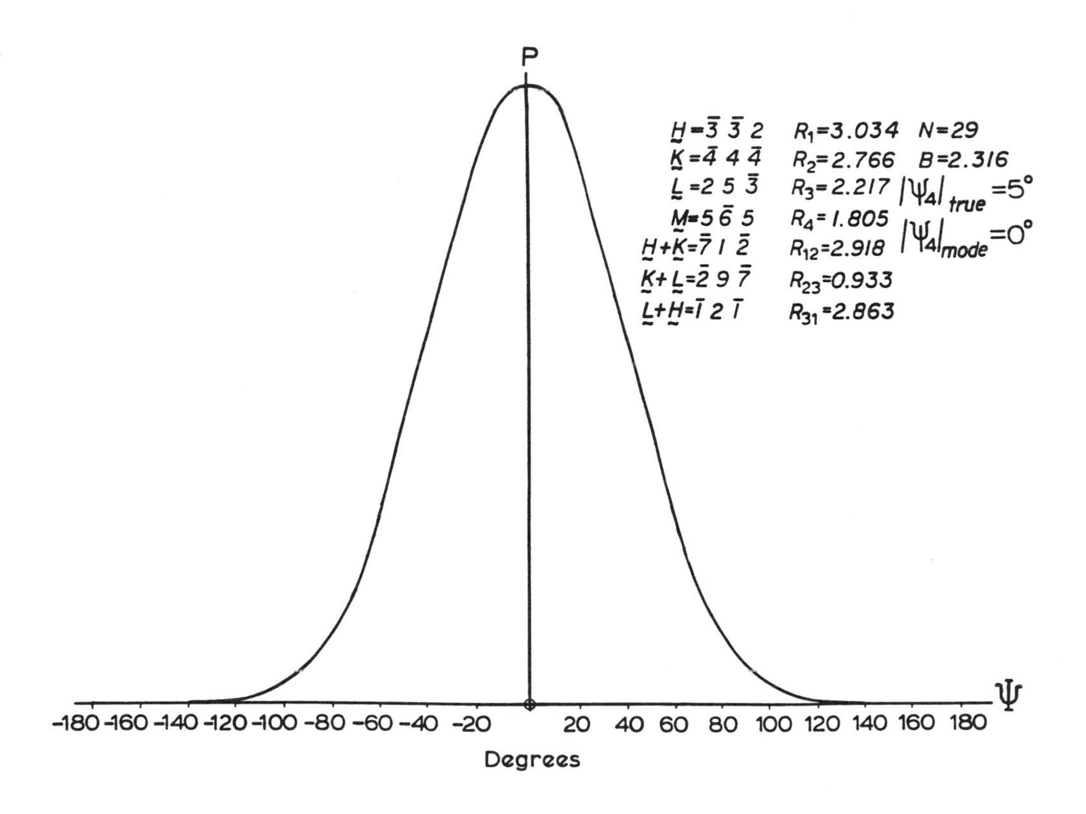

Figure 5. The distribution $P_{1/7}$, Eq. (26), for the values of the seven parameters, Eqs. (27) and (28), shown. The mode of $P_{1/7}$ is 0.

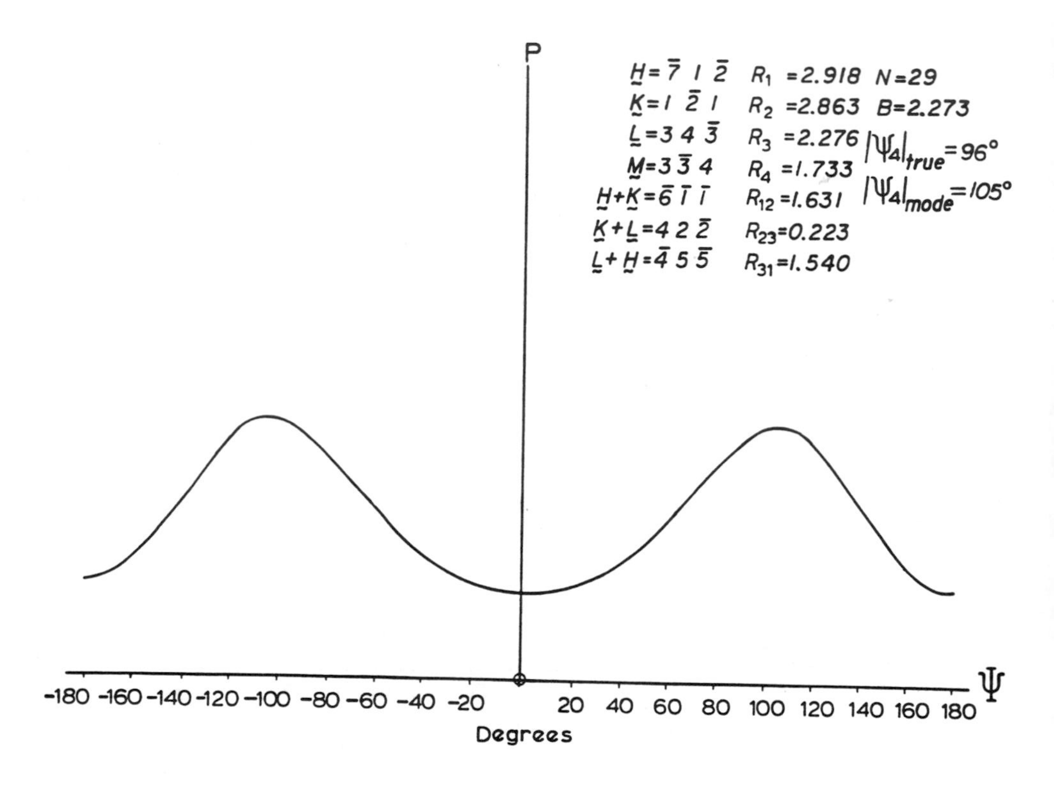

Figure 6. The distribution $P_{1/7}$, Eq. (26) for the values of the seven parameters, Eqs. (27) and (28), shown. The mode of $P_{1/7}$ is $\pm 105°$.

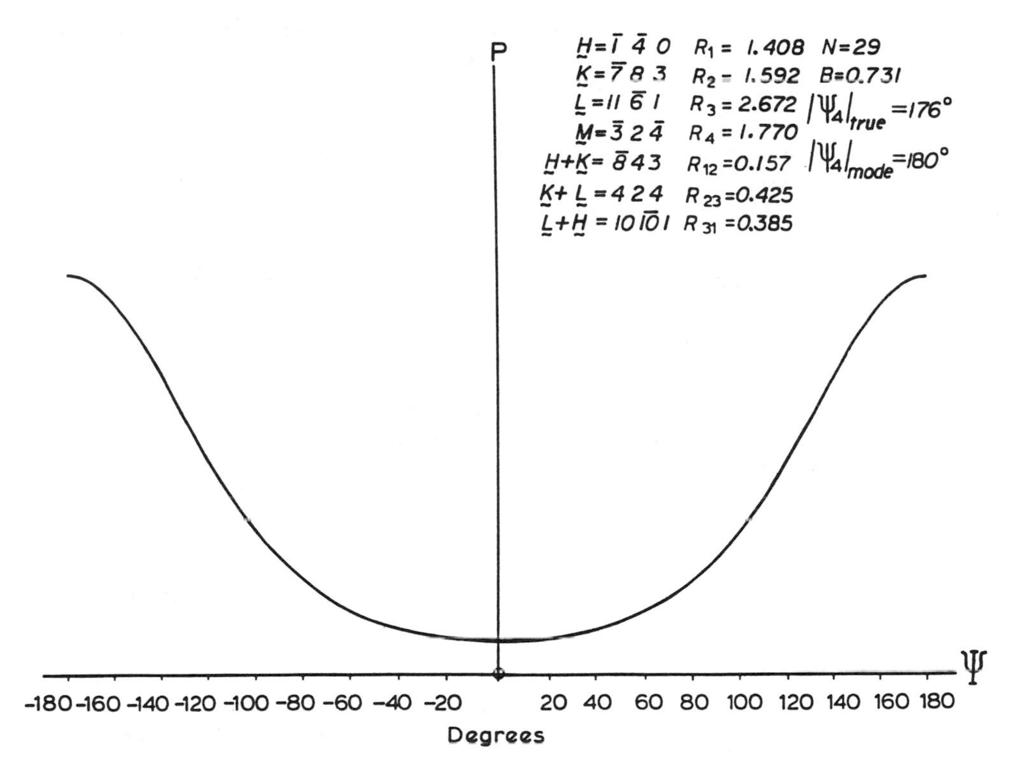

Figure 7. The distribution $P_{1/7}$, Eq. (26) for the values of the seven parameters, Eqs. (27) and (28), shown. The mode of $P_{1/7}$ is 180°.

10. Summary. The problem of determining crystal and molecular structures by the techniques of X-ray crystallography is the problem of determining the positions of the atoms in the crystal when only the intensities of the diffraction maxima are available from experiment. However the associated phases, which are also needed if one is to deduce the structure unambiguously from the experimental observations, are lost in the diffraction experiment. By exploiting prior structural information, usually the atomicity property of real crystal structures and the nonnegativity of the corresponding electron density function, it can be shown that the lost phase information is contained in the measured diffraction intensities and can be recovered provided that the molecular structure is not too large. To this end one introduces special linear combinations of the phases, the so-called structure invariants, whose values are uniquely determined by the structure alone, independently of the choice of origin. Estimates of the structure invariants in terms of measured intensities then lead unambiguously to the values of the individual phases. The method is to derive the conditional probability distributions of the structure invariants assuming as known certain suitably chosen sets of intensities and to use the mode as an estimate of the corresponding structure invariant.

The estimate (zero) of the triplet Ψ_3 (Eq. 24) and the related tangent formula (Eq. 25) are the cornerstones of most computer programs *(8, 9)* used for the direct solution of crystal structures. However, distributions of the quartets, in particular Eq. (30), used to identify those quartets whose values are close to π (the so-called negative quartets because their cosines are negative), as well as of the higher order structure invariants, often play an important (sometimes indispensable) role *(10–12)*, particularly for complex structures when diffraction data may be limited in number and quality *(13–23)*.

Major emphasis has been placed on the neighborhood principle and the important role played by the structure invariants. The conditional probability distribution of a structure invariant ψ, given the magnitudes $|E|$ in any of its neighborhoods, yields a reliable estimate for ψ in the favorable case that the variance of the distribution happens to be small. Since the structure invariants are the essential link between magnitudes $|E|$ and phases ϕ, probabilistic methods are seen to play the central role in the solution of the phase problem.

REFERENCES

[1] H. HAUPTMAN, *Crystal Structure Determination: The Role of the Cosine Seminvariants*, Plenum, New York (1972).

[2] H. HAUPTMAN, Acta Cryst. A31 (1975), 671.

[3] H. HAUPTMAN, Acta Cryst. A31 (1975), 680.

[4] H. HAUPTMAN AND J. KARLE, *Solution of the Phase Problem I. The Centrosymmetric Crystal*, American Crystallographic Association Monograph 3, Polycrystal Book Service, Western Springs, IL (1953).

[5] W. COCHRAN, Acta Cryst. 8 (1955), 473.

[6] J. KARLE AND H. HAUPTMAN, Acta Cryst. 9 (1956), 635.

[7] J. HAUPTMAN, Acta Cryst. 32 (1976), 877.

[8] P. MAIN ET. AL., *MULTAN 80: A System of Computer Programs for the Automatic Solution of Crystal Structures from X-ray Diffraction Data*, University of York, York, England, and University of Louvain, Louvain, Belgium (1980).

[9] I.L. KARLE AND J. KARLE, Acta Crystallogr. 16 (1963), 969.

[10] G.T. DETITTA, J.W. EDMONDS, D.A., LANGS, H. HAUPTMAN, Acta Crystallogr. Sect. A 31 (1975), 472.

[11] C.J. GILMORE, *A Computer Program for the Automatic Solution of Crystal Structures from X-ray DAta*, University of Glasgow, Glasgow (1983).

[12] D.A. LANGS AND G.T. DETITTA, Acta Crystallogr. Sect. A 31 (1975) (Abstr. 02.2–14), 516.

[13] I.L. KARLE, H. HAUPTMAN, J. KARLE, A.R. WING, Acta Crystallogr. 11 (1958), 257.

[14] G. BLANK, M. RODRIGUES, J. PLETCHER, M. SAX, Acta Crystallogr. Sect. B 32 (1976), 2970.

[15] R. BONNETT, J.E. DAVIES, M.B. HURSTHOUSE, G.M. SHELDRICK, Proc. R. Soc. London Ser. B 202 (1978), 249.

[16] B. BUSETTA, Acta Crystallogr. Sect. A 32 (1976), 139.

[17] A.A. FREER AND C.J. GILMORE, J. Chem. Soc. Chem. Commun. (1977), 296.

[18] C.J. GILMORE, Acta Crystallogr. Sect. A 33 (1977), 712.

[19] C.J. GILMORE, A.D.U. HARDY, D.D. MACNICOL, D.R. WILSON, J. Chem. Soc. Perkin Trans. 2 (1977), 1427.

[20] C. GLIDEWELL, D.C. LILES, D.J. WALTON, G.M. SHELDRICK, Acta Crystallogr. Sect. B 35 (1979), 500.

[21] M. SAX, M. RODRIGUES, G. BLANK, M.K. WOOD, J. PLETCHER, ibid. 32 (1976) 1953.

[22] J.V. SILVERTON AND T. AKIYAMA, Acta Crystallogr. Assoc. Program Abst. Ser. 2 (1978) (Abstr. PB15).

[23] J.V. SILVERTON AND C. KABUTO, Acta Crystallogr. Sect. B 34 (1978), 588.

INVERSION OF THE X-RAY TRANSFORM FROM DATA IN A LIMITED ANGULAR RANGE

STEVEN H. IZEN*

Abstract. The x-ray transform in three dimensions is shown to be applicable to an inversion problem arising from optical diagnostics of supersonic gas flow. The reconstruction of the index of refraction of the gas becomes a fully three-dimensional, limited solid angle parallel beam tomography problem. Two algorithms for the reconstruction and some preliminary numerical results on the ill-posedness of such a reconstruction are presented.

Introduction. The optical diagnostic technique of Diffuse Illumination Heterodyne Holographic Interferometry can provide data which has the potential of allowing a three-dimensional reconstruction of the density of a supersonic gas flow. The data measured can be interpreted as line integrals of the index of refraction, which in turn, are directly proportional to the line integrals of density. The holographic nature of the measurement allows the collection of line integral data from different views simultaneously. However, physical constraints restrict the range of viewing. From a mathematical viewpoint the quantity measured is a restricted view x-ray transform of the density, where the data are available for a parallel beam geometry. Previous fully three-dimensional applications of the x-ray transform have been mainly in the medical setting where different geometries apply. For example, see [HS]. The motivation for this work is to determine whether sufficient three-dimensional information can be extracted using this technique to justify the expenditure of significant amounts of money on the experimental development and applications for this technique. Two ways of exploiting the parallel geometry are introduced in this paper with the goal of obtaining a practical reconstruction from experimental data and of quantifying the ill-posedness of such a reconstruction. One of these methods takes advantage of the full knowledge of the x-ray transform along planes by using the projection-slice theorem to move the problem into Fourier space. The other method uses the complete knowledge of the x-ray transform along planes to enable the use of orthogonality relations. In both reconstruction methods orthogonal function series expansions are used to represent the function to be recovered. The rest of this paper is organized as follows: First, we present a brief summary of the experimental apparatus. Next, we present the mathematical formulation for the problem. Then the orthogonality approach and the Fourier approach to inverting the x-ray transform are presented. Next, some preliminary numerical results on the ill-posedness are given. Lastly, some comments are made about on-going work on this problem. The author would like to thank Dr. Arthur J. Decker of the NASA-Lewis Research Center for introducing him to this application for the x-ray transform, and the NASA-Lewis Research Center for their support.

*Case Western Reserve University

Experimental description. The experimental technique of Diffuse Illumination Heterodyne Holographic Interferometry is used to obtain experimental data corresponding to the x-ray transform of the density of a flowing gas. A simplified description of the technique follows. A laser beam is split by a beam splitter. One of the two resulting beams, the reference beam, is immediately directed onto a photographic plate. The other beam is directed onto a diffuser. The light scattered off the diffuser passes through the transparent object and recombines with the reference beam on the photographic plate, forming a holographic image. The diffuse illumination can be thought of as having object beams passing through the object from all directions. The hologram is exposed twice, once with the object in place as described above, and once with the object not present. It is assumed that there are only small variations of the optical path lengths of the object beams between exposures. The developed hologram is illuminated by a duplicate of the reference beam. An interference pattern results from the interference of the two reconstructions. A lens is used to form an image of the interference pattern. The resulting planar interference pattern is the projection parallel to the optical axis of the lens. This planar interference pattern corresponds to the difference, between exposures, of the optical path length of object beams parallel to the viewing direction. By moving the lens, projections from different directions are obtained. The difference in optical path length is proportional to the line integral of the change in index of refraction along the path of the light. [DeS] This is, for a gas, proportional to the line integral of density change between exposures. When the density distribution is known for one exposure, line integrals of the density of the object are obtained. The angular range of views available will be limited by the physical size of the holographic plate. The application to which this technique will be applied is the measurement of gas flows between compressor blades inside jet engines. In that case the compressor blades will also restrict the viewing directions. For simplicity, we assume that the hologram can be viewed for directions lying inside the narrow spherical angle $0 \leq \varphi \leq \phi$, $0 \leq \theta \leq 2\pi$. Here (φ, θ) are spherical coordinates. The maximum angular deflection ϕ is at most $20°$. A more realistic value for ϕ is $10°$. The experimental work described above is being done by Dr. A. J. Decker of the NASA-Lewis Research Center.

Mathematical formulation. Let f be a function in a suitable function space for the representation of an object which is to be imaged. The x-ray transform of f is defined by

$$Pf(\omega, x) = \int_{-\infty}^{\infty} f(x + t\omega)dt,$$

where $\omega \in S^2$, $x \in \omega^{\perp}$. This is the projection of f along the line with direction given by ω which intersects the orthogonal plane through the origin at x. The data corresponding to a view of the object along a fixed direction ω is $P_\omega f$, which is defined by

$$P_\omega f = Pf(\omega, x).$$

The mathematical problem can be stated as follows: *Given $P_\omega f$ only for ω restricted to the interior of a narrow cone, reconstruct f, and/or quantify the ill-*

posedness of such a reconstruction. Note that the limited view tomography problem presented here has the range of views much more severely restricted than the limited view tomography problems normally arising in a medical setting.

For the rest of this paper we make the assumption that f is supported on the unit ball B in $\mathbf{R}^3$ and is square integrable. More specifically,

$$f \in L^2(B, w(x)),$$

where $w(x) = (1 - |x|^2)^{-(\alpha - 3/2)}$. With these assumptions f can be expanded in terms of orthogonal functions.

$$f(r,\varphi,\theta) = \sum_{s=0}^{\infty} \sum_{\substack{m_0=0 \\ s\equiv m_0(2)}}^{m_0} \sum_{m_1=-m_0}^{m_0} a_s^{m_0,m_1} V_s^{m_0,m_1}(r,\varphi,\theta), \tag{1}$$

where (r,φ,θ) are polar coordinates, and the $V_s^{m_0,m_1}$ form an orthonormal basis for $L^2(B, w(x))$. They are given by [Iz2]

$$V_s^{m_0,m_1}(r,\varphi,\theta) = R_s^{m_0}(r) S(m_0, m_1; \varphi, \theta),$$

where

$$R_s^{m_0}(r) = c_1^{s,m_0,\alpha} r^{m_0} (w(r))^{-1} P_{(s-m_0)/2}^{(m_0+1/2,\alpha-3/2)}(1-2r^2),$$
$$S(m_0, m_1; \varphi, \theta) = c_2^{m_0,m_1} P_{m_0}^{m_1}(\cos\varphi) e^{im_1\theta}.$$

Here $P_n^{(\beta,\gamma)}(x)$ are Jacobi polynomials, and $P_\ell^m(x)$ are associated Legendre functions. The constants $c_1^{s,m_0,\alpha}$ and $c_2^{m_0,m_1}$ are listed in table 2. For m_0 fixed, the $S(m_0, m_1; \varphi, \theta)$ form a normalized basis for spherical harmonics of degree m_0 on S^2. If the first sum is terminated at a finite value S instead of ∞, the sum of the remaining terms will be the same as the Taylor series for $w(r)f$ about the origin, truncated at degree S. It has been shown [Ma] that the a truncation of (1) gives the approximation of degree S to f minimizing the L^2 measure of estimation error.

Reconstruction methods. We can restrict our attention to the recovery of the unknown coefficients $a_s^{m_0,m_1}$. There are two practical methods presented here for the computation of the $a_s^{m_0,m_1}$ from the measured x-ray transform data. In the first method full advantage is taken of the orthogonality relations available for the transform of the basis. Here the calculations are performed in the x-ray transform space. The orthogonality relations allow the decoupling of the problem for different values for s, which reduces the size of the linear systems which need to be solved. The second method takes advantage of the projection slice theorem to convert the x-ray transform data to the Fourier transform data. Then the calculations are performed in the Fourier domain. In this case, there is no decoupling in the variable s, but it may be possible to effect such a decoupling by using an alternate basis for $L^2(B, w(x))$.

A more general, in depth, treatment of the first method can be found in [Iz1] and [Iz2]. The following is an outline of this method. The numerical implementation of this method is under development.

In the special case when ω points along the pole ω_0 of the basis for the spherical harmonics $S(m_0, m_1; \varphi, \theta)$, $P_{\omega_0} V_s^{m_0,m_1}$ is computable directly. For this case $\omega_0^\perp$ is the equatorial plane of the unit ball.

$$\text{(2a)} \qquad P_{\omega_0} V_s^{m_0,m_1}(p,\tau) = c_3^{s,m_0,m_1,\alpha} p^{m_1}(1-p^2)^{\alpha-1} P_{(s-m_1)/2}^{(m_1,\alpha-1)}(1-2p^2)e^{im_1\tau}$$

for $m_0 - m_1$ even, and

$$\text{(2b)} \qquad P_{\omega_0} V_s^{m_0,m_1}(p,\tau) = 0,$$

for $m_0 - m_1$ odd. (p,τ) are polar coordinates on $\omega_0^\perp$. We use the notation $c_3^{2,m_0,m_1,\alpha} U_s^{m_1}$, where $c_3^{2,m_0,m_1,\alpha}$ is given in table 2, as a shorthand for the right hand side of equation (2). This result has been obtained (in a more general form) independently by both P. Maaß and the author.

Applying equation (2) to (1) gives

$$\text{(3)} \qquad P_{\omega_0} f(p,\tau) = \sum_{s=0}^{\infty} \sum_{\substack{m_0=0 \\ s\equiv m_0(2)}}^{s} \sum_{m_1=-m_0}^{m_0} a_s^{m_0,m_1} c_3^{2,m_0,m_1,\alpha} U_s^{m_1}$$

To apply the orthogonality of the Jacobi polynomials and the circular harmonics, fix s and m_1. Then

$$\text{(4)} \qquad \begin{aligned} &\int_0^1 \int_0^{2\pi} (P_{\omega_0} f)(p,\tau) e^{-im_1\tau} p^{m_1+1} P_{(s-m_1)/2}^{(m_1+1,\alpha-1)}(1-2p^2) d\tau dp \\ &= \sum_{\substack{s\geq m_0 \geq m_1 \\ s\equiv m_0(2)}} a_s^{m_0,m_1} c_4^{s,m_0,m_1,\alpha}. \end{aligned}$$

The constant $c_4^{s,m_0,m_1,\alpha}$ is given in table 2. Consider equation (4) for all m_1 such that $0 \leq m_1 \leq s$ and $m_1 \equiv s(2)$. Equation (4) then gives $s+1$ linear equations relating the $(s+1)(s+2)/2$ unknowns $a_s^{m_0,m_1}$. In order to obtain the additional linear relations needed, data from other projection directions must be used.

Unfortunately, it is not possible to directly compute $P_{\tilde{\omega}} f$ for arbitrary directions $\tilde{\omega}$. The trick used is to compute $P_{\tilde{\omega}} f$ with respect to a different basis for the spherical harmonics. This basis is determined by the choice of an $O \in SO(3)$ which maps ω_0 to $\tilde{\omega}$. The new basis consists of spherical harmonics for which the pole points along $\tilde{\omega}$. $\tilde{S}(m_0, \tilde{m}_1, \tilde{\varphi}, \tilde{\theta})$ is defined in the same way as $S(m_0, m_1, \varphi, \theta)$ except (φ, θ) have been replaced by the spherical coordinates $(\tilde{\varphi}, \tilde{\theta})$ induced by the transformation O. The two families of harmonics are related by

$$\text{(5)} \qquad S(m_0, m_1, \varphi, \theta) = \sum_{\tilde{m}_1=-m_0}^{m_0} g_{m_0,m_1,\tilde{m}_1}(\tilde{\omega}) \tilde{S}(m_0, \tilde{m}_1 . \tilde{\varphi}, \tilde{\theta})$$

The coefficients $g_{m_0,m_1,\tilde{m}_1}$ are directly computable by integration.[EM2]

Define

$$\tilde{V}_s^{m_0,\tilde{m}_1} = R_s^{m_0}\tilde{S}(m_0,\tilde{m}_1,\tilde{\varphi},\tilde{\theta})$$

and $\tilde{U}_s^{\tilde{m}_1}$ by

$$P_{\tilde{\omega}}\tilde{V}_s^{m_0,\tilde{m}_1}(p,\tilde{\tau}) = c_3^{s,m_0,\tilde{m}_1,\alpha}\tilde{U}_s^{\tilde{m}_1}.$$

where p and $\tilde{\tau}$ are polar coordinates on $\tilde{\omega}^{\perp}$. Expand f in terms of the basis $\{\tilde{V}_s^{m_0,\tilde{m}_1}\}$, then apply equations (5) and (2).

$$P_{\tilde{\omega}}f(p,\tilde{\tau}) = \sum_{s=0}^{\infty}\sum_{\substack{m_0=0\\ s\equiv m_0(2)}}^{s}\sum_{m_1=-m_0}^{m_0} a_s^{m_0,m_1}\sum_{\tilde{m}_1=-m_0}^{m_0} g_{m_0,m_1,\tilde{m}_1}(\tilde{\omega})c_3^{s,m_0,m_1,\alpha}\cdot \tilde{U}_s^{\tilde{m}_1}(p,\tilde{\tau}).$$

Fix s and $\tilde{m}_1$ and again apply orthogonality relations to obtain

$$\begin{aligned}\int_0^1\int_0^{2\pi}&(P_{\tilde{\omega}}f)(p,\tilde{\tau})e^{-i\tilde{m}_1\tilde{\tau}}p^{\tilde{m}_1+1}P_{(s-\tilde{m}_1)/2}^{(\tilde{m}_1+1,\alpha-1)}(1-2p^2)d\tilde{\tau}dp\\ &= \sum_{\substack{s\geq m_0\geq m_1\\ s\equiv m_0(2)}}\sum_{m_1=-m_0}^{m_0} a_s^{m_0,m_1}g_{m_0,m_1,\tilde{m}_1}(\tilde{\omega})c_4^{s,m_0,\tilde{m}_1}.\end{aligned} \tag{6}$$

For each direction $\tilde{\omega}$, equation (6), like equation (4), will give $s+1$ relations among the $(s+1)(s+2)/2$ unknowns $a_s^{m_0,m_1}$. Thus we have the following result.

THEOREM. *At least $(s+2)/2$ directions are required to recover f up to degree s.*

For each s we obtain the linear system

$$Q_s\vec{a}_s = \vec{b}_s, \tag{7}$$

where $\vec{a}_s$ is the column vector consisting of all coefficients $a_s^{m_0,m_1}$ of degree s, $\vec{b}_s$ is the column vector consisting of the left hand side of the equations obtained from (4) and (6) as $\tilde{m}_1$ and m_1 vary, and $\tilde{\omega}$ ranges over the directions at which data is sampled. Q_s is the corresponding matrix from the terms $g_{m_0,m_1,\tilde{m}_1}(\tilde{\omega})c_4^{s,m_0,\tilde{m}_1}$ which multiply $a_s^{m_0,m_1}$ in the right hand side of (4) and (6). The matrix Q_s depends on the geometry of the directions for which data are available, and $\vec{b}_s$ depends on the sampled data. Note that the use of the orthogonality relations has decoupled the recovery of the coefficients by the degree s. This will provide a large reduction in the sizes of the linear systems (7) to be solved.

The ill-posedness of the recovery of f can be determined empirically as a function of the available directions by performing numerical singular value decompositions of the matrices Q_s obtained for different data sampling geometries. One numerical difficulty in applying this method is the discretization of equation (6). It will be necessary to have enough sample points so that the representation of the integrals in (6) as sums is a reasonable approximation. Though it appears that this requirement, combined with the large amounts of data needed to combat the ill-posedness would be prohibitively expensive in computational resources, the size reduction given by the decoupling in s keeps the linear systems small enough to allow the computations to be done on scientific workstations.

The second reconstruction method is Fourier reconstruction. In this technique, the projection slice theorem is used to transform the x-ray transform data into the Fourier domain. The basis functions $V_s^{m_0,m_1}$ are transformed also, becoming a basis for L^2 functions in the Fourier domain which are transforms of L^2 functions supported on the unit ball in the spatial $\mathbf{R}^3$. For each ω, the x-ray transform data $P_\omega f$ is mapped to the Fourier transform of f, sampled on the plane $\omega^\perp$. Since ω is only available in the range $0 \leq \varphi \leq \phi$, $0 \leq \theta \leq 2\pi$, f is known in the Fourier domain only on the complement of the cone, $\pi/2 - \phi \leq \varphi \leq \pi/2$ and $0 \leq \theta \leq 2\pi$, in Fourier $\mathbf{R}^3$. Thus the limited view manifests itself as an incompleteness of the Fourier domain on which f is known. If, for all ω in a dense set of directions $\{\omega_j \mid \omega_j \text{ is in the solid cone } 0 \leq \varphi \leq \phi, 0 \leq \theta \leq 2\pi\}$, $P_\omega f$ is available for all $x \in \omega^\perp$, then the missing data is uniquely determined by an analytic continuation argument. However, practically, 1) $P_\omega f$ is only discretely and finitely sampled, and 2) A dense set $\{\omega_j\}$ is not available since ω is also discretely and finitely sampled. Thus the missing data cannot be directly recovered, and other techniques must be used to recover the Fourier transform of the object function. We accomplish this by using a least squares fit for the coefficients $a_s^{m_0,m_1}$. Also, in this framework, the singular value decomposition can be used to combat and measure ill-posedness.

The Projection-Slice Theorem relates the two dimensional Fourier transform of the x-ray transform data to three-dimensional Fourier transform of a function. For minor technical reasons, we use the inverse Fourier transform, defined for $y \in \mathbf{R}^n$ by

$$F_n^{-1} f(y) = (2\pi)^{-n} \int_{\mathbf{R}^n} e^{ix\cdot y} f(x) dx,$$

instead of the forward Fourier transform.

THEOREM. *(Projection slice theorem) For* $\eta \in \omega^\perp$,

$$(2\pi)(F_3^{-1} f)(\eta) = F_2^{-1}(P_\omega f)(\eta). \tag{8}$$

See [Iz1] or [Na] for a proof.

Because of the discrete sampling of $P_\omega f$, a discrete approximation to the right hand side (8) is computed using a two dimensional FFT algorithm. This translates on the left hand side of equation (8) to having $F^{-1}f$ known on sample points η_i on planes through the origin lying inside of the cone complement. A least squares fit is then performed to determined $F^{-1}f$, and hence, f up to degree S. Transforming equation (1) and evaluating it at the sample point η_i in Fourier $\mathbf{R}^3$ gives

$$F^{-1} f(\eta_i) = \sum_{s=0}^{S} \sum_{\substack{m_0=0 \\ s\equiv m_0(2)}}^{s} \sum_{m_1=-m_0}^{m_0} a_s^{m_0,m_1} F^{-1} V_s^{m_0,m_1}(\eta_i), \tag{9}$$

where the transform of the basis functions are computed analytically [Iz2] and can be expressed in spherical coordinates (q, μ, τ) as

$$(10) \qquad F^{-1}V_s^{m_0,m_1}(q,\mu,\tau) = c_5^{s,m_0,\alpha} q^{-\alpha} J_{\alpha+s}(q) S(m_0, m_1; \mu; \tau),$$

where J_ν is the Bessel function of the first kind and $c_5^{s,m_0,\alpha}$ is listed in table 2.

The left hand side of equation (9) is computed from the experimentally obtained x-ray transform data, and the basis functions $F^{-1}V_s^{m_0,m_1}$ can be evaluated numerically at each sample location η_i. Thus equation (9) can be written is matrix form as

$$(11) \qquad Q\vec{a} = \vec{b}.$$

Note that the linear system of equation (11) is much larger than the systems of equation (7). In equation (11) there is no decoupling in the degree. That is, the column vector $\vec{a}$ consists of all $(S+1)(S+2)(S+3)/6$ coefficients $a_s^{m_0,m_1}$ of degree $s \leq S$. If $P_\omega f$ is sampled for J directions ω, and N^2 points in $\omega^\perp$ are sampled for each ω, then $\vec{b}$, the vector computed from the sampled data will be of dimension JN^2. That is, $\vec{b}$ will have one entry for each sample point. Q is the $JN^2 \times (S+1)(S+2)(S+3)/6$ matrix of the $(S+1)(S+2)(S+3)/6$ basis functions evaluated at JN^2 points in Fourier space. For even moderately small degrees S and sparse sampling geometries the linear system to be solved can be prohibitively large, requiring the use of supercomputers. Contrast this with the equation (7). The largest of the $S+1$ systems defined by (7) has $\vec{a}_S$ of size $(S+1)(S+2)/2$, and $\vec{b}$ of size $J(S+1)$, which means makes Q_S of size $J(S+1) \times (S+1)(S+2)/2$. For any practical reconstruction,

$$(S+1) \ll N^2(S+3)/3,$$

so the largest system from (7) is considerably smaller than that of (11). An $(S+3)/3$ reduction in the number of columns came from the use of orthogonality relations to decouple in s. An additional size reduction came from collapsing the data by equation (6) from the N^2 sample points in each $\omega^\perp$ to $(S+1)$ linear relations in the decoupled $a_s^{m_0,m_1}$. Even though $S+1$ linear systems must be solved, each one will be manageable in size, so tremendously powerful computing facilities are unlikely to be necessary. A disadvantage of the smaller method is that a sufficiently dense set of sample points on each $\omega^\perp$ must be taken in order to have a discrete approximation to equation (6) be valid. An advantage to the larger approach is that it much easier to implement numerically, and therefore is a quick way to obtain results on the ill-posedness of the reconstruction as a function of the available view and sampling geometry. It appears, however, that for any situation where repeated reconstructions will be performed, the first method should be used.

Numerical Results. We now present some results of computer experiments on the ill-posedness as a function of the available views of the recovery of the object function from its x-ray transform. It should be strongly emphasized that these numerical results are preliminary, and that as such should be quoted only with a disclaimer to that effect.

Singular value decompositions were performed numerically on a variant of equation (11). Instead of for each S performing a singular value decomposition on the full matrix Q, a singular value decomposition was done on the submatrix of Q consisting of columns corresponding to $a_s^{m_0,m_1}$ for a fixed value for s (as opposed to for all s such that $s \leq S$). The calculated singular values are interpreted as giving a comparison between the relative ease with which coefficients $a_s^{m_0,m_1}$ of the same degree s can be reconstructed. There is no information obtained from this experiment which can be used to compare coefficients of different degrees.

It was assumed that for each projection, 256 data points arranged in a evenly spaced square 16×16 array would be available ($N = 16$). Also, 64 projections ($J = 64$) were assumed to be available at directions given by (μ_i, τ_j), where $\tau_j = 2\pi j/8$ for $j = 0, \dots, 7$ and for $\mu_i = i\phi/8$ for $i = 1, \dots, 8$. The degree S was allowed to range from 0 through 10. The experiments were repeated for $\phi = 10°$, $\phi = 90°$. The $\phi = 90°$ case allowed to range from 0 through 10. The experiments were repeated for $\phi = 10°$, $\phi = 20°$, and $\phi = 90°$. The $\phi = 90°$ case corresponds to the view not being limited. This sampling geometry was chosen for ease in programming, and we make no claims as to its optimality. As the results listed in table 1 indicate, when the view is extremely limited, the reconstruction is highly ill-posed, and the ill-posedness decreases as the viewing angle increases. For $\phi = 90°$, the reconstruction is a fairly well posed problem. Also, as expected, the coefficients which are poorly determined for small angles ϕ, are the ones corresponding to the basis functions $V_s^{m_0,m_1}$ which are of higher degree in the azimuthal and radial variables. That is, the variables which contain the most information in the direction of the axis of the viewing cone.

The singular value decompositions were performed in complex double precision by the LINPACK routine ZSVDC on a VAX 8800 at the NASA-Lewis Research Center. For each run, a matrix of size $(S+1)(S+2)/2 \times 16384$ was analyzed. The largest case was for $S = 10$, when approximately 17 megabytes (66×16384 entries times 16 bytes per entry) of storage were needed. The CPU time required in that case was one and a half hours. The storage costs are already prohibitive, and for a full Fourier reconstruction (equation (11)) up to degree 10, the size of the matrix Q would be about 71 megabytes ($286 \times 16384 \times 16$).

The corresponding analysis performed using the first method exposed here (equation(7)) would be much smaller. For example, for $S = 10, 10$ singular value decompositions would be done, the largest of which would be on Q_{10}, a matrix requiring only approximately 0.75 Megabytes ($66 \times 11 \times 64 \times 16$) of memory. Work is underway to implement this method.

Acknowledgement. This work was partially supported by NASA grant NAG 3-832 and by the NASA-ASEE Summer Faculty Fellowship Program at the NASA-Lewis Research Center in Cleveland, Ohio.

TABLE 1

S		0	1	2	3	4	5	6	7	8	9	10
Number of coefficients of degree S.		1	3	6	10	15	21	28	36	45	54	66
	$\rho = 0.1$	1	3	5	7	9	11	13	15	18	23	26
$\phi = 10°$	$\rho = 0.01$	1	3	6	9	12	15	18	22	28	32	38
	$\rho = 0.001$	1	3	6	10	14	18	21	25	33	40	45
	$\rho = 0.1$	1	3	5	7	12	15	18				
$\phi = 20°$	$\rho = 0.01$	1	3	6	9	14	18	22				
	$\rho = 0.001$	1	3	6	10	15	20	25				
	$\rho = 0.1$	1	3	6	10	15	21					
$\phi = 90°$	$\rho = 0.01$	1	3	6	10	15	21					
	$\rho = 0.001$	1	3	6	10	15	21					

The table lists the number of singular values within a factor of ρ from the largest singular value when reconstructing the coefficients for the object function of degree S. Here 2ϕ is the angle of opening of the cone of available viewing directions. That all the singular values for $\phi = 90°$ are within a factor of .1 from the largest singular value indicates that the reconstruction in this case is not ill-posed. The relative scarcity of singular values in that range for $\phi = 10°$ and $\phi = 20°$ shows the relative ill-posedness of the corresponding reconstructions.

TABLE 2

$$c_1^{s,m_0,\alpha} = \left((2s+3)\frac{\Gamma((s+m_0+2\alpha)/2)\Gamma((s-m_0+2)/2)}{\Gamma((s+m_0+3)/2)\Gamma((s-m_0+2\alpha-1)/2)}\right)^{1/2}$$

$$c_2^{m_0,m_1} = (-1)^{m_1}(2\pi)^{-1/2}(m_0+1/2)^{1/2}\left(\frac{(m_0-m_1)!}{(m_0+m_1)!}\right)^{1/2}$$

$$c_3^{s,m_0,m_1,\alpha} = c_1^{s,m_0,\alpha}(-1)^{m_0}2^{m_1}\pi^{1/2}\Gamma(m_1+1/2)\left(\frac{(m_0-m_1)!}{(m_0+m_1)!}\right)^{1/2}\cdot$$

$$\cdot\frac{\Gamma((s-m_0+2\alpha-1)/2)\Gamma((s-m_1+2)/2)}{\Gamma((s-m_0+2)/2)\Gamma((s-m_1+2\alpha)/2)}$$

$$c_4^{s,m_0,m_1,\alpha} = c_1^{s,m_0,\alpha}c_3^{s,m_0,m_1,\alpha}\frac{\Gamma((s+m_1+2)/2)\Gamma((s-m_1+2\alpha-2)/2)}{2(s+\alpha)\Gamma((s-m_1+2)/2\Gamma((s+m_1+2\alpha)/2)}$$

$$c_5^{s,m_0,\alpha} = c_1^{s,m_0,\alpha}\frac{i^{m_0}(2\pi)^{-3/2}2^{\alpha-3/2}\Gamma((s-m_0+2\alpha-1)/2)}{\Gamma((s-m_0+2)/2)}$$

This is a table of various constants referred to in the text.

[DeA] A. J. Decker, J. Stricker, *Comparison of Electronic Heterodyne Holographic Interferometry and Electronic Heterodyne Moiré Deflectometry for Flow Measurements*, S.A.E. Trans., paper #851896 (1985).

[DeS] S. Deans, *The Radon Transform and Some of its Applications*, Wiley, New York, 1983.

[EM1] A. Erdelyi, W. Magnus, F. Oberhettinger, F. G. Tricomi, *Higher Transcendental Functions*, Volumes I and II, McGraw-Hill, 1953.

[EM2] A. Erdelyi, W. Magnus, F. Oberhettinger, F. G. Tricomi, *Tables of Integral Transforms*, Volumes I and II, McGraw-Hill, 1954.

[HS] C. Hamaker, K. T. Smith, D. C. Solmon, S. L. Wagner, *The Divergent Beam X-ray Transform*, Rocky Mtn. J. Math., 10 (1980), pp. 253–283.

[Iz1] S. H. Izen, *A Series Inversion for the X-ray Transform in n Dimensions*, Inverse Problems, 4 (1988), pp. 725–748.

[Iz2] S. H. Izen, *Inversion of the k-plane Transform by Orthogonal Function Series Expansions*, Inverse Problems, 5 (1989), pp. 181–202.

[Lo] A. K. Louis, *Orthogonal Function Series Expansion and the Null Space of the Radon Transform*, SIAM J. Math. Anal., 15 (1984), pp. 621–633.

[Ma] P. Maass, *The X-ray Transform: Singular Value Decomposition and Resolution*, Inverse Problems, 3 (1987), pp. 729–741.

[Na] F. Natterer, *The Mathematics of Computerized Tomography*, John Wiley and Sons, New York, 1986.

THE EIKONAL APPROXIMATION IN ULTRASOUND COMPUTER TOMOGRAPHY

ALFRED K. LOUIS†

Abstract. The eikonal approximation was proposed to compute the far field pattern for a given potential in scattering theory. Here we study this method for treating the inverse problem; i.e., the determination of a potential from a given far field pattern. The ill-conditioned behaviour is discussed and the steps in the inversion formula are investigated.

Key words. inverse scattering, eikonal approximation

AMS(MOS) subject classifications. 35R30,65N30,76Q05

1. Introduction. In many applications as medical imaging or exploration geophysics the object to be studied cannot directly be measured. Hence from observations influenced by the object under consideration we compute the desired information. To this end a mathematical model describing the mapping from the searched-for parameters to the observations has to be formulated and the inverse problem of identifying the parameters has to be solved.

When we send a plane wave to an object and record the scattered wave we are led, assuming time-harmonic waves, to the Helmholtz equation which we write down in the form of the Schrödinger equation, namely

$$(\Delta + k^2)u = Vu \text{ in} I\!R^3. \tag{1.1}$$

We assume that we have homogeneous background in our experiment. Then V describes the changes relative to the background, and when the scatterer has finite extend, the potential V is compactly supported. Furthermore in our model absorption should be included, which means that V is complex valued with

$$\Im V \leq 0. \tag{1.2}$$

In the acoustic applications the potential depends on the wave number, we have

$$V = k^2 V_A$$

where V_A is independent of k. In quantum mechanics we use

$$V = \frac{2m}{\hbar^2} V_Q.$$

†Fachbereich Mathematik, TU Berlin, D-1000 Berlin 12, FRG. The work of the author was supported by the Deutsche Forschungsgemeinschaft under grant Lo 310/2-4.

The solution of problem (1.1) is the sum of the incoming plane wave $\exp(\imath k < \theta, x >)$ and the scattered wave u^s which essentially consists of a spherical wave. In $I\!R^3$ this means

$$u(x) = e^{\imath k<\theta,x>} + \frac{e^{\imath k|x|}}{|x|} f(k,\theta,\frac{x}{|x|}) + o(|x|^{-1}) \tag{1.3}$$

where $\theta \in S^2$ is the direction of the incoming plane wave. The complex-valued function f is the so-called far field pattern or the scattering amplitude. For more details see for example [2,3,14,15].

In the following we address the problem of determining the potential V when the far field pattern f is given for fixed wave number k on $S^2 \times S^2$. It suffices to know f only on a restricted range $S_1^2 \times S_2^2$ because of the analyticity of f. We then can determine extensions of f to all of $S^2 \times S^2$ using spherical harmonics similar to the situation in limited angle problems for x-ray computer tomography, see [8].

If V is assumed to be real and compactly supported it is shown in [15] that it is uniquely determined by f. Obviously the problem of solving

$$AV = f \tag{1.4}$$

is extremely ill-posed, because the image f is an analytical function, hence much smoother than V. This behaviour is inherited by any reasonable approximation of the problem. In the low frequency case; i.e., small k, which is treated in [3,7] a regularization is achieved by representing the solution with not too many parameters, hence the solution is projected onto a subspace of small dimension.

In the other extreme case, very large k, linearizations are used as Born or Rytov approximation, see for example [4,6].

In this paper we adapt a method which was developed for treating the forward problem, the so-called eikonal approximation, see [5,12,14]. We approximate the mapping A from (1.4) and decompose it into simpler operations including the x-ray transform. In Section 2 we describe a derivation of the approximation, and then we discuss the inverse problem in Section 3. Finally questions concerning the resolution for finitely many data are addressed.

2. The Eikonal Approximation. In the above mentioned references, [5,12,14] the eikonal approximation is derived by using arguments concerning stationary phase, see [17]. Here we derive the formula in a way such that the approximation, which occurs at just one place, is obvious.

We start out from the Lippmann-Schwinger equation, an integral equation for the solution of (1.1), which utilizes the fundamental solution of the Helmholtz equation in three dimensions. It is

$$u(x) = e^{\imath k<\theta,x>} - \frac{1}{4\pi} \int_{I\!R^3} |y|^{-1} e^{\imath k|y|} V(x-y) u(x-y)\, dy. \tag{2.1}$$

Putting

$$\varphi(x) = e^{-\imath k<\theta,x>} u(x) = 1 + \frac{u^s(x)}{u^i(x)},$$

where we denoted the incoming wave by

$$u^i(x) = e^{\imath k<\theta,x>},$$

and the scattered wave by

$$u^s(x) = \frac{e^{\imath k|x|}}{|x|} f(k, \theta, \frac{x}{|x|}) + o(|x|^{-1}).$$

This results in an equation for φ

$$\varphi(x) = 1 - \frac{1}{4\pi} \int_{\mathbb{R}^3} |y|^{-1} \exp(\imath k(|y| - < \theta, y >)) V(x-y)\varphi(x-y)\, dy.$$

By θ' we denote the direction where we observe the scattered wave, and by ω we denote the direction in the middle of θ and θ'

$$\omega = \frac{\theta + \theta'}{|\theta + \theta'|}.$$

Next we use polar coordinates with north pole ω, the angle between ω and y is called μ and defined as

$$\cos\mu = \frac{< \omega, y >}{|y|},$$

and the zonal angle is α. Furthermore we abbreviate

$$\psi(r, \mu, \alpha) := V(x-y)\varphi(x-y).$$

Then we can rewrite the relation for the function φ as

$$\varphi(x) = 1 - \frac{1}{4\pi} \int_0^\infty \int_0^{2\pi} \int_0^\pi r e^{\imath kr(1-\cos\mu)} \psi(r, \mu, \alpha) \sin\mu \, d\mu \, d\alpha \, dr.$$

Because of

$$r \sin\mu e^{\imath kr(1-\cos\mu)} = \frac{1}{\imath k} \frac{d}{d\mu} e^{\imath kr(1-\cos\mu)}$$

we can evaluate the inner integral using partial integration:

$$\int_0^\pi r \sin\mu e^{\imath kr(1-\cos\mu)} \psi(r, \mu, \alpha)\, d\mu$$

$$= \frac{1}{\imath k} e^{\imath kr(1-\cos\mu)} \psi(r, \mu, \alpha)|_0^\pi - \frac{1}{\imath k} \int_0^\pi e^{\imath kr(1-\cos\mu)} \frac{d}{d\mu} \psi(r, \mu, \alpha)\, d\mu.$$

The polar coordinates give

$$\psi(r, 0, \alpha) = \psi(r, 0, 0) = V(x - r\omega)\varphi(x - r\omega),$$

$$\psi(r, \pi, \alpha) = \psi(r, \pi, 0) = V(x + r\omega)\varphi(x + r\omega).$$

Hence these terms are independent of α and the integration over α just gives the factor 2π. Totally we get

(2.2)
$$\varphi(x) = 1 - \frac{\imath}{2k}\int_0^\infty V(x - r\omega)\varphi(x - r\omega)\, dr$$

$$(2.3) \qquad + \frac{\imath}{2k}\int_0^\infty e^{2\imath kr} V(x + r\omega)\varphi(x + r\omega)\, dr$$

$$(2.4) \qquad - \frac{\imath}{4\pi k}\int_0^\infty \int_0^\pi e^{\imath kr(1-\cos\mu)} \frac{\partial}{\partial\mu}\int_0^{2\pi} \psi(r,\mu,\alpha)\, d\alpha\, d\mu\, dr.$$

For large values of k the terms (2.3) and (2.4) are Fourier coefficients for high frequencies. The functions $V\varphi$ are shown in [1] to be in $C \cap H^2_{loc}$, hence we can hope that these two terms are relatively small compared with to (2.2). This even holds in the acoustic case. Therefore, and this is the aforementioned approximation, we neglect these two terms. Using the coordinate transform

$$x = b + \rho\omega \ , \ b \in \omega^\perp,$$

we find the approximation

$$\varphi(b + \rho\omega) \approx 1 - \frac{\imath}{2k}\int_{-\infty}^\rho V(b + r\omega)\varphi(b + r\omega)\, dr.$$

Differentiating both sides with respect to ρ we get an ordinary differential equation which is easily solvable leading to

$$(2.5) \qquad \varphi(b + \rho\omega) \approx \exp(-\frac{\imath}{2k}\int_{-\infty}^\rho V(b + r\omega)dr).$$

Finally we find the following approximation for the solution u of equation (1.1)

$$(2.6) \qquad u(b + \rho\omega) \approx \exp \imath[k < b + \rho\omega, \theta > -\frac{1}{2k}\int_{-\infty}^\rho V(b + r\omega)dr].$$

For the scattering amplitude holds the following relation

$$f(k,\theta,\theta') = -\frac{1}{4\pi}\int_{I\!R^3} e^{-\imath k<\theta',y>} V(y)u(y)\, dy.$$

Using again $y = b + \rho\omega$ and the above approximation for u we find with

$$< \omega, \theta + \theta' >= \frac{1}{|\theta + \theta'|} < \theta + \theta', \theta - \theta' >= \frac{1}{|\theta + \theta'|}(|\theta|^2 - |\theta'|^2) = 0$$

the relation

$$f(k,\theta,\theta') \approx -\frac{1}{4\pi}\int_{\omega^\perp} e^{-\imath k<b,\theta-\theta'>}\int_{-\infty}^\infty V(b+\rho\omega)\exp(-\frac{\imath}{2k}\int_{-\infty}^\rho V(b+r\omega)dr)d\rho db.$$

The integrand in the ρ - integral is the derivative with respect to ρ of the exponential, hence this integration can easily be performed resulting in the searched for relation between far field and potential, namely

$$f(k,\theta,\theta') \approx -\frac{\imath k}{2\pi}\int_{\omega^\perp} e^{-\imath k<b,\theta-\theta'>}\Big(\exp(-\frac{\imath}{2k}\int_{-\infty}^\infty V(b + r\omega)dr) - 1\Big)\, db \qquad (2.7)$$
$$= g(k,\theta,\theta').$$

This is the relation which was derived by [12] in order to compute an approximation to the far field pattern.

3. The Inverse Scattering Problem. In order to describe the approximation given in (2.7), we introduce some operators. First we denote by

$$\mathcal{P}V(\omega, b) = \int_{I\!R} V(b + t\omega)dt \tag{3.1}$$

the three-dimensional x-ray transform, see e.g. [10,11,13]. It assings a function in $I\!R^3$ its integrals over all lines. Therefore $\mathcal{P}V$ is defined on the tangent bundle $T = \{(\omega, b) : \omega \in S^2, b \in \omega^\perp\}$. The use of ω and the coupling of θ' and ω allows to sample the data on a plane; i.e., we get the data in the so-called parallel geometry. Then we use the nonlinear operator defined for complex numbers

$$\mathcal{E}z = \exp(\frac{-\imath}{2k}z) - 1.$$

We immediately see, that $\mathcal{E}0 = 0$. Next we describe by

$$\mathcal{F}^2_{\omega^\perp} h(\xi) = \frac{1}{2\pi} \int_{\omega^\perp} h(b)e^{-\imath b\xi} db$$

the two-dimensional Fourier transform on $\omega^\perp$. The variable ξ in formula (2.7) is

$$\xi = k(\theta - \theta').$$

Because of $\theta, \theta' \in S^2$ this variable is restricted to the ball with radius $2k$. Therefore we introduce

$$\mathcal{B}_{2k}h(\eta) = \begin{cases} h(\eta), & \text{if } |\eta| \leq 2k, \\ 0, & \text{if } |\eta| > 2k. \end{cases}$$

the bandlimiting operator restricting the function onto the ball with radius $2k$. Then we can write the approximation as

$$g = \mathcal{A}_k V \tag{3.2}$$

with

$$\mathcal{A}_k = -\imath k \mathcal{B}_{2k} \mathcal{F}^2_{\omega^\perp} \mathcal{E}\mathcal{P}. \tag{3.3}$$

In the following we study properties of the operator $\mathcal{A}_k$.

We denote by Ω the unit ball in $I\!R^3$ and

$$L_2^-(\Omega) = \{f \in L_2(\Omega) : \Im f \leq 0\}.$$

THEOREM 3.1. *The operator $\mathcal{A}_k$ is pointwise continuous as mapping from $L_2^-(\Omega) \to L_2(S^2 \times S^2)$.*

Proof. Let $V, W \in L_2^-(\Omega)$, then

$$\|\mathcal{A}_k V - \mathcal{A}_k W\|^2_{L_2(S^2 \times S^2)} = \int_{S^2} \int_{S^2} |\mathcal{A}_k V(k, \theta, \theta') - \mathcal{A}_k W(k, \theta, \theta')|^2 d\theta \, d\theta'$$

$$\leq 4k^2 \int_{S^2} \int_{\omega^\perp} |\exp(\frac{-\imath}{2k}\mathcal{P}V(\omega,b)) - \exp(\frac{-\imath}{2k}\mathcal{P}W(\omega,b))|^2 \, db \, d\omega.$$

The operator $\mathcal{P}$ is real, $\mathcal{P}\bar{V} = \overline{\mathcal{P}V}$, and positive, which means that if $h \geq 0$ then also $\mathcal{P}h \geq 0$. Because of $\Im V \leq 0$ we have $\Re(-\imath\mathcal{P}V) \leq 0$, hence we have the estimate

$$|\mathcal{E}PV - \mathcal{E}PW| \leq |\mathcal{P}V - \mathcal{P}W|.$$

The continuity of $\mathcal{P}$ as mapping from $L_2(\Omega) \to L_2(T)$ gives

$$\begin{aligned} \|\mathcal{A}_k V - \mathcal{A}_k W\| &\leq \|\mathcal{P}(V - W)\|_{L_2(T)} \\ &\leq \|V - W\|_{L_2(\Omega)} \end{aligned}$$

which completes the proof. □

THEOREM 3.2. *The Operator $\mathcal{F}^2_{\omega^\perp}\mathcal{E}P$ maps $L_2^-(\Omega)$ into analytic functions of $\theta - \theta'$.*

Proof. If V is compactly supported in Ω, then $\mathcal{P}(\omega, \cdot)$ is compactly supported in the unit ball in $\omega^\perp$. Furthermore the Fourier transform of a compactly supported function is analytic.

Remark. This smoothing property of the operator A_k results in an extreme ill-conditionedness of the problem, see [9].

THEOREM 3.3. *Let $V \in L_2^-(\Omega)$ be such that*

$$\sup |\frac{1}{k}\Re V| \leq \pi.$$

Then V is uniquely determined by $\mathcal{A}_k V$.

Proof. The condition on the real part of V guarantees that the phase of $\mathcal{E}PV$ is less than π in modulus, hence the nonlinear problem has a unique solution. □

Remark. If the above condition on the real part of V is not met, then we can use the compact support and the smoothness of $\mathcal{P}V$ for uniquely determining $\mathcal{P}V$ by starting from the boundary of the support and avoiding jumps in the function $\mathcal{P}V$.

THEOREM 3.4. *Let $V \in L_2^-(\Omega)$ be independent of k; i.e., we consider especially the quantum mechanical situation. Then*

$$\lim_{k\to\infty} \mathcal{A}_k V = -\sqrt{2\pi}\hat{V} \text{ on } \omega^\perp.$$

Proof. If V is independent of k then also $\mathcal{P}V$ is independent of k. With

$$\lim_{k\to\infty} (-\imath k(e^{-\frac{\imath}{2k}z} - 1)) = -\frac{1}{2}z$$

and the fact that $\mathcal{P}V \in H^{1/2}(T)$ we get with the Lebesgue dominated convergence theorem

$$\lim_{k\to\infty} \mathcal{A}_k V(\omega,\eta) = -\mathcal{F}^2_{\omega^\perp}\mathcal{P}V(\omega,\eta) = -\sqrt{2\pi}\hat{V}(\eta) \ , \ \eta \in \omega^\perp.$$

In the last step we have used the projection theorem for the x-ray transform, see [13]. □

Remark. In the quantum mechanical case we have as high-frequency limit $k \to \infty$ the Born approximation.

THEOREM 3.5. *In order to have an optimal resolution in the reconstruction it suffices to use $2k$ different incoming plane waves.*

Proof. This is a consequence of Lemma 10 in [11] where the resolution of the x-ray transform for a finite number of directions is discussed. □

Remark. The above decomposition of the operator A_k indicates a numerical algorithm for solving the inverse scattering problem. The question of achievable resolution is settled by the results on the x-ray transform.

REFERENCES

[1] S. AGMON, *Spectral Properties of Schrödinger Operators and Scattering Theory*, Ann. Scuol. Norm. Sup. Pisa, 2 (1975), pp. 151.

[2] K. CHADAN AND P.C. SABATIER, *Inverse Problems in Quantum Scattering Theory 2nd ed.*, Springer, Berlin, 1989.

[3] D. COLTON AND R. KRESS, *Integral Equation Methods in Scattering Theory*, Wiley, New York, 1983.

[4] A.J. DEVANEY, *Reconstructive Tomography with Diffracting Wavefields*, Inverse Problems, 2 (1986), pp. 161 - 183.

[5] R. GLAUBER, *High Energy Collision Theory*, in *W.E. Brittin, L.G. Dunhams (eds.) Lectures in Theoretical Physics, Vol.1*, Interscience, New York, 1959, pp. 315 - 414.

[6] N. GORENFLO, *Inversion Formulae for First - Order Approximations in Fixed Energy Scattering by Compactly Supported Potentials*, Inverse Problems, (1988), pp. 1025–1035.

[7] A. KIRSCH, *IMA Journal Appl. Math.*.

[8] A.K. LOUIS, *Picture Reconstruction from Projections in Restricted Range*, Math. Meth. Appl. Sci., 2 (1980), pp. 209 - 220.

[9] ————, *Inverse und schlecht gestellte probleme*, Teubner, Stuttgart, 1989.

[10] A.K. LOUIS AND F. NATTERER, *Mathematical Problems of Computerized Tomography*, Proc. IEEE, 71 (1983), pp. 379 - 389.

[11] P. MAASS, *The X-Ray Transform : Singular Value Decomposition and Resolution*, Inverse Problems, 3 (1987), pp. 729 - 741.

[12] G. MOLIÈRE, Z. für Naturforschung, 2A (1947), pp. 133.

[13] F. NATTERER, *The Mathematics of Computerized Tomography*, Teubner - Wiley, Stuttgart - New York, 1986.

[14] R.G. NEWTON, *Scattering Theory of Waves and Particles*, 2nd ed, Springer, Berlin, 1986.

[15] A.G. RAMM, *Scattering by Obstacles*, Reidel, Dorderecht, 1986.

[16] ————, *Multidimensional Inverse Problems and Completeness of the Product of Solutions to PDE*, J. Math. Anal. Appl., 134 (1988).

[17] M. REED AND B. SIMON, *Methods of Modern Mathematical Physics, III : Scattering Theory*, Academic Press, New York, 1979.

RADAR SIGNAL CHOICE AND PROCESSING FOR A DENSE TARGET ENVIRONMENT

HAROLD NAPARST*

Abstract. Consider a dense group of reflecting radar objects moving with different velocities at different ranges. the problem is to determine the density of targets at every range and velocity. The problem of how to choose the outgoing signals and how to process the echoes of those signals from the targets so as to determine the density function is discussed. The problem is a classical inverse problem. The object is to reconstruct a function of two variables (range and velocity) from limited information. Two schemes are given. The first of these methods modifies the method of Klauder and Wilcox to the case of signals with a large range of frequency components (wideband signals). The second is an improvement on the first which uses a formula of Khalil from affine group theory. It will be seen that this work is closely related to the "wavelet" work recently completed by Daubechies & Meyer. Numerical simulations support the conclusions.

This paper is a summary of the authors Ph.D. dissertation, which may be ordered from University Microfilms, Inc.

Contents

*Fidelity Management Co. 82 Devonshire St. #I40B, Boston, MA 02109

April 14, 1988

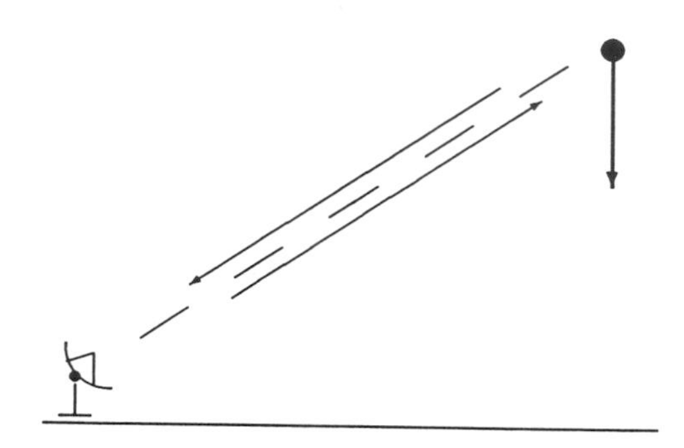

Figure 1: A radar setup with one moving target

1 Introduction. Suppose that there is an object of some sort which is moving relative to us. We would like to learn as much as possible about this object as we can just by sending pulses of energy to it and observing the returned echo. The type of energy might be electromagnetic or acoustic. See Figure 1.

Suppose that a signal $s(t)$ is sent out, and an echo $e(t)$ is observed. In this paper we are going to assume that:

$$e(t) = \sqrt{y}s(y(t-x)) \tag{1}$$

where x is the distance to the target (divided by the speed of propagation), and y is the compression factor:

$$y = \frac{1 - v/c}{1 + v/c}$$

where v is the velocity of the target in the line of sight, and c is the speed of propagation. This is the exact "wideband" return, rather than the commonly used "narrowband" approximation. For further details, please consult the authors thesis [21].

We point out that it is equivalent to assume

$$e(t) = \sqrt{y}s(yt + x)$$

as a change of variables will give. We will use these forms interchangeably.

The type of environment which we will study is a dense-target environment, which means that there are a lot of targets, and what is worse, the ranges and velocities of those objects are very close to each other. This type of environment has not received extensive treatment in the literature to my knowledge. The most complete discussion is given in Section 10.6 of Cook & Bernfeld [2]. A cursory discussion is given by Rihaczek [23], and it appears that the first reference to the dense-target environment was given by Fowle, Kelly, and Sheehan [12].

Suppose then, that the "density" of targets at distance x and velocity y is $D(x, y)$. The echo from this assemblage of targets will be:

$$e(t) = \int_{0}^{\infty} \int_{-\infty}^{\infty} D(x, y) \sqrt{y} s(y(t - x)) dx\, dy \tag{2}$$

This is to say that every place in the range-velocity plane where the targets could possibly be contributes to the echo.

Equation 2 is based on the assumption that all targets are illuminated equally by the radar beam. This may not be true at all. For example, if there is any attenuation of the radar beam, it will not be true. It may be possible to compensate for attenuation in the processing as is done in tomography [22]. We shall not do this. We will assume that the wavelength has been chosen so the echo will be in the form of Equation 2.

There is one additional assumption that we will make which really has nothing to do with the physics per se. That is, we will want to send multiple signals and record the echoes. During the time of the experiment, the target environment will change, especially if the velocities of the targets are high. We will ignore any change in the target environment over the period of the experiment. This constrains the length of the experiment to be a function of the "width" of $D(x, y)$ in the range-velocity plane and also the maximum velocity.

1.1 Statement of the Problem. Assume we are in the presence of a dense-target environment which satisfies the preceding assumptions, so that the target environment can be described by a continuous function $D(x, y)$. By observing only echoes from the target(s), we would like to determine $D(x, y)$. Each signal that we choose to send out will yield information, and the problem becomes twofold:

- How to choose the signals to send out.
- What to do with the echoes to determine $D(x, y)$.

In practice we will only want to determine $D(x, y)$ approximately. This will lead to problems of how to *best* implement the processing. That is, we will not only want to be able to know that we can find $D(x, y)$, but we will want to find an approximation to $D(x, y)$ as *efficiently* as possible. This will prove to be a real challenge.

1.2 Examples of Distributed Targets. It might be thought that the model of a continuous distribution of targets is artificial, since all target environments are really discrete. This is not true. Here are some ideas for possible realizations of distributed target environments.

- A rain cloud. Clearly one is not interested in imaging the individual water droplets which form the elementary scatterers, so this is a perfect example of a continuous target distribution. The real problem may come in satisfying the conditions of negligible signal attenuation and negligible multiple reflections.

- The wake of a re-entry vehicle such as the space shuttle. As an orbiting object re-enters the earth's atmosphere, the wake of particles in the ionosphere may be detectable by radar.[1]
- The surface of a large moving object such as a close airplane, the moon, or the hull of a ship underwater using sonar. This is the subject of inverse scattering and Inverse Synthetic Aperture Radar (ISAR) [10].
- Imaging ground clutter. As a missile flies over the ground, it may realize the ground as a distributed moving target and apply these techniques. This is the basis of Synthetic Aperture Radar (SAR).
- Discrete Targets. Although we have been describing the concept of a continuous distribution of targets, the techniques we will develop will be just as applicable to "almost discrete" target environments. Each technique will have its own limitations, but they are by no means going to be limited to smooth functions. The techniques we will describe are not restricted to be applicable only in the presence of a continuous distribution of targets; rather they are designed to deal with dense-target environments when conventional processing cannot and to replace conventional processing when their cost is less.

1.3 Structure of Paper. We are going to present two methods for solving our general problem. Naturally, we devote a section to each method.

In the next section we present a discussion of the work of Wilcox [26] & Klauder [18] and an extension to the wideband case. Also in that section we present a new theorem regarding the completeness and orthogonality of wideband cross-ambiguity functions. We provide a computer simulation which points to the uses and limitations of this method.

In the following section we present a very exciting and completely new method of reconstruction based on Lie Group theory. The formula that we use as the basis of our method is due to Khalil [17]. Very encouraging simulations are given.

Finally, we will sum up the methods in a concluding section. We will indicate directions for future research as well.

2 Expansion by Cross-Ambiguity Functions. A well-known way of expressing a vector $\vec{v}$ in a Hilbert space $\mathcal{H}$ is in terms of its components along an orthonormal basis $\{\vec{v}_i \in \mathcal{H} | i = 1, 2, \ldots \infty\}$:

$$\vec{v} = \sum_{i=1}^{\infty} c_i \vec{v}_i \tag{3}$$

[1] suggested by Marvin Bernfeld.

In this work, the "vector" we wish to describe is the function $D(x, y)$. The analog of (3) would be to find an orthonormal basis $\{\psi_{nm}(x, y)\}$ and constants $\{c_{nm}\}$ so that:

$$D(x,y) = \sum_{n,m=0}^{\infty} c_{nm}\psi_{nm}(x,y) \tag{4}$$

Wilcox [26, Lemma 4.1] accomplished this in the narrowband case. The functions $\psi_{nm}(x, y)$ were narrowband cross-ambiguity functions:

$$\psi_{nm}(x,y) = \int_{-\infty}^{\infty} s_m\left(t - \frac{x}{2}\right)\overline{s_n}\left(t + \frac{x}{2}\right)e^{-jyt}dt \tag{5}$$

The set $\{s_n\}$ was assumed to be a basis for an appropriate signal space, and the space $\mathcal{H}$ was chosen to be $L^2(\mathcal{R}^2)$.

In this section we propose to do the same thing for the wideband case. The practical importance of this result is as follows:

Suppose signals $\{s_1, s_2, \dots\}$ are sent out and their echoes $\{e_1, e_2, \dots\}$ are recorded. Suppose we form

$$\begin{aligned}\langle e_n, s_m\rangle & \int_0^{\infty}\int_{-\infty}^{\infty} D(x,y)\underbrace{\sqrt{y}\int_{-\infty}^{\infty} s_n(y(t-x))\overline{s}_m(t)\;dt}_{A_{nm}(x,y)}\;dx\;dy \\ &= \int_0^{\infty}\int_{-\infty}^{\infty} D(x,y)A_{nm}(x,y)\;dx\;dy \end{aligned} \tag{6}$$

If we can find signals $\{s_1, s_2, \dots\}$ such that their cross-ambiguity functions $\{A_{nm}(x, y)|n, m > 0\}$ are orthonormal with respect to some inner product and complete in a Hilbert space $\mathcal{H}$ which contains $D(x, y)$, then the numbers $\langle e_n, s_m\rangle$ will allow reconstruction of $D(x, y)$ as follows:

$$D(x,y) = \sum_{n,m=0}^{\infty} \langle e_n, s_m\rangle A_{nm}(x,y) \tag{7}$$

Although the theory for the narrowband case was founded by Wilcox, I have not seen a numerical simulation of the preceding technique. If such a simulation were done, there is a theoretical certainty and practical possibility that the narrowband approximation would be violated. Of course, there is not such problem in the wideband case.

In this section we shall develop the theory necessary to demonstrate the preceding technique. We will also give a numerical simulation of the technique.

2.1 The Theorems.

DEFINITION 1. Let $s_n(t)$ and $s_m(t)$ be complex valued functions on the real line. Define for $y > 0, x \in \mathcal{R}$

$$A_{nm}(x,y) = \sqrt{y} \int_{-\infty}^{\infty} sn(y(t-x))\overline{s_m(t)}dt \tag{8}$$

for those x, y for which the integral exists. Call $A_{nm}(x,y)$ the cross-ambiguity function of s_n and s_m.

The domain of $A_{nm}(x,y)$ may be empty or all of $\mathcal{R}^2_+ = \{(x,y)|x \in \mathcal{R}, y > 0\}$, depending on the properties of s_n and s_m.

DEFINITION 2. *Let $\mathcal{S}$ be the class of C-valued functions $s(t)$ for which:*

1. $s(t) \in L^2(\mathcal{R}, dt) \cap L^1(\mathcal{R}, dt)$
2. $S(\omega) \in L^2(\mathcal{R}, \sqrt{|2/\omega|}d\omega)$

Notice that property 1 implies that $S(\omega)$ exists for all ω and is in $L^2(\mathcal{R}, d\omega)$. Property 2 further restricts S to those functions whose moduli are square-integrable with respect to the measure $\sqrt{|2/\omega|}d\omega$ as well.

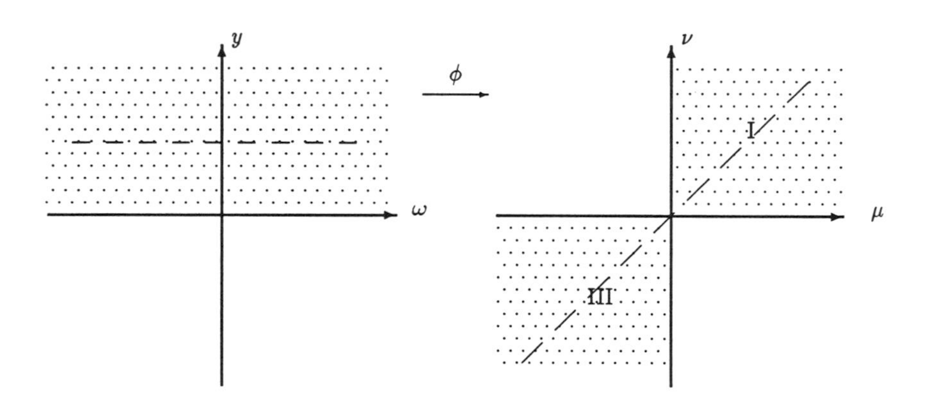

Figure 2: A Change of variables

PROPOSITION 1. *If s_n and s_m are elements of L^2, then $A_{nm}(x,y)$ exists on the upper half-plane $\mathcal{R}^2_+$. If s_n, s_m are elements of $\mathcal{S}$, then $A_{nm}(x,y)$ is an element of $L^2(\mathcal{R}^2_+, y^{-1/2}dx\ dy)$.*

Proof. Since s_n is in $L^2(\mathcal{R}, dt)$, so is $\sqrt{y}s_n(y(t-x))$. Therefore $A_{nm}(x,y)$ exists, since it is the inner product of two elements of L^2.

Since S_n and S_m are square-integrable with respect to $\sqrt{|2/\omega|}d\omega$,

(9)

$$\int_0^\infty |S_n(\nu)|^2 \frac{d\nu}{\nu^{1/2}} \int_0^\infty |S_m(\mu)|^2 \frac{d\mu}{\mu^{1/2}} + \int_{-\infty}^0 |S_n(\nu)|^2 \frac{d\nu}{|\nu|^{1/2}} \int_{-\infty}^0 |S_m(\mu)|^2 \frac{d\mu}{|\mu|^{1/2}} < \infty$$

The map $\phi : (\omega, y) \mapsto (\mu, \nu)$ given by

$$\phi(\omega, y) = (\omega, \omega/y) \tag{10}$$

sends the region $\mathcal{R}^2_+$ into the first and third quadrants as depicted in Figure 2, and transforms the area element $d\omega dy$ as follows:

$$d\omega \; dy = \left| \frac{\partial(\omega, y)}{\partial(\mu, \nu)} \right| d\mu \; d\nu$$

$$d\omega \; dy = \frac{\nu^2}{|\mu|} d\mu \; d\nu$$

Although this map is not bijective, it only fails to be bijective on a set of measure zero. Thus, we can do integrations as if it were bijective. Equation 9 therefore becomes:

$$\int_0^\infty \int_0^\infty |S_n(\nu)|^2 |S_m(\mu)|^2 \overbrace{\left(\frac{\nu}{\mu}\right)^{3/2}}^{y^{-3/2}} \overbrace{\left(\frac{\mu}{\nu^2}\right) d\mu \; d\nu}^{d\omega dy} +$$

$$\int_{-\infty}^0 \int_{-\infty}^0 |S_n(\nu)|^2 |S_m(\mu)|^2 \left(\frac{\nu}{\mu}\right)^{3/2} \left(\frac{|\mu|}{\nu^2}\right) d\mu \; d\nu < \infty$$

$$\int_0^\infty \int_{-\infty}^\infty y^{-3/2} |S_n(\omega/y)|^2 |S_m(\omega)|^2 \; d\omega \; dy < \infty$$

$$\int_0^\infty \int_{-\infty}^\infty \sqrt{y}\frac{1}{y} S_n\left(\frac{\omega}{y}\right) \overline{S_m(\omega)} \; \overline{\sqrt{y}\frac{1}{y} S_n\left(\frac{\omega}{y}\right) S_m(\omega)} \quad d\omega \frac{dy}{y^{1/2}} < \infty$$

Since $S_n, S_m \in L^2, S_n\left(\frac{\omega}{y}\right) \overline{S_m(\omega)} \in L^1(\mathcal{R}, d\omega)$. Therefore, the inverse Fourier transform of $\sqrt{y}\frac{1}{y} S_n\left(\frac{\omega}{y}\right) \overline{S_m(\omega)}$ exists and Parseval's theorem applies. The above integral becomes:

$$\int_0^\infty \int_{-\infty}^\infty \sqrt{y} \int_{-\infty}^\infty s_n(y(t-x))\overline{s_m(t)} dt \overline{\sqrt{y} \int_{-\infty}^\infty s_n(y(t-x))\overline{s_m(t)} \; dt} \frac{dx \; dy}{y^{1/2}} < \infty$$

$$\iint_{\mathcal{R}^2_+} |A_{nm}(x, y)|^2 \frac{dx \; dy}{y^{1/2}} < \infty$$

which was to be shown.

□

In the above theorem, the measure $y^{-1/2}dx\ dy$ is unsettling. From a group theoretic point of view, it is neither the left nor the right Haar measure. One wonders why we are using it at all, and where it came from. The point of using the strange measure $y^{-1/2}dx\ dy$ is as follows:

THEOREM 1 (ORTHOGONALITY). *Let $\mathcal{J}$ be an index set. Suppose that $\{s_n(t), n \in \mathcal{J}\}$ is a subset of $\mathcal{S}$ with the properties that:*

$$n, m \in \mathcal{J} \Longrightarrow \int_0^\infty S_n(\omega)\overline{S_m(\omega)}\frac{d\omega}{\sqrt{\omega}} = \frac{1}{\sqrt{2}}\delta_{mn} \tag{11}$$

and

$$n, m \in \mathcal{J} \Longrightarrow \int_{-\infty}^0 S_n(\omega)\overline{S_m(\omega)}\frac{d\omega}{\sqrt{\omega}} = \frac{1}{\sqrt{2}}\delta_{mn} \tag{12}$$

Then the cross-ambiguity functions $\{A_{nm}(x,y), n, m \in \mathcal{J}\}$ are pairwise orthonormal in $L^2(\mathcal{R}^2_+ \to \mathcal{C}, y^{-1/2}dx\ dy)$.

Proof. In what follows we use the map $(\omega, y) \mapsto (\mu, \nu)$ given by Equation 10.

$$\langle A_{n,m}A_{n',m'}\rangle = \int_0^\infty \int_{-\infty}^\infty \sqrt{y} \int_{-\infty}^\infty s_n(y(t-x))\overline{s_m(t)}dt$$

$$\sqrt{y} \int_{-\infty}^\infty \overline{s_{n'}(y(t'-x))}s_{m'}(t')dt'\frac{dx\ dy}{y^{1/2}}$$

$$= \int_0^\infty \int_{-\infty}^\infty y^{-3/2} S_n\left(\frac{\omega}{y}\right) \overline{S_{n'}\left(\frac{\omega}{y}\right)} \overline{S_m(\omega)} S_{m'}(\omega) d\omega\ dy$$

$$= \iint_{I+III} S_n(\nu)\overline{S_{n'}(\nu)S_m(\mu)}S_{m'}(\mu)\left(\frac{\nu}{\mu}\right)^{3/2}\left(\frac{|\mu|}{\nu^2}\right)\ d\mu\ d\nu$$

$$= \int_0^\infty S_n(\nu)\overline{S_{n'}(\nu)}\frac{d\nu}{\nu^{1/2}} \int_0^\infty \overline{S_m(\mu)}S_{m'}(\mu)\frac{d\mu}{\mu^{1/2}}$$

$$+ \int_{-\infty}^0 S_n(\nu)\overline{S_{n'}(\nu)}\frac{d\nu}{|\nu|^{1/2}} \int_{-\infty}^0 \overline{S_m(\mu)}S_{m'}(\mu)\frac{d\mu}{|\mu|^{1/2}}$$

$$= \delta_{mm'}\delta_{nn'}$$

This was to be shown.

□

The previous theorem is related to work done by Duflo and Moore [7]. The reader may also want to consult [27, 25]. The results obtained here are perhaps a more symmetric form of their equations.

COROLLARY 1. *Let $\mathcal{J}$ be an index set. Suppose that $\{s_n(t), n \in \mathcal{J}\}$ is a subset of $\mathcal{S}$ with the properties that:*

$$n, m \in \mathcal{J} \Longrightarrow \int_0^\infty S_n(\omega)\overline{S_m(\omega)}\frac{d\omega}{\sqrt{\omega}} = \frac{1}{\sqrt{2}}\delta_{mn} \tag{13}$$

and $\forall n \in \mathcal{J}$, $s_n(t)$ is even. Then the cross-ambiguity functions $\{A_{nm}(x,y), n, m \in \mathcal{J}\}$ are pairwise orthonormal in $L^2(\mathcal{R}^2_+, y^{-1/2}dx\ dy)$.

Proof. Because the signals $\{s_n(t)\}$ are even, their Fourier transforms are even, so (13) automatically implies (12). The fact that the cross-ambiguity functions are real follows immediately from the definition since the signals $s_n(t)$ are real.

□

In the above Corollary we could have taken all the signals $\{s_n(t)\}$ to be odd instead of even, but we could not have had a mix of even and odd functions. The reader should carefully check this.

DEFINITION 3. Let $\mathcal{S}_e$ be the subset of $\mathcal{S}$ whose members are real and even.

LEMMA 1. *If $s_n(t)$ and $s_m(t)$ are in $\mathcal{S}_e$, then $A_{nm}(x,y)$ is real and even in x.*

Proof. Left to the reader.

□

THEOREM 2 (COMPLETENESS). *Let $\{s_n(t) | n \in \mathcal{J}\}$ span $\mathcal{S}$. Then their cross-ambiguity functions $\{A_{nm}(x,y) | n, m \in \mathcal{J}\}$ span $L^2(\mathcal{R}^2_+, y^{-1/2}dx\ dy)$.*

Proof. Because the set $\{S_n(\omega)\}$ spans $L^2(\mathcal{R}, \sqrt{|2/\omega|}\ d\omega)$, the product functions $\{S_n(\nu)\overline{S_m(\mu)}\}$ span $L^2(I + III, |\mu\nu|^{-1/2}\ d\mu\ d\nu)$. Applying the map (10) shows that the functions $\{y^{-1/2}S_n(\omega/y)\overline{S_m(\omega)}\}$ span $L^2(\mathcal{R}^2_+, y^{-1/2}d\omega\ dy)$. Taking Fourier transforms, the functions $\{A_{nm}(x,y)\}$ span $L^2(\mathcal{R}^2_+, y^{-1/2}dx\ dy)$.

□

COROLLARY 2. *Let $\{s_n | n \in \mathcal{J}\}$ span $\mathcal{S}_e$. Then their cross-ambiguity functions*

$$\{A_{nm}(x,y) | n, m \in \mathcal{J}\}$$

span the subset of $L^2(\mathcal{R}^2_+ \to \mathcal{R}, y^{-1/2}dx\ dy)$ consisting of real functions even in x.

Proof. The proof of this Corollary is identical to the proof of the previous theorem. It is only necessary to note that the extra properties are also satisfied. This is left to the reader.

□

2.2 The Method. Our immediate goal is a simulation of the expansion of Equation 7. First we will give a choice of the set of signals $\{s_n(t)\}$, and we will summarize the method again. Then, we will present the results of a numerical simulation using the proposed functions and a phantom target environment.

Theorems 1 and 2 provide a way of reconstructing $D(x,y)$ in the event that it is either not real or not even in x. Since x is proportional to distance — a positive quantity — $D(x,y)$ will have support in the right half-plane. In order to reconstruct a function with support in the right half-plane it suffices to have an orthonormal basis of the *even* subset of $L^2(\mathcal{R}^2_+, y^{-1/2}dx\ dy)$. After generating the even reconstruction, one just ignores the left half-plane. Although there may be no justification for assuming that $D(x,y)$ is real valued, we will make this assumption just to simplify the exhibition of the method. It is simply easier to plot real functions.

2.2.1 Choice of Signals. In practice, the choice of signals $\{s_n(t)\}$ greatly influences the performance of this method. In the narrowband case, Wilcox decided to choose the signals so as to make the auto-ambiguity functions $\{A_{nn}(x,y)\}$ circularly symmetric. It is not clear that circular symmetry means anything deep when the x-axis represents distance and the y-axis represents velocity, but the intention is clear. If the units are chosen so that a one unit change in the x-direction is as important as one unit change in the y-direction, then circularly symmetric auto-ambiguity function will resolve equally in both variables, where the word *equally* means that it is probably resolving in the way we want it to.

Let us denote the signals that Wilcox proposed to use by $w_n(t)$. We will describe $w_n(t)$ by giving the Fourier transform $W_n(\omega)$. Let ω_0 be any positive real number. Then

$$W_n(\omega) = \frac{H_n(\omega-\omega_0)}{\pi^{1/4}2^{n/2}\sqrt{n!}}e^{-(\omega-\omega_0)^2/2} + \frac{H_n(\omega+\omega_0)}{\pi^{1/4}2^{n/2}\sqrt{n!}}e^{-(\omega+\omega_0)^2/2} \tag{14}$$

where $H_n(\omega)$ is the nth Hermite polynomial. It is tempting to use the above functions in our simulation, but there is a problem. For small n and large ω_0, it is true that the $W_n(\omega)$ are nearly orthogonal. But as n is increased, the overlap of the two terms in (14) becomes significant, the near-orthogonality is lost, and the narrowband approximation is violated. In order to fix this problem we propose to separate the two terms. Let $p : (0,\infty) \to \mathcal{R}$ be a C^1-bijection such as

$$p(\omega) = \begin{Bmatrix} \omega_0 - \cot\left(\frac{\pi}{2}\frac{\omega}{\omega_0}\right) & \omega < \omega_0 \\ \omega & \omega \geq \omega_0 \end{Bmatrix} \tag{15}$$

For $\omega > 0$, let

$$U_n(\omega) = \sqrt{p'(\omega)}\frac{H_n(p(\omega)-\omega_0)}{\pi^{1/4}2^{n/2}\sqrt{n!}}e^{-(p(\omega)-\omega)^2/2}$$

For $\omega < 0$, we let $U_n(\omega) = U_n(-\omega)$. Now $\{U_n(\omega)\}$ is orthogonal for all n, regardless of the choice of ω_0. The choice of $p(\omega)$ was rather arbitrary, chosen more for ease of implementation that anything else. It is true that the choice of $p(\omega)$ will affect the "circular symmetry" of the auto-ambiguity function. The method in which this happens has not been investigated. It will be seen that the choice of p is not a critical factor, though.

The carrier frequency ω_0 provides only one parameter to control the resolution characteristics of the ambiguity functions $A_{nm}(x,y)$. In order to gain more control, we modify $U_n(\omega)$ by introducing a positive parameter α as follows:

$$U_n(\omega) = \alpha\sqrt{p'(\omega)}\frac{H_n(\alpha^2(p(\omega)-\omega_0))}{\pi^{1/4}2^{n/2}\sqrt{n!}}e^{-\alpha^4(p(\omega)-\omega_0)^2/2} \qquad (\omega > 0)$$

by varying α and ω_0, the functions $A_{nm}(x,y)$ will be distorted. We may choose α and ω_0 to affect the desired distortion. The only deficiency with the above functions $U_n(\omega)$ is that they are orthogonal not with respect to $\sqrt{|2/\omega|}d\omega$, but with respect to $d\omega$. To fix this, we simply multiply by $|\omega|^{1/4}$. Thus, define

$$S_n(\omega) = \omega^{1/4}U_n(\omega) \tag{16}$$

The $s_n(t)$, just defined through their Fourier transforms, form a basis of $\mathcal{S}_e$, orthogonal with respect to $\sqrt{|2/\omega|}d\omega$. It is these signals that we will use in the simulation.

2.2.2 Summary. As a final step before the simulation, let us summarize the method that we will employ to reconstruct the real-valued function $D(x,y)$, which is even in x.

1. Let $\alpha,\ \omega_0 > 0$. Let $p : (0,\infty) \to \mathcal{R}$ be a C^1-bijection such as is given in Equation 15. For $\omega > 0$ let

$$U_n(\omega) = \alpha\sqrt{p'(\omega)}\frac{H_n(\alpha^2(p(\omega)-\omega_0))}{\pi^{1/4}2^{n/2}\sqrt{n!}}e^{-\alpha^4(p(\omega)-\omega_0)^2/2}$$

   For $\omega < 0$, let $U_n(\omega) = U_n(-\omega)$.

2. Let $S_n(\omega) = |\omega|^{1/4}U_n(\omega)$. Normalize the $S_n(\omega)$ in the space $L^2(\mathcal{R}_+, \sqrt{|2/\omega|}d\omega)$. Le $s_n(t)$ be the inverse Fourier transform of $S_n(\omega)$. Use the signals $\{s_n(t), n = 0,1,2,\dots\}$ for the radar pulses. Denote the received echoes by $\{e_n(t), n = 0,1,2,\dots\}$.

   From the inner products[2]

$$c_{nm} = \int_{-\infty}^{\infty} e_n(t)\overline{s_m(t)}dt$$

$$\iint_{\mathcal{R}_+^2} \sqrt{y}D(x,y)A_{nm}(x,y)\frac{dx\,dy}{\sqrt{y}} \qquad (m,n < N)$$

3. Then,

$$D(x,y) \approx \frac{1}{\sqrt{y}}\sum_{m,n=0}^{N} c_{nm}A_{nm}(x,y)$$

[2]The signals $s_n(t)$ are real. The bar denoting conjugation is not necessary.

2.3 The Simulation. With the choice of signals we have made, the functions $A_{nm}(x, y)$ are fairly concentrated near $x = 0,\ y = 1$. In fact, they were chosen by Klauder [18] to have exactly this property. Such $A_{nm}(x, y)$, will not be suitable for expansion of densities $D(x, y)$ which are large far away from $x = 0,\ y = 1$. In the interest of showing some possible value of this technique, let us try to reconstruct

$$D(x, y) = \delta(x)\delta(y - 1)$$

Certainly this function is concentrated near $x = 0,\ y = 1$! Choosing such a function may therefore be considered cheating by some. As we said, though, we are interested first in seeing if we can get *some* value from this technique.

Over the next pages, we present the reconstructions we obtained with N equal to 1 and 4. The reconstructions are all fairly circularly symmetric, showing that our choice of $p(\omega)$ didn't affect things too much. However, the results are not very encouraging even for this ideal target density. Of course, they are an improvement over previously reported reconstructions, which used only one signal instead of our four, but the method clearly won't work well if the object is not at the point $x = 0,\ y = 1$.

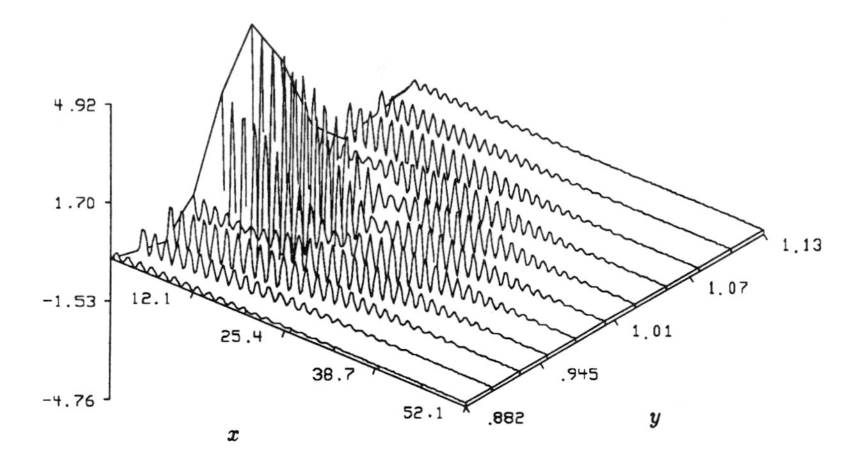

Figure 3: Approximation of a δ-function with $N = 1$

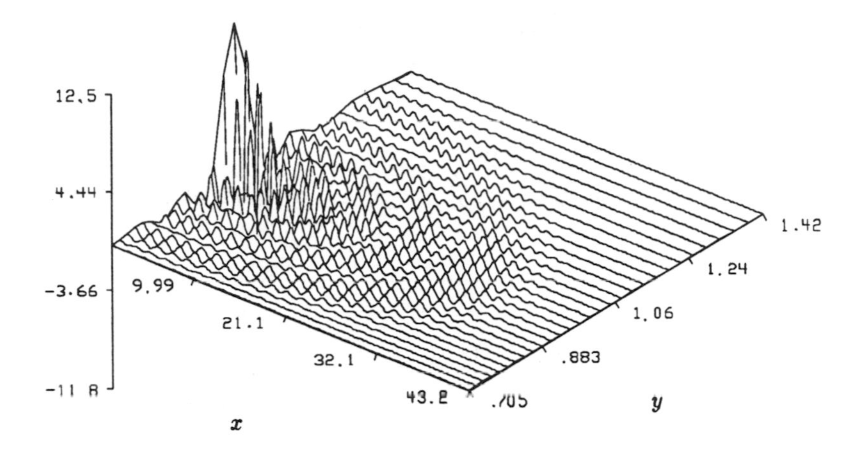

Figure 4: Approximation of a δ-function with $N = 4$

The amount of computation required, added to the probable failure of this method for functions which are not concentrated near $x = 0$, $y = 1$, does not exactly encourage us. Perhaps with a different set of functions $\{s_n(t)\}$ this method will be more useful, but we are not optimistic. This is unfortunate, because it is (was) our first try with a scheme which showed some practical promise. Even from a theoretical point of view, this method can be viewed as an implementation of the Peter–Weyl theorem for the affine group. Nonetheless, in this incarnation at least, it doesn't work very well. Let us now try to learn from this failure and improve.

3 Affine Group Methods. In the last section we expanded the target density $D(x, y)$ by means of cross-ambiguity functions:

$$D(x,y) = \sum_{n,m=0}^{\infty} c_{nm} A_{nm}(x,y) \tag{17}$$

where

$$A_{nm}(x,y) = \sqrt{y} \int_{-\infty}^{\infty} s_n(yt+x)\overline{s_m}(t)dt \tag{18}$$

If (x, y) is not *close* to $(0, 1)$, $|A_{nm}(x, y)|$ will be small, and the expansion (17) will generally be poor for that reason. As we have seen, the results given in the previous section are not impressive. Now we will try to improve our results by reconsidering conventional processing methods. Then we will compare the two methods side by side.

One usually cross-correlates the echo with the signal rather than taking the inner product as in (18). Let us define the cross-correlation operator $\star$ by:

$$(f \star g)(x,y) = \sqrt{y} \int_{-\infty}^{\infty} f(yt+x)\overline{g(t)}dt$$

If we do this, (18) is replaced by:

$$A_{nm}(x,y,x'y') = \sqrt{yy'} \int_{-\infty}^{\infty} s_n(yt+x)\overline{s_m}(y't+x')dt \tag{19}$$

Let

$$\begin{aligned} r_{nm}(x',y') &= e_n(t) \star s_m(t) \\ &= \iint_{\mathcal{R}_+^2} D(x,y)A_{nm}(x,y,x',y')dx\, dy \end{aligned}$$

We might ask if observation of the functions $r_{nm}(x',y')$ is sufficient to determine $D(x,y)$. In fact, we don't even need that much information. We already know from Section 2 that we could observe just $r_{nm}(0,1)$ and still reconstruct D—although poorly.

Looking for another way to reduce the amount of information — and therefore the computation — in (20), we might try sending out only one signal and observing

$$\begin{aligned} r_{nm}(x',y') &= e(t) \star s(t) \\ &= \iint_{\mathcal{R}_+^2} D(x,y)A(x,y,x',y')dx\, dy \end{aligned}$$

This is certainly no good. It is well known [6, 5, 4, 3, 15, 16, 14, 13, 20] that $r(x',y')$ cannot determine $D(x,y)$. In fact, in the references just cited, the kernel of $\Phi : D \to r$ is discussed.

We have just described two ways of deleting some of the four parameters $n, m, x'y'$. The first method, that of §2, deleted x' and y' and didn't work well. The second method, the conventional one, deleted n and m and didn't work at all. A compromise is in order. Let us just delete m and keep everything else. That is, let us try observing

$$r_n(x',y') = \iint_{\mathcal{R}_+^2} D(x,y)A_{nn}(x,y,x',y')dx\, dy \tag{20}$$

We shall see that this is a good idea. With one modification, it is very easy to recover $D(x,y)$ from the functions $r_n(x',y')$.

3.1 Fourier Analysis on the Affine Group. The affine group is the group of translations and compressions of the real line. That is, it is the set of maps:

$$\Phi_{x,y} : t \longmapsto yt + x \qquad x \in \mathcal{R} \qquad y > 0$$

The Doppler effect describes an action of this group G on the signal space. If we let x and y be the coordinates on the affine group and let $g = (x, y)$ be an element of G, the echo $e(t)$ is

$$e(t) = \int_G \sqrt{y} D(g) s(gt) dg$$

We will leave the measure dg unspecified for the moment. Then

$$\begin{aligned} e(t) \star s(g't) &= \int_G D(g)\sqrt{yy'} \int_{-\infty}^{\infty} s(gt)\overline{s}(g't)dt\ dg \\ &= \int_G D(g)\sqrt{y/y'} \int_{-\infty}^{\infty} s(gg'^{-1}t)\overline{s}(t)dt\ dg \\ &= \int_G D(g)A(gg'^{-1})dg \qquad (21) \end{aligned}$$

Equation 21 is a convolution on the affine group. Evidently the inversion problem is reduced to Fourier analysis on this group. Originally our intention was to take the "Fourier transform" of (21) and "divide" to obtain D. However, as we present the Fourier theory for G, a more natural solution will become evident. This happens mainly because the function $A(g)$ is a very special type of function—a matrix element of an irreducible unitary representation of G.

There are exactly two non-equivalent infinite dimensional irreducible unitary representations of G. Call them π_+ and π_-. We will describe four concrete realizations of them, depending on the representation space $\mathcal{H}$ chosen:

1. If $\mathcal{H}_1 = L^2\left(\mathcal{R}_+, \frac{dt}{t}\right)$, then define

$$\pi^1_\pm(x, y)\phi(t) = e^{\pm jxt}\phi(yt) \qquad (22)$$

   To check unitarity note that,

$$\langle \pi^1_\pm(x, y)\phi, \pi^1_\pm(x, y)\psi\rangle = \int_0^\infty \phi(yt)\overline{\psi(y^t)}\frac{dt}{t} = \langle \phi, \psi\rangle$$

2. If we apply the map $\phi \mapsto \phi/\sqrt{t}$ to the above representation, and let $\mathcal{H}_2 = L^2(\mathcal{R}_+, dt)$ we obtain equivalent representations:

$$\pi^2_\pm(x, y)\phi(t) = e^{\pm jxt}\sqrt{y}\phi(yt) \qquad (23)$$

To check unitarity,

$$\langle \pi^2_{\pm}(x,y)\phi, \pi^2_{\pm}(x,y)\psi\rangle = \int_0^{\infty} \phi(yt)\overline{\psi(yt)}d(yt) = \langle \phi, \psi\rangle$$

3. **Kohari** [19]: Here we let the representation space be $\mathcal{H}_3 = L^2(\mathcal{R}, dt)$. We let $\pi^3_{\pm}$ act on the subset of $\mathcal{H}_3$ which consists of functions supported only on the right half-line. Denote such a function by ϕ_+. π^3_- will act on the orthogonal complement, which are functions supported on the left half-line. We define

$$\pi^3_{\pm}(x,y)\phi_{\pm}(t) = e^{jxt}\sqrt{y}\phi_{\pm}(yt) \tag{24}$$

It is easy to see that this representation is equivalent to the previous one.

4. If we take the Fourier transform of the above representation, we obtain a very familiar expression. Let $\mathcal{H}^{\pm}_4$ be the set of functions $\phi_{\pm}$ whose Fourier transforms have support on the right (left) half-line respectively. Let

$$\pi^4_{\pm}(x,y)\phi_{\pm}(t) = \frac{1}{\sqrt{y}}\phi_{\pm}(t/y + x) \tag{25}$$

We hope the reader will recognize this as the action of the affine group in time that we have been using all along.

Representations are very important in the study of Lie groups. It is clear from the last representation π^4 that the ambiguity functions are just diagonal matrix elements of this representation. Let us define the Fourier transform of a function $f(x,y)$ on the affine group to be the pair of operators $(\mathcal{F}_+(f), \mathcal{F}_-(f))$ defined by:

$$\mathcal{F}_{\pm}(f) = \int_G dg f(g)\pi_{\pm}(g) \tag{26}$$

where $dg = y^{-2}dy\ dx$ is the left-invariant measure on G. If we define convolution by

$$f * h(g) = \int_G f(g'^{-1}g)h(g')dg' \tag{27}$$

then convolution is mapped into composition:

$$\mathcal{F}(f * h) = \mathcal{F}(h) \circ \mathcal{F}(f) \tag{28}$$

Clearly one gets different results by defining convolution differently. The *amazing* thing, however, is that with this definition of $\mathcal{F}$, one has the inversion formula of Khalil [17]:

$$f(g) = \sum_{\pm} \text{tr}\ (\pi^*_{\pm}(g) \circ \mathcal{F}_{\pm}(f) \circ \delta_1) \tag{29}$$

where the operator δ_1 is defined by

$$\delta_1 \phi(t) = t\phi(t)$$

if the concrete form of the representation is chosen to be the first one, π^1. It is interesting to see what δ_1 is in every realization we have given. For example, in π^4, it corresponds to taking the derivative and then taking the Hilbert transform.

The formula (29) can be verified by direct calculation. We use the first form of the representation to do it, but it could clearly be done in any of the equivalent representations we have given. Instead of supposing that the functions in the representation space are functions of t, let us make a small change and denote the independent variable by ω. This makes more sense really, because the action of G is most naturally viewed in the frequency domain. Because the representations are unitary, $\pi^*(g) = \pi(g^{-1})$. Let

$$P = \pi(g^{-1}) \circ \int_G f(g')\pi(g')dg' \circ M(\omega) \tag{30}$$

By $M(\omega)$ as an operator, we mean multiplication by ω. Then,

$$\begin{aligned} P_{\pm}s(\omega) &= \pi_{\pm}(g^{-1}) \int_G f(g')e^{\pm jx'\omega}\ y'\omega s(y'\omega)dg' \\ &= \int_G f(g')e^{\mp j\frac{x}{y}\omega}e^{\pm j\frac{x'}{y}\omega}\frac{y'}{y}\omega\ s\left(\frac{y'}{y}\omega\right)\frac{dy'dx'}{y'^2} \end{aligned}$$

Let $\widetilde{\omega} = y'\omega/y$. Then,

$$\begin{aligned} P_{\pm}s(\omega) &= \int_G f(g')e^{\pm j\frac{\omega}{y}(x'-x)}\frac{\omega}{y}dx'\ s(\tilde{\omega})\frac{dy'}{y'} \\ &= \int_0^\infty K_{\pm}(\omega,\tilde{\omega})s(\tilde{\omega})\frac{d\tilde{\omega}}{\tilde{\omega}} \end{aligned}$$

where

$$\begin{aligned} K_{\pm}(\omega,\tilde{\omega}) &= \frac{\omega}{y}e^{\mp j\frac{\omega}{y}x}\int_{-\infty}^{\infty} f(x',y')e^{\pm\frac{\omega}{y}x'}\ dx' \\ &= \frac{\omega}{y}e^{\mp j\frac{\omega}{y}x}\mathfrak{F}_1 f\left(\pm\frac{\omega}{y},y'\right) \\ &= \frac{\omega}{y}e^{\mp j\frac{\omega}{y}x}\mathfrak{F}_1 f\left(\pm\frac{\omega}{y},\frac{\tilde{\omega}y}{\omega}\right) \end{aligned}$$

From an exercise in Dunford & Schwartz [8, Chapter 11, # 49],

$$\begin{aligned} \mathrm{tr}P &= \sum_{\pm}\int_0^\infty K_{\pm}(\omega,\omega)d\omega \\ &= \sum_{\pm}\int_0^\infty e^{\mp j\frac{\omega}{y}x}\mathfrak{F}_1 f\left(pm\frac{\omega}{y},a\right)\frac{\omega}{y}\frac{d\omega}{\omega} \end{aligned}$$

Now let $\omega' = \frac{\omega}{y}$. Then

$$\begin{aligned}
\mathrm{tr}P &= \sum_{\pm} \int_0^\infty e^{\pm j\omega' x} \mathcal{F}_1 f(\pm\omega', y) d\omega' \\
&= \int_0^\infty e^{-j\omega' x} \mathcal{F}_1 f(\omega', y) d\omega' + \int_0^\infty e^{+j\omega' x} \mathcal{F}_1 f(-\omega', y) d\omega' \\
&= \int_0^\infty e^{-j\omega' x} \mathcal{F}_1 f(\omega', y) d\omega' + \int_{-\infty}^0 e^{-j\omega' x} \mathcal{F}_1 f(+\omega', y) d\omega' \\
&= f(x, y)
\end{aligned}$$

which proves (29).

3.2 Implications of Khalil's Formula. Khalil's formula (29) is just a mathematical theorem like the Fourier inversion theorem. The usefulness of Fourier's theorem likes in the physical importance of the Fourier transform. Likewise we shall now see that Khalil's formula will give us a reconstruction algorithm as a composition of simple operations.

First, if P is any trace class operator in a Hilbert space $\mathcal{H}$, and $\mathcal{B} = \{s_n(t)\}$ is a normalized basis of $\mathcal{H}$, the trace of P is independent of $\mathcal{B}$ and is given by

$$\mathrm{tr}P = \sum_{n=0}^{\infty} \langle Ps_n, s_n \rangle \tag{31}$$

In our case, we have two operators, two Hilbert spaces, and two bases. Let us denote them by $P_\pm(g), \mathcal{H}_\pm$, and $\{s_n^\pm(t)\}$, respectively. Define

$$P_\pm(g) = \pi_\pm^*(g) \circ \mathcal{F}_\pm(f) \circ \delta_1 \tag{32}$$

Then Khalil's formula becomes

$$\begin{aligned}
f(g) &= \sum_{\pm} \sum_{n=0}^{\infty} \langle \pi_\pm^*(g) \circ \mathcal{F}_\pm(f) \circ \delta_1 s_n^\pm(t), s_n^\pm(t) \rangle \\
&= \sum_{\pm} \sum_{n=0}^{\infty} \langle \mathcal{F}_\pm(f)(\delta_1, s_n^\pm(t)), \pi_\pm(g) s_n^\pm(t) \rangle
\end{aligned}$$

We know that if $\sigma(t)$ is a signal, $\mathcal{F}(f)\sigma(t)$ is the echo from the target density $f(x, y)$. Also, taking the inner product with $\pi(g)s_n(t)$ is exactly cross-correlation. The reader will now see that the only "modification" that is necessary to the idea discussed at the beginning of this section is the inclusion of δ_1.

3.3 Choice of Signals. Although the value of the sum (31) is independent of $\mathcal{B}$, in practice the choice of $\mathcal{B}$ will have a great impact on the rate of convergence of the sum. In order to send as few signals as possible, we would like to make the first few terms as large as possible. If P is an Hermitian operator, we could use the power method to find the eigenfunction $s_0(t)$ corresponding to the largest eigenvalue as follows:

1. Let $\sigma_0(t)$ be any $L^2(\mathcal{R})$ function.
2. Let

$$\sigma_n(t) = \frac{P\sigma_{n-1}}{\|P\sigma_{n-1}\|} \qquad n = 1, 2, \ldots \tag{33}$$

3. Let

$$s_0(t) = \lim_{n \to \infty} \sigma_n(t) \tag{34}$$

One could then find the second eigenfunction by restricting P to the orthogonal complement of the one dimensional space spanned by $s_0(t)$ and applying the power method again. Continuing in this way, we could find as many $s_n(t)$ as desired.

The reader may protest that P depends on f and g. Thus, the optimal set $\{s_n(t)\}$ will also depend on f and g. This is true. However, let us make some observations:

1. An excellent choice of $\mathcal{B}$ for one value of g will probably represent a pretty good choice for other values of g.
2. Of course the choice of $\mathcal{B}$ depends on the target environment $f(g)$. The proposal is to observe a few echoes as in Equation 33 and determine $\mathcal{B}$ that way. In practice we may be observing similar target environments, and a fixed $\mathcal{B}$ may be possible.

The previous scheme is so nice, it's a pity that our P is not usually Hermitian. We have made an initial study of when P is Hermitian, and have found that $f(x, y)$ and g may have to satisfy ridiculously strict conditions. We state the following proposition without proof:

PROPOSITION 2. *Sufficient (but not necessary) conditions for P to be Hermitian are:*

1. $x = 0$
2. $f(x, y)$ *can be written as a product:* $f(x, y) = f_1(x) f_2(y)$
3. $f_2(-x) = f_2^*(x)$
4. $f_1(y\gamma) = f_1(y/\gamma)$ *for all* γ.

For example, if $g = (0, 1)$ then the following target density works:

$$f(x, y) = e^{-x^2}\, e^{-(\log y)^2} \tag{35}$$

Regarding the previous example, we found something rather curious. The computed eigenfunctions bear a similarity to the functions $U_n(\omega)$ that we used in Chapter 2. Based on this observation, we have used these functions $U_n(\omega)$ in our simulation. the question of how to choose the parameters α and ω_0 is important, however.

Even if P is not Hermitian, and we have seen that it is usually not, it may still be possible to gain information about how to choose $\mathcal{B}$ by observing the functions $\sigma_n(t)$ for a few iterations. In practice we see that although the guess $\sigma_0(t)$ might be poor, $\sigma_1(t)$ "jumps" to about the right position. Thus, the parameters α and ω_0 are chosen by observing the σ_n for what in practice may amount to only one iteration.

The subject of how to choose the signals $\{s_n(t)\}$ is truly fascinating. From a purely mathematical point of view, is it possible to analytically determine the eigenfunctions corresponding to the target density in Equation 35? We don't know. More practically, is there a *much* better way of choosing $\{s_n(t)\}$ than the one we have described? Again, we don't know. We hope to see further work on these questions.

But now, let us move on to a summary of the method we will use in our numerical simulation.

3.4 The Method. It is possible to implement the method that we have just discussed with no knowledge of the subject of Lie groups. Let $D(x, y)$ be a complex valued target density function. To determine $D(x, y)$:

1. For $n = 0, 1, 2, \ldots$, let

$$S_n^+(\omega = \begin{cases} \alpha\sqrt{p'(\omega)}\frac{H_n(\alpha^2[p(\omega)-\omega_0])}{\pi^{1/4}2^{n/2}\sqrt{n!}}e^{-\alpha^4(p(\omega)-\omega_0)^2/2} & \omega > 0 \\ 0 & \omega \leq 0 \end{cases}\Bigg\}$$

   and

$$S_n^-(\omega) = S_n^+(-\omega)$$

   The parameters α and ω_0 are arbitrary and positive.

2. Let

$$\widetilde{S}_n^{\pm}(\omega) = |\omega| S_n^{\pm}(\omega)$$

3. Let $\widetilde{s}_n^{\pm}(t)$ denote the inverse Fourier transform of $\widetilde{S}_n^{\pm}(\omega)$.

4. Let $e_n^{\pm}(t)$ denote the echoes from the signals $\tilde{s}_n^{\pm}(t)$¿

5. Let

$$r_n^{\pm}(x, y) = \int_{-\infty}^{\infty} e_n^{\pm}(t)\sqrt{y}\overline{s_n^{\pm}(yt + x)}dt$$

6. Then

$$D(x, y) = \sum_{\pm}\sum_{n=0}^{\infty} r_n^{\pm}(x, y)$$

3.5 The Numerical Simulation. The first target environment which will simulate is shown in figure 5. It consists of six Gaussians fairly well separated. The reconstruction is shown in figure 6. We used 10 signals to obtain this. Obviously this target environment is more complex than that of Chapter 2, and the results are clearly more impressive.

Now that we know that the method works well in a simple case, where will it have trouble? The choices of α and ω_0 were made with a particular idea of what the target's variances would be. That is, the signal set $\{s_n(t)\}$ was chosen with *a priori* knowledge of what the target environment was likely to be. If the target environment consists of objects of widely different shapes, the method may fail. To test this possibility, we used the target density shown in figure 7. It consists of six Gaussians, as in the previous case, but of more varying widths. by looking at the reconstruction (figure 8), we see that we didn't try hard enough to make the method fail. The method picked out all the peaks. Apparently this method is fairly robust.

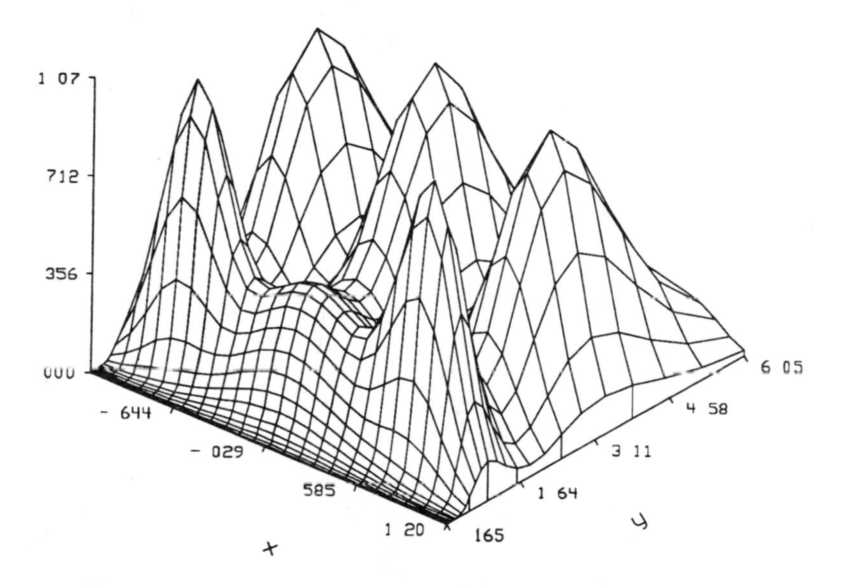

Figure 5: A Phantom Environment

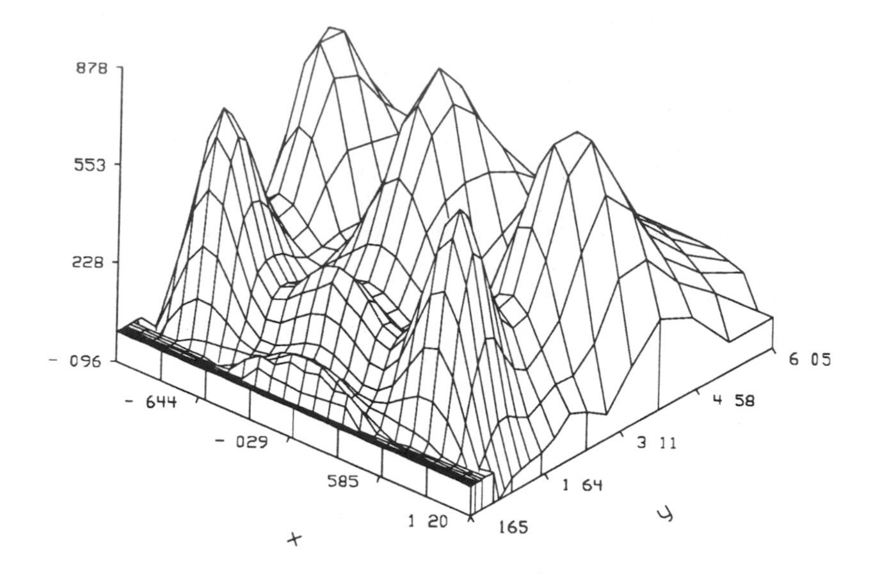

Figure 6: Reconstruction with 10 signals

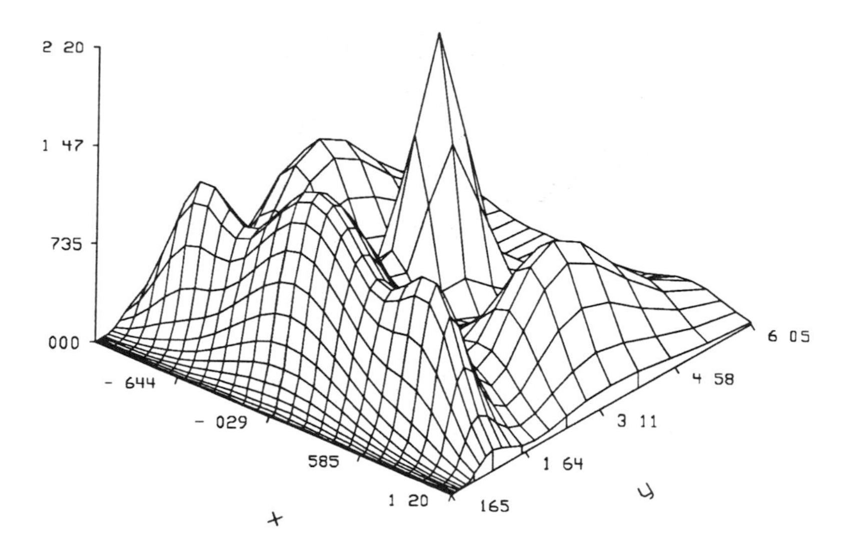

Figure 7: Another Phantom Environment

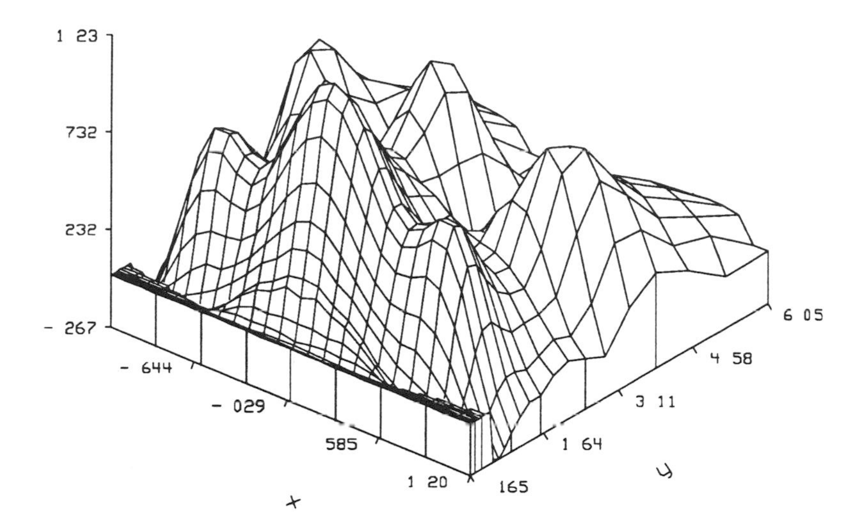

Figure 8: The Reconstruction with 10 signals

3.6 What's the Difference? For mathematicians, the problem of decomposition of functions on Lie groups is classical. The main result in this area is the Peter–Weyl theorem, which says that if G is a compact group, the matrix elements of the irreducible unitary representations form a basis of $L^2(G)$ orthogonal with respect to the bi-invariant Haar measure. By using the homomorphic properties of the representations, the Peter–Weyl theorem can often be written in a simpler "trace" form. We maintain that the difference between the methods of Chapters 2 and 3 is exactly this. That is, they are different ways of expressing the same decomposition of a function on G. To illustrate, let us carry through some computations with a compact group, $SO(3)$.

For positive integer $n, SO(3)$ has a representation $\sigma_n(g)$ of dimension $2n+1$. Denote the matrix elements by $\sigma_n^{ij}(g)$ where $-n \leq i,j \leq n$. These matrix elements are precisely the spherical harmonics. The trace formula for G is [9]:

$$f(g) = \sum_{n=0}^{\infty}(2n+1)\mathrm{tr}\left\{\underbrace{\left[\int_G f(g')\sigma_n^*(g')dg'\right][\sigma_n(g)]}_{M}\right\} \tag{36}$$

$$= \sum_{n=0}^{\infty}(2n+1)\mathrm{tr}\left\{\underbrace{\sum_{k=-n}^{n}\left[\int_G f(g')\sigma_n^{jk}(g')dg'\right][\sigma_n^{ki}(g)]}_{M_{ji}}\right\}$$

$$= \sum_{n=0}^{\infty}(2n+1)\sum_{j,k=-n}^{n}\int_G f(g')\sigma_n^{jk}(g')dg'\sigma_n^{kj}(g) \tag{37}$$

The last formula is the classical Peter–Weyl theorem. This shows how the trace formula is related to it. Notice that we never used the homomorphic property of σ. If instead we use it, (36) can be written:

$$f(g) = \sum_{n=0}^{\infty}(2n+1)\mathrm{tr}\int_G f(g')\sigma_n(g'^{-1}g)dg'$$

$$= \sum_{n=0}^{\infty}(2n+1)\sum_{i=-n}^{n}\int_G f(g')\sigma_n^{ii}(g'^{-1}g)dg' \tag{38}$$

The computation of $f(g)$ by (38) will evidently be easier, since it only involves the diagonal matrix elements, whereas (37) involves the off-diagonal matrix elements as well. This situation will occur whenever the representations are not one dimensional, which is to say whenever G is not abelian. The affine group is non-abelian, of course, so we have the same sort of phenomenon. It is more complex to describe, however because the representations are infinite-dimensional.

The reader can see how the two methods are different. We maintain that in every target environment, the method of this section will be at least as good as the "Peter–Weyl" expansion method of section 2. This follows because the functions are orthogonal. In this way, one is assured that the L^2 error using the method of section 3 is less than that of section 2.

This brings us to an end of our study of the affine group methods for the solution of our problem. While we have probably raised more questions than we have answered, it is appropriate now to summarize.

4 Conclusions. In this paper we have considered the problem of resolving a continuous distribution of targets in range and doppler simultaneously. The object is to determine the target density function $D(x, y)$ as a function of range (x), and velocity (y).

We provided two solutions. In both cases a basis of the signal space is chosen. Call the signals $\{s_0(t), s_2(t), \dots\}$. In section 2 we showed how $D(x,y)$ could be expanded into a sum of cross-ambiguity functions:

$$D(x,y) = \sum_{n,m=0}^{\infty} c_{nm} A_{nm}(x,y) \tag{39}$$

where

$$A_{nm}(x,y) = \sqrt{y} \int_{-\infty}^{\infty} s_n(yt+x)\overline{s_m(t)}dt \tag{40}$$

This is a conceptually simple approach, but unfortunately it did not work well in our simulations. This method may potentially be more useful in other target environments than we have shown it to be here. Our target distributions were very widely spread across the range-doppler plane. We have not conducted extensive tests of how it would behave in different target environments or noisy target environments. Also, we have not extensively considered different choices for the signal set.

Further analysis may be academic, however, because in section 3 we showed how group theory due to Khalil [7] gave another reconstruction scheme which yielded $D(x,y)$ with less computational effort. We obtained reconstructions of complex target distributions with 10 signals. At the end of section 3 we showed how this "trace" method is related to the expansion method of section 2. Our conclusion is that the two are related but that the "trace" method provides a quicker algorithm for computation of $D(x,y)$.

We hope that the results given in this paper will spark other researchers to investigate these methods, both theoretically and practically. We have barely scratched the surface, and the possibilities for future research are endless. Among the many areas needing attention, the most important ones are:

- Methods of choosing the signals based on a priori knowledge of $D(x,y)$.
- Conversely, how a choice of signals affects the rate of convergence of these methods.
- Addition of noise.

The methods just discussed have the disadvantage that the optimal signals to use with them might be difficult to implement in hardware. One might ask if there is another method which uses standard radar signals, and might therefore be implementable on existing hardware. In fact there is a very inspired method of reconstruction which attempts to use the mathematics of tomography in this radar problem. This method has received a bit of attention recently. It was discovered independently by three sets of researchers [1, 11, 24]. Although no one to my knowledge has ever published a complete simulation study of this method, there are rumors that a radar has actually been built (!!) which uses this method. We have not seen any proof of this, however.

REFERENCES

[1] MARVIN BERNFELD, *Chirp doppler radar*, Proceedings of the IEEE, 72 (4) (April 1984), 540–541.

[2] CHARLES E. COOK AND MARVIN BERNFELD, *Radar Signals*, Academic Press, New York (1967).

[3] INGRID DAUBECHIES, *Time-frequency localization operators: A geometric phase space approach*, I don't know if this has been published.

[4] INGRID DAUBECHIES, *Wavelets and applications*, I don't know if this has been published.

[5] INGRID DAUBECHIES, *Discrete sets of coherent states and their use in signal analysis*, In International Conference on Differential Equations and Mathematical Physics, Birmingham, Alabama (1986).

[6] INGRID DAUBECHIES, A. GROSSMAN, AND Y. MEYER, *Painless nonorthogonal expansions*, Journal of Mathematical Physics, 27 (5) (May 1986), 1271–1283.

[7] M. DUFLO AND C.C. MOORE, *On the regular representation of a nonunimodular locally compact group*, Journal of Functional Analysis 21 (1976), 209–243.

[8] NELSON DUNFORD AND JACOB T. SCHWARTZ, *Linear Operators*, Interscience Publishers, New York (1958).

[9] H. DYM AND H.P. MCKEAN, *Fourier Series and Integrals*, Academic Press, New York (1972).

[10] N.H. FARHAT, T.H. CHU, AND C.L. WERNER, *Tomographic and projective reconstruction of 3-d image detail in inverse scattering*, Proceedings of the Society of Photo-Optical Instrumentation Engineers 422 (1984), 82–88.

[11] EPHRAIM FEIG AND F. ALBERTO GRÜNBAUM, *Tomographic methods in range-doppler radar*, Inverse Problems, 2(2) (1986), 185–195.

[12] E.N. FOWLE, E.J. KELLY, AND J.A. SHEEHAN, *Radar system performance in a dense-target environment*, In IRE International Convention Record, number 4 (1961), 136–145.

[13] A. GROSSMANN, M. HOLSCHNEIDER, R. KRONLAND-MARTINET, AND J. MORLET, *Detection of abrupt changes in sound signals with the help of wavelet transforms*, I don't know if this has been published (April 1987).

[14] A. GROSSMANN AND J. MORLET, *Decomposition of Hardy functions into square integrable wavelets of constant shape*, SIAM Journal of Mathematical Analysis, 15 (1984), 723–726.

[15] A. GROSSMANN, J. MORLET, AND T. PAUL, *Transforms associated to square integrable group representations. Part 1: General results*, Journal of Mathematical Physics, 26(10) (October 1985), 2473–2479.

[16] A. GROSSMAN, J. MORLET, AND T. PAUL, *Transforms associated to square integrable group representations. Part 2: Examples*, Ann. Inst. Henri Poincaré, 45(3) (1986), 293–309.

[17] IDRISS KHALIL, *Sur l'analyse harmonique du groupe affine de la droite*, Studia Mathematica, LI(2) (1974), 139–167.

[18] J.R. KLAUDER, *The design of radar signals having both high range resolution and high velocity resoltuion*, The Bell System Technical Journal (July 1960), 808–819.

[19] AKIHIRO KOHARI, *Harmonic analysis on the group of linear transformations of the straight line*, Proceedings of the Japanese Academy, 37 (1961), 250–254.

[20] R. KRONLAND-MARTINET, J. MORLET, AND A. GROSSMAN, *Analysis of sound patterns through wavelet transforms*, I don't know if this has been published, (January 1987).

[21] HAROLD L. NAPARST, *Radar Signal Choice and Processing for a Dense Target Environment*, PhD thesis, University of California, Berkeley (1988). Published by University Microfilms, Inc., Ann Arbor, Michigan.

[22] FRANK NATTERER, *On the inversion of the attenuated Radon transform*, Numerische Mathematik, 32 (1979), 431–438.

[23] A.W. RIHACZEK, *Radar resolution of moving targets*, IEEE Transactions on Information Theory, IT-13(1) (1967), 51–56.

[24] DONALD L. SNYDER AND HARPER J. WHITEHOUSE, *Delay-doppler radar imaging using chirp-rate modulation. Tenth Colloquium of GRETSI, Nice, France (1985).*

[25] N. Tatsuuma, *Plancherel formula for nonunimodular locally compact groups*, J. Math. Kyoto University, 12 (1972), 179–261.

[26] Calvin H. Wilcox, *The synthesis problem for radar ambiguity functions*, MRC Technical Summary Report 157, Mathematics Research Center, United States Army, University of Wisconsin, Madison, Wisconsin (April 1960).

[27] Joseph A. Wolf, *Fourier inversion Problems on Lie Groups and a class of Pseudodifferential operators*, volume 48 of Lecture Notes in Pure and Applied Math, Mercel Dekker, New York (1979), 293–305.

BASIC ALGORITHMS IN TOMOGRAPHY

F. NATTERER* AND A. FARIDANI*

Abstract. After a short survey on reconstruction algorithms in tomography we give a detailed discussion of the filtered backprojection algorithm. We describe all the necessary discretizations and truncations and study the errors resulting from these approximations. We also point out deficiencies of the filtered backprojection algorithm for fan-beam data and show how they can be overcome by using direct algebraic methods.

1. Introduction. The basic problem in computerized tomography (CT) is the reconstruction of a function from its line integrals. Applications come from diagnostic radiology [1], astronomy [2], electron microscopy [3], seismology [4], radar [5], [6], [7], plasma physics [8], nuclear medicine [9] and from many other fields.

We restrict ourselves to the two-dimensional problem even though there are also interesting applications in three dimensions, see e.g. [10]. We use the following notation. f denotes the function in $\mathbf{R}^2$ which has to be reconstructed. We always assume that f is supported in the unit disk, i.e. $f(x) = 0$ for $|x| > 1$. For any unit vector Θ and any real number s we denote by $L(\Theta, s)$ the straight line $x \cdot \Theta = s$ where the dot indicates the dot product. Sampling the line integrals gives rise to an integral transform, namely the Radon transform

$$Rf(\Theta, s) = \int_{L(\Theta,s)} f(x)dx.$$

In the jargon of CT, $Rf(\Theta, \cdot)$ is called a projection.

We shall frequently use the n-dimensional Fourier transform

$$\hat{f}(\xi) = (2\pi)^{-n/2} \int e^{-ix\cdot\xi} f(x)dx.$$

There are essentially 4 types of reconstruction algorithms.

(i) Fourier reconstruction is a direct implementation of the projection slice theorem

$$\widehat{Rf}(\Theta, \sigma) = (2\pi)^{1/2} \hat{f}(\sigma\Theta), \quad \sigma \in \mathbf{R}.$$

Here, $\widehat{Rf}$ stands for the one-dimensional Fourier transform of Rf with respect to the second variable and $\hat{f}$ is the two-dimensional Fourier transform of f. Thus a discrete Fourier transform of the parallel data $Rf(\Theta_j, s_l)$ (see (1)) yields an approximation to $\hat{f}$ at points $\sigma_k\Theta_j$ lying on a polar grid. In order to do the inverse Fourier transform by FFT, one interpolates these values to a rectangular grid. In the early days of tomography this interpolation in Fourier space caused serious loss in accuracy. Meanwhile, reliable implementations of Fourier algorithms have been found, see

*Institut für Numerische und instrumentelle Mathematik Universität Münster, Einsteinstr. 62 D-4400 Münster Federal Republik of Germany.

[11], [12], [13], [14]. Among all reconstruction techniques, Fourier reconstruction is the most efficient one.

(ii) Algebraic reconstruction technique (ART) is an SOR-type iterative method applied to the discretized problem, see [15]. Its success depends mainly on a proper ordering of the projections and/or a good choice of the SOR-parameter ω, see [16]. Among all the reconstruction techniques it is the least efficient one. It needs for a single iteration step as many operations as the whole filtered backprojection algorithm (see *(iv)*).

(iii) Direct algebraic methods, see [16]. They compute regularized minimum norm solutions without any discretization errors. They require highly symmetric scanning geometries and sophisticated programming, but provide excellent reconstructions even in cases in which other methods behave poorly, compare section 3 of this paper. They are as efficient as the filtered backprojection algorithm (see *(iv)*).

(iv) Filtered backprojection is an implementation of Radon's 1917 inversion formula (see e.g. [16, p. 20])

$$f = \frac{1}{4\pi} R^* H \frac{\partial}{\partial s} Rf.$$

Here, H is the Hilbert transform, i.e.

$$\widehat{Hg}(\sigma) = -i \operatorname{sgn}(\sigma)\hat{g}(\sigma),$$

and

$$(R^* g)(x) = \int_{S^1} g(\Theta, x \cdot \Theta) d\Theta$$

is the backprojection operator. This formula is obtained from the projection slice theorem by introducing polar coordinates in the two dimensional Fourier inversion integral, thus avoiding the difficulties with the interpolations in algorithm (i). At least in the medical field, filtered backprojection is the standard algorithm.

The purpose of this paper is to give a detailed error analysis of the filtered backprojection algorithm. Much of this analysis has been done before, but we also arrive at some new conclusions.

We do our analysis in a Fourier transform framework.

We call a function (essentially) b-band-limited if $\hat{f}(\xi)$ is (almost) zero for $|\xi| \geq b$. As is customary in communication theory we think of a b-band-limited function as being capable of representing details of size down to $2\pi/b$. Shannon's sampling theorem states that it is possible to recover a b-band-limited function from its samples if the sampling distance h is not bigger then one half of the smallest detail contained in f, i.e. $h \leq \pi/b$ (Nyquist condition). All we need to know about Fourier analysis and sampling is contained in the survey article [17] or in [16].

We consider two scanning geometries. In parallel scanning, Rf is sampled for

$$\begin{aligned} \Theta_j &= \begin{pmatrix} \cos\varphi_j \\ \sin\varphi_j \end{pmatrix}, \quad && \varphi_j = \pi j/p, \quad && j = 0, \dots, p-1 \\ s_l &= lh, && h = 1/q, && l = -q, \dots, q, \end{aligned} \tag{1}$$

i.e. p projections with $2q+1$ parallel lines each are taken, the detector spacing in each projection being h. In fan-beam scanning we make the substitution

$$\Theta = \begin{pmatrix} \cos(\alpha+\beta-\pi/2) \\ \sin(\alpha+\beta-\pi/2) \end{pmatrix}, \qquad s = r \, \sin\beta$$

and sample Rf for

$$(2) \qquad \begin{aligned} \alpha_j &= 2\pi j/p, \quad j = 0,\ldots,p-1, \\ \beta_l &= hl, \qquad h = \arcsin(\frac{1}{r})/q, \quad l = -q,\ldots,q. \end{aligned}$$

This corresponds to p fan-beam projections emanating from sources equally distributed on a circle of radius $r > 1$, the (angular) detector spacing in each fan being h.

It is well known [2] that the parallel data (1) determine an essentially b-band-limited function provided that

$$(3) \qquad p > b, \quad h \leq \pi/b.$$

For the fan-beam data (2) we have found the conditions

$$(4) \qquad p > 2b, \quad rh \leq \pi/b,$$

see [16, p. 84]. In the light of the sampling theorem, the conditions on h are easily understood. The conditions on p are less obvious.

The case of parallel data, which we analyse in section 2, is fairly well understood. We not only discuss the effect of violating (3) but also explain conditions on placing detectors which so far have been poorly understood. In section 3 we consider the fan-beam case. The failure of the algorithm when the sources are close to the reconstruction region (i.e. $r \sim 1$) is explained. It is shown by examples that this difficulty can be overcome by using direct algebraic methods.

2. The filtered backprojection algorithm for parallel data. This is a computer implementation of Radon's inversion formula

$$(5) \qquad f = \frac{1}{4\pi} R^* H \frac{\partial}{\partial s} g, \quad g = Rf.$$

The numerical implementation of (5) for the parallel geometry (1) consists of three steps. In the convolution step we replace $\frac{1}{4\pi} H \frac{\partial}{\partial s} g$ by a discrete convolution

$$w \overset{h}{*} g(s) = h \sum_l w(s-hl) g(hl)$$

with a suitable function w. In a second step, an interpolation operator I_h with step-size h is applied to $w \overset{h}{*} g$. In principle, this interpolation is not necessary. However, it reduces the complexity of the algorithm. Finally, the discrete backprojection operator

$$R_p^* g(x) = \frac{2\pi}{p} \sum_{j=0}^{p-1} g(\Theta_j, x \cdot \Theta_j)$$

is applied. Summarizing, the algorithm computes the approximation

$$f_{FB} = R_p^* I_h w \overset{h}{*} g$$

to f.

In the following we analyse the impact of the various approximations.

(i) Convolution

The convolution with w has to mimic the action of $\frac{1}{4\pi} H \frac{\partial}{\partial s}$. Looking at this operator from the Fourier transform side suggests to take

$$\hat{w}(\sigma) = \frac{1}{2}(2\pi)^{-3/2}|\sigma|$$

but we use a regularized version thereof, namely

$$\hat{w}_b(\sigma) = \frac{1}{2}(2\pi)^{-3/2}|\sigma| F(|\sigma/b|)$$

where F is a filter function which vanishes (or is small in some sense) outside $[0,1]$. Thus w_b is a b-band-limited (or essentially b-band-limited) function. In order to avoid ringing artifacts it is advisable to use filter functions which are sufficiently smooth. The simplest choice for F is

$$F(\sigma) = \begin{cases} 1 & 0 \le \sigma \le 1 \\ 0 & \text{otherwise.} \end{cases}$$

A straight forward calculation yields

$$(7) \quad w_b(s) = \frac{b^2}{4\pi^2} u(bs), \quad u(s) = \text{ sinc }(s) - \frac{1}{2}(\text{ sinc }(\frac{s}{2}))^2, \quad \text{sinc }(s) = \frac{\sin(s)}{s},$$

see fig. 1. This is the kernel suggested in [18]. Choosing

$$F(\sigma) = \begin{cases} \text{sinc }(\frac{\sigma\pi}{2}), & 0 \le \sigma \le 1 \\ 0 & \text{otherwise} \end{cases}$$

yields the kernel

$$w_b(s) = \frac{b^2}{2\pi^3} u(bs), \quad u(s) = \frac{\frac{\pi}{2} - s \sin s}{(\frac{\pi}{2})^2 - s^2},$$

see fig. 2. Both kernels simplify considerably when evaluated at $s = l\pi/b$, l an integer. We get

$$w_b(l\frac{\pi}{b}) = \frac{b^2}{2\pi^2} \begin{cases} \frac{1}{4}, & l = 0, \\ 0, & l \ne 0 \quad \text{even}, \\ -\frac{1}{\pi^2 l^2}, & l \quad \text{odd} \end{cases} \quad \text{for (7)}$$

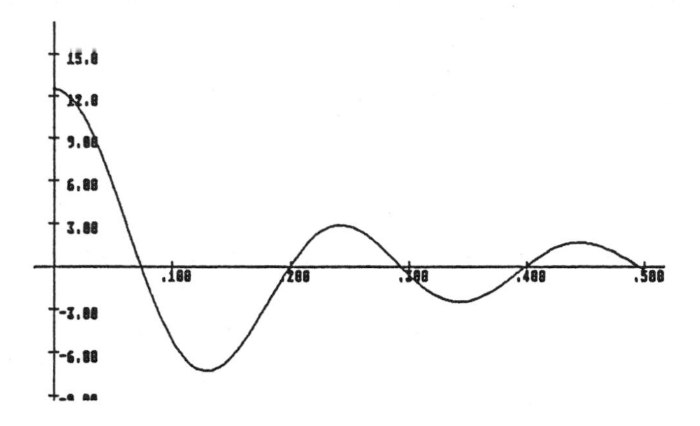

Figure 1

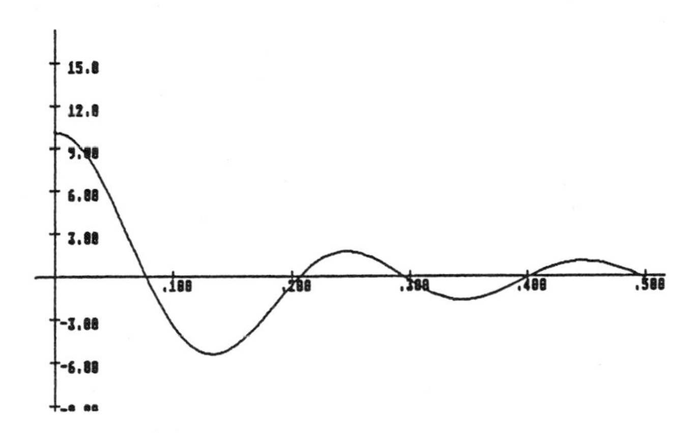

Figure 2

$$w_b(l\frac{\pi}{b}) = \frac{b^2}{\pi^4}\frac{1}{1-4l^2} \qquad \text{for (8)}$$

Many other kernels have been considered, see [19], [20].

In order to find out the effect w_b has on the reconstruction we make use of

$$R^*(w_b * Rf) = W_b * f, \quad W_b = R^* w_b,$$

see [20]. For w_b as in (6) we have

$$\hat{W}_b(\xi) = \frac{1}{2\pi}F(\frac{|\xi|}{b}),$$

see [16, chapt. V.1]. Thus, W_b is a low-pass filter. We conclude that replacing $\frac{1}{4\pi}H\frac{\partial}{\partial s}$ by a convolution with w_b amounts to low-pass filtering f, the band-width of the filter being essentially b. Thus, disregarding all discretizations, the parameter b controls the resolution of the algorithm.

Now we come to the discretization errors. Since w_b has essential band-width b, a necessary condition for $w_b \stackrel{h}{*} g$ to be close to $w_b * g$ is that the detector spacing h satisfies

$$h \leq \pi/b. \tag{9}$$

This condition is also sufficient if f, hence g, is essentially b-band-limited. Violating (9) leads to serious artifacts in the reconstruction.

From reasons to become clear in *(ii)* one often wants to work with kernels which are obtained from band-limited kernels w_b such as (7), (8) via interpolation. Let I_k, k the interpolation stepsize, be the operator

$$I_k w_b(s) = \sum_l B(s/k - l) w_b(kl)$$

where B is the B-spline of order r, see [21], i.e.

$$\hat{B}(\sigma) = (2\pi)^{-1/2}(\mathrm{sinc}(\frac{\sigma}{2}))^r. \tag{10}$$

We are only interested in the cases $r = 1$ (nearest neighbour interpolation) and $r = 2$ (broken line interpolation). For the Fourier transform we obtain from Poisson's formula (see e.g. [16, p. 184])

$$(I_k w_b)^\wedge(\sigma) = k\, \hat{B}(k\sigma) \sum_l e^{-ikl\sigma} w_b(kl) = (2\pi)^{1/2} \hat{B}(k\sigma) \sum_l \hat{w}_b(\sigma - 2\pi l/k).$$

Now assume $k \leq \pi/b$. Then $\hat{w}_b(\sigma) = 0$ for $|\sigma| \geq \pi/k$, and for $k\sigma/\pi$ not an odd integer the sum reduces to one single term with $l = l(\sigma)$ where $l(\sigma)$ is the uniquely defined integer statisfying

$$-\frac{\pi}{k} < \sigma - \frac{2\pi l(\sigma)}{k} \leq \frac{\pi}{k}.$$

Hence,

$$(I_k w_b)^\wedge(\sigma) = (2\pi)^{1/2} \hat{B}(k\sigma) \hat{w}_b(\sigma - 2\pi l(\sigma)/k). \tag{11}$$

Now we study the discrete convolution $(I_k w_b) \stackrel{h}{*} g$ where $h \leq \pi/b$. Its Fourier transform is (see [16, p. 58])

$$\begin{aligned} ((I_k w_b) \stackrel{h}{*} g)^\wedge(\sigma) &= (2\pi)^{1/2} (I_k w_b)^\wedge(\sigma) \sum_l \hat{g}(\sigma - 2\pi l/h) \\ &= 2\pi \hat{B}(k\sigma) \hat{w}_b(\sigma - 2\pi l(\sigma)/k) \sum_l \hat{g}(\sigma - 2\pi l/h). \end{aligned} \tag{12}$$

We compare the discrete convolution with the exact convolution whose Fourier transform is

$$(13)\qquad \begin{aligned}((I_k w_b) * g)^\wedge(\sigma) &= (2\pi)^{1/2}(I_k w_b)^\wedge(\sigma)\hat{g}(\sigma)\\ &= 2\pi\hat{B}(k\sigma)\hat{w}_b(\sigma - 2\pi l(\sigma)/k)\hat{g}(\sigma).\end{aligned}$$

Now assume that h is chosen according to the Nyquist criterion, i.e. $\hat{g}(\sigma)$ negligible for $|\sigma| \geq \pi/h$. Then, for $|\sigma| \leq \pi/h$, $\hat{g}(\sigma)$ and $\sum_l \hat{g}(\sigma - 2\pi l/h)$ are close to each other, and so are (12) and (13). This is not necessarily the case for $|\sigma| \geq \pi/h$ because $\sum_l \hat{g}(\sigma - 2\pi l/h)$ has period $2\pi/h$. In most applications, $\hat{g}$ is fairly concentrated around 0. This means that $\sum_l \hat{g}(\sigma - 2\pi l/h)$ is far from being small in the vicinity of $\sigma = 2\pi l/h$, $l = \pm 1, \pm 2, \ldots$. Thus, (12) and (13) differ greatly for these values of σ. This causes severe artifacts unless $\hat{B}(k\sigma)$ is small for these values of σ, i.e. unless k is a multiple of h. Thus we come to the following important conclusion.

When working with interpolated kernels it is wise to choose the stepsize in the interpolation as a multiple of the stepsize of the discrete convolution, i.e. to put a detector at each interpolation point.

As a special case we consider the kernel which is obtained from (8) by interpolation with stepsize $k = \pi/b$. Alternatively one obtains this kernel by putting

$$F(\sigma) = (\text{sinc}(\frac{\sigma\pi}{2}))^r |\text{sinc}\frac{\sigma\pi}{2}|$$

in (6), as is easily seen from (11). For $r = 2$ this kernel has been suggested in [22]. From fig. 2 we see that this kernel assumes its minimum at $\sigma = k = \frac{\pi}{b}$. Our rule for interpolated kernels says that a detector must sit at this minimum. This rule has been found empirically in [20].

We emphasise that this rule holds only for interpolated kernels. For instance, it is not valid for the kernels (7), (8). From the graph of w_b (see fig. 1, 2) we see that $\frac{\pi}{b}$ lies well to the left of the first minimum of w_b in $[0, \infty)$. Thus, for this kernel, (9) implies that the detector spacing h must be significantly smaller than the σ-value for which $w_b(\sigma)$ is minimal.

(ii) Interpolation

We use the B-spline operator I_k introduced above. The ensueing error is well understood, see e.g. the analysis in [16 p. 107] and [23]. Up to a high frequency error with spatial frequencies above π/k it causes an other filtering of f with the low-pass filter whose Fourier transform is

$$\begin{cases} (2\pi)^{-n/2}(\text{sinc}(\frac{k|\xi|}{2}))^r, |\xi| \leq \pi/k \\ \qquad 0 \qquad\qquad\qquad \text{otherwise.}\end{cases}$$

The high frequency error has been analyzed further in [23]. It has been shown there, that ringing artifacts generated by a jump discontinuity of the filter function at $\sigma = b$ can be suppressed by choosing $k = h$. We remark that for $k = h$ (and, more generally, for h a multiple of k)

$$I_k(w_b \overset{h}{*} g) = (I_k w_b) \overset{h}{*} g,$$

i.e. the interpolation step can be interpreted as interpolating the kernel. This is why we studied interpolated kernels.

(iii) Backprojection

For the analysis of this step we assume that the previous steps (i) and (ii) have been done properly, i.e. with negligible error. Then it suffices to discuss under which circumstances

$$f_p = R_p^* w_b * g$$

is close to f. It is well known that this is the case for w_b b-band-limited provided that $p > b$ and b sufficiently big. In order to see this we have to remember that R_p^* is just the trapeziodal rule in $[0, \pi]$ for the integral in R^* and that the trapeziodal rule integrates accurately periodic functions whose Fourier coefficients of order $> p$ vanish. Thus we have to consider the Fourier coefficients

$$c_l = \int_0^\pi e^{-2il\varphi}(w_b * g)(\Theta, x \cdot \Theta) d\varphi, \quad \Theta = \begin{pmatrix} \cos\varphi \\ \sin\varphi \end{pmatrix}.$$

After some algebra (see e.g. [16, p. 105]) one can use Debye's asymptotic formula for the Bessel functions to conclude that c_l is negligible for $l > b$, hence the result.

We convince ourselves that the condition $p > b$ is the right one by looking at the point spread function W_p for which

$$f_p = W_p * f, \quad W_p(x) = \frac{2\pi}{p} \sum_{j=0}^{p-1} w_b(x \cdot \Theta_j),$$

see [19]. For the kernel w_b from (7) with $b = 15.7$ (i.e. $\pi/b = 0.2$) we computed $W_p(x_1, 0)$, $0 \le x_1 \le 2$ for $p = 16$ (fig. 3) and $p = 12$ (fig. 4). For W_{16} the condition $p > b$ is satisfied, and we expect W_{16} to look like a typical point spread function. From fig. 3 we see that this is in fact the case. For W_{12} the condition $p > b$ is not satisfied. From fig. 4 we see that W_{12} is virtually identical to W_{16} for $x_1 \le 1.5$ but assumes significantly positive values for $x_1 \ge 1.5$. This means that W_{12} is a good point spread function in $|x| \le 1.5$ only, permitting good reconstructions in $|x| \le 0.50$ while serious artifacts are to be expected outside this circle.

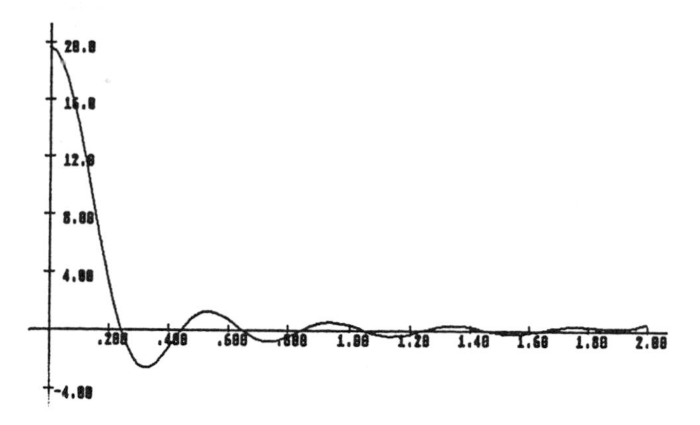

Figure 3

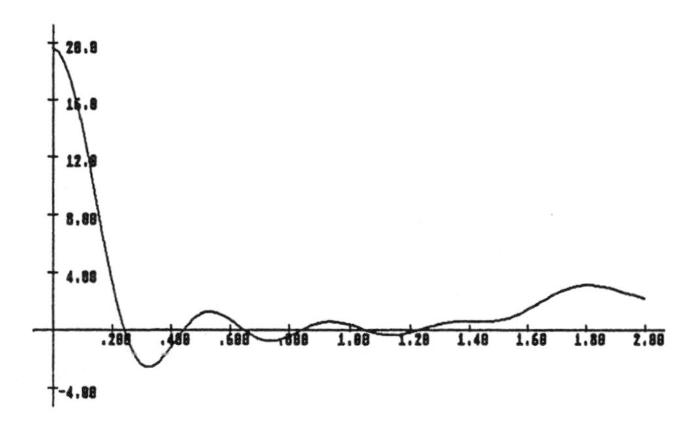

Figure 4

Note that the conditions (9) and $p > b$ are identical to (3). This means that the filtered backprojection algorithm achieves the resolution which is theoretically possible. However, one can get the same resolution by deleting every second data point in (1), i.e. by working only on the data with $j - l$ even, see [24]. Versions of the filtered backprojection algorithm which achieve the best possible resolution on the reduced data set have been given in [25] and [23].

The smallest values of p, q for which (3) holds fulfil approximately

$$p = \pi q. \tag{14}$$

In fig. 5 we demonstrate the effect of violating this condition. The phantom (original not shown) consists of a large circle of diameter 2 with function value 1 and 4 small circles of diameter 0.094, 0.126, 0.188, 0.376, respectively, with function value 2. We took $p = 32$ directions which should suffice to reconstruct details of size $2\pi/32 = 0.196$. According to (14) the correct choice of q would be $q = 10$. The reconstruction with these values of p, q and $b = 32$ is shown on the left hand side of fig. 5. From the horizontal cross section through the circles with diameter 0.094 and 0.188 we see that the larger one is recovered with its correct function value, while the reconstruction of the smaller one shows serious damping. This is in agreement with our theory since 0.094 is less than the theoretical resolution of 0.196 but 0.188 is at the resolution limit.

Now we try to improve the resolution by increasing q while keeping p fixed, violating (14). If b is left at its old value, increasing q has no effect whatsoever except for some averaging which may help to reduce the noise in the data. On the other hand, adjusting b to the new value of q (i.e. $b = \pi q$) in fact increases the resolution but introduces disturbing oscillations. This can be seen from fig. 5. The reconstruction with $q = 20$ and $b = 64$ is shown on the right hand side of fig. 5. We see that the smallest circle comes out a little better, but at the expense of significantly increasing the oscillations in the smooth part of the picture. Thus we see that any attempt to improve the resolution by increasing q beyond its values set by (14) is problematic. The reconstructions have been done with the kernel (7), but other kernels gave similar results.

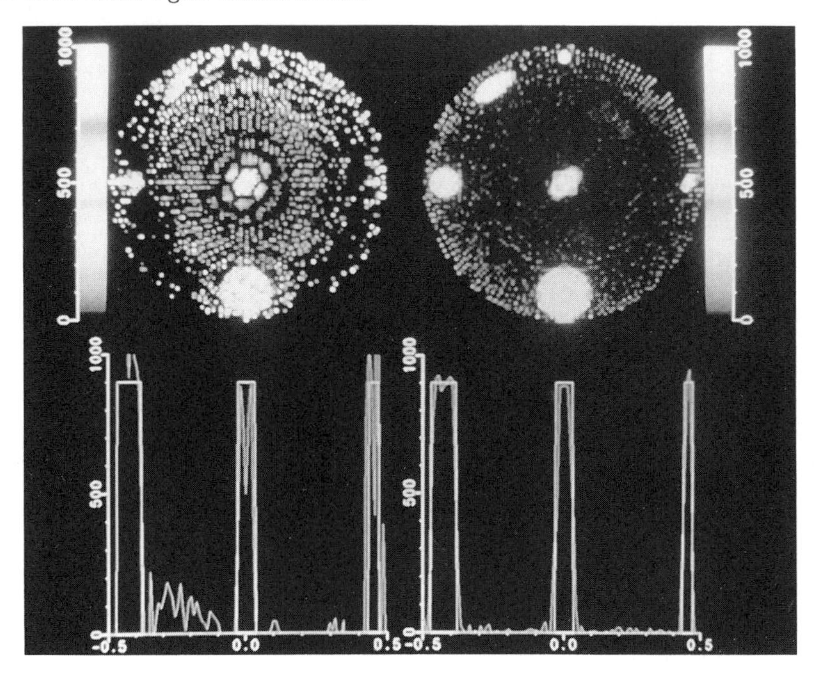

Figure 5

3. The filtered backprojection algorithm for fan-beam data.

For handling fan-beam data (2) we start out from

$$W_b * f = R^*(w_b * Rf), \tag{15}$$

see [20] with w_b, W_b as in the parallel case. The variables Θ and s in (5) are related to the fan-beam coordinates α, β by

$$\Theta = \begin{cases} \cos(\alpha + \beta - \pi/2) \\ \sin(\alpha + \beta - \pi/2) \end{cases}, \quad s = r \sin \beta.$$

Substituting this turns (15) into

$$W_b * f(x) = \int_0^{2\pi} \int_{-\pi/2}^{\pi/2} w_b(|x-a|\sin(\gamma - \beta))g(\alpha,\beta)\cos\varphi \, d\beta \, d\alpha$$

where $g(\alpha, \beta) = Rf(\Theta, s)$ and

$$\cos(\pm\gamma) = \frac{(a-x,a)}{|a-x||a|}, \quad \text{'+' for } a^\perp \cdot x \le 0, \quad \text{'−' for } a^\perp \cdot x \ge 0$$

with $a^\perp = r \begin{pmatrix} \cos(\alpha + \pi/2) \\ \sin(\alpha + \pi/2) \end{pmatrix}$, see [16, p. 112]. We see that the inner integral is almost a convolution, except for the factor $|x - a|$ in the argument of w_b. In order to remove this factor we remark that every kernel w_b with (6) satisfies

$$w_b(\rho s) = \rho^{-2} w_{\rho b}(s).$$

Using this with $\rho = |x - a|$ and $s = \sin(\gamma - \beta)$ we get

$$W_b * f(x) = \int_0^{2\pi} |x-a|^{-2} \int_{-\pi/2}^{\pi/2} w_{b|x-a|}(|x-a|\sin(\gamma-\beta))g(\alpha,\beta)\cos\beta \, d\beta \, d\alpha.$$

Here we make an approximation which is not well understood. We replace the variable cut-off frequency $b|x - a|$ of the kernel by a constant c:

$$W_b * f(x) = \int_0^{2\pi} |x-a|^{-2} \int_{-\pi/2}^{\pi/2} w_c(|x-a|\sin(\gamma-\beta))g(\alpha,\beta)\cos\beta \, d\beta \, d\alpha.$$

This is a good approximation if $b|x - a|$ is almost constant, i.e. if $r >> 1$, in which case we can put $c = br$. However, for r close to 1, the approximation is very poor.

Apart from this problem the implementation of (17) is analogous to the parallel case. We do not describe the details.

However we want to demonstrate that the algorithm performs poorly for $r \sim 1$. In fig. 6 we show reconstructions of a simple object (not shown) consisting of circles and ellipses. We used $p = 132$ fan-beam projections with $r = 1.1$ and $q = 37$. This choice agrees with (4) and should suffice to reconstruct all the details of the object. The reconstruction with the filtered backprojection with $b = 66$ given in the left half of fig. 6 demonstrate clearly that this algorithm does not achieve the best possible resolution. This is one of the situations in which direct algebraic methods are clearly superior. The reconstruction with such an algorithm is displayed in the right half of fig. 6. Its superiority is obvious. We also see that direct algebraic algorithms do achieve the optimal resolution.

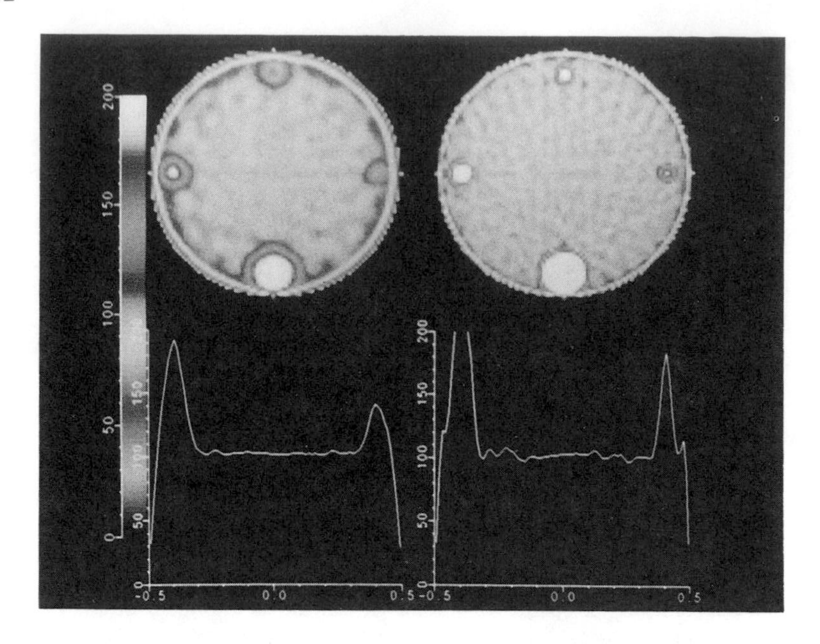

Figure 6

REFERENCES

[1] SCUDDER, H. J., *Introduction to computer aided tomography.*, Proceedings of the IEEE, 66 (1978), pp. 628-637.

[2] BRACEWELL, R. N., *Strip integration in radio astronomy*, Aus. J. Phys., 9 (1956), pp. 198-217.

[3] HOPPE, W. AND HEGERL, R., *Three-dimensional structure determination by electron microscopy, in Hawkes, P. W. (ed.)*, Computer Processing of Electron Microscope Images, Springer. (1980).

[4] DINES, K. A. AND LYTLE, R. J., *Computerized geophysical tomography*, Proceedings of the IEEE, 67 (1979), pp. 1065-1073.

[5] DAS, Y. AND BOERNER, W. M., *On radar target shape estimate using algorithms for reconstruction from projections*, IEEE Trans. Antenna Propagation, 26 (1978), pp. 274-279.

[6] FEIG, E. AND GRÜNBAUM, F. A., *Tomography methods in range - Doppler radar*, Inverse Problems, 2 (1986), pp. 185-195.

[7] HELLSTEN, H. AND ANDERSSON, L. E., *An inverse method for the processing of synthetic aperture radar data*, Inverse Problems, 3 (1987), pp. 111-124.

[8] PIKALOV, V. V. AND PREOBRAZHENSKY, N. G., *Tomographic reconstruction in gas dynamics and plasma physics*, (Russian). Novosibirsk Nauka (1987).

[9] BUDINGER, T. F., GULLBERG, G. T. AND HUESMAN, R. H., *Emission computed tomography, in Herman, G. T. (ed.)*,, Image Reconstruction from Projections. Springer (1979).

[10] GRANGEAT, P., *Analyse d'un système d'imagerie 3D par reconstruction á partir de radiographies X en géométrie conique. Thèse de doctorat, présenteé à l'école nationale superieure des telecommunication.*

[11] STARK, H., WOODS, J. W., PAUL, I. P. AND HINGORANI, R., *An investigation of computerized tomography by direct Fourier inversion and optimum interpolation*, IEEE Trans. Biomed. Eng., BME-28 (1981), pp. 496-505..

[12] NATTERER, F., *Fourier reconstruction in tomography*, Numer. Math., 47 (1985), pp. 343-353.

[13] O'SULLIVAN, J. D., *A fast sinc function gridding algorithm for Fourier inversion in computer tomography*, IEEE Trans. Med. Imaging, MI-4 (1985), pp. 200-207.

[14] KAVEH, M., *Image Reconstruction, in Lacoume, J. L., Durrani, T. S. and Stara, R. (ed.), Treatment du signal. Elsevier Science Publishers B.V. 1987.*

[15] HERMAN, G. T., *Image Reconstruction from Projections. The Fundamentals of Computerized Tomography*, Academic Press (1980).

[16] NATTERER, F., *The Mathematics of Computerized Tomography*, John Wiley & Sons and B.G. Teubner (1986).

[17] JERRI, A. J., *The Shannon sampling theorem - its various extensions and applications: a tutorial review*, Proc. IEEE, 65 (1977), pp. 1565-1596.

[18] RAMACHANDRAN, G. N. AND LAKSHMINARAYANAN, A. V., *Three-dimensional reconstruction from radiographs and electron micrographs: application of convolutions instead of Fourier transforms*, Proc. Nat. Acad. Sci. US, 68 (1971), pp. 2236-2240.

[19] ROWLAND, S. W., *Computer implementation of image reconstruction formulas, in Herman, G.T. (ed.)*, Image Reconstruction from Projections (1979), Springer.

[20] SMITH, K. T. AND KEINERT, F., *Mathematical foundations of computed tomography*, Applied Optics, 24 (1985), pp. 3950-3957.

[21] DE BOOR, C., *A Practical Guide to Splines*, Springer (1978).

[22] SHEPP, L. A. AND LOGAN, B. F., *The Fourier reconstruction of a head section*, IEEE Trans. Nucl. Sci., NS-21 (1974), pp. 21-43.

[23] FARIDANI, A., *Abtastbedingungen und Auflösung in der Beugungstomographie*, Dissertation, Fachbereich Mathematik der Universität Münster (1988).

[24] LINDGREN, A, G. AND RATTEY, P. A., *Sampling the 2-D Radon transform with parallel- and fan-beam projections. Department of Electrical Engineering, University of Rhode Island, Kingston, RIO2881*, Techn. Rep. 5-33285-01, August 1981 (1981).

[25] KRUSE, H., *Resolution of Reconstruction Methods in Computerized Tomography*, SIAM J. on Scient. and Statist. Computing, 10 (1989).

HIGH RESOLUTION RADAR IMAGING USING SPECTRUM ESTIMATION METHODS*

JOSEPH A. O'SULLIVAN† AND DONALD L. SNYDER†

Abstract. This paper summarizes a new approach to high resolution radar imaging based on modern spectrum estimation techniques. First a statistical model of the radar reflections which properly accounts for the randomness of reflections by targets which are rough on the order of a wavelength of the carrier frequency is introduced. The model for the radar return signal is valid for all transmitted narrowband radar signals. Equations which generate maximum likelihood estimates for the reflectivity power as a function of delay and Doppler coordinates are derived.

Introduction. This paper presents recent research results in high resolution delay-Doppler radar imaging based on statistical models obtained from the underlying physics of spotlight mode radar reflections. We present solution equations which may be used to process radar return signals to form images. These equations bear a strong resemblance to the equations derived in [1] for the generic Toeplitz covariance estimation problem. This is a result of the fact that the reflectivity process which characterizes the target is a random process which is stationary at each delay. Thus the samples of the reflectivity process have Toeplitz covariances and the problem of estimating the parameters of the underlying spectra reduces to a Toeplitz covariance estimation problem. We take circulant extensions of the covariance matrices and estimate spectrum samples. There are several aspects of this problem which differ significantly from the generic problem, but the underlying issue is to estimate Toeplitz covariance matrices.

The model of the reflectivity process is described first. Since the processing is performed digitally, the discrete form of this model is examined in detail. Next, the manner in which the transmitted signal interacts with the reflectivity to form the radar return signal is presented. The imaging problem is reduced to the problem of estimating the spectrum underlying the reflectivity process given samples of the radar return signal. A necessary condition for the maximum likelihood solution is obtained and an EM algorithm approach to solving for the maximum is taken. The results extend those in [1] in three important ways. First, the samples of the process with a Toeplitz covariance are not available. Instead, the stationary process is multiplied by a signal matrix. Second, the model includes an additive white Gaussian noise. Third, the process of interest is a function of two variables. For each value of the delay variable, the process is Toeplitz. Thus the model from [1] is extended to spectra which change as a function of an independent variable.

*This work was supported by contract number N00014-86-K-0370 from the Office of Naval Research.

†Both authors are with the Electronic Systems and Signals Research Laboratory, Department of Electrical Engineering, Washington University, St. Louis, MO 63130

Reflectivity Process. The target is described in terms of its reflectivity. It is assumed that the radar transmitted signal is a plane wave at the target so points on the target at a cross-section perpendicular to the line of sight sum up to contribute to the same return signal. This sum changes as a function of time and is denoted by $y(t,\tau)$. The variable τ is the distance of points on the target from the transmitter given in time units as the time it takes for a wave to propagate to the target and back to the transmitter (two-way delay). From this model, it is not apparent that separate points at the same two-way delay τ may be differentiated to obtain an image, but they may be if their velocities relative to the transmitter are different. In particular, if the target is a rigid body and is rotating about a point along the line of sight, then the velocity of a point in the direction of the transmitter is proportional to the distance of that point from the line of sight. Since the Doppler shift introduced by a point on the target is proportional to this directed velocity, a delay-Doppler image of a rotating rigid body is equivalent to a range-crossrange image. The problem is to determine the power reflected from these points as a function of delay and Doppler and then to display this power function as an image of the target. Even if this rigid body assumption for the target is not valid, a delay-Doppler image can be a useful image of the target region.

Since the reflectivity $y(t,\tau)$ is the superposition of all reflectivities at delay τ times the Doppler shift terms they introduce, it may be expressed as

$$y(t,\tau) = \int_{-\infty}^{\infty} c(f,\tau)e^{j2\pi ft}df, \tag{1}$$

where $c(f,\tau)$ is the reflectivity of the points on the target at two-way delay τ which introduce a Doppler shift f. The target is assumed to have finite extent. This implies that both $y(t,\tau)$ and $c(f,\tau)$ are zero for τ outside of some fixed interval. It also implies that $c(f,\tau)$ is zero for f outside of some finite interval because the Doppler variable corresponds to crossrange extent of the target. The processing is assumed to be performed digitally so that discretized versions of y and c are used. Suppose that the resolution cells of f and τ are Δf and $\Delta\tau$ respectively. Let there be I_R bins in the delay or range direction (in delay coordinates, the target is of length $I_R\Delta\tau/2$) and let there be I_{CR} bins in the Doppler or cross range direction (so the target is of width $I_{CR}\Delta f$). If samples of the radar return signal are taken every Δt seconds, Equation (1) may be approximated by

$$y(k\Delta t, n\Delta\tau) = \sum_{m=-(I_{CR}-1)/2}^{(I_{CR}-1)/2} c(m,n)e^{j2\pi mk\Delta f\Delta t}, \tag{2}$$

where

$$c(m,n) = c(m\Delta f, n\Delta\tau)\Delta f. \tag{3}$$

The literature on radar reflections [2,3] describes statistical models for the reflectivity when the target is rough on the order of a wavelength of the carrier signal.

The model states that the reflectivity of a patch on the target is a complex valued Gaussian random variable with zero mean and is uncorrelated from patch to patch. For our model this corresponds to the assumptions on the reflectivity of patches of the target.

$$E[c(i,k)c(m,n)] = 0$$
$$E[c(i,k)c^*(m,n)] = \sigma(i,k)\delta_{i,m}\delta_{d,n}. \tag{4}$$

Here, $\sigma(i,k)$ is the real, nonnegative covariance of the reflectivity of the patch at delay k and Doppler i. The scatterers described by this model are called wide-sense stationary uncorrelated scatterers (WSSUS) by Van Trees [2] because the assumptions imply that

$$E[y(i,k)y^*(m,l)] = K_G(i-m,k)\delta_{k,l}, \tag{5}$$

where $y(i,k) = y(i\Delta t - k\Delta\tau/2, \Delta\tau)$ and

$$K_G(n,k) = \sum_{m=-(I_{CR}-1)/2}^{(I_{CR}-1)/2} \sigma(m,k)e^{j2\pi mn\Delta f\Delta t}. \tag{6}$$

Let $\mathbf{y}_G(k)$ be the vector of samples from delay k

$$\mathbf{y}_G(k)^t = [y(0,k)\ y(1,k)\ y(2,k)\cdots y(G-1,k)]. \tag{7}$$

The covariance matrix for $\mathbf{y}_G(k)$ is a Toeplitz matrix $\mathbf{K}_G(k)$ whose i,m element is $K_G(i-m,k)$. The restriction imposed by the constraint of a target of finite extent is represented by Equation (6). That is, the only $\sigma(m,k)$ which can be nonzero are for $-(I_{CR}-1)/2 \leq m \leq (I_{CR}-1)/2$ and $0 \leq k \leq I_R - 1$. This constraint on m is only meaningful if $\Delta f\ \Delta t I_{CR}$ is less than one and it limits the $\mathbf{K}_G(k)$ possible. If $\Delta f\ \Delta t = 1/N$, then the $\sigma(m,k)$ may be thought of as samples of the spectrum of the periodic extension of the covariance matrix at delay k. Only $I_{CR}(I_{CR} < N)$ of the spectrum samples at each delay are nonzero.

We now introduce larger vectors and matrices so that the results which follow may be written compactly. The $\mathbf{y}_G(k)$ are loaded into one $GI_R \times 1$ vector $\mathbf{y}_G$. The covariance of $\mathbf{y}_G$ is block diagonal matrix $\mathbf{K}_G$ whose k^{th} diagonal block is $\mathbf{K}_G(k)$. Define $\Sigma(k)$ to be the $I_{CR} \times I_{CR}$ diagonal matrix

$$\Sigma(k) = diag[\sigma(0,k)\ \sigma(1,k)\ \sigma(2,k)\cdots\sigma(-2,k)\ \sigma(-1,k)], \tag{8}$$

and define Σ to be the $I_R I_{CR} \times I_R I_{CR}$ block diagonal matrix whose k^{th} block is $\Sigma(k)$. Then we have

$$\mathbf{K}_G(k) = \mathbf{W}_G^\dagger \mathbf{J}_{CR}^t \Sigma(k)\mathbf{J}_{CR}\mathbf{W}_G, \tag{9}$$

where $\mathbf{W}$ is an $N \times N$ normalized DFT matrix $(N > G, N > I_{CR}), I_{CR}), \mathbf{W}_G$ consists of the first G columns of $\mathbf{W}$, and

$$\mathbf{J}_{CR} = \begin{bmatrix} \mathbf{I}_1 & 0 & 0 \\ 0 & 0 & \mathbf{I}_2 \end{bmatrix}, \tag{10}$$

where $\mathbf{J}_{CR}$ is $I_{CR} \times N$, $\mathbf{I}_1$ is an identity matrix of dimension $(I_{CR}+1)/2$, and $\mathbf{I}_2$ is an identity matrix of dimension $(I_{CR}-1)/2$. $\mathbf{J}_{CR}$ implements the assumption that some spectrum samples are zero. Having now defined several matrices, with a little more notation we can give a more precise relationship between the matrix Σ and the matrix $\mathbf{K}_G$. Let $\mathbf{M}_{CR}$ be the $I_R I_{CR} \times I_R N$ block diagonal matrix each of whose I_R blocks is $\mathbf{J}_{CR}$. Let $\mathbf{M}_R$ be the $I_R N \times I_R G$ block diagonal matrix each of whose I_R blocks is $\mathbf{J}_R$($\mathbf{J}_R$ is $N \times G$ and equals $[\mathbf{I}\ 0]^t$). Finally, let $\widehat{\mathbf{W}}$ be the $NI_R \times NI_R$ block diagonal matrix each of whose I_R blocks equals $\mathbf{W}$. Then

$$\mathbf{K}_G = \mathbf{M}_R^t \widehat{\mathbf{W}}^\dagger \mathbf{M}_{CR}^t \Sigma \mathbf{M}_{CR} \widehat{\mathbf{W}} \mathbf{M}_R. \tag{11}$$

Except for constraining some spectrum samples to be zero, if samples of $\mathbf{y}_G(k)$ from each delay k were directly available, this problem would be very similar to that from [1] and the approach to the solution almost identical. Instead, for our problem, we have the data $\mathbf{r}$ which is described below.

Radar data. In a radar system, each signal is typically described as the product of a baseband signal times a complex exponential at the carrier frequency. All of the interactions of interest for narrow band radar systems may be described in terms of these complex valued baseband signals which will be called complex envelopes of the signals or simply the signals.

Let $S_T(t)$ and $S_R(t)$ be the complex envelopes of the transmitted and received signals, respectively. The distance of a point on the target from the transmitter/receiver is measured in time units as the two-way delay. The received signal is the superposition of the reflectivity from all two-way delays times the appropriate transmitted signals,

$$S_R(t) = \int_{-\infty}^{\infty} S_T(t-\tau) y(t-\tau/2, \tau) d\tau. \tag{12}$$

In Equation (12), the $\tau/2$ arises because it takes that long for the transmitted signal to get to a point at two-way delay τ. The available data are the sum of the radar return signal and additive white Gaussian noise

$$r(t) = S_R(t) + w(t)$$

$$r(t) = \int_{-\infty}^{\infty} S_T(t-\tau) y(t-\tau/2, \tau) d\tau + w(t). \tag{13}$$

This equation forms the basis for our model. We now assume that samples of the data are available and we wish to estimate the covariance of the reflectivities from patches on the target. The sampled version of Equation (13) is

$$r(k) = \sum_{n=0}^{I_R-1} S_T(k\Delta t - n\Delta\tau) y(k\Delta t - n\Delta\tau/2, n\Delta\tau) + w(k)\ . \tag{14}$$

Given G samples $r(k), 0 \leq k \leq G-1$, we may write Equation (14) in vector form as

$$\mathbf{r} = \mathbf{S}_T^\dagger \mathbf{y} + \mathbf{w} \ , \tag{15}$$

where $\mathbf{r}$ is the vector of samples $r(k)$; $\mathbf{w}$ is a white noise vector with covariance $N_0\mathbf{I}$; $\mathbf{y}$ is the $GI_R \times 1$ vector of samples of the reflectivity of the target described above; and $\mathbf{S}_T^\dagger$ is a $G \times GI_R$ matrix composed of I_R $G \times G$ submatrices each of which is diagonal:

$$\mathbf{S}_T^\dagger = [\mathbf{S}_0 \ \mathbf{S}_1 \ \mathbf{S}_2 \cdots \mathbf{S}_{I_R-1}], \tag{16}$$

where

$$\mathbf{S}_n = diag[S_T(-n\Delta\tau) \ S_T(\Delta t - n\Delta\tau) \ S_T(2\Delta t - n\Delta\tau) \cdots S_T((G-1)\Delta t - n\Delta\tau)].$$

Note that Equation (15) could also be written in terms of samples of $c(f, \tau)$ AS

$$\mathbf{r} = \Gamma^\dagger \mathbf{c} + \mathbf{w}, \tag{17}$$

where $\mathbf{c}$ is an $I_R I_{CR} \times 1$ vector of samples of the reflectivity of the target arranged as I_R subvectors each of length I_{CR} of the samples from each delay and $\Gamma^\dagger$ is a $G \times I_R I_{CR}$ matrix each element of which is a product of a sample of S_T times a complex exponential. More explicitly, Γ is given by

$$\Gamma = \mathbf{M}_{CR} \widehat{\mathbf{W}} \mathbf{M}_R \mathbf{S}_T. \tag{18}$$

There are obviously issues associated with the appropriate or desired sampling rates of $r(t)$ and of $c(f, \tau)$. Some of these issues are addressed in [4]. The quality of the image obtained and the resolution achievable are intimately related to the sampling issues.

Since $\mathbf{r}$ is the result of linear operations on Gaussian random variables, $\mathbf{r}$ is a zero mean Gaussian random variable with covariance

$$\mathbf{K}_R = \mathbf{S}_T^\dagger \mathbf{K}_G \mathbf{S}_T + N_0 \mathbf{I} = \Gamma^\dagger \Sigma \Gamma + N_0 \mathbf{I}, \tag{19}$$

Maximum Likelihood Solution. The loglikelihood function for the data is

$$L(\mathbf{K}_R; \mathbf{r}) = -\ln(\det \mathbf{K}_R) - \mathbf{r}^\dagger (\mathbf{K}_R)^{-1} \mathbf{r}. \tag{20}$$

As shown first by Burg, et al. [5] a necessary condition for a matrix to maximize the loglikelihood function is the trace condition

$$tr[(\mathbf{K}_R^{-1} \mathbf{r}\mathbf{r}^\dagger \mathbf{K}_R^{-1} - \mathbf{K}_R^{-1}) \delta \mathbf{K}_R] = 0. \tag{21}$$

The matrix $\delta\mathbf{K}_R$ is a variational matrix which takes values in all possible additive variations of the matrix $\mathbf{K}_R$. The set of possible $\mathbf{K}_R$ is described by Equation (19). If we rewrite the trace condition in terms of $\mathbf{K}_G$, then we get

(22)

$$tr[\mathbf{S}_T(\mathbf{S}_T^\dagger\mathbf{K}_G\mathbf{S}_T + N_0\mathbf{I})^{-1}(\mathbf{r}\mathbf{r}^\dagger - \mathbf{S}_T^\dagger\mathbf{K}_G\mathbf{S}_T - N_0\mathbf{I})(\mathbf{S}_T^\dagger\mathbf{K}_G\mathbf{S}_T + N_0\mathbf{I})^{-1}\mathbf{S}_t^\dagger\delta\mathbf{K}_G] = 0$$

Equation (22) shows the three ways in which the present spectrum estimation problem differs from that in [1]. First, there is the signal matrix $\mathbf{S}_T$ appearing in (22) multiplying $\mathbf{K}_G$ wherever it appears. Second, the additive noise manifests its influence in the equation above through the appearance of $N_0\mathbf{I}$. Third, the matrix $\mathbf{K}_G$ in Equation (22) is a block diagonal matrix with Toeplitz blocks, not merely Toeplitz. Despite these differences an EM algorithmic approach to the solution can be derived in much the same way as in [1].

For each delay k, define the complete data vector $\mathbf{y}_N(k)$ to be the periodic extension of the data vector $\mathbf{y}_G(k)$

$$\mathbf{y}_N(k)^t = [\mathbf{y}_G(k)^t \; \mathbf{y}_A(k)^t] \tag{23}$$

where $\mathbf{y}_A(k)$ is an $(N-G)\times 1$ vector which augments $\mathbf{y}_G(k)$ to obtain a full period sample of the periodic process. Let $\mathbf{y}_N$ be the $NI_R \times 1$ vector made by stacking the $\mathbf{y}_N(k)$ to form one long vector. The complete data loglikelihood is

$$L(\mathbf{K}_N;\mathbf{y}_N) = -(\ln(\det\mathbf{K}_N)) - \mathbf{y}_N^\dagger(\mathbf{K}_N)^{-1}\mathbf{y}_N, \tag{24}$$

where $\mathbf{K}_N$ is a block diagonal matrix with each of the I_R blocks being a circulant Toeplitz matrix. Premultiplying $\mathbf{K}_N$ by $\widehat{\mathbf{W}}$ and postmultiplying by $\widehat{\mathbf{W}}^\dagger$ yields a diagonal matrix. This diagonal matrix consists of the samples of the spectrum, some of which are constrained to be zero by the assumption of a target of finite extent. This constraint results in

$$\mathbf{K}_N = \widehat{\mathbf{W}}^\dagger\mathbf{M}_{CR}^t\Sigma\mathbf{M}_{CR}\widehat{\mathbf{W}}. \tag{25}$$

What this implies is that after rotating the data $\mathbf{y}_N$ using the orthogonal matrix $\widehat{\mathbf{W}}$, some of the entries of the resulting vector are zero. These elements may be removed from consideration by multiplying the resulting data by $\mathbf{M}_{CR}$, giving a vector of uncorrelated samples of a process whose covariance matrix is given by Σ.

$$\mathbf{c} = \mathbf{M}_{CR}\,\widehat{\mathbf{W}}\mathbf{y}_N \tag{26}$$

$$E[\mathbf{c}\mathbf{c}^\dagger] = \Sigma\,. \tag{27}$$

This vector $\mathbf{c}$ is the same as in Equation (17). Under the assumptions stated, the matrix $\mathbf{K}_N$ is not invertible. The reflectivity, however, almost surely does not have a component in the null space of $\mathbf{K}_N$ so we can make some sense of the complete data loglikelihood (24). A more correct way to write the complete data loglikelihood is in terms of the rotated coordinates $\mathbf{c}$ and its covariance Σ:

$$L(\Sigma;\mathbf{c}) = -\sum_{k=0}^{I_R-1}\;\sum_{i=-(I_{CR}-1)/2}^{(I_{CR}-1)/2}[\ln\sigma(i,k) + \frac{|c(i,k)|^2}{\sigma(i,k)}]\,. \tag{28}$$

The EM algorithm is an iterative algorithm which at each step updates the estimate for the Σ by maximizing the conditional expected value of the complete data loglikelihood over Σ. The E-step of the EM algorithm performs the expected value of (28) given the incomplete data and the previous estimate for Σ. The M-step consists of maximizing the result of this expectation over the $\sigma(i,k)$. Since the complete data loglikelihood in rotated coordinates separates into the sum of independent samples in (28), the result of taking the maximum over the spectral values at step $p+1$ is just

$$\sigma^{(p+1)}(i,k) = E[|c(i,k)|^2|\Sigma^{(p)},\mathbf{r}]. \tag{29}$$

The estimate of the covariance of $\mathbf{y}_N$ at step $p+1$ is found by transforming back to those coordinates

$$\mathbf{K}_N^{(p+1)}(l-n,k) = \frac{1}{N}\sum_{m=0}^{N-1} E[y_N(m,k)y_N^*(\langle m+l-n\rangle_N,k)|\mathbf{r},\mathbf{K}_N^{(p)}]. \tag{30}$$

This equation makes sense intuitively. It says that to find the maximum likelihood estimate over the constrained set of Toeplitz covariances, augment the covariance matrix at each delay with the conditional mean and mean square estimates of the missing lags. At the convergence point of the algorithm, the covariance estimates equal the conditional mean estimates of the lag products.

Returning to Equation (27), the estimate at step $p+1$ of Σ is given by the diagonal elements of the conditional expectation of $\mathbf{cc}^\dagger$ or

$$\Sigma^{(p+1)} \overset{d}{=} E[\mathbf{cc}^\dagger|\Sigma^{(p)},\mathbf{r}], \tag{31}$$

where $\overset{d}{=}$ means that the off diagonal terms on the left are zero and the diagonal terms on the left equal the diagonal terms on the right. The computation indicated in (31) is a standard problem in estimation theory. This equation may be written as [4]

$$\Sigma^{(p+1)} \overset{d}{=} \Sigma^{(p)}+\Sigma^{(p)}\Gamma(\Gamma^\dagger\Sigma^{(p)}\Gamma+N_0\mathbf{I})^{-1}(\mathbf{rr}^\dagger-\Gamma^\dagger\Sigma^{(p)}\Gamma-N_0\mathbf{I})(\Gamma^\dagger\Sigma^{(p)}\Gamma+N_0\mathbf{I})^{-1}\Gamma^\dagger\Sigma^{(p)}. \tag{32}$$

Equation (32) defines the iteration sequence used by the computations. The diagonal elements of Σ are the values which are displayed as the scattering function image of the target. Thus, at each stage an image is calculated and may be displayed. Some appropriate stopping criterion is used to terminate the algorithm. It should be pointed out that at each stage of the algorithm the conditional mean estimate of the reflectance is also generated. At step p, this estimate is

$$E[\mathbf{c}|\Sigma^{(p)},\mathbf{r}] = \Sigma^{(p)}\Gamma(\Gamma^\dagger\Sigma^{(p)}\Gamma+N_0\mathbf{I})^{-1}\mathbf{r}. \tag{33}$$

The magnitude or the magnitude squared of $\mathbf{c}$ are commonly viewed as images of the target. Thus both types of radar image commonly viewed are generated by our algorithm. We feel this is a unique feature of our algorithm.

Let the corresponding sequence of covariance matrices for the data $\mathbf{r}$ be denoted $\mathbf{K}_R^{(p)} = \Gamma^\dagger \Sigma^{(p)} \Gamma + N_0 \mathbf{I}$. Since this iteration is an EM algorithm, it has all the properties associated with this type of algorithm. In particular, the incomplete data loglikelihood is nondecreasing in the sequence of covariance matrices $\mathbf{K}_R^{(p)}$.

This section has presented a derivation of the equations used to produce target images. This approach starts with a model which accurately accounts for the random nature of radar reflections and adopts the maximum likelihood method of statistics to estimate delay-Doppler high resolution images of radar targets. Questions associated with uniqueness of spectrum estimates and convergence of the EM algorithm are discussed in the following section.

Convergence Issues. This section addresses some of the convergence questions associated with the EM algorithms proposed in earlier sections. These results are stated so that they apply to both the radar imaging problem studied here and the Toeplitz estimation problem from [1]. Let the integer M stand for $I_R I_{CR}$ in the radar problem. The matrix Γ refers to the Γ from the last section or simply to W_G if we are discussing the original Toeplitz problem. Also, N_0 is zero for the original Toeplitz problem.

One question of importance is the uniqueness of estimates of parameters of interest. This question is addressed by looking at the Cramer-Rao bounds on the variance of estimates. The Cramer-Rao bounds are obtained by inverting the Fisher information matrix. When the Fisher information matrix is singular, these bounds are infinite. It is shown how singularity of the Fisher information matrix corresponds to nonuniqueness of parameter estimates.

DEFINITION. Let γ_k denote the kth row of the $M \times G$ matrix Γ. The $M \times G^2$ matrix $\mathbf{F}$ has k^{th} row given by $\gamma_k \otimes \gamma_k^*$, where $\otimes$ denotes the Kronecker product.

THEOREM 1. *The Fisher information matrix for estimating Σ given data $\mathbf{r}$ is equal to*

$$\mathbf{F}(\mathbf{K}_R^{-1} \otimes \mathbf{K}_R^{-\dagger})\mathbf{F}^\dagger. \tag{34}$$

Proof. The Fisher information matrix is just the negative of the expected value of the second derivative of the log-likelihood function. This second derivative is evaluated in the appendix of [4] and taking the expected value yields the above expression.

THEOREM 2. *Suppose the Fisher information matrix in (34) is singular and that the matrix $\mathbf{K}_R$ is positive definite. Then there does not exist a Σ which is positive definite which yields a unique maximum of the log-likelihood.*

Proof. Since the rank of the matrix $\mathbf{K}_R^{-1} \otimes \mathbf{K}_R^{-\dagger}$ equals G^2, and by the form of the matrix, the Fisher information matrix is singular if and only if the matrix $\mathbf{F}$ has rank less than M if and only if there exists a real vector $\mathbf{s}$ such that $\mathbf{F}^\dagger \mathbf{s} = 0$. Such an $\mathbf{s}$ exists if and only if there exists a real diagonal matrix $\mathbf{D}$ ($\mathbf{D} = diag(\mathbf{s})$) such that

$$\Gamma^\dagger(\Sigma + \alpha \mathbf{D})\Gamma = \Gamma^\dagger \Sigma \Gamma \tag{35}$$

for all real α. If Σ is positive definite and maximizes the log-likelihood, then there exists a β such that for all $0 \leq \alpha \leq \beta$ the matrix $\Sigma + \alpha\mathbf{D}$ is nonnegative definite and yields the same covariance matrix and hence the same value for the log-likelihood.

COROLLARY 1. *For the spectrum estimation problem from [1], there does not exist a positive definite Σ which yields a unique maximum of the log-likelihood if $N > 2G - 1$.*

Proof. The matrix $\mathbf{F}$ constructed above has rank less than or equal to $2G - 1$.

Note that this theorem does not say that the estimate of the Toeplitz covariance matrices generated by the algorithm are not unique. The theorem and its corollary relate to the uniqueness of the spectrum samples. For some problems the parameters of interest are in the covariance matrix $\mathbf{K}_R$ or in the Toeplitz matrix $\mathbf{K}_G$. There could be (and indeed are for M large enough) many Σ which yield the same estimate for $\mathbf{K}_R$. Theorem 2 can be applied to problems where one desires to know how big to make N. In general, it is desired to have N as short as possible in order to reduce the number of parameters to be estimated. If it can be shown that with a given N there exists a positive definite matrix which yields the maximum likelihood estimate for the covariance matrix and the conditions of the theorem are satisfied, then one might consider using a smaller N to reduce the size of Σ.

For some problems, including the radar problem, it is the matrix Σ which is of interest. For these problems, it is very important to know when the estimate is unique. In the radar problem, the result which is displayed as an image is an array whose elements are the diagonal entries from Σ. In order to be able to generate a unique image, the conditions of theorem 2 must be satisfied.

Some of the issues associated with convergence of the EM algorithm for the problems described are addressed next. The following material is adapted from the material in [1] to be applicable to the radar problem as well.

DEFINITION. Let $\mathbf{K}_R^a$ be the set of positive definite matrices $\mathbf{K}_R$ given by Equation (19) whose entries are bounded by some number a. Let $\overline{\mathbf{K}}_R^a$ be the closure of $\mathbf{K}_R^a$, the set of nonnegative definite matrices of this parameterized form.

The important issues for the following theorems are not the matrices which go into Γ. What is important is that any $\mathbf{K}_R \in \mathbf{K}_R^a$ may be written as

$$\mathbf{K}_R = \sum_{k=0}^{M-1} \sigma(k)\gamma_k^\dagger \gamma_k + N_0\mathbf{I}, \tag{36}$$

where γ_k is the kth row of Γ which is fixed once the model is specified. The only parameters which must be found are the $\sigma(k)$ which are specified to be greater than or equal to zero. The case where N_0 equals zero is an important one for which we wish to guarantee that the estimate of $\mathbf{K}_R$ is nonsingular. Clearly for any fixed a the set $\overline{\mathbf{K}}_R^a$ is compact.

THEOREM 3. *Let Γ be any fixed $M \times G$ matrix with complex entries. Let $\mathbf{r}$ be an observation of a $G \times 1$ zero-mean Gaussian random vector whose covariance is some positive definite hermitian symmetric matrix. Then*

a) There does not exist a singular $\mathbf{K}_R \in \overline{\mathbf{K}}_R^a$ such that $\mathbf{r}$ is in the range space of $\mathbf{K}_R$, with probability one.

b) The log-likelihood function is bounded from above over the set $\overline{\mathbf{K}}_R^a$, with probability one.

Proof. a) Suppose that any G rows of Γ are linearly independent. Since $\mathbf{K}_R$ is given by (36), a singular matrix in this class must be given by

$$\mathbf{K}_R^s = \sum_{k \in J} \sigma(k) \gamma_k^\dagger \gamma_k, \tag{37}$$

where $J \subset \{0, 1, 2, \ldots, M-1\}$ consists of $G-1$ or fewer integers which correspond to the nonzero diagonal entries of Σ. Since the true covariance for $\mathbf{r}$ is nonsingular, the probability that $\mathbf{r}$ lies in the subspace spanned by $\{\gamma_k^\dagger | k \in J\}$ is zero. Since there are a finite number of such spaces and the probability that $\mathbf{r}$ is in any one of them is zero, the probability that $\mathbf{r}$ is in the range of any singular $\mathbf{K}_R$ in the set is zero. If any G rows are not independent, then singular matrices may be written as the sum of more than $G-1$ outer products $\gamma_k^\dagger \gamma_k$. But the data would still have to lie in a subspace spanned by fewer than G independent vectors and thus the probability of this is zero and this part of the theorem follows.

b) This part follows from [6] where it is shown that the log-likelihood function is bounded above when the data are not in the range space of a singular covariance matrix in the set in question. The proof is based on the following facts. First, if $\mathbf{K}_R$ is nonsingular and its eigenvalues are bounded from above and below, the log-likelihood is bounded. Second, if $\mathbf{K}_R$ is singular and the data are in its range, the log-likelihood is unbounded above; but this is a zero probability event. Third, if $\mathbf{K}_R$ is singular (with rank n) and the data are not in its range, the log-likelihood is unbounded from below. This is shown by writing $\mathbf{K}_R$ as the limit as $\epsilon \to 0$ of $\mathbf{K}_R + \Gamma^\dagger(\epsilon \mathbf{I})\Gamma$. We examine the loglikelihood (20) when $\mathbf{r}$ is not in the range of $\mathbf{K}_R$. The term $-\ln(\det(\mathbf{K}_R + \epsilon\Gamma^\dagger\Gamma))$ has bounded terms and an unbounded component of the form $-(G-n)\ln(\epsilon)$. The term $-\mathbf{r}^\dagger(\mathbf{K}_R + \epsilon\Gamma^\dagger\Gamma)^{-1}\mathbf{r}$ has bounded terms and an unbounded component of the form $-c/\epsilon$, where c is some positive number. Then,

$$\lim_{\epsilon \to 0} -(G-n)\ln(\epsilon) - \frac{c}{\epsilon} = -\infty\ . \tag{38}$$

There are more facts needed to prove convergence. One fact is that the log-likelihood is increasing at each step of the algorithm. Another is that the iterates stay in a bounded set so that the above theorems apply. The theorems before this point apply to the problem no matter what algorithm is used to find the maximum likelihood estimate, while the theorems that follow apply for our particular algorithm.

THEOREM 4. *The iterates defined by the EM algorithm (32) produce a sequence of log-likelihoods which are nondecreasing,*

$$L(\mathbf{K}_R^{(p+1)};\mathbf{r}) - L(\mathbf{K}_R^{(p)};\mathbf{r}) \geq Q(\mathbf{K}_N^{(p+1)}|\mathbf{K}_N^{(p)}) - Q(\mathbf{K}_N^{(p)}|\mathbf{K}_N^{(p)}) \geq 0 ,$$

where $L(.;.)$ is the log-likelihood for the problem.

Proof. This is just a result of the sequence being generated by an EM algorithm [7,8].

THEOREM 5. $L(\mathbf{K}_R^{(p+1)};\mathbf{r}) = L(\mathbf{K}_R^{(p)};\mathbf{r})$ *if and only if* $\Sigma^{(p+1)} = \Sigma^{(p)}$.

Proof. This is a result of the concavity of the complete data log-likelihood. Take the second derivative with respect to the variable being maximized, Σ. For each diagonal entry of Σ this derivative is either positive or zero. It's zero if and only if the previous corresponding entry of Σ is zero, and this entry would remain zero. Thus the maximizer is unique and is given by $\Sigma^{(p+1)}$. By the inequality from the theorem 4, theorem 5 follows.

The one last theorem we would like to have is that iterates remain in a bounded set. It has been our experience in computations that the iterates do remain bounded, but we have had trouble proving this in the general case. We have observed in computations that Σ may tend to a singular limit. This is not precluded by any of the above theorems. In fact, for our radar imaging problem we do not wish to exclude this possibility since a zero estimate of the power reflected from a point simply means that there is no target at that point.

Conclusions. We have presented an algorithm for generating images of radar targets in the delay-Doppler plane. The approach has been estimation-based because of our assumption of targets which are rough on the order of a wavelength of the carrier frequency. Some of the theoretical properties of this approach and uniqueness of estimates have been discussed. Presently we are implementing the proposed algorithm and performing computational studies. We are also addressing several theoretical issues. One issue of particular importance is the incorporation of specular components in the algorithm. These points would have a different statistical characterization than the diffuse components considered here and a correspondingly altered loglikelihood to be maximized. Computationally, we are examining the convergence of our algorithm, its computational complexity, and comparing the performance of the algorithm to other approaches.

REFERENCES

[1] M.I. MILLER AND D.L. SNYDER, *The Role of Likelihood and Entropy in Incomplete-Data Problems: Applications to Estimating Point-Process Intensities and Toeplitz-Constrained Covariances*, Proceedings of the IEEE, 75 (July 1987), pp. 892–907.

[2] H.L. VAN TREES, *Detection, Estimation, and Modulation Theory, Part III*, Wiley and Sons, New York, NY (1971).

[3] J. SHAPIRO, B. A. CAPRON, AND R.C. HARNEY, *Imaging and Target Detection with a Heterodyne-Reception Optical Radar*, Applied Optics, 20 (October 1981), pp. 3292–3313.

[4] D.L. SNYDER, J.A. O'SULLIVAN, AND M.I. MILLER, *The Use of Maximum-Likelihood Estimation for Forming Images of Diffuse Radar-Targets from Delay Doppler Data*, IEEE Transactions on Information Theory, 35 (May 1989), pp. 536–548.

[5] J.P. BURG, D.G. LUENBERGER, AND D.L. WENGER, *Estimation of Structured Covariance Matrices*, Proceedings of the IEEE, 70 (September 1982), pp. 963–974.

[6] D.R. FUHRMANN AND M.I. MILLER, *On the Existence of Positive Definite Maximum-Likelihood Estimates of Structured Covariance Matrices*, IEEE Transactions on Information Theory, 34 (July 1988).

[7] A.D. DEMPSTER, N.M. LAIRD, AND D.B. RUBIN, *Maximum Likelihood from Incomplete Data via the EM Algorithm*, J. Royal Statistical Society, B39 (1977), pp. 1–37.

[8] C.F.J. WU, *On the Convergence Properties of the EM Algorithm*, The Annals of Statistics, 11 (1983), pp. 95–103.

LIMITED DATA TOMOGRAPHY IN NON-DESTRUCTIVE EVALUATION*

ERIC TODD QUINTO†

Abstract. X-ray computed tomography is effective for the non-destructive evaluation of rocket exit cones and bodies, objects with centers that do not need to be analyzed. The author's algorithm has been modified to reconstruct these objects without using X-rays passing through their centers. Defects such as delaminations in exit cones and separations in rocket body gaskets are easily seen from such data and perhaps small scanners could be made to acquire data on site. The algorithm is outlined and reconstructions of mathematical phantoms of exit cones are given in this article.

AMS(MOS) subject classifications. 92A07, 44A05

1. Introduction. X-ray tomography is a natural technique for non-destructive evaluation of rocket parts [Shepp and Srivastava]. Because defects are shown precisely in CT reconstructions, engineers can easily assess the magnitude and nature of failures. The planar crossection of a rocket cone or rocket body generally consists of a ring, or annulus, surrounding a hollow center or rocket fuel. Defects in the object such as delaminations in rocket exit cones or deteriorations of rubber gaskets in rocket bodies[1] are important to detect, but often the centers of these objects are of no interest diagnostically. Normal CT scanners use data along X-rays through the center, because standard inversion algorithms need these data to reconstruct the ring. However, since the crossection of the object to be tested is annular, many fewer X-rays can be used; X-rays that do not pass through the center of the ring are sufficient to reconstruct the ring.

The mathematical goal of this article is to reconstruct a function outside of a disc in the plane (the ring outside the center of the rocket part) from X-ray data (integrals) over lines not intersecting that disc, from so called *exterior Radon data.* The hole theorem [Cormack 1963, Helgason, Ludwig] asserts this is possible for functions of compact support, the objects that occur in practice. This problem is also of interest in medical CT [Shepp and Kruskal] and in astronomical studies of the corona of the sun [Altschuler].

The author has developed an algorithm for this problem [Quinto 1983, 1988] that reconstructs the ring itself using only X-rays not passing through the center. The algorithm employs a singular value decomposition [Perry, Quinto 1983 for $\mathbf{R}^n$] and *a priori* information about the shape of the ring. In this article we adapt the algorithm to reconstruct rocket exit cones and rocket bodies and then test the algorithm on mathematical phantoms of exit cones that have common defects.

With exterior data a CT scanner needs only to rotate around the object, not translate the object. It would be easy to build a compact scanner that, in some cases, would be usable on site. Recent tragedies demonstrate the utility of such

---

*The author was partially supported by NSF grant MCS 8701415

†Department of Mathematics, Tufts University, Medford, MA 02155

[1] Obviously the materials must be appropriate for X-ray tomography.

on site scanners. Acquisition time for exterior data can be much lower than with normal CT data. On a rocket exit cone with outside diameter 23 cm. and thickness 1 cm., only 9 % of the data for normal CT scans are needed with the proposed method.

Common defects are seen well from the given data. Shepp and Srivastava [p. 78] argue that CT is ideally suited to detecting delaminations in rocket exit cones because of the *tangent casting effect* associated with sharp material density differences that occur tangent to X-rays in the data set. Simply put, an X-ray that travels tangent to a defect will detect it easily. Exterior data are tangent to delaminations as well as separations in rocket body gaskets and therefore, detect these common defects clearly (see §3).

The standard algorithms [Natterer 1986, Lewitt] do not solve this problem. They need "complete" data, including data over X-rays through the center even to reconstruct the ring. As with other limited data problems [Davison, Louis], this problem is much more highly ill-posed (that is, much more difficult and more sensitive to noise) than standard tomography. Moreover, examples [Finch] can be used to show that reconstruction from exterior data is continuous in no range of Sobolev norms. Therefore inversion of this transform for functions of compact support is much more unstable than inversion of the Radon transform with complete data (the inverse is continuous of order $+1/2$ in Sobolev norms with complete data).

Cormack [1963] developed an inversion method for this problem using integral equations, but the numerical results are not good (private communication from Allan Cormack, see also [Hansen]), but Lewitt and Bates [1978], Louis, and Natterer [1980] give algorithms that are numerically better. The excluded region in all of these algorithms, the set over which data are not taken, is assumed to be a disc. However, any inversion algorithm for circular excluded regions can be adapted to elliptical excluded regions by linear transformation with appropriate multiplicative factors for the line integrals.

Section 2 of this article provides the definitions and the inversion procedure. Reconstructions and analysis are in §3.

2. The algorithm. First let $\cdot$ denote the standard inner product on $\mathbf{R}^2$; let $|\ \ |$ be the induced norm, and let dx be Lebesgue measure on $\mathbf{R}^2$. At the same time let $\theta \in [0, 2\pi]$, and let $p \in \mathbf{R}$. Now let $d\theta$ and dp denote the standard measures on $[0, 2\pi]$ and $\mathbf{R}$, respectively. In order to define the Radon transform let $L(\theta, p) = \{x \in \mathbf{R}^2 | x \cdot \overline{\theta} = p\}$, the line with normal vector $\overline{\theta} = (\cos\theta, \sin\theta)$ and directed distance p from the origin. The points (θ, p) and $(\theta + \pi, -p)$ parametrize the same line $L(\theta, p)$, so we will always assume $p \geq 0$ in this article. Let dx_L be arc length, the measure on $L(\theta, p)$ induced from Lebesgue measure on $\mathbf{R}^2$. The classical Radon transform is defined for an integrable function f on $\mathbf{R}^2$ by

$$Rf(\theta, p) = \int_{L(\theta,p)} f(x) dx_L. \tag{2.1}$$

$Rf(\theta, p)$ is just the integral of f over the line $L(\theta, p)$.

Let E be the exterior of the unit disc in $\mathbf{R}^2, E = \{x \in \mathbf{R}^2 \,|\, 1 \le |x|\}$, and let $E' = [0, 2\pi] \times [1, \infty)$. E' corresponds to the set of lines $L(\theta, p)$ that are contained in $\bar{E}$.

The *exterior Radon transform* is the transform R as a map from integrable functions on E to integrable functions on E'. The problem posed in §1, recovering a function defined outside a disc from integrals over lines not intersecting that disc, is solved by inverting the exterior Radon transform.

Let $L^2(E)$ be the Hilbert space of functions on E defined by weight $|x|(1 - |x|^{-2})^{1/2}/\pi dx$ and let $L^2(E')$ be the Hilbert space of functions on E' defined by weight $(\pi p)^{-1} d\theta dp$. One part of the inversion method is the following singular value decomposition (see [Quinto, 1988 Propositions 3.1-2] for precise statements).

PROPOSITION 2.1. *There are orthonormal bases $\{f_m | m \in \mathbb{Z}\}$ of $L^2(E)$ and $\{g_m | m \in \mathbf{N}\}$ of $L^2(E')$ such that the exterior Radon transform $R : L^2(E) \to L^2(E')$ satisfies*

$$Rf_m(\theta, p) = 0 \quad \text{for} \quad m \le 0 \tag{2.2}$$

and,

$$Rf_m(\theta, p) = R_m g_m(\theta, p) \quad \text{for} \quad m > 0 \tag{2.3}$$

where the constants R_m satisfy

$$R_m = \mathcal{O}(1/\sqrt{m}) \quad \text{for} \quad m > 0. \tag{2.4}$$

Each basis function $f_m(r\overline{\theta})$ is an orthogonal polynomial in $1/r$ multiplied by $e^{i\ell\theta}$ for some $\ell \in \mathbb{Z}$; thus each f_m has only one non-zero term in its polar Fourier expansion (the Fourier expansion in the polar coordinate θ).

Proposition 2.1 is R. M. Perry's decomposition for the exterior Radon transform on $\mathbf{R}^2$ and is proved using identities in [Perry] involving orthogonal polynomials. This is *not* an inversion method because of the presence of the non-trivial null space (2.2). Thus, any successful reconstruction algorithm must use more than just the singular value decomposition.

The SVD is generalized in [Quinto 1983] to the Radon transform integrating over hyperplanes in $\mathbf{R}^n$, and the author's basic inversion method is outlined there. The algorithm uses the singular value decomposition and the null space characterization of (2.2) to give an exact inversion method for functions of compact support. Let $f(x) \in L^2(E)$ be the density to be reconstructed, then the SVD gives f_R, the projection of f onto the orthogonal complement of the null space of the exterior transform. Because of (2.4) this procedure is only as mildly ill-posed as inversion of the Radon transform with complete data (compare with the singular values in [Cormack 1964]). Now $f = f_R + f_N$, where f_N, the projection of f onto the null space of the exterior transform, is not recovered by the SVD.

One of the keys to the inversion method is the null space characterization (see [Quinto, 1988]):

PROPOSITION 2.2. *A function $f_N \in L^2(E)$ is in the null space of $R : L^2(E) \to L^2(E')$* **iff** *for each $\ell \in \mathbb{Z}$ the ℓth polar Fourier coefficient, $(f_N)_\ell(r)$, is a polynomial in $1/r$ of the same parity as ℓ, of degree less than $|\ell|$, and with lowest order term in $1/r$ of degree at least 2.*

The other key is the simple but useful observation that rocket exit cones have known inner and outer radii. So, given the function f to be reconstructed, there are *a priori* known inner radius (**set equal to 1.05 in this article**) and outer radius K such that

$$f(x) = 0 \quad \text{for} \quad |x| \in [1, 1.05] \cup [K, \infty) \tag{2.5}$$

Because of (2.5),

$$f_N(x) = -f_R(x) \quad \text{for} \quad |x| \in [1, 1.05] \cup [K, \infty). \tag{2.6}$$

This uniquely determines $f_N(x)$ for $x \in E$, (that is $|x| \geq 1$) [Quinto, 1988].

The basic algorithm to recover f is to use the SVD to recover $f_R(x)$ and then use equation (2.6) and the null space restriction, Proposition 2.2, to find $f_N(x)$. The unstable part of the algorithm is the recovery of f_N. Because the basis functions are expressed in terms of trigonometric monomials, the algorithm is performed on each Fourier coefficient. For the ℓth coefficient, f_ℓ, a polynomial of degree less than $|\ell|$ is least-square fit to the data, $-(f_R)_\ell(r)$, for $r \in [1, 1.05] \cup [K, \infty)$. This polynomial, the reconstruction for $(f_N)_\ell$, is interpolated to $[1.05, K]$ using the explicit knowledge of its coefficients. This gives the reconstruction of the Fourier coefficient f_ℓ, and the reconstruction of f is a smoothed sum of the Fourier coefficients (see [Quinto 1988] for details).

The author's original inversion method [Quinto 1983] was an *extrapolation* procedure assuming only that $f(x) = 0$ for $|x| \geq K$ rather than (2.5). The *interpolation* procedure using (2.6) is more stable and more accurate for higher Fourier coefficients than the original method but the values of f need to be known near $|x| = 1$. For this reason we have set the inner radius of the exit cone equal to 1.05, a number greater than one.

Equations (2.5-6) can be modified for rocket body shells. For gasket defects, one can scale radii so that the gasket is at radius greater than 1.05 and then apply the algorithm with 0 replaced by the known inside density from radius 1.0 to 1.05 in (2.5-6). In many cases the rocket shell density, f, is fairly constant near $r = 1$ (perhaps with a rescaling of radii) so high Fourier coefficients of f are approximately zero near $r = 1$ and the interpolation, performed on Fourier coefficients, is simpler. Reconstructions of phantoms for which f is not known *a priori* near $r = 1$ are given in [Quinto 1988].

3. Reconstructions and analysis. Figure 1a is a phantom with outer radius $K = 1.15$ and inner radius 1.05 (the third quadrant is shown; reconstructions of the other quadrants are at least as good). The delamination is an off-center annulus (tangent to the inside and outside boundaries of the large annulus at antipodal points) with zero density and thickness 0.002, 1/50 that of the large annulus. Figure 1b is the reconstruction with 1% L^∞ noise (the reconstruction without noise looks better). Fifty Fourier coefficients are recovered ($0 \leq |\ell| < 50$) and 40 are given near the inner boundary, $r = 1$. For each coefficient, 180 polynomials in $1/r$ are reconstructed. Data are collected for 170 values of p and 512 values of θ.

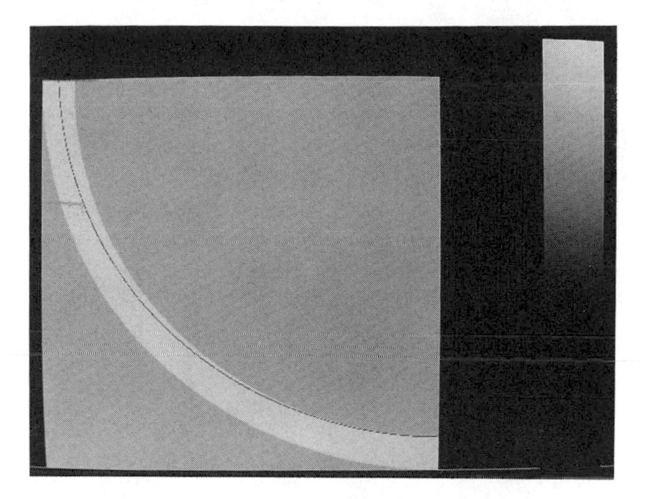

Figure 1a.

Figure 1b.

Figure 1. (a) Phantom with outer radius 1.15; (b) Reconstruction with 1% noise. Fifty Fourier coefficients are recovered, (40 near $r = 1.05$)

Figure 2a.

Figure 2b.

Figure 2. (a) Phantom with outer radius 1.32; (b) Reconstruction with 1% noise. Thirty-two Fourier coefficients are recovered (23 near $r = 1.05$)

Figure 2a is a phantom with outer radius $K = 1.32$ and inner radius 1.05 (third quadrant is shown). The delamination is an off-center annulus with zero density and thickness 0.0054, 1/50 that of the large annulus. Figure 2b is the reconstruction with 1% L^∞ noise. Thirty-two Fourier coefficients are used (23 near the inner boundary, $r = 1$) and, for each, 180 polynomials in $1/r$ are reconstructed. Data are collected for 230 values of p and 512 of θ.

The phantoms were chosen to be similar to real defects and to be difficult for the author's algorithm. The delamination is a very thin off-center annulus; it has high polar Fourier coefficients, ones that are less well reconstructed by the algorithm. The third quadrant is shown because reconstructions with this method are generally less accurate near the inner boundary of the annulus and in quadrant three the delamination lies near this boundary. Summing the same number of polar Fourier coefficients of the phantom's density itself gives similar pictures to figures 1b and 2b. (These reconstructions are better than the ones shown at the IMA Signal Processing Conference because a program bug was found and fixed after that conference.)

An object to be reconstructed, f, consists of a cross section of a region consisting of material and air space with well defined boundaries, and a reconstruction is good if it precisely shows these boundaries. An analytic property of the Radon transform predicts which boundaries of f will be detected well by the data:

(3.1) *The Radon transform detects boundaries of f that are tangent to lines in the data set, but not boundaries in other directions.*

Thus reconstructions using limited data can be expected to show boundaries tangent to lines in the data set more clearly than other boundaries. So delaminations are easy for exterior data to detect because boundaries of delaminations are generally along the curve of the annulus–they are tangent to the lines in the exterior data set and, therefore, they are easy to detect.

This basic idea has been noted by several researchers; for example, (3.1) explains the tangent casting of [L. A. Shepp and S. Srivastava]. The rigorous mathematical reason for (3.1) is that R is an elliptic Fourier integral operator that detects singularities when the singularities are "tangent" to the line being integrated over but not otherwise (this can be made precise and proven using wave front sets, see [Quinto, 1980, 1988]).

Acknowledgements: The author is indebted to Merilee Goldberg and Jon Polito for their practical help as well as to Alfred Louis, Frank Natterer, and Larry Shepp for their valuable information. We thank Allan Cormack for his constant support, as well as for suggesting this problem. This research was supported in part by the Institute for Mathematics and its Applications at the University of Minnesota with funds provided by the National Science Foundation.

REFERENCES

[1] M. D. ALTSCHULER, *Reconstruction of the global-scale three-dimensional solar corona*, in *Image Reconstruction from Projections, Implementation and Applications*, (G. T. Herman,

Ed.); pp. 105-145, Topics in Applied Physics, Vol. 32, Springer-Verlag, New York/Berlin, 1979.

[2] A. M. CORMACK, *Representation of a function by its line integrals with some radiological applications*, J. Appl. Phys., 34 (1963), pp. 2722–2727.

[3] A. M. CORMACK, *Representation of a function by its line integrals with some radiological applications II*, J. Appl. Phys, 35 (1964), pp. 2908-2913.

[4] M. DAVISON, *The ill-conditioned nature of the limited angle tomography problem*, SIAM J. Appl. Math., 43 (1983), pp. 428–448.

[5] D. V. FINCH, *Cone beam reconstruction with sources on a curve*, SIAM J. Appl. Math., 45 (1985), pp. 665–673..

[6] E. W. HANSEN, *Circular harmonic image reconstruction: Experiments,*, Appl. Optics, 20 (1981), pp. 2266–2274.

[7] R. M. LEWITT, *Overview of inversion methods used in computed tomography*, in *Advances in Remote Sensing Retrieval Methods*, Deepak Publishing, Hampton, VA 1988 (in press).

[8] R. M. LEWITT AND R. H. T. BATES, *Image reconstruction from projections*, **II**, Projection completion methods (theory), *Optik* **50** (1978), 189-204, **III**: Projection completion methods (computational examples), Optik, 50 (1978), pp. 269–278..

[9] A. LOUIS, *Incomplete data problems in X-ray computerized tomography 1. Singular value decomposition of the limited angle transform*, Numer. Math., 48 (1986), pp. 251-262.

[10] F. NATTERER, *Efficient implementation of "optimal" algorithms in computerized tomography*, Math. Meth. Appl. Sci., 2 (1980), pp. 545–555..

[11] F. NATTERER, *The mathematics of computerized tomography*, Wiley, New York, 1986.

[12] R. M. PERRY, *On reconstruction a function on the exterior of a disc from its Radon transform*, J. Math. Anal. Appl., 59 (1977), pp. 324-341.

[13] E.T. QUINTO, *The dependence of the generalized Radon transform on defining measures*, Trans. Amer. Math. Soc., 257 (1980), pp. 331-346..

[14] E. T. QUINTO, *The invertibility of rotation invariant Radon transforms*, J. Math. Anal. Appl., 91 (1983,), pp. 510–522.

[15] E. T. QUINTO, *Tomographic reconstructions from incomplete data–numerical inversion of the exterior Radon transform.*, Inverse Problems, 4 (1988,), pp. 867–876.

[16] L. A. SHEPP AND J. B. KRUSKAL, *Computerized tomography: The new medical X-ray technology*, Amer. Math. Monthly, 85 (1978), pp. 420-439.

[17] L. A. SHEPP AND S. SRIVASTAVA, *Computed tomography of PKM and AKM exit cones*, A. T & T. Technical Journal, 65 (1986), pp. 78–88.

SPEECH RECOGNITION BASED ON PATTERN RECOGNITION APPROACHES

LAWRENCE R. RABINER*

Abstract. Algorithms for speech recognition can be characterized broadly as pattern recognition approaches and acoustic phonetic approaches. To date, the greatest degree of success in speech recognition has been obtained using pattern recognition paradigms. Thus, in this paper, we will be concerned primarily with showing how pattern recognition techniques have been applied to the problems of isolated word (or discrete utterance) recognition, connected word recognition, and continuous speech recognition. We will show that our understanding (and consequently the resulting recognizer performance) is best for the simplest recognition tasks and is considerably less well developed for large scale recognition systems.

1. Introduction. The ultimate goal of most research is speech recognition is to develop a machine that had the ability to understand fluent, conversational speech, with unrestricted vocabulary, from essentially any talker. Although the promise of such a capable machine is as yet unfulfilled, the field of automatic speech recognition has made significant advances in the past decade [1–3]. This is due, in part, to the great advances made in VLSI technology, which has greatly lowered the cost and increased the capability of individual devices (e.g. processors, memory), and in part due to the theoretical advances in our understanding of how to apply powerful mathematical modelling techniques to the problems of speech recognition.

When setting out to define the problems associated with implementing a speech recognition system, one finds that there are a number of general issues that must be resolved before designing and building the system. One such issue is the size and complexity of the user vocabulary. Although useful recognition systems have been built with as few as two words (yes, no), there are at least four distinct ranges of vocabulary size of interest. Very small vocabularies (on the order of 10 words) are most useful for control tasks – e.g. all digit dialing of telephone numbers, repertory name dialing, access control etc. Generally the vocabulary words are chosen to be highly distinctive words (i.e. of low complexity) to minimize potential confusions. The next range of vocabulary size is moderate vocabulary systems having on the order of 100 words. Typical applications include spoken computer languages, voice editors, information retrieval from databases, controlled access via spelling etc. For such applications, the vocabulary is generally fairly complex (i.e. not all pairs of words are highly distinctive), but word confusions are often resolved by the syntax of the specific task to which the recognizer is applied. The third vocabulary range of interest is the large vocabulary system with vocabulary sizes on the order of 1000 words. Vocabulary sizes this large are big enough to specify fairly comfortable subsets of English and hence are used for conversational types of applications – e.g. the IBM laser patent text, basic English, etc. [4, 5]. Such vocabularies are inherently very complex and rely heavily on task syntax to resolve recognition ambiguities between similar sounding words. Finally the last range of vocabulary

*Head, Speech Research Department, AT&T Bell Laboratories, Murray Hill, New Jersey 07974

size is the very large vocabulary system with 10,000 words or more. Such large vocabulary sizes are required for office dictation/word processing applications.

Although the vocabulary size and complexity is of paramount importance in specifying a speech recognition system, several other issues can also greatly affect the performance of a speech recognizer. The system designer must decide if the system is to be speaker trained, or speaker independent; the format for talking must be specified (e.g. isolated inputs, connected inputs, continuous discourse); the amount and type of syntactic and semantic information must be specified; the speaking environment and transmission conditions must be considered; etc. The above set of issues, by no means exhaustive, gives some idea as to how complicated it can be to talk about speech recognition by machine.

There are two general approaches to speech recognition by machine, the statistical pattern recognition approach, and the acoustic-phonetic approach. The statistical pattern recognition approach is based on the philosophy that if the system has "seen the pattern, or something close enough to it, it can recognize it". Thus a fundamental element of the statistical pattern recognition approach is pattern training. The units being trained, be they phrases, words, or subword units, are essentially irrelevant, so long as a good training set is available, and a good pattern recognition model is applied. On the other hand, the acoustic-phonetic approach to speech recognition has the philosophy that speech sounds have certain invariant (acoustic) properties, and that if one could only discover these invariant properties, continuous speech could be decoded in a sequential manner (perhaps with delays of several sounds). Thus, the basic techniques of the acoustic-phonetic approach to speech recognition are feature analysis (i.e. measurement of the invariants of sounds), segmentation of the feature contours into consistent groups of features, and labelling of the segmented features so as to detect words, sentences, etc.

To date, the greatest successes in speech recognition have been achieved using the pattern recognition approach. Hence, for the remainder of this paper, we will restrict our attention to trying to explain how the model works, and how it has been applied to the problems of isolated word, connected word, and continuous speech recognition.

II. The Statistical Pattern Recognition Model. Figure 1 shows a block diagram of the pattern recognition model used for speech recognition. The input speech signal, $s(n)$, is analyzed (based on some parametric model) to give the test pattern, T, and then compared to a prestored set of reference patterns, $\{R_v\}$, $1 \leq v \leq V$ (corresponding to the V labelled patterns in the system) using a pattern classifier (i.e. a similarity procedure). The pattern similarity scores are then sent to a decision algorithm which, based upon the syntax and/or semantics of the task, chooses the best transcription of the input speech.

There are two types of reference patterns which can be used with the model of Fig. 1. The first type, called nonparametric reference patterns, are patterns created from one or more real world tokens of the actual pattern. The second type, called statistical reference models, are created as a statistical characterization (via a fixed type of model) of the behavior of a collection of real world tokens. Ordinary template approaches [6], are examples of the first type of reference patterns; hidden Markov models [7, 8] are examples of the second type of reference patterns.

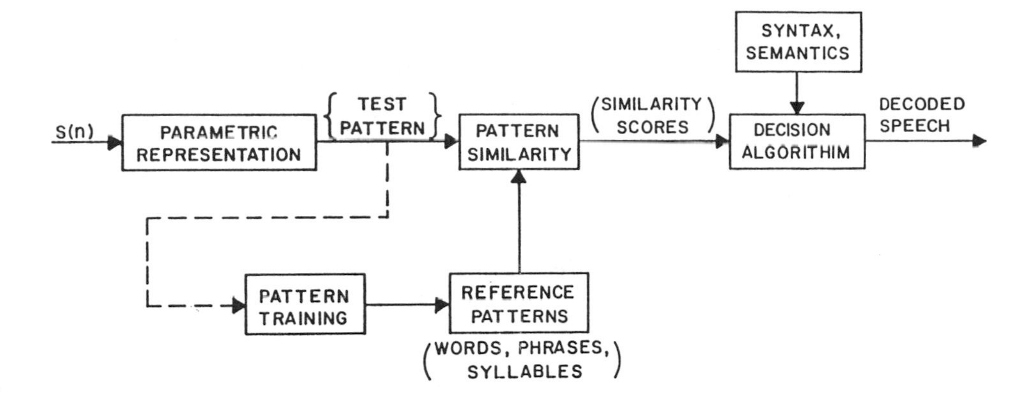

FIGURE 1. Pattern Recognition Model for Speech Recognition

The model of Fig. 1 has been used (either explicitly or implicitly) for almost all commercial and industrial speech recognition systems for the following reasons:

1. It is invariant to different speech vocabularies, users, feature sets, pattern similarity algorithms, and decision rules
2. It is easy to implement in either software or hardware
3. It works well in practice.

For all of these reasons we will concentrate on this model throughout this paper. In the remainder of this paper we will discuss the elements of the pattern recognition model and show how it has been used for isolated word, connected word, and for continuous speech recognition. Because of the tutorial nature of this paper we will minimize the use of mathematics in describing the various aspects of the signal processing. The interested reader is referred to the appropriate references [e.g. 6–15].

2.1 Parametric Representation. Parametric representation (or feature measurement, as it is often called) is basically a data reduction technique whereby a large number of data points (in this case samples of the speech waveform recorded at an appropriate sampling rate) are transformed into a smaller set of features which are equivalent in the sense that they faithfully describe the salient properties of the acoustic waveform. For speech signals, data reduction rates from 10 to 100 are generally practical.

For representing speech signals, a number of different feature sets have been proposed ranging from simple sets, such as energy and zero crossing rates (usually in selected frequency bands), to complex, complete representations, such as the short-time spectrum or a linear predictive coding (LPC) model. For recognition systems, the motivation for choosing one feature set over another is often complex and highly dependent on constraints imposed on the system (e.g. cost, speed, response time, computational complexity etc). Of course the ultimate criterion is overall system performance (i.e. accuracy with which the recognition task is performed). However, this criterion is also a complicated function of all system variables.

The two most popular parametric representations for speech recognition are the short-time spectrum analysis (or bank of filters) model, and the LPC model. The bank of filters model is illustrated in Figure 2.

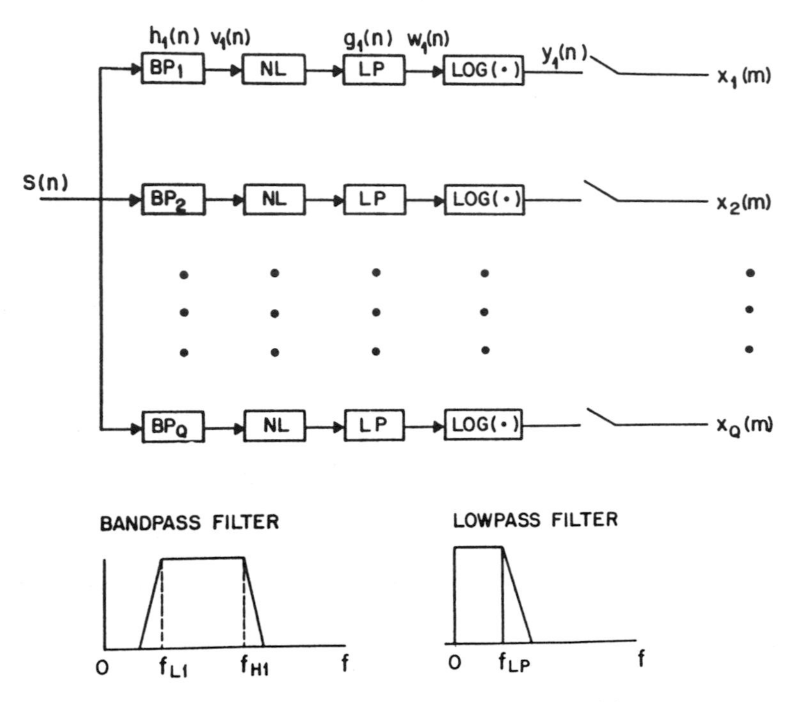

FIGURE 2. Bank of Filters Analysis Model

The speech signal is passed through a bank of Q bandpass filters covering the speech band from 100 Hz to some upper cutoff frequency (typically between 3000 and 8000 Hz). The number of bandpass filters used varies from as few as 5 to as many as 32. The filters may or may not overlap in frequency. Typical filter spacings are linear until about 1000 Hz and logarithmic beyond 1000 Hz [9].

The output of each bandpass filter is generally passed through a nonlinearity (e.g. a square law detector or a full wave rectifier) and lowpass filtered (using a 20–30 Hz width filter) to give a signal which is proportional to the energy of the speech signal in the band. A logarithmic compressor is generally used to reduce the dynamic range of the intensity signal, and the compressed output is resampled (decimated) at a low rate (generally twice the lowpass filter cutoff) for efficiency of storage.

The LPC feature model for recognition is shown in Figure 3.

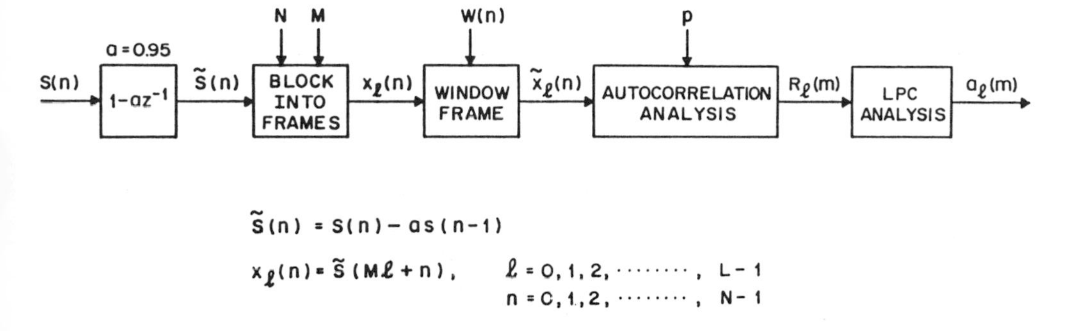

FIGURE 3. LPC Analysis Model

Unlike the bank of filters model, this system is a block processing model in which a frame of N samples of speech is processed, and a vector of features is computed. The steps involved in obtaining the vector of LPC coefficients, for a given frame of N speech samples, are as follows:

1. preemphasis by a first order digital network in order to spectrally flatten the speech signal
2. frame windowing, i.e. multiplying the N speech samples within the frame by an N-point Hamming window, so as to minimize the endpoint effects of chopping an N-sample section out of the speech signal
3. autocorrelation analysis in which the windowed set of speech samples is autocorrelated to give a set of $(p+1)$ coefficients, where p is the order of the desired LPC analysis (typically 8 to 12)
4. LPC analysis in which the vector of LPC coefficients is computed from the autocorrelation vector using a Levinson or a Durbin recursive method [10].

New speech frames are created by shifting the analysis window by M samples (typically $M < N$) and the above steps are repeated on the new frame until the entire speech signal has been analyzed.

The LPC feature model has been a popular speech representation because of its ease of implementation, and because the technique provides a robust, reliable, and accurate method for characterizing the spectral properties of the speech signal.

As seen from the above discussion, the output of the feature measurement procedure is basically a time-frequency pattern – i.e. a vector of spectral features is obtained periodically in time throughout the speech.

2.2 Pattern Training. Pattern training is the method by which representative sound patterns are converted into reference patterns for use by the pattern similarity algorithm. There are several ways in which pattern training can be performed, including:

1. casual training in which each individual training pattern is used directly to create either a non-parametric reference pattern or a statistical model. Casual training is the simplest, most direct method of creating reference patterns.
2. robust training in which several (i.e. two or more) versions of each vocabulary entry are used to create a single reference pattern or statistical model. Robust training gives statistical confidence to the reference patterns since multiple patterns are used in the training.
3. clustering training in which a large number of versions of each vocabulary entry are used to create one or more reference patterns or statistical models. A statistical clustering analysis is used to determine which members of the multiple training patterns are similar, and hence are used to create a single reference pattern. Clustering training is generally used for creating speaker independent reference patterns, in which case the multiple training patterns of each vocabulary entry are derived from a large number of different talkers.

The final result of the pattern training algorithm is the set of reference patterns used in the recognition phase of the model of Fig. 1.

2.3 Pattern Similarity Algorithm. A key step in the recognition algorithm of Fig. 1 is the determination of similarity between the measured (unknown) test pattern, and each of the stored reference patterns. Because speaking rates vary greatly from repetition to repetition, pattern similarity determination involves both time alignment (registration) of patterns, and once properly aligned, distance computation along the alignment path.

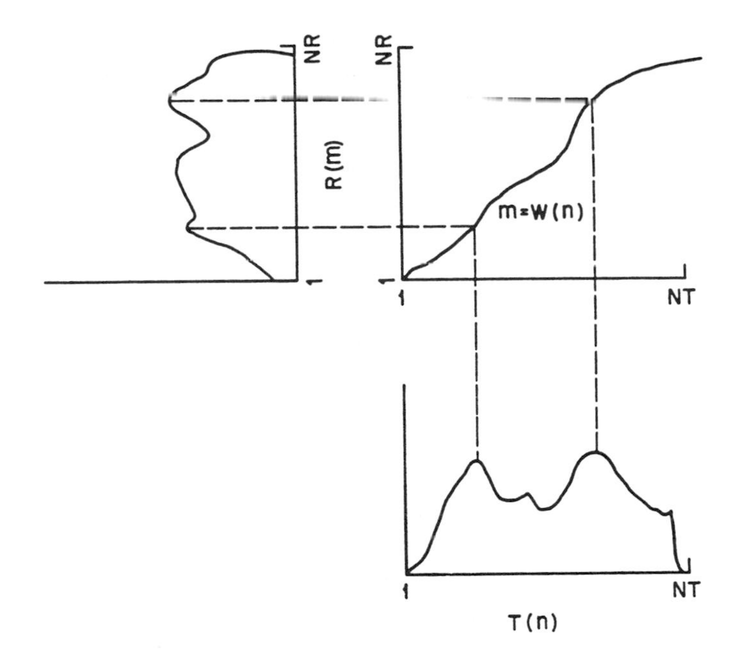

FIGURE 4. Example of Time Registration of a Test and Reference Pattern

Figure 4 illustrates the problem involved in time aligning a test pattern, $T(n), 1 \leq n \leq NT$ (where each $T(n)$ is a spectral vector), and a reference pattern $R(m)$, $1 \leq m \leq NR$. Our goal is to find an alignment function, $m = w(n)$, which maps R onto the corresponding parts of T. The criterion for correspondence is that some measure of distance between the patterns be minimized by the mapping w. Defining a local distance measure, $d(n,m)$, as the spectral distance between vectors $T(n)$ and $R(m)$, then the task of the pattern similarity algorithm is to determine the optimum mapping, w, to minimize the total distance

$$D^* = \min_{w(n)} \sum_{i=1}^{NT} d(i, w(i)) \tag{1}$$

The solution to Eq. (1) can be obtained in an efficient manner using the techniques of dynamic programming. In particular a class of procedures called dynamic time warping (DTW) techniques, has evolved for solving Eq. (1) efficiently [6].

The above discussion has shown how to time align a pair of templates. In the case of aligning statistical models, an analogous procedure, based on the Viterbi algorithm, can be used [7,8,16].

2.4 Decision Algorithm. The last step in the statistical pattern recognition model of Fig. 1 is the decision which utilizes both the set of pattern similarity scores (distances) and the system knowledge, in terms of syntax and/or semantics, to decode the speech into the best possible transcription. The decision algorithm can (and generally does) incorporate some form of nearest neighbor rule to process the distance scores to increase confidence in the results provided by the pattern similarity procedure. The system syntax helps to choose among the candidates with the lowest distance score by eliminating candidates which don't satisfy the syntactic constraints of the task, or by deweighting extremely unlikely candidates. The decision algorithm can also have the capability of providing multiple decodings of the spoken string. This feature is especially useful in cases in which multiple candidates have indistinguishably different distance scores.

2.5 Summary. We have now outlined the basic signal processing steps in the pattern recognition approach to speech recognition. In the next sections we illustrate how this model has been applied to problems in isolated word, connected word, and continuous speech recognition.

III. Results on Isolated Word Recognition. Using the pattern recognition model of Fig. 1, with an 8th order LPC parametric representation, and using the non-parametric template approach for reference patterns, a wide variety of tests of the recognizer have been performed with isolated word inputs in both speaker dependent (SD) and speaker independent (SI) modes. Vocabulary sizes have ranged from as small as 10 words (i.e. the digits zero-nine) to as many as 1109 words. Table I gives a summary of recognizer performance under the conditions discussed above. It can be seen that the resulting error rates are not strictly a function of vocabulary size, but also are dependent on vocabulary complexity. Thus a simple vocabulary of 200 polysyllabic Japanese city names had a 2.7% error rate (in an SD mode), whereas a complex vocabulary of 39 alphadigit terms (in both SD and SI modes) had error rates of on the order of 4.5 to 10.5%.

Table 1 also shows that in case where the same vocabulary was used in both SD and SI modes (e.g. the alphadigits and the airline words), the recognizer gave reasonably comparable performances. This result indicates that the SI mode clustering analysis, which yielded the set of SI templates, was capable of providing the same degree of representation of each vocabulary word as either casual or robust training for the SD mode. Of course the computation of the SI mode recognizer was comparably higher than that required for the SD mode since a larger number of templates were used in the pattern similarity comparison.

Vocabulary	Mode	Error Rate (%)
10 Digits	SI	0.8
37 Dialer Words	SD	0
39 Alphadigits	SD	4.5
	SI	10.5
54 Computer Terms	SI	3.5
129 Airline Words	SD	1.0
	SI	2.9
200 Japanese Cities	SD	2.7
1109 Basic English	SD	4.3

Table I

Performance of Template-Based Isolated Word Systems

The results in Table I are based on using word templates created from isolated word training tokens. Studies have shown that when adequate training data is available, the performance of isolated word recognizers based on statistical models is comparable to or better than that of recognizers based on templates. The main issue here is the amount of training data available relative to the number of parameters to be estimated in the statistical model. For small amounts of training data, very unreliable parametric estimates result, and the template approach is generally superior to the statistical model approach. For moderate amounts of training data, the performance of both types of models is comparable. However, for large amounts of training data, the performance of statistical models is generally superior to that of template approaches because of their ability to accurately characterize the tails of the distribution (i.e. the outliers in terms of the templates).

IV. Connected Word Recognition Model. The basic approach to connected word recognition from discrete reference patterns is shown in Fig. 5.

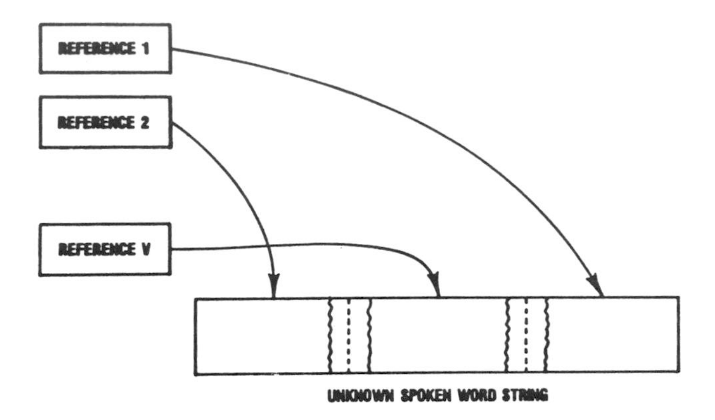

FIGURE 5. Illustration of Connected Word Recognition from Word Templates

Assume we are given a test pattern $\mathbf{T}$, which represents an unknown spoken word string, and we are given a set of V reference patterns, $\{R_1, R_2, \ldots, R_v\}$ each representing some word of the vocabulary. The connected word recognition problem consists of finding the "super" reference pattern, $\mathbf{R}^s$, of the form

$$\mathbf{R}^s = R_{q(1)} \oplus R_{q(2)} \cdots R_{q(L)}$$

which is the concatenation of L reference patterns, $R_{q(1)}, R_{q(2)}, \ldots, R_{q(L)}$, which best matches the test string, $\mathbf{T}$, in the sense that the overall distance between $\mathbf{T}$ and $\mathbf{R}^s$ is minimum over all possible choices of $L, q(1), q(2), \ldots, q(L)$, where the distance is an appropriately chosen distance measure.

There are several problems associated with solving the above connected word recognition problem. First we don't know L, the number of words in the string. Hence our proposed solution must provide the best matches for all reasonable values of L, e.g. $L = 1, 2, \ldots, L_{MAX}$. Second we don't know nor can we reliably find word boundaries, even when we have postulated L, the number of words in the string. The implication is that the word recognition algorithm must work without direct knowledge of word boundaries; in fact the estimated word boundaries will be shown to be a byproduct of the matching procedure. The third problem with a template matching procedure is that the word matches are generally much poorer at the boundaries than at frames within the word. In general this is a weakness of word matching schemes which can be somewhat alleviated by the matching procedures which can apply lessor weight to the match at template boundaries than at frames within the word. A fourth problem is that word durations in the string are often grossly different (shorter) than the durations of the corresponding reference patterns. To alleviate this problem one can use some time prenormalization procedure to warp the word durations accordingly, or rely on reference patterns extracted from embedded word strings. Finally the last problem associated with matching word strings is that the combinatories of matching strings exhaustively (i.e. by trying all combinations of reference patterns in a sequential manner) is prohibitive.

A number of different ways of solving the connected word recognition problem have been proposed which avoid the plague of combinatorics mentioned above. Among these algorithms are the 2-level DP approach of Sakoe [11], the level building approach of Myers and Rabiner [12], the parallel single stage approach of Bridle et al. [13], and the nonuniform sampling approach of Gauvain and Mariani [14]. Although each of these approaches differs greatly in implementation, all of them are similar in that the basic procedure for finding $\mathbf{R}^s$ is to solve a time-alignment problem between $\mathbf{T}$ and $\mathbf{R}^s$ using dynamic time warping (DTW) methods.

The level building DTW based approach to connected word recognition is illustrated in Fig. 6.

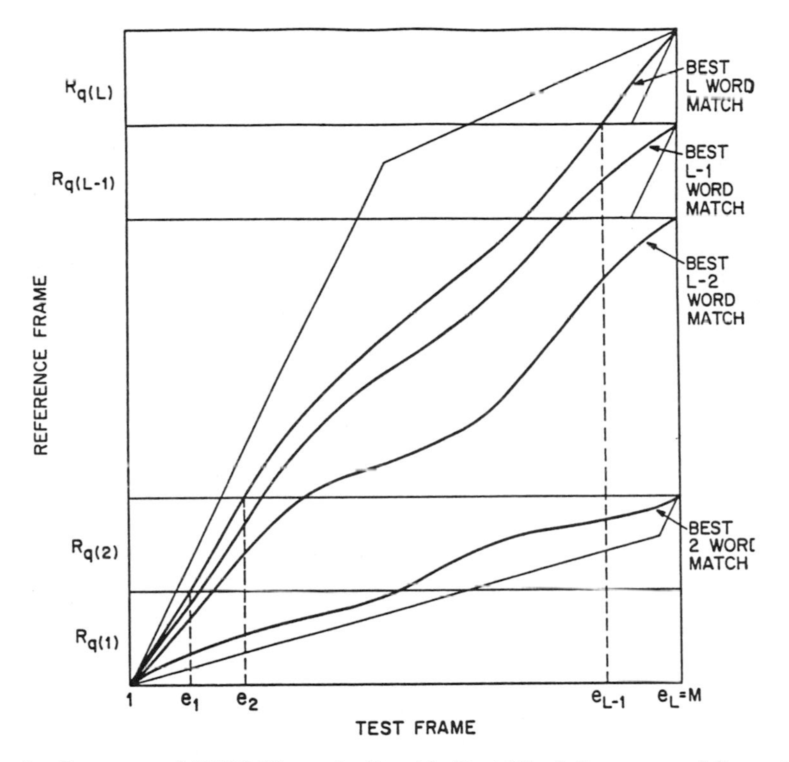

FIGURE 6. Sequence of DTW Warps to Provide Best Word Sequences of Several Different Lengths

Shown in this figure are the warping paths for all possible length matches to the test pattern, along with the implicit word boundary markers $(e_1, e_2, \ldots, e_{L-1}, e_L)$ for the dynamic path of the L-word match. The level building algorithm has the property that it builds up all possible L-word matches one level (word in the string) at a time. For each string match found, a segmentation of the test string into appropriate matching regions for each reference word in $\mathbf{R}^s$ is obtained. In addition, for every string length L, the best Q matches (i.e. the Q lowest distance L-word strings) can be found. The details of the level building algorithm are available elsewhere [12], and will not be discussed here.

Typical performance results for connected word recognizers, based on a level building implementation, are shown in Table II. For a digits vocabulary, string accuracies of 98-99% have been obtained. For name retrieval, by spelling, from a 17,000 name directory, string accuracies of from 90% to 96% have been obtained. Finally, using a moderate size vocabulary of 127 words, the accuracy of sentences for obtaining information about airlines schedules is between 90% and 99%. Here the average sentence length was close to 10 words. Many of the errors occurred in sentences with long strings of digits.

VOCABULARY	MODE	WORD ACCURACY	TASK	STRING (TASK) ACCURACY
Digits (10 Words)	Speaker Dependent or Speaker Independent	>99% SI >99% SD	1–7 Digit Strings 1–7 Digit Strings	98.5% SI* 99%SD*
Letters of the Alphabet (26 words)	Speaker Dependent or Speaker Independent	≈90%	Directory Listing Retrieval (17,000 Name Directory)	96% SD 90% SI
Airline Terms (129 words)	Speaker Dependent or Speaker Independent	99% SD 97% SI	Airline Information and Reservations	99% SD 90% SI

* Known string length.

Table II

Performance of Connected Word Recognizers on Specific Recognition Tasks

V. Continuous, Large Vocabulary, Speech Recognition. The area of continuous, large vocabulary, speech recognition refers to systems with at least 1000 words in the vocabulary, a syntax approaching that of natural English (i.e. an average branching factor on the order of 100), and possibly a semantic model based on a given, well defined, task. For such a problem, there are three distinct sub-problems that must be solved, namely choice of a basic recognition unit (and a modelling technique to go with it), a method of mapping recognized units into words (or, more precisely, a method of scoring words from the recognition scores of individual word units), and a way of representing the formal syntax of the recognition task (or, more precisely, a way of integrating the syntax directly into the recognition algorithm).

For each of the three parts of the continuous speech recognition problem, there are several alternative approaches. For the basic recognition unit, one could consider whole words, half syllables such as dyads, demisyllables, or diphones, or sound units as small as phonemes or phones. Whole word units, which are attractive because of our knowledge of how to handle them in connected environments, are totally impractical to train since each word could appear in a broad variety of contexts. Therefore the amount of training required to capture all the types of word environments is unrealistic. For the sub-word units, the required training is extensive, but could be carried out using a variety of well known, existing training procedures. A full system would require between 1000 and 2000 half syllable speech units. For the phoneme-like units, only about 30–100 units would have to be trained.

The problem of representing vocabulary words, in terms of the chosen speech unit, has several possible solutions. One could create a network of linked word unit models for each vocabulary word. The network could be either a deterministic (fixed) or a stochastic structure. An alternative is to do lexical access from a dictionary in which all word pronounciation variants (and possibly part of speech information) are stored, along with a mapping from pronunciation units to speech representation units.

Finally the problem of representing the task syntax, and integrating it into the recognizer, has several solutions. The task syntax, or grammar, can be represented as a deterministic state diagram, as a stochastic model (e.g. a model of word trigram statistics), or as a formal grammar. There are advantages and disadvantages to each of these approaches.

To illustrate the state of the art in continuous speech recognition, consider the office dictation system discussed in Reference [16]. This system uses phoneme-like units in a statistical model to represent words, where each phoneme-like unit is a statistical model based on vector-quantized spectral outputs of a speech spectrum analysis. A third statistical model is used to represent syntax; thus the recognition task is essentially a Bayesian optimization over a triply embedded sequence of statistical models. The computational requirements are very large, but a system has been implemented using isolated word inputs for the task of automatic transcription of office dictation. For a vocabulary of 5000 words, in a speaker trained mode, with 20 minutes of training for each talker, the average *word* error rates for 5 talkers are 2% for prerecorded speech, 3.1% for read speech, and 5.7% for spontaneously spoken speech [16].

VI. Summary. In this paper we have reviewed and discussed the general pattern recognition framework for machine recognition of speech. We have discussed some of the signal processing and statistical pattern recognition aspects of the model and shown how they contribute to the recognition.

The challenges in speech recognition are many. As illustrated above, the performance of current systems is barely acceptable for large vocabulary systems, even with isolated word inputs, speaker training, and favorable talking environment. Almost every aspect of continuous speech recognition, from training to systems implementation, represents a challenge in performance, reliability, and robustness.

REFERENCES

[1] N.R. Dixon and T.B. Martin, Eds., *Automatic Speech and Speaker Recognition*, New York: IEEE Press (1979).

[2] W. Lea, Ed., *Trends in Speech Recognition*, Englewood Cliffs, NJ: Prentice-Hall (1980).

[3] G.R. Doddington and T.B. Schalk, *Speech Recognition: Turning Theory into Practice*, IEEE Spectrum, 18, 9 (Sept. 1981), pp. 26–32.

[4] L.R. Bahl, F. Jelinek and R.L. Mercer, *A Maximum Likelihood Approach to Continuous Speech Recognition*, IEEE Trans. on Pattern Analysis and Machine Intelligence, PAMI-5, 2 March (1983), pp. 179–190.

[5] A.E. Rosenberg, L.R. Rabiner, J.G. Wilpon, and D. Kahn, *Demisyllable-Based Isolated Word Recognition*, IEEE Trans. on Acoustics, Speech, and Signal Processing, ASSP-31, 3 (June 1983), pp. 713-726.

[6] F. Itakura, *Minimum Prediction Residual Principle Applied to Speech Recognition*, IEEE Trans. on Acoustics, Speech, and Signal Processing, ASSP-23 (Feb. 1975), 67–72.

[7] F.Jelinek, *Speech Recognition by Statistical Methods*, Proc. IEEE, 65, (April 1976), pp. 532–556.

[8] S.E. Levinson, L.R. Rabiner, and M.M. Sondhi, *An Introduction to the Application of the Theory of Probabilistic Functions of a Markov Process to Automatic Speech Recognition*, Bell System Tech. Jour., 62, 4 (April 1983), pp. 1035–1074.

[9] B.A. Dautrich, L.R. Rabiner, and T.B. Martin, *On the Effects of Varying Filter Bank Parameters on Isolated Word Recognition*, IEEE Trans. on Acoustics, Speech, and Signal Processing, ASSP-31, 4 (Aug. 1983), pp. 793–807.

[10] L.D. Markel and A.H. Gray, Jr., *Linear Prediction of Speech*, New York: Springer-Verlag (1976).

[11] H. Sakoe, *Two Level DP Matching — A Dynamic Programming Based Pattern Matching Algorithm for Connected Word Recognition*, IEEE Trans. on Acoustics, Speech, and signal Processing, ASSP-27 (Dec. 1979), pp. 588–595.

[12] C.S. Myers and L.R. Rabiner, *Connected Digit Recognition Using a Level Building DTW Algorithm*, IEEE Trans. on Acoustics, Speech, and Signal Processing, ASSP-29, 3 (June 1981), pp. 351–363.

[13] J.S. Bridle, M.D. Brown and R.M. Chamberlain, *An Algorithm for Connected Word Recognition*, Automatic Speech Analysis and Recognition, J.P. Haton, Ed. (1982), pp. 191–204.

[14] J.L. Gauvain and J. Mariani, *A Method for Connected Word Recognition and Word Spotting on a Microprocessor*, Proc. 1982 ICASSP (May 1982), pp. 891–894.

[15] L.R. Rabiner, *A Tutorial on Hidden Markov Models and Selected Applications in Speech Recognition*, Proc. IEEE., 77, 2 (Feb. 1989), pp. 257–286.

[16] F. Jelinek, *The Development of an Experimental Discrete Dictation Recognizer*, Proc. IEEE, Vol. 73, No. 11 (Nov. 1985), pp. 1616-1624.

ESPRIT - Estimation of Signal Parameters via Rotational Invariance Techniques

R. ROY AND T. KAILATH

Information Systems Laboratory
Stanford University
Stanford, CA 94305

Abstract

High-resolution signal parameter estimation is a problem of significance in many signal processing applications. Such applications include *direction-of-arrival* (DOA) estimation, system identification, and time series analysis. A novel approach to the general problem of signal parameter estimation is described. Though discussed in the context of direction-of-arrival estimation, ***ESPRIT*** can be applied to a wide variety of problems. It exploits an underlying *rotational invariance* among signal subspaces induced by an array of sensors with a *translational invariance* structure. A few simulation results are presented and directions of current and future research are discussed.

CONTENTS

I. Introduction

IN MANY PRACTICAL signal processing problems, the objective is to estimate, using measurements from a sensor array, a set of *constant* parameters upon which the signals being received depend. Problems of this type include high resolution direction-of-arrival (DOA) estimation in sensor systems such as radar, sonar, and electronic surveillance, and high resolution spectral analysis of time series. In such problems, the functional form of the underlying signals can often be assumed to be known (*e.g.*, narrowband plane waves, cisoids). The quantities to be estimated are parameters (*e.g.*, frequencies and DOAs of plane waves, cisoid frequencies) upon which the sensor outputs depend, and these parameters are assumed to be *constant.*[1]

There have been several approaches to such problems including the so-called maximum likelihood (ML) method of Capon (1969) and Burg's (1967) maximum entropy (ME) method. Though often successful and widely used, these methods have certain fundamental limitations (*esp.* bias and sensitivity in the parameter estimates), largely because they do not explicitly use the actual underlying model of the measurements. Pisarenko (1973) was one of the first to exploit the structure of the data model, doing so in the context of estimation of parameters of cisoids in additive noise using a covariance approach. Schmidt (1977) and independently Bienvenu (1979) were the first to correctly exploit the measurement model in the case of sensor arrays of arbitrary form. Schmidt, in particular, accomplished this by first deriving a complete geometric solution in the absence of noise, then cleverly extending the geometric concepts to obtain a *reasonable* approximate solution in the presence of noise. The resulting algorithm was called **MUSIC** (MUltiple SIgnal Classification) and has been widely studied. In a detailed evaluation based on thousands of simulations, MIT's Lincoln Laboratory concluded that, among currently accepted high-resolution algorithms, **MUSIC** was the most promising and a leading candidate for further study and actual hardware implementation. However, though the performance advantages of **MUSIC** are substantial, they are achieved at a considerable cost in computation (searching over parameter space) and storage (of array calibration data, *cf.* Section III).

In the sequel. a recently developed algorithm (***ESPRIT***) that dramatically reduces the aforementioned computation and storage costs is described. In the context of DOA estimation, the reductions are achieved by requiring that the sensor array possess a *displacement invariance*, *i.e.*, that sensors occur in matched pairs with identical *displacement* vectors (*cf.* Figure 3). Fortunately, there are many practical problems in which these conditions are or can easily be satisfied. Linear arrays of equi-spaced identical sensors are commonplace in sonar applications, as are regular rectangular arrays of identical elements (*e.g.*, phased-arrays) in radar applications. In addition to obtaining signal parameter estimates efficiently, *optimal signal copy vectors* for reconstructing the signals are also provided by ***ESPRIT***. ***ESPRIT*** is also manifestly

[1] Extensions to situations in which the parameters may be time varying can be made, however they rely on an inherent *time-scale* or *eigenvalue separation* between the parameter dynamics and the dynamics of the signal process. Fundamentally, the assumption is made that over time intervals long enough to collect sufficient information from which to obtain accurate parameter estimates, the parameters have not changed significantly.

less sensitive to array imperfections than previous techniques including **MUSIC** [1], a desirable property when system implementation is considered.

Notation and Mathematical Preliminaries

To make the presentation as clear as possible, an attempt is made to adhere to a somewhat standard notational convention. Lower case **boldface** characters will generally refer to vectors. Upper case **BOLDFACE** characters will generally refer to matrices. For either real or complex-valued matrices, $(\cdot)^*$ will be used to denote the Hermitian conjugate (or complex-conjugate transpose) operation. Eigenvalues of square Hermitian matrices are assumed to be ordered in decreasing magnitude, as are the singular values of non-square matrices. Knowledge of the fundamental theorems of matrix algebra dealing with eigendecompositions and singular value decompositions (SVD) is assumed (*cf.* [2]).

II. The Data Model

Though ***ESPRIT*** is generally applicable to a wide variety of problems, for illustrative purposes the discussions herein focus on direction-of-arrival (DOA) estimation. In many practical signal processing applications, data from an array of sensors are collected and the objective is to locate point sources assumed to be radiating energy that is detectable by the sensors as depicted in Figure 1. Mathematically, such problems are quite simply, though abstractly, modeled using *Green's functions* for the particular differential operator that describes the physics of radiation propagation from the sources to the sensors. For the intended applications however, a few *reasonable* assumptions can be invoked to simplify the model.

The transmission medium is assumed to be isotropic and non-dispersive so that the radiation propagates in *straight lines*, and the sources are assumed to be in the *far-field* of the array. Consequently, the radiation impinging on the array is in the form of a sum of *plane waves*. For simplicity, it is assumed that the problem is planar and that all sources lie in a half-space, thus reducing the location parameter space to a single-dimensional subset of $\Re$, *i.e.*, $\theta_i \in [0, \pi]$, where θ_i is the direction-of-arrival (DOA) of the ith source.

The signals are assumed to be *narrowband* processes, and can be considered to be sample functions of a stationary stochastic process or deterministic functions of time. In terms of estimating the signals themselves, these two assumptions lead to entirely different estimation algorithms. However, as far as estimation of signal parameters such as DOA is concerned, both assumptions lead to the same (suboptimal) algorithm under certain assumptions (*e.g., persistent excitation*).

Since the narrowband signals are assumed to have the same *known* center frequency[2] (ω_0), the ith signal can be written as

$$\tilde{s}_i(t) = u_i(t)\cos(\omega_0 t + v_i(t)), \tag{1}$$

[2]In the multiple signal environment, this condition is often termed *co-channel interference* in communication applications. If the center frequencies are not the same, the problem can be greatly simplified by first separating the signals in the frequency domain.

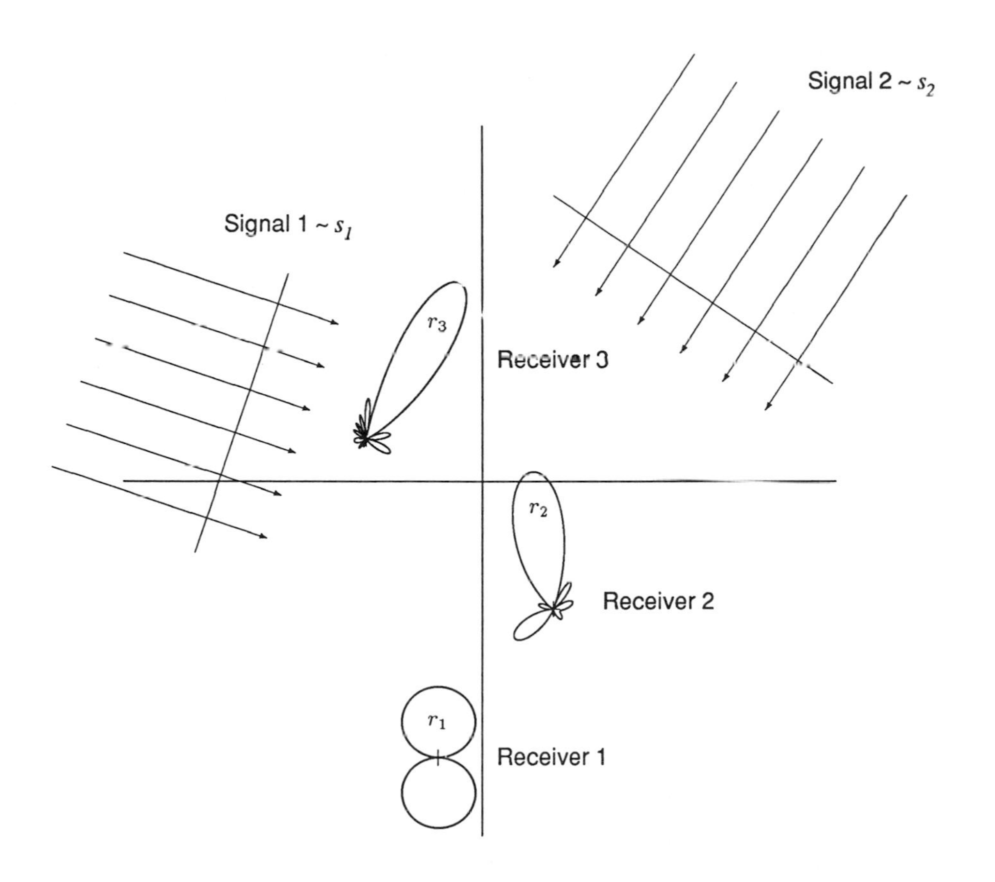

Figure 1: Passive Sensor Array Geometry

where $u_i(t)$ and $v_i(t)$ are slowly varying[3] functions of time that define the amplitude and phase of $\tilde{s}_i(t)$ respectively. For such signals, it is often more convenient to use the so-called *complex envelope* representation [3] in which $\tilde{s}_i(t) = \text{Re}\{s(t)\}$, where $s(t) = u(t)e^{j(\omega_0 t + v(t))}$. Noting that the narrowband assumption implies $u(t) \approx u(t-\tau)$ and $v(t) \approx v(t-\tau)$ for all possible propagation delays τ, the effect of a time-delay on the received waveforms is simply a phase shift, *i.e.*,

$$s(t-\tau) \approx s(t)e^{-j\omega_0\tau} . \tag{2}$$

The result is that $x_k(t)$, the complex signal output of the kth sensor at time t, can be written as

$$x_k(t) = \sum_{i=1}^{d} a_k(\theta_i)s_i(t-\tau_k(\theta_i)) = \sum_{i=1}^{d} a_k(\theta_i)s_i(t)e^{-j\omega_0\tau_k(\theta_i)} , \tag{3}$$

where $\tau_k(\theta_i)$ is the propagation delay between a reference point and the kth sensor for the ith wavefront impinging on the array from direction θ_i, $a_k(\theta_i)$ is the corresponding sensor element complex response (gain and phase) at frequency ω_0, and there are assumed to be d point sources present.

Employing vector notation for the outputs of the m sensors, the *data model* becomes

$$\mathbf{x}(t) = \sum_{i=1}^{d} \mathbf{a}(\theta_i)s_i(t) , \tag{4}$$

where

$$\mathbf{a}(\theta_i) = [a_1(\theta_i)e^{-j\omega_0\tau_1(\theta_i)}, \ldots, a_m(\theta_i)e^{-j\omega_0\tau_m(\theta_i)}]^T, \tag{5}$$

often termed the *array steering vector* for direction θ_i. Setting

$$\mathbf{A}(\boldsymbol{\theta}) = [\mathbf{a}(\theta_1) \vdots \cdots \vdots \mathbf{a}(\theta_d)], \tag{6}$$

$$\mathbf{s}(t) = [s_1(t), \ldots, s_d(t)]^T, \tag{7}$$

and adding measurement noise $\mathbf{n}(t)$, the *measurement model* for the passive sensor array narrowband signal processing problem becomes

$$\mathbf{z}(t) = \mathbf{A}(\boldsymbol{\theta})\mathbf{s}(t) + \mathbf{n}(t) , \tag{8}$$

where $\mathbf{z}(t), \mathbf{n}(t) \in \mathbb{C}^m$, $\mathbf{s}(t) \in \mathbb{C}^d$, and $\mathbf{A}(\boldsymbol{\theta}) \in \mathbb{C}^{m \times d}$, and it is assumed that $d < m$.

By employing the simplifying assumptions above, the DOA estimation problem, which is in general a complex problem in two independent variables, space and time, is reduced to a parameter estimation problem in one independent variable, time. The resulting measurement model is equivalent to one commonly employed in time series analysis. Therein, given a sequence of samples (indexed by a single real independent variable, usually time) from a series presumed to be composed of a superposition of

[3]The definition of slowly varying is taken to mean that the approximation $\tilde{s}_i(t-\tau_k(\theta_i)) \approx u_i(t)\cos(\omega_0(t-\tau_k(\theta_i)) + v_i(t))$, is valid, *i.e.*, the amplitude and phase variations as functions of spatial position for fixed t are *negligible* over the extent of the array.

sinusoids with possibly growing or decaying envelopes, the objective is to obtain estimates of the sinusoidal frequencies, the envelopes, and the relative strengths (powers) of all the components. Mathematically, the measurement model can be expressed as follows:

$$z(t) = \sum_{i=1}^{d} a_i e^{\sigma_i t} \cos(\omega_i t + \phi_i) + n(t) , \tag{9}$$

where a_i is the residue at the ith pole $\gamma_i = \sigma_i + j\omega_i$, and ϕ_i is the relative phase of the ith cisoidal component. The term $n(t)$ represents *additive* noise present in the measurements and accounts for such sources as thermal noise in amplifiers and quantization noise introduced in analog-to-digital (A/D) conversion. In a manner entirely analogous to the development of the DOA data model, the complex representation can be employed to simplify the notation. Assuming uniform sampling, any vector of m consecutive samples can be written as

$$\tilde{\mathbf{z}}(t) = \sum_{i=1}^{d} \mathbf{a}(\gamma_i)\tilde{s}_i(t) + \tilde{\mathbf{n}}(t) , \tag{10}$$

where $\tilde{s}_i(t) = a_i e^{(\gamma_i t + j\phi_i)}$ and $\mathbf{a}(\gamma_i) = [1 \ , \ e^{-\gamma_i D} \ , \ldots, \ e^{-\gamma_i (m-1)D}]$, a Vandermonde vector. Combining the column vectors, $\mathbf{a}(\gamma_i)$, into a matrix $\mathbf{A}(\boldsymbol{\gamma})$,

$$\tilde{\mathbf{z}}(t) = \mathbf{A}(\boldsymbol{\gamma})\tilde{\mathbf{s}}(t) + \tilde{\mathbf{n}}(t), \tag{11}$$

which is exactly the same form as equation (8). Recall that in the DOA estimation application, however, the columns of $\mathbf{A}$ are arbitrary array steering vectors. They are Vandermonde vectors only in the special case of a uniform linear array of sensors.

III. The Geometric Approach

In 1977, Schmidt [4] developed the **MUSIC** (**MU**ltiple **SI**gnal Classification) algorithm by taking a geometric view of the signal parameter estimation problem. As aforementioned, one of the major breakthroughs afforded by the **MUSIC** algorithm was the ability to handle arbitrary arrays of sensors. Until the mid-1970's, direction finding techniques required knowledge of the array directional sensitivity pattern in analytical form, and the task of the antenna designer was to build an array of antennae with a pre-specified sensitivity pattern. Inevitable imperfections and interelement interactions make this a difficult task. The work of Schmidt essentially relieved the designer from such constraints by exploiting the reduction in analytical complexity that could be achieved by *calibrating* the array. Thus, the highly nonlinear problem of calculating the array response to a signal from a given direction was reduced to that of measuring and storing the response. Although **MUSIC** did not mitigate the computational complexity of solution to the DOA estimation problem, it did extend the applicability of high-resolution DOA estimation to *arbitrary* arrays of sensors.

A. Array Manifolds and Signal Subspaces

To introduce the concepts of the *array manifold* and the *signal subspace*, recall the noise-free data model (*cf.* equation (4)):

$$\mathbf{x}(t) = \mathbf{A}(\boldsymbol{\theta})\mathbf{s}(t).$$

The vectors $\mathbf{a}(\theta_i) \in \mathbb{C}^m$, the columns of $\mathbf{A}(\boldsymbol{\theta})$, are elements of a set (not a subspace), termed the **array manifold**[4] ($\mathcal{A}$), composed of all array response (*steering*) vectors obtained as θ ranges over the entire parameter space. $\mathcal{A}$ is completely determined by the sensor directivity patterns and the array geometry, and can sometimes be computed analytically. For example, as discussed in Section II, in DOA estimation using uniform linear arrays as well as in time series analysis using a uniform tapped-delay line, the elements of $\mathcal{A}$ are Vandermonde vectors of the form

$$\mathbf{a}(\omega) = [1, \lambda, \lambda^2, \ldots, \lambda^{m-1}].$$

where the parameter λ is a (nonlinear) function of the signal parameters of interest (DOAs or cisoid frequencies). However, for complex arrays that defy analytical description, $\mathcal{A}$ can be obtained by calibration (*i.e.*, physical measurements). For azimuth only DOA estimation, the array manifold is a one-parameter manifold that can be viewed as a *rope* weaving through $\mathbb{C}^m$. For azimuth and elevation DOA estimation, the manifold is a *sheet* in $\mathbb{C}^m$. Note that to avoid ambiguities, it is necessary to assume that the map from θ to $\mathbf{a}(\theta)$ is one-to-one, a property that can be ensured by proper array design.[5]

The key observation is that if $\mathbf{x}(t) = \mathbf{a}(\theta)s_\theta(t)$ is an appropriate data model for a single signal, the data are confined to a *one-dimensional subspace* of $\mathbb{C}^m$ characterized by the vector $\mathbf{a}(\theta)$. For d signals, the observed data vectors $\mathbf{x}(t) = \mathbf{A}(\boldsymbol{\theta})\mathbf{s}(t)$ are constrained to the d-dimensional subspace of $\mathbb{C}^m$, termed the **signal subspace** ($\mathcal{S}_Z$), that is spanned by the d vectors $\mathbf{a}(\theta_i)$, the columns of $\mathbf{A}(\boldsymbol{\theta})$.[6]

B. Intersections as Solutions

The concepts of an observed signal subspace and a calibrated array manifold permit an immediate visualization of the solution. In the absence of noise, the outputs of the sensor array lie in a d-dimensional subspace of $\mathbb{C}^m$, the *signal subspace* ($\mathcal{S}_Z$) spanned by the columns of $\mathbf{A}(\boldsymbol{\theta})$. Once d independent vectors have been observed,[7] $\mathcal{S}_Z$ is known, and intersections between the observed subspace and the array manifold yield the set of vectors from the array manifold that span the observed signal subspace. A three-sensor, two-source example is graphically depicted in Figure 2. Assuming that the

[4]Technically, a *k-dimensional manifold* in $\mathbb{C}^m$ is a subset of points in $\mathbb{C}^m$ satisfying certain local continuity and differentiability conditions. The physics of sensor arrays guarantee the continuity and differentiability properties will be satisfied. Associated with each point on the manifold is a vector to that point from the origin in $\mathbb{C}^m$ using the standard basis.

[5]This condition is necessary, but not sufficient, to guarantee unique solutions in the multiple source environment. Higher order ambiguities can exist as discussed in detail in [1].

[6]It is also convenient to define the noise subspace ($\mathcal{S}_X^\perp$) as the orthogonal complement of the signal subspace, $\mathcal{S}_Z$, in $\mathbb{C}^m$.

[7]The problem of degenerate signal spaces, *i.e.*, fully correlated signals, is discussed in [1]. For the purposes of this discussion, it is assumed that the signals are not fully correlated.

sensor array has been designed such that the map from parameters to array manifold vectors is unique, the parameters are immediately determined.

Problems arise when only noisy measurements $\mathbf{z}(t) = \mathbf{A}(\theta)\mathbf{s}(t) + \mathbf{n}(t)$ of the array output are available, since $\mathcal{S}_Z$ must be estimated. Imposing the constraint that the estimate $\widehat{\mathcal{S}}_Z$ be spanned by elements from $\mathcal{A}$ and assuming unknown deterministic signals and Gaussian noise, a maximum-likelihood (ML) estimator can be formulated as described in Appendix D. However, the ML solution (of obtaining a set of vectors from the array manifold that *best fits* all the measurements) is computationally prohibitive in most practical applications.

Schmidt's idea was to employ a suboptimal *two-step* procedure instead. First, an *unconstrained* set of d vectors that best fits all the measurements is found. Then points of closest approach of the space spanned by those vectors to the array manifold are sought. This procedure, though clearly suboptimal, retains some of the key properties of the ML solution, including the fact that the *exact* answer is obtained asymptotically as the number of measurements goes to infinity.

C. Estimating the Signal Subspace

To obtain an unconstrained estimate of the signal subspace, the *least-squares* (LS) criterion is most often employed. The idea is to find a set of d vectors that span a subspace of $\mathbb{C}^m$ that *best fits*, in a LS sense, the observed data. Assuming the signals and noise are zero-mean, a method can be derived by first examining the covariance[8] matrix of the measurements. If the signals are modeled as stationary stochastic processes, they are assumed to be uncorrelated with the noise and possess a positive definite covariance matrix $\mathbf{R}_{SS} > 0$. If, on the other hand, a deterministic signal model is chosen, a persistent excitation condition, *i.e.*,

$$\mathbf{R}_{SS} \stackrel{\text{def}}{=} \lim_{N\to\infty} \frac{1}{N} \sum_{t=1}^{N} \mathbf{s}(t)\mathbf{s}^*(t) > 0,$$

is assumed. Without loss of generality, the covariance matrix of the measurement noise[9] ($\mathbf{R}_{NN}$) can be assumed to be $\sigma^2\mathbf{I}$, where σ may be unknown.

Under the conditions given above, the covariance matrix of the measurements is given by

$$\mathbf{R}_{ZZ} \stackrel{\text{def}}{=} E\{\mathbf{z}\mathbf{z}^*\} = \mathbf{A}\mathbf{R}_{SS}\mathbf{A}^* + \sigma^2\mathbf{I}. \tag{12}$$

The objective is to find a set of d linearly independent vectors that is contained in $\mathcal{S}_Z = \mathcal{R}\{\mathbf{A}\}$, the subspace spanned[10] by the columns of $\mathbf{A}$. One such set of vectors is easily obtained as the set of eigenvectors $\{\mathbf{e}_i, i = 1, \ldots, d\}$ of $\mathbf{R}_{ZZ}$ corresponding to the d largest eigenvalues, a fact that follows from elementary linear algebra as follows. First, any $m \times m$ unitary[11] matrix $\mathbf{E}$ can be taken as a matrix of eigenvectors for

[8]The sample covariance is easily seen to be a *sufficient statistic* for this least-squares estimation problem.

[9]The more general situation where the noise covariance $\mathbf{R}_{NN} = \sigma^2\Sigma_n$, where $\Sigma_n > 0$, leads naturally to generalized eigenproblems that can be avoided by *pre-whitening* the measurements by first filtering the data using $\Sigma_n^{-\frac{1}{2}}$.

[10]This subspace is also referred to as the *range* of $\mathbf{A}$.

[11]A matrix $\mathbf{E} \in \mathbb{C}^{m\times m}$ is unitary if and only if $\mathbf{E}\mathbf{E}^* = \mathbf{I}$.

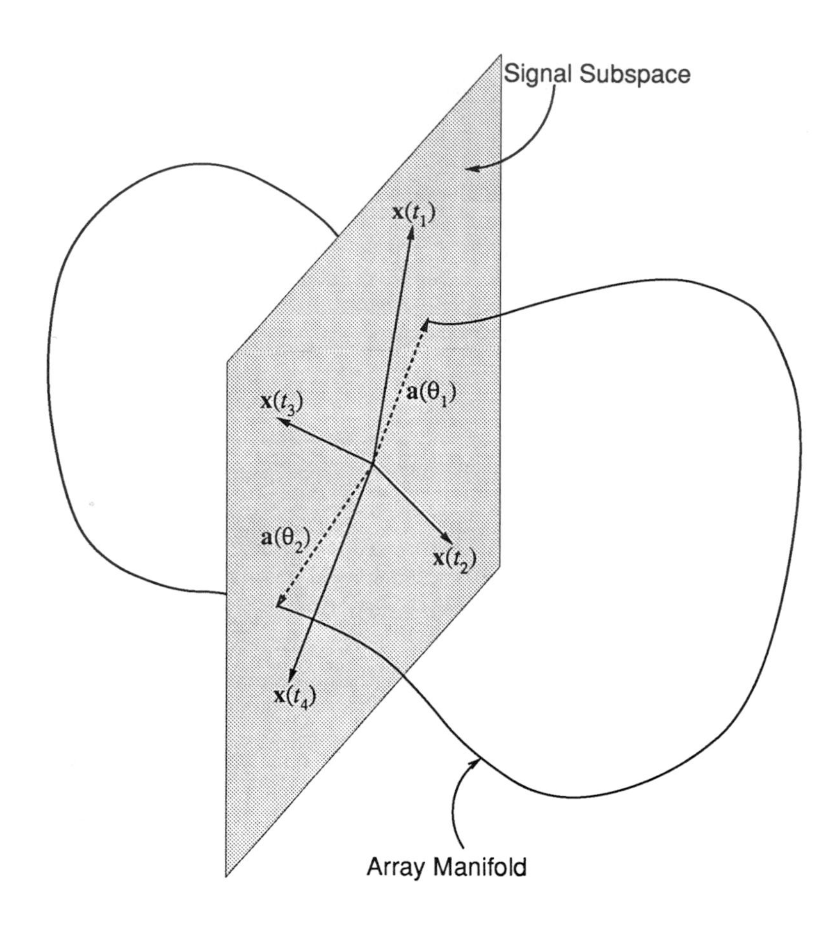

Figure 2: The Geometry of MUSIC for a Three-Sensor/Two-Source Example (No Noise)

$\sigma^2\mathbf{I}$ since the identity matrix commutes with any matrix, *i.e.,* $\mathbf{EIE}^* = \mathbf{I}$ for unitary $\mathbf{E}$. Thus, any set of eigenvectors of $\mathbf{AR}_{SS}\mathbf{A}^*$ is also a set[12] of eigenvectors of $\mathbf{AR}_{SS}\mathbf{A}^*+\sigma^2\mathbf{I}$. Since $\mathbf{AR}_{SS}\mathbf{A}^*$ is rank d and positive semi definite by construction, there are d positive eigenvalues and $m-d$ zero eigenvalues. Thus, the d largest eigenvalues of $\mathbf{R}_{ZZ}$ are simply the d non-zero eigenvalues of $\mathbf{AR}_{SS}\mathbf{A}^*$ augmented by σ^2, and the associated eigenvectors lie in the subspace spanned by the columns of $\mathbf{A}$. Note also that the $m-d$ smallest eigenvalues of $\mathbf{R}_{ZZ}$ are all equal to σ^2, a fact that can be used to find d and σ. Thus, $\mathcal{R}\{\mathbf{E}_z\} = \mathcal{S}_z$ where $\mathbf{E}_z \stackrel{\text{def}}{=} [\mathbf{e}_1 \mid \cdots \mid \mathbf{e}_d]$. Similarly defining $\mathbf{E}_n \stackrel{\text{def}}{=} [\mathbf{e}_{d+1} \mid \cdots \mid \mathbf{e}_m]$, $\mathcal{R}\{\mathbf{E}_n\} = \mathcal{S}_X^{\perp}$.

In most situations, the covariance matrices required above are not known and must be estimated from measurements. Practically, the noise covariance can be estimated from measurements made when signals of interest are not present. Measurements made when signals of interest are present are used to estimate $\mathbf{R}_{ZZ}$. The standard estimate

$$\widehat{\mathbf{R}}_{ZZ} = \frac{1}{N}\sum_{t=1}^{N} \mathbf{z}(t)\mathbf{z}^*(t),$$

is most often employed, and in terms of the $m \times N$ data matrix $\mathbf{Z}$,

$$\widehat{\mathbf{R}}_{ZZ} = \frac{1}{N}\mathbf{ZZ}^*.$$

Note that this estimate can be scaled by $N/(N-m)$ to obtain an unbiased estimate, but the scaling only effects the magnitudes of the eigenvalues, not the eigenvectors, so the subspace estimates remain unaffected.

An alternative to first forming the measurement covariance matrix and then performing an eigendecomposition is to operate directly on the measurements using the singular value decomposition. In addition to avoiding *squaring* the data, this approach has a nice geometric interpretation. Letting $\mathbf{Z} \in \mathbb{C}^{m\times N}$ denote the data matrix, the objective is to obtain a set of vectors spanning the column space of the rank d matrix, $\hat{\mathbf{Z}} \in \mathbb{C}^{m\times N}$, that best approximates $\mathbf{Z}$ in a least-squares sense. The solution is given by the d left singular vectors of $\mathbf{Z}$ corresponding to the d largest singular values.[13]

It is easily demonstrated that the eigendecomposition and the SVD yield the same subspace estimate. If the SVD of $\mathbf{Z}/\sqrt{N}$ is given by $\mathbf{U\Sigma V}^*$,

$$\frac{1}{N}\mathbf{ZZ}^* = \mathbf{U\Sigma}^2\mathbf{U}^* = \widehat{\mathbf{R}}_{ZZ},$$

since $\mathbf{\Sigma}$ is diagonal and real, and $\mathbf{U}$ and $\mathbf{V}$ are unitary. Thus, the left singular vectors, $\mathbf{U}$, of $\mathbf{Z}$ are the right eigenvectors of $\widehat{\mathbf{R}}_{ZZ}$, the sample covariance matrix. Thus, in the absence of finite precision effects in computing the decompositions, the subspace estimates from both techniques are identical.

[12] Since there are eigenvalues with multiplicity greater than one, the eigenvectors are not unique. However, the invariant subspaces spanned by eigenvectors corresponding to distinct eigenvalues are unique, and this is all that is required.

[13] A discussion of this problem can be found in [2, pp. 19-20], though therein the 2-norm is employed. The extension to the Frobenius norm (*cf.* total least-squared error) is straightforward.

There are, however, significant computational differences. The computation of the full SVD of $\mathbf{Z}$ is of order mN^2 which can be significantly larger than the $O(m^3)$ operations required for an eigendecomposition of $\widehat{\mathbf{R}}_{ZZ}$. The increase in computation is due to the fact that the full SVD is also obtaining a set of d vectors, the first d columns of $\mathbf{V} \in \mathbb{C}^{N \times N}$, that span the d-dimensional subspace of N-dimensional space spanned by the d signal vectors, vectors in $\mathbb{C}^N$ whose components are the samples of the underlying signals.[14] If the information of the full SVD is not required, *partial* SVD algorithms that compute only the left singular vectors and singular values can be employed resulting in substantial computational savings [5].

If N is the number of measurements to be processed, and m is the number of elements in each sample vector, forming the sample covariance matrix requires on the order of Nm^2 operations. Eigendecompositions of matrices in $\Re^{m \times m}$ require on the order of $10m^3$ operations, whereas a standard SVD of $\mathbf{Z} \in \Re^{m \times N}$ requires $\approx 2Nm^2 + 4m^3$ (*cf.* [2, page 175]) if only the singular values and left singular vectors are required. The computational effort is therefore on the same order for both the SVD and eigenvector approaches to reducing the measurements to the statistic $\mathbf{E}_z$ used by ***ESPRIT*** and **MUSIC**. Again, the major advantage of the SVD over the eigendecomposition is that the measurements are processed directly without *squaring* them. Thus, numerical problems associated with ill-conditioned matrices are mitigated to some extent by using the SVD.

When the covariances must be estimated from finite data matrices (or the data matrices used directly), the $m - d$ smallest eigenvalues (singular values) are clustered around, but not all equal to σ^2 (σ). In this case, special statistical techniques based on likelihood ratio (*LR*) tests (*cf.* Appendix E) can be used to obtain an estimate $\hat{d}$ of the signal subspace dimension. In any case, in the presence of a finite amount of noisy measurements, $\mathcal{R}\{\mathbf{E}_z\} = \widehat{\mathcal{S}}_Z$ and $\mathcal{R}\{\mathbf{E}_n\} = \widehat{\mathcal{S}}_Z^{\perp}$, the *estimated signal and noise subspaces.*

D. Estimating the Signal Parameters

In the absence of noise, parameter estimates can be obtained by finding *intersections* of $\mathcal{A}$ with $\widehat{\mathcal{S}}_Z$, or equivalently finding elements of $\mathcal{A}$ that are orthogonal to $\widehat{\mathcal{S}}_Z^{\perp}$. At this point in the suboptimal signal subspace algorithms, the real computational effort[15] begins. Even with perfect knowledge of the signal subspace, searching the array manifold for d intersections with $\mathcal{S}_Z$ can be quite costly, especially for multi-dimensional parameter spaces (*e.g.,* azimuth, elevation, and range).

The problem is further complicated in the presence of noise since, with probability one, $\widehat{\mathcal{S}}_Z \cap \mathcal{A} = \emptyset$; there are no intersections. Consequently there are no elements of $\mathcal{A}$ that are orthogonal to $\widehat{\mathcal{S}}_Z^{\perp}$. Referring to Figure 2, it would seem that intersections could almost always be found. In this respect, the figure is somewhat misleading. For three sensors and two sources, the signal subspace is a two-dimensional *complex* subspace

[14] In the presence of noise, these d vectors are biased estimates due to the fact that for each measurement taken, another signal vector must be estimated. Thus, there is no *averaging of the noise.*

[15] Only for extremely large arrays such as phased array radars with hundreds of elements is the eigendecomposition of the measurement covariance matrix computationally significant in comparison to the search over $\mathcal{A}$.

of three-dimensional *complex* space. In the real field, the estimated signal subspace is actually a four-dimensional subspace of six-dimensional space, and it need not intersect the one-dimensional manifold at all. Obviously, elements of $\mathcal{A}$ that are *closest* to $\widehat{\mathcal{S}}_Z$ should be considered as potential solutions, but the issue of an appropriate measure of *closeness* remains.

Schmidt [4] proposed the following function as one possible measure[16] of the *closeness* of an element of $\mathcal{A}$ to $\widehat{\mathcal{S}}_Z$:

$$P_M(\theta) = \frac{\mathbf{a}^*(\theta)\mathbf{a}(\theta)}{\mathbf{a}^*(\theta)\mathbf{E}_N\mathbf{E}_N^*\mathbf{a}(\theta)}. \tag{13}$$

In the absence of noise, this measure, termed the **MUSIC** *spectrum*, is infinite for elements of $\mathcal{A}$ belonging to $\mathcal{S}_Z$. In the presence of noise, this measure is clearly peaked near points of closest approach of $\mathcal{A}$ to $\widehat{\mathcal{S}}_Z$, points where $\mathcal{A}$ is nearly orthogonal to $\widehat{\mathcal{S}}_Z^{\perp}$. This property is used to obtain parameter estimates as those parameters, or parameter vectors in the case of multi-dimensional array manifolds, that yield the d largest *peaks in the spectrum.*

Though conceptually simple, the *one-dimensional*[17] **MUSIC** measure has several drawbacks. Primarily, problems in the finite measurement case arise from the fact that since d signals are known to be present, d parameter estimates, $\{\theta_1, \ldots, \theta_d\}$, should be sought *simultaneously* by maximizing an appropriate functional (*cf.* [1]) rather than obtaining estimates one at a time as is done in the search over $P_M(\theta)$. However, multi-dimensional searches are exponentially more expensive than one-dimensional searches. The price paid for the computational reduction achieved by employing a *one-dimensional search* for d parameters is that the method is finite-sample-biased in the multiple source environment (*cf.* [1]).

Problems also arise in low SNR scenarios, and in situations where even small sensor array errors are present. The ability of the conventional **MUSIC** *spectrum* to *resolve* closely spaced sources (*i.e.*, observe multiple peaks in the measure) is severely degraded (*cf.* [1]). Fundamentally, the problem is the same as the cisoidal frequency resolution problem in time series analysis using DFTs. As discussed in detail in [1], this *lack of resolution* is a consequence of the one-dimensional nature of the measure. As with the finite-sample-bias, the problem is mitigated by proper choice of a *multi-dimensional measure* of closeness of d elements of $\mathcal{A}$ to $\widehat{\mathcal{S}}_Z$. Nevertheless, it should be emphasized that in spite of these drawbacks, **MUSIC** has been shown to outperform previous techniques (*cf.* [6]).

Finally, as indicated above, **MUSIC** asymptotically yields unbiased parameter estimates since as the amount of data becomes infinite, errors in the estimate $\widehat{\mathcal{S}}_Z$ vanish. If the noise is spatially Gaussian and temporally independent, the distribution of the eigenvectors of the sample covariance is asymptotically Gaussian, with mean equal to the true eigenvectors (assuming distinct eigenvalues) and covariances that go

[16] In [4], the numerator was not explicitly included in the measure since it was assumed that the array manifold vectors were all *normalized in some suitable fashion.*

[17] The conventional **MUSIC** measure is herein referred to as a *one-dimensional* measure, though the search over $\mathcal{A}$ is potentially a multi-dimensional one, the dimension being that of the parameter vector (*e.g.*, three for a parameter vector consisting of range, azimuth and elevation).

to zero [7, 8]. Thus, the estimated signal subspace *converges in mean-square* to the true signal subspace, and the parameter estimates converge to the true values as well.

E. Summary of the MUSIC *Algorithm*

The following is a summary of the **MUSIC** algorithm based on the covariance[18] approach described above.

1. Collect the data and estimate $\mathbf{R}_{ZZ} = E\{\mathbf{z}\mathbf{z}^*\} = \mathbf{A}\mathbf{R}_{SS}\mathbf{A}^* + \sigma^2\mathbf{I}$ denoting the estimate $\widehat{\mathbf{R}}_{ZZ}$.

2. Solve for the eigensystem;
$$\widehat{\mathbf{R}}_{ZZ}\bar{\mathbf{E}} = \bar{\mathbf{E}}\mathbf{\Lambda},$$
where $\mathbf{\Lambda} = \text{diag}\{\lambda_1, \ldots, \lambda_m\}$, $\lambda_1 \geq \cdots \geq \lambda_m$, and $\bar{\mathbf{E}} \stackrel{\text{def}}{=} [\mathbf{e}_1 \,\vdots\, \cdots \,\vdots\, \mathbf{e}_m]$.

3. Estimate the multiplicity of $\lambda_{\min}$, *i.e.*, estimate $m - d$.

4. Evaluate
$$P_M(\theta) = \frac{\mathbf{a}^*(\theta)\mathbf{a}(\theta)}{\mathbf{a}^*(\theta)\mathbf{E}_N\mathbf{E}_N^*\mathbf{a}(\theta)},$$
where $\mathbf{E}_n = [\mathbf{e}_{d+1} \,\vdots\, \cdots \,\vdots\, \mathbf{e}_m]$.

5. Find the d (largest) peaks of $\mathbf{P}(\theta)$ to obtain estimates of the parameters.

IV. *ESPRIT*

Though **MUSIC** was the first of the high-resolution algorithms to correctly exploit the underlying data model of narrowband signals in additive noise, the algorithm has several limitations including the fact that complete knowledge of the array manifold is required, and that the search over parameter space is computationally very expensive. In this section, an approach (***ESPRIT***) to the signal parameter estimation problem that exploits sensor array invariances is described.[19] ***ESPRIT*** is similar to **MUSIC** in that it correctly exploits the underlying data model, while manifesting significant advantages over **MUSIC** as described in Section I.

A. Array Geometry

ESPRIT retains most of the essential features of the *arbitrary* array of sensors, but achieves a significant reduction in computational complexity by imposing a constraint on the structure of the sensor array, a constraint most easily described by an example. Consider a planar array of arbitrary geometry composed of m sensor *doublets* as shown in Figure 3. The elements in each doublet have identical sensitivity patterns and are

[18] Alternatively, a signal subspace estimate can be obtained by performing a (generalized) SVD of the data matrix $\mathbf{Z} = [\mathbf{z}(t_1) \,\vdots\, \cdots \,\vdots\, \mathbf{z}(t_N)]$. The resulting subspace estimate is the same; the choice of which approach to use is made based on external factors such as numerical and hardware implementation considerations.

[19] A patent has been issued on the sensor array design and concepts embodied in ***ESPRIT*** [9].

translationally separated by a known constant displacement vector $\mathbf{\Delta}$. Other than the obvious requirement that each sensor have non-zero sensitivity in all directions of interest, the gain, phase, and polarization sensitivity of the elements in the doublet are arbitrary. Furthermore, there is no requirement that any of the doublets possess the same sensitivity patterns, though as discussed in [1, 10], there are advantages to employing arrays with such characteristics.

B. The Data Model

Assume that there are $d \leq m$ narrowband sources[20] centered at frequency ω_0, and that the sources are located sufficiently far from the array such that in homogeneous isotropic transmission media, the wavefronts impinging on the array are planar. As before, the sources may be assumed to be stationary zero-mean random processes or deterministic signals. Additive noise is present at all $2m$ sensors and is assumed to be a stationary zero-mean random process with a *spatial* covariance $\sigma^2\mathbf{I}$. There is no loss of generality in comparison with the assumption that the covariance is $\sigma^2\mathbf{\Sigma}_n$. As long as $\mathbf{\Sigma}_n$ is known, *pre-whitening* of the measurement noise can be performed.

To describe mathematically the effect of the *translational invariance* of the sensor array, it is convenient to describe the array as being comprised of two subarrays, Z_X and Z_Y, identical in every respect although physically displaced (not rotated) from each other by a known displacement vector $\mathbf{\Delta}$. The signals received at the ith doublet can then be expressed as

$$
\begin{aligned}
x_i(t) &= \sum_{k=1}^{d} s_k(t) a_i(\theta_k) + n_{x_i}(t)\,, \\
y_i(t) &= \sum_{k=1}^{d} s_k(t) e^{j\omega_0 \Delta \sin\theta_k / c} a_i(\theta_k) + n_{y_i}(t)\,,
\end{aligned}
\tag{14}
$$

where θ_k is now the direction of arrival of the kth source relative to the direction of the translational displacement vector $\mathbf{\Delta}$.

Since the sensor gain and phase patterns are arbitrary and since ***ESPRIT*** does not require any knowledge of the sensitivities, the subarray displacement vector $\mathbf{\Delta}$ sets not only the scale for the problem, but the *reference direction* as well. The DOA estimates obtained are angles-of-arrival with respect to the direction of the vector $\mathbf{\Delta}$, and they are obtained as nonlinear functions of the ***ESPRIT*** *electrical* angles, angles that are linear functions of $\Delta = \|\mathbf{\Delta}\|_2$. A natural consequence of this fact is the necessity for a corresponding displacement vector for each *dimension* in which parameter estimates are desired.

Combining the outputs of each of the sensors in the two subarrays, the received data vectors can be written as follows:

$$\mathbf{x}(t) = \mathbf{A}\mathbf{s}(t) + \mathbf{n}_x(t)\,, \tag{15}$$

$$\mathbf{y}(t) = \mathbf{A}\mathbf{\Phi}\mathbf{s}(t) + \mathbf{n}_y(t)\,, \tag{16}$$

[20] **MUSIC** imposes the requirement $d < 2m$, and can therefore can handle roughly twice as many sources as ***ESPRIT*** in general. For uniform linear arrays, however, ***ESPRIT*** can handle as many sources as **MUSIC** [1] by employing *overlapping* subarrays.

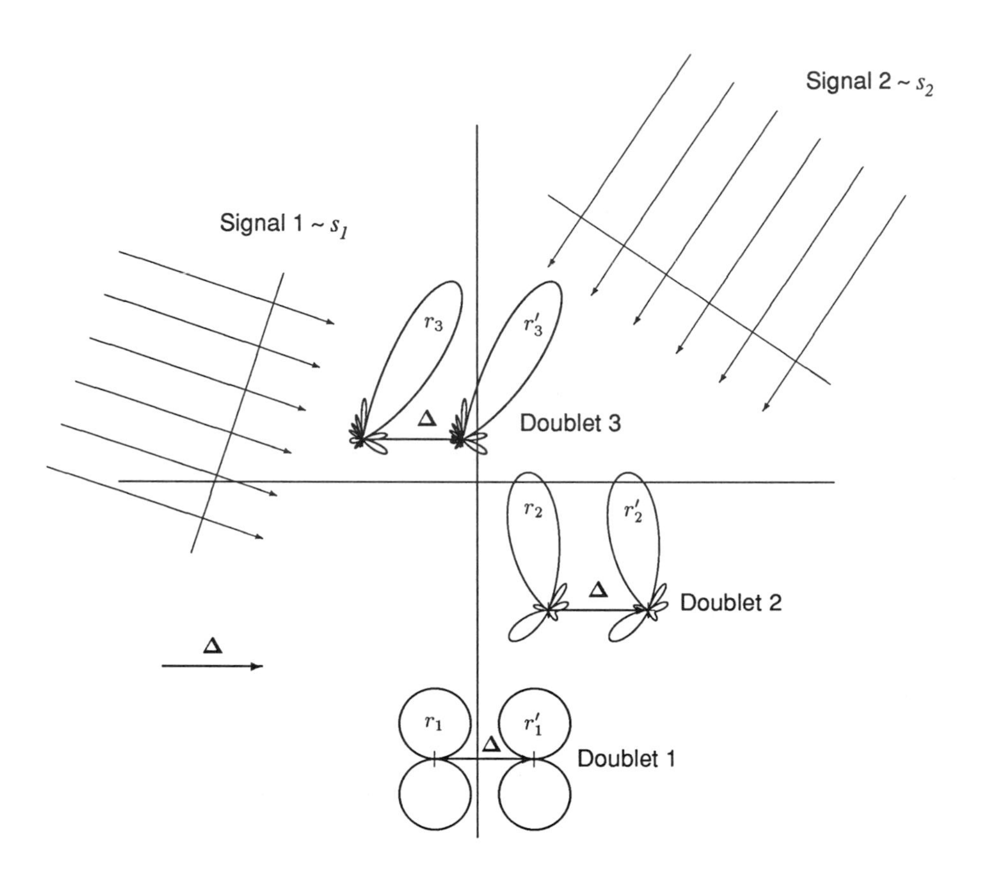

Figure 3: Sensor Array Geometry for Multiple Source DOA Estimation Using ***ESPRIT***

where the vector $\mathbf{s}(t)$ is the $d \times 1$ vector of impinging signals (wavefronts) as observed at the reference sensor of subarray X. The individual signals can be temporally correlated, *i.e.*, $E\{s_i(t)s_j^*(t)\} \neq 0$, $i \neq j$, though the case of coherent (or fully correlated) sources is not considered herein (*cf.* [1, Section 7.11] for a complete discussion). The matrix $\mathbf{\Phi}$ is a diagonal $d \times d$ matrix of the phase delays between the doublet sensors for the d wavefronts, and is given by

$$\mathbf{\Phi} = \text{diag}[\, e^{j\gamma_1}, \ldots, e^{j\gamma_d} \,], \tag{17}$$

where

$$\gamma_k = \omega_0 \Delta \sin \theta_k / c. \tag{18}$$

$\mathbf{\Phi}$ is a unitary matrix (operator) that relates the measurements from subarray Z_X to those from subarray Z_Y. In the complex field, $\mathbf{\Phi}$ is a simple scaling operator. However, it is isomorphic to the real two-dimensional rotation operator and is herein referred to as a *rotation*[21] operator. The unitary nature of $\mathbf{\Phi}$ is a consequence of the narrowband planewave assumption, an assumption that leads to unit-modulus cisoidal *signals* in the spatial domain. In time series analysis, the diagonal elements of $\mathbf{\Phi}$ are potentially arbitrary complex numbers in which case $\mathbf{\Phi}$ could be an *expansive* or *contractive* operator.

Defining the total array output vector as $\mathbf{z}(t)$, the subarray outputs can be combined to yield

$$\mathbf{z}(t) = \begin{bmatrix} \mathbf{x}(t) \\ \mathbf{y}(t) \end{bmatrix} = \bar{\mathbf{A}}\mathbf{s}(t) + \mathbf{n}_z(t), \tag{19}$$

$$\bar{\mathbf{A}} = \begin{bmatrix} \mathbf{A} \\ \mathbf{A\Phi} \end{bmatrix}, \quad \mathbf{n}_z(t) = \begin{bmatrix} \mathbf{n}_x(t) \\ \mathbf{n}_y(t) \end{bmatrix}. \tag{20}$$

It is the structure of $\bar{\mathbf{A}}$ that is exploited to obtain estimates of the diagonal elements of $\mathbf{\Phi}$ *without having to know* $\mathbf{A}$.

Scale Invariance

From equation (19), it is easily seen that the estimation problem posed is *scale-invariant* in the sense that absolute signal powers are not observable. For any nonsingular diagonal matrix, $\mathbf{D}$, the data model is invariant with respect to the transformations $\mathbf{s}(t) \to \mathbf{D}^{-1}\mathbf{s}(t)$ and $\bar{\mathbf{A}} \to \bar{\mathbf{A}}\mathbf{D}$. Thus, estimates of the signals and the associated array manifold vectors derived herein are to be interpreted modulo an arbitrary scale factor unless knowledge of the gain pattern of *one* of the sensors is available.

C. *ESPRIT* – *The Invariance Approach*

The basic idea behind ***ESPRIT*** is to exploit the *rotational* invariance of the underlying signal subspace induced by the translational invariance of the sensor array. The relevant signal subspace is the one that contains the outputs from the two subarrays

[21] This is the origin of the term *rotational* in the acronym ***ESPRIT***.

described above, Z_X and Z_Y. Simultaneous sampling of the output of the arrays leads to two sets of vectors, $\mathbf{E}_X$ and $\mathbf{E}_Y$, that span the same signal subspace (ideally, that spanned by the columns of $\mathbf{A}$).

The ***ESPRIT*** algorithm is based on the following results for the case in which the underlying $2m$-dimensional signal subspace containing the entire array output is known. In the absence of noise, the signal subspace can be obtained as before by collecting a sufficient number of measurements and finding *any set of d linearly independent vectors* that span the same column space, the subspace spanned by $\bar{\mathbf{A}}$. The signal subspace can also be obtained from knowledge of the covariance of the measurements $\mathbf{R}_{ZZ} = \bar{\mathbf{A}}\mathbf{R}_{SS}\bar{\mathbf{A}}^* + \sigma^2\mathbf{I}$. As shown in Section C., if $d \leq m$ (an assumption required in later developments), the $2m - d$ smallest eigenvalues of $\mathbf{R}_{ZZ}$ are equal to σ^2. The d eigenvectors ($\mathbf{E}_Z$) corresponding to the d largest eigenvalues satisfy $\mathcal{R}\{\mathbf{E}_Z\} = \mathcal{R}\{\bar{\mathbf{A}}\}$, the signal subspace. Also as shown in Section C., the *same* set of vectors $\mathbf{E}_Z$ can be obtained by performing a singular value decomposition (SVD) of the data matrix $\mathbf{Z}$.

Since $\mathcal{R}\{\mathbf{E}_Z\} = \mathcal{R}\{\bar{\mathbf{A}}\}$, there must exist a unique (recall $d \leq m$), nonsingular $\mathbf{T}$ such that

$$\mathbf{E}_Z = \bar{\mathbf{A}}\mathbf{T}, \tag{21}$$

Furthermore, the invariance structure of the array implies $\mathbf{E}_Z$ can be decomposed into $\mathbf{E}_X \in \mathbb{C}^{m\times d}$ and $\mathbf{E}_Y \in \mathbb{C}^{m\times d}$ (*cf.* X and Y subarrays) such that

$$\mathbf{E}_X = \mathbf{A}\mathbf{T}, \tag{22}$$

$$\mathbf{E}_Y = \mathbf{A}\mathbf{\Phi}\mathbf{T}. \tag{23}$$

which implies that

$$\mathcal{R}\{\mathbf{E}_X\} = \mathcal{R}\{\mathbf{E}_Y\} = \mathcal{R}\{\mathbf{A}\}.$$

Since $\mathbf{E}_X$ and $\mathbf{E}_Y$ share a common column space, the rank of

$$\mathbf{E}_{XY} \stackrel{\text{def}}{=} [\,\mathbf{E}_X \,\vdots\, \mathbf{E}_Y\,] \tag{24}$$

is d, which implies there exists a unique (recall $d \leq m$) rank d matrix[22] $\mathbf{F} \in \mathbb{C}^{2d\times d}$ such that

$$\mathbf{0} = [\,\mathbf{E}_X \,\vdots\, \mathbf{E}_Y\,]\,\mathbf{F} = \mathbf{E}_X\mathbf{F}_X + \mathbf{E}_Y\mathbf{F}_Y\,, \tag{25}$$

$$= \mathbf{A}\mathbf{T}\mathbf{F}_X + \mathbf{A}\mathbf{\Phi}\mathbf{T}\mathbf{F}_Y\,. \tag{26}$$

$\mathbf{F}$ spans the *null-space* of $[\,\mathbf{E}_X \,\vdots\, \mathbf{E}_Y\,]$. Defining

$$\mathbf{\Psi} \stackrel{\text{def}}{=} -\mathbf{F}_X[\mathbf{F}_Y]^{-1}\,, \tag{27}$$

equation (26) can be rearranged to yield[23]

$$\mathbf{A}\mathbf{T}\mathbf{\Psi} = \mathbf{A}\mathbf{\Phi}\mathbf{T} \Longrightarrow \mathbf{A}\mathbf{T}\mathbf{\Psi}\mathbf{T}^{-1} = \mathbf{A}\mathbf{\Phi}\,. \tag{28}$$

---

[22] This derivation, though somewhat more lengthy than at first glance seems necessary, will prove useful when noisy estimates of $\mathbf{E}_X$ and $\mathbf{E}_Y$ are available.

[23] The same argument used in deriving equation (21) can be used to derive equation (28) directly. However, the derivation only insures the existence and uniqueness of such a full-rank $\mathbf{\Psi}$. The advantage of the preceding derivation is that implicitly a prescription for obtaining $\mathbf{\Psi}$ is given. Note that the existence and uniqueness of a full-rank $\mathbf{\Psi}$ guarantees the invertibility of $\mathbf{F}_Y$.

Assuming $\mathbf{A}$ to be full rank implies

(29)
$$\boxed{\mathbf{T}\mathbf{\Psi}\mathbf{T}^{-1} = \mathbf{\Phi}.}$$

Therefore, the eigenvalues of $\mathbf{\Psi}$ must be equal to the diagonal elements of $\mathbf{\Phi}$ and the columns of $\mathbf{T}$ are the eigenvectors of $\mathbf{\Psi}$. This is the key relationship in the development of TLS ***ESPRIT*** and its properties. The signal parameters are obtained as nonlinear functions of the eigenvalues of the operator $\mathbf{\Psi}$ that maps (rotates) one set of vectors ($\mathbf{E}_X$) that span an m-dimensional signal subspace into another ($\mathbf{E}_Y$).

D. *Estimating the Subspace Rotation Operator*

In practical situations where only a finite number of noisy measurements are available, $\mathbf{E}_Z$ is estimated from the covariance matrices of the measurements, $\mathbf{R}_{ZZ}$, or equivalently from the data matrix $\mathbf{Z}$. The result is that $\mathcal{R}\{\mathbf{E}_Z\}$ is only an estimate of $\mathcal{S}_Z$, and with probability one, $\mathcal{R}\{\mathbf{E}_Z\} \neq \mathcal{R}\{\bar{\mathbf{A}}\}$. Furthermore, $\mathcal{R}\{\mathbf{E}_X\} \neq \mathcal{R}\{\mathbf{E}_Y\}$. Thus the objective of finding a $\mathbf{\Psi}$ such that $\mathbf{E}_X\mathbf{\Psi} = \mathbf{E}_Y$ is no longer achievable. A criterion for obtaining a suitable estimate must be formulated. The most commonly employed criterion for problems of this nature is the *least-squares* (LS) criterion.

The standard LS criterion applied to the model $\mathbf{AX} = \mathbf{B}$ to obtain an estimate of $\mathbf{X}$ assumes $\mathbf{A}$ is known and the error is to be attributed to $\mathbf{B}$. Assuming the set of equations is overdetermined, the columns of $\mathbf{A}$ are linearly independent, and the noise in $\mathbf{B}$ is zero mean with covariance proportional to $\mathbf{I}$, the LS solution is

$$\widehat{\mathbf{X}} = [\mathbf{A}^*\mathbf{A}]^{-1}\mathbf{A}^*\mathbf{B}.$$

It is easily verified that the estimate is unbiased and minimum variance. If both $\mathbf{A}$ and $\mathbf{B}$ are *noisy* however, the LS solution is known to be biased.

Since it is not difficult to argue that the estimates $\mathbf{E}_X$ and $\mathbf{E}_Y$ are *equally* noisy, the LS criterion is clearly inappropriate. A criterion that takes into account noise on both $\mathbf{A}$ and $\mathbf{B}$ is the *total least-squares* (TLS) criterion. The TLS criterion can be stated [2] as finding *residual* matrices $\mathbf{R_A}$ and $\mathbf{R_B}$ of minimum Frobenius norm, and $\hat{\mathbf{X}}$ such that

(30)
$$[\mathbf{A} + \mathbf{R_A}]\dot{\mathbf{X}} = \mathbf{B} + \mathbf{R_B}.$$

This criterion is equivalent (*cf.* Appendix A) to replacing the zero matrix in equation (25) by a matrix of errors, the Frobenius norm (*i.e., total* least-squared error) of which is to be minimized. Appending a nontriviality constraint $\mathbf{F}^*\mathbf{F} = \mathbf{I}$ to eliminate the zero solution and applying standard Lagrange techniques (*cf.* Appendix B) leads to a solution for $\mathbf{F}$ given by the eigenvectors corresponding to the d smallest eigenvalues of $\mathbf{E}_{XY}^*\mathbf{E}_{XY}$. The eigenvalues of $\mathbf{\Psi}$ as defined above and calculated from the estimates $\mathbf{F}_X$ and $\mathbf{F}_Y$ are taken as estimates of the diagonal elements of $\mathbf{\Phi}$.

E. *Summary of the TLS* **ESPRIT** *Covariance Algorithm*

The TLS ***ESPRIT*** algorithm based on a covariance formulation can be summarized as follows.

1. Obtain an estimate of $\mathbf{R}_{ZZ}$, denoted $\widehat{\mathbf{R}}_{ZZ}$, from the measurements $\mathbf{Z}$.

2. Compute the eigen-decomposition of $\widehat{\mathbf{R}}_{ZZ}$

$$\widehat{\mathbf{R}}_{ZZ}\bar{\mathbf{E}} = \bar{\mathbf{E}}\mathbf{\Lambda}\,,$$

where $\mathbf{\Lambda} = \mathrm{diag}\{\lambda_1, \ldots, \lambda_{2m}\}$, $\lambda_1 \geq \cdots \geq \lambda_{2m}$, and $\bar{\mathbf{E}} = [\mathbf{e}_1 \,\vdots\, \cdots \,\vdots\, \mathbf{e}_{2m}]$.

3. Estimate the number of sources $\hat{d}$ (*cf.* Appendix E).

4. Obtain the signal subspace estimate $\widehat{\mathcal{S}}_Z = \mathcal{R}\{\mathbf{E}_Z\}$ where

$$\mathbf{E}_Z \stackrel{\text{def}}{=} [\mathbf{e}_1 \,\vdots\, \cdots \,\vdots\, \mathbf{e}_{\hat{d}}] = \begin{bmatrix} \mathbf{E}_X \\ \mathbf{E}_Y \end{bmatrix}.$$

5. Compute the eigendecomposition $(\lambda_1 > \ldots > \lambda_{2\hat{d}})$,

$$\mathbf{E}_{XY}^*\mathbf{E}_{XY} \stackrel{\text{def}}{=} \begin{bmatrix} \mathbf{E}_X^* \\ \mathbf{E}_Y^* \end{bmatrix} [\mathbf{E}_X \,\vdots\, \mathbf{E}_Y] = \mathbf{E}\mathbf{\Lambda}\mathbf{E}^*,$$

and partition $\mathbf{E}$ into $\hat{d} \times \hat{d}$ submatrices,

$$\mathbf{E} \stackrel{\text{def}}{=} \begin{bmatrix} \mathbf{E}_{11} & \mathbf{E}_{12} \\ \mathbf{E}_{21} & \mathbf{E}_{22} \end{bmatrix}.$$

6. Calculate the eigenvalues of $\mathbf{\Psi} = -\mathbf{E}_{12}\mathbf{E}_{22}^{-1}$,

$$\widehat{\phi}_k = \lambda_k(-\mathbf{E}_{12}\mathbf{E}_{22}^{-1}), \;\; \forall\, k = 1, \ldots, \hat{d}\,.$$

7. Estimate $\hat{\theta}_k = f^{-1}(\widehat{\phi}_k)$; *e.g.*, for DOA estimation, $\hat{\theta}_k = \sin^{-1}\{c \arg(\widehat{\phi}_k)/(\omega_0\Delta)\}$.

As alluded to earlier, in many instances it is preferable to avoid forming covariance matrices, and instead to operate directly on the *data*. This approach leads to singular value decompositions (SVDs) of data matrices, and an SVD variant of ***ESPRIT*** summarized in Appendix C.

From the key relation, equation (29), several other quite striking results can be derived. For example, not only is knowledge of the array manifold not required, but the elements thereof associated with the estimated signal parameters (DOAs) can be estimated if desired. The same is true of the source correlation matrix, knowledge of which is not needed in ***ESPRIT***.

F. Array Calibration

Using the TLS formulation of ***ESPRIT***, the array manifold vectors associated with each signal (parameter) can be estimated (to within an arbitrary scale factor). From equation (29), the right eigenvectors of $\mathbf{\Psi}$ are given by $\mathbf{E}_{\Psi} = \mathbf{T}^{-1}$. This result can be used to obtain estimates of the array manifold vectors as

$$(31) \qquad \mathbf{E}_z \mathbf{E}_{\Psi} = \bar{\mathbf{A}}\mathbf{T}\mathbf{T}^{-1} = \bar{\mathbf{A}}\,.$$

No assumption concerning the source covariance is required.

Though simple to compute, this estimate will not in general conform to the invariance structure of the array in the presence of noise. In low SNR scenarios, the deviation from the assumed structure $\bar{\mathbf{A}} = [\mathbf{A}^T(\mathbf{A\Phi})^T]^T$ may be significant. In such situations, improved estimates of the array manifold vectors can be obtained by employing the TLS formulation discussed in Appendix A. Therein, the equivalence of the TLS ***ESPRIT*** solution and the following minimization problem is demonstrated.

$$(32) \qquad \min_{\{\mathbf{B},\mathbf{\Psi}\}} J = \min_{\{\mathbf{B},\mathbf{\Psi}\}} \|[\mathbf{E}_X - \mathbf{B} \,\vdots\, \mathbf{E}_Y - \mathbf{B\Psi}]\|_F^2 \,,$$

The minimizing $\mathbf{B}$, the TLS estimate of $\mathbf{AT}$, is given by

$$(33) \qquad \mathbf{B} = \mathbf{E}_{XY}\bar{\mathbf{\Psi}}^* \left[\bar{\mathbf{\Psi}}\bar{\mathbf{\Psi}}^*\right]^{-1} \,,$$

where $\bar{\mathbf{\Psi}} = [\mathbf{I} \,\vdots\, \mathbf{\Psi}]$. Alternative forms for computing $\mathbf{B}$ from elements of the SVD of $\mathbf{E}_{XY}$ are given in Appendix A. Since in the absence of noise, $\mathbf{E}_X = \mathbf{AT}$ and $\mathbf{E}_Y = \mathbf{A\Phi T} = \mathbf{AT\Psi}$, the TLS estimate of $\bar{\mathbf{A}}$ with the assumed invariance structure is given by

$$(34) \qquad \bar{\mathbf{A}} = \begin{bmatrix} \mathbf{B} \\ \mathbf{B\Psi} \end{bmatrix} \mathbf{T}^{-1} \,,$$

where $\mathbf{T}^{-1}$ is the matrix of right eigenvectors of $\mathbf{\Psi}$.

G. Signal Copy

In many practical applications, not only the signal parameters, but the signals themselves are of interest. Estimation of the signals as a function of time from an estimated DOA is termed *signal copy.* The basic objective is to obtain estimates $\hat{\mathbf{s}}(t)$ of the signals $\mathbf{s}(t)$ from the array output, $\mathbf{z}(t) = \bar{\mathbf{A}}\mathbf{s}(t) + \mathbf{n}(t)$. Employing a linear estimator, a squared-error cost criterion in the metric of the noise[24] (which is ML if the noise is Gaussian), and conditioning on knowledge of $\bar{\mathbf{A}}$, leads to the estimate $\hat{\mathbf{s}}(t)$, the vector of coefficients resulting from the *oblique projection* of $\mathbf{z}(t)$ onto the space spanned by the columns of $\bar{\mathbf{A}}$ (*cf.* Appendix D). The resulting weight matrix $\mathbf{W}$ (*i.e.,* the *linear* estimator) whose i^{th} column is a weight vector that can be used to obtain an estimate of the signal from the ith estimated DOA and reject those from the other DOAs is given by

$$(35) \qquad \mathbf{W} = \mathbf{\Sigma}_n^{-1}\bar{\mathbf{A}}[\bar{\mathbf{A}}^*\mathbf{\Sigma}_n^{-1}\bar{\mathbf{A}}]^{-1}.$$

[24] Herein a non-identity noise covariance $\mathbf{\Sigma}_n$ is retained for complete generality. In this case, $\mathbf{E}_Z = \mathbf{\Sigma}_n[\mathbf{e}_1 \,\vdots\, \cdots \,\vdots\, \mathbf{e}_d]$.

In terms of quantities already available, equation (35) can be written as

$$\mathbf{W} = \Sigma_n^{-1}\mathbf{E}_z[\mathbf{E}_z^*\Sigma_n^{-1}\mathbf{E}_z]^{-1}\mathbf{E}_\Psi^{-*}, \tag{36}$$

using equation (31) to estimate $\bar{\mathbf{A}}$. This equivalence is easily established since from equation (29) it follows that the right eigenvectors of $\mathbf{\Psi}$ equal $\mathbf{T}^{-1}$. Combining this fact with $\mathbf{E}_z = \bar{\mathbf{A}}\mathbf{T}$ and substituting in (36) yields

$$\mathbf{W}^* = \mathbf{E}_\Psi^{-1}[\mathbf{E}_z{}^*\Sigma_n^{-1}\mathbf{E}_z]^{-1}\mathbf{E}_z{}^*\Sigma_n^{-1} = [\bar{\mathbf{A}}^*\Sigma_n^{-1}\bar{\mathbf{A}}]^{-1}\bar{\mathbf{A}}^*\Sigma_n^{-1}. \tag{37}$$

Note that the *optimal* copy vector is a vector that is Σ_n^{-1} orthogonal to all but one of the vectors in the columns of $\bar{\mathbf{A}}$ since $\mathbf{W}^*\bar{\mathbf{A}} = \mathbf{I}$.

There is, of course, a *total least-squares* alternative to conditioning on knowledge of $\bar{\mathbf{A}}$. Since only estimates of $\bar{\mathbf{A}}$ are available, in low SNR scenarios where accurate signal estimates are desired, the TLS approach yields improved estimates at the cost of increased computation. Though not derived herein, $\hat{\mathbf{s}}(t)$ can be obtained by performing a (generalized) singular value decomposition of $[\bar{\mathbf{A}} \,\vdots\, \mathbf{z}(t)]$. The right singular vector corresponding to the *smallest* singular value yields $\hat{\mathbf{s}}(t)$ as the first d elements after normalizing the last element to unity.

H. Source Covariance Estimation

There are several approaches that can be used to estimate the source covariance matrix $\mathbf{R}_{SS}$. Since the problem as posed is scale invariant as discussed in Section B., the source covariance is only estimable to within an arbitrary diagonal scaling matrix $\mathbf{D}$. If estimates of the signals have been obtained, $\mathbf{D}\mathbf{R}_{SS}\mathbf{D}^*$ can be estimated by

$$\mathbf{D}\mathbf{R}_{SS}\mathbf{D}^* = \frac{1}{N}\sum_{t=1}^{N} \hat{\mathbf{s}}(t)\hat{\mathbf{s}}^*(t).$$

If signal copy has not been performed, the most straightforward is to simply note that the *optimal signal copy* matrix $\mathbf{W}$ obtained above removes the *spatial correlation* in the observed measurements (*cf.* (36)). Thus, $\mathbf{W}^*\mathbf{C}_{ZZ}\mathbf{W} = \mathbf{D}\mathbf{R}_{SS}\mathbf{D}^*$ where $\mathbf{R}_{SS}$ is the source correlation (not covariance) matrix, $\mathbf{C}_{ZZ} = \mathbf{R}_{ZZ} - \sigma^2\Sigma_n = \mathbf{A}\mathbf{R}_{SS}\mathbf{A}^*$, and the diagonal factor $\mathbf{D}$ accounts for arbitrary normalization of the columns of $\mathbf{W}$. Note that when $\mathbf{C}_{ZZ}$ must be estimated, a manifestly rank d estimate

$$\hat{\mathbf{C}}_{ZZ} = \mathbf{E}_z[\Lambda_Z^{(d)} - \hat{\sigma}^2\mathbf{I}_d]\mathbf{E}_Z^*$$

can be used, where $\Lambda_Z^{(d)} = \text{diag}\{\lambda_1, \ldots, \lambda_d\}$ and λ_i is a generalized eigenvalue of $(\mathbf{R}_{ZZ}, \Sigma_n)$. Combining this with $\mathbf{E}_z = \bar{\mathbf{A}}\mathbf{T}$ gives

$$\mathbf{D}\mathbf{R}_{SS}\mathbf{D}^* = \mathbf{T}[\Lambda_Z^{(d)} - \hat{\sigma}^2\mathbf{I}_d]\mathbf{T}^*. \tag{38}$$

The same result is obtained if $\mathbf{E}_z$ is replaced by its TLS estimate $[\mathbf{B}^T \,\vdots\, (\mathbf{B}\mathbf{\Psi})^T]^T$ and the corresponding TLS copy vector used as well. Finally, if the gain pattern of one of the sensors is known, specifically if the gain $g_1(\theta_k)$ is known for all θ_k associated with sources whose power is to be estimated, then source power estimation is possible since the array manifold vectors can now be obtained with proper scaling. Further details can be found in [1].

V. Simulation Results

Many simulations have been conducted exploring different aspects of ***ESPRIT*** and making comparisons with other techniques (*cf.* [1]). Herein, only one of the scenarios, but one that addresses several issues that arise in a practical implementation of ***ESPRIT***, is presented. Thus, sensor gain and phase errors, as well as sensor spacing errors are included.[25] Finally, unequal source powers and a high degree of source correlation are assumed.

The array chosen was a ten element array with doublet spacing $\lambda/4$ and the five doublets randomly spaced on a line resulting in an aperture of approximately 4λ. Two sources were located at 24° and 28° and were of unequal powers, 20 dB and 15 dB respectively. Sensor response errors were introduced by including zero-mean normal random additive errors, independent of angle, with sigmas of 0.1 dB in amplitude and 2° in phase. Additive sensor position errors were similarly included with sigmas of 0.5% of the nominal separation in each doublet. An identity noise covariance was used, *i.e.*, $\mathbf{R}_{NN} = \mathbf{I}$.

The sources were initially 50% temporally correlated. The results are given in Figures 4 through 6. The indicated failure rate for **MUSIC** of 6% indicates the percentage of the number of trials in which the conventional MUSIC spectrum either exhibited less than two peaks, or either of the final DOA estimates was outside the interval $[20°, 32°]$. The sample means and variances of the ***ESPRIT*** and MUSIC estimates (excluding the failures) are given in Table 1. Notice in Figure 6 that there is

Estimator	*Source DOA Estimates*	
	$\hat{\theta}_1 \quad (\theta_1 = 24°)$	$\hat{\theta}_2 \quad (\theta_2 = 28°)$
MUSIC*	$24.25° \pm 0.23°$	$27.69° \pm 0.27°$
ESPRIT	$23.97° \pm 0.89°$	$28.01° \pm 1.01°$

* Based on 4713 successes out of 5000 trials.

Table 1: ***ESPRIT*** and **MUSIC** DOA Sample Means and Sigmas — Random 10 Element Linear Array, SNR [20,15] dB, Source Correlation 50%, $\Delta = \lambda/4$.

clearly a small overlap in the distributions of the ***ESPRIT*** parameter estimates. Since a simple *nearest-to-the-true-location* assignment scheme was employed in the calculation of the sample statistics,[26] the resulting mean and variance estimates are biased, though for this scenario the effect is not large. Ideally, a 5-parameter (two means, two variances, and a correlation coefficient) fit of a two-dimensional Gaussian distribution to the estimates should be performed.

[25] While the array is linear, the elements are not uniformly spaced rendering root-**MUSIC** inapplicable.

[26] This does not imply that the statistics are compiled by *splitting the histogram* down the middle and computing the center of mass and second moments of the truncated distributions.

The results indicate the presence of a finite sample bias in the conventional **MUSIC** estimates, the source of which is described in detail in [1, pp. 98-100]. They also indicate that knowledge of the array manifold is certainly valuable, at least in situations where the ***ESPRIT*** subarray separation is small. As the subarray separation increases, however, the relative value of the knowledge of the subarray manifold decreases as is clearly demonstrated in the third of the three scenarios to be described herein.

To investigate the sensitivity to source correlation, that parameter was increased from 50% to 90% while holding all other parameters fixed. The results are shown in Figures 7 through 9. The means and sigmas of the DOA estimates are given in Table 2. The results indicate the relative lack of sensitivity of ***ESPRIT*** to the dramatic increase

Estimator	*Source DOA Estimates*	
	$\hat{\theta}_1 \quad (\theta_1 = 24°)$	$\hat{\theta}_2 \quad (\theta_2 = 28°)$
MUSIC*	$24.35° \pm 0.28°$	$27.48° \pm 0.38°$
ESPRIT	$23.93° \pm 1.07°$	$28.06° \pm 1.37°$

* Based on 3175 successes out of 5000 trials.

Table 2: ***ESPRIT*** and **MUSIC** DOA Sample Means and Sigmas — Random 10 Element Linear Array, Source Powers [20,15] dB, Source Correlation 90%, $\Delta = \lambda/4$.

in correlation, while **MUSIC** degrades significantly. The **MUSIC** failure rate has increased to 37% as manifest in the spectra shown in Figure 7, and the finite sample bias in the estimate of the source DOA for the lower power source at 28° has increased significantly. Note that in computing the statistics, only the trials that resulted in *successes* are included. Note that as in the previous scenario, the ***ESPRIT*** mean and variance estimates are approximate as a two-dimensional Gaussian distribution fit was eschewed in favor of a simple angle ordering assignment procedure.

To investigate the effect of increasing the subarray separation, Δ was increased from $\lambda/4$ to 4λ. The results are given in Figures 10 through 12, and the sample means and sigmas in Table 3. The conventional **MUSIC** algorithm did not fail once in the 1000 trials. This is a consequence of the increase in the magnitude of the gradient of the array manifold vectors evaluated at the nominal angles resulting from the increase in array aperture. As alluded to earlier, as the subarray separation increases, knowledge of the subarray manifold becomes relatively less important. In this scenario, ***ESPRIT*** is performing as well as, if not better than **MUSIC**, and at a fraction of the computational cost!

VI. Discussion

Herein, a new technique (TLS ***ESPRIT***) for signal parameter estimation has been described. The algorithm differs from its predecessors in that a *total least-squares* rather than a standard least-squares (LS) criterion is employed.

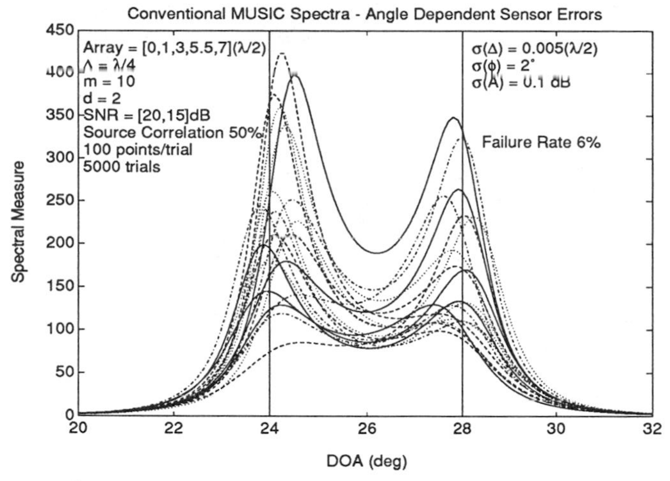

Figure 4: Conventional **MUSIC** Spectra — Random 10 Element Linear Array, Source Correlation 50%, Small Array Aperture.

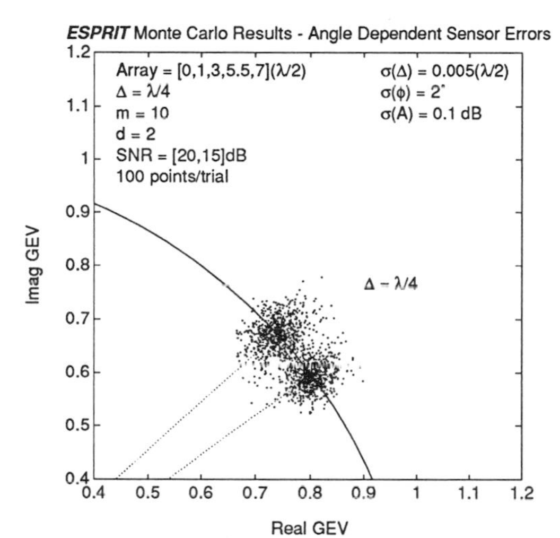

Figure 5: ***ESPRIT*** GEVs — Random 10 Element Linear Array, Source Correlation 50%, Small Array Aperture ($\Delta = \lambda/4$).

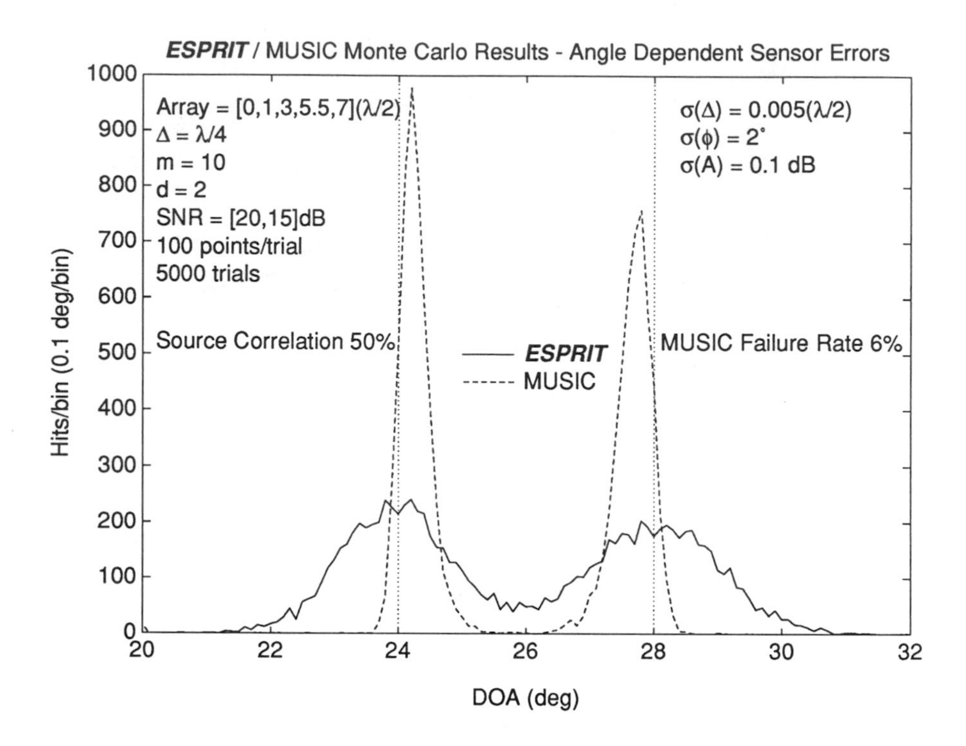

Figure 6: Histogram of **MUSIC** and ***ESPRIT*** Results — Random 10 Element Linear Array, Source Correlation 50%, Small Array Aperture ($\Delta = \lambda/4$).

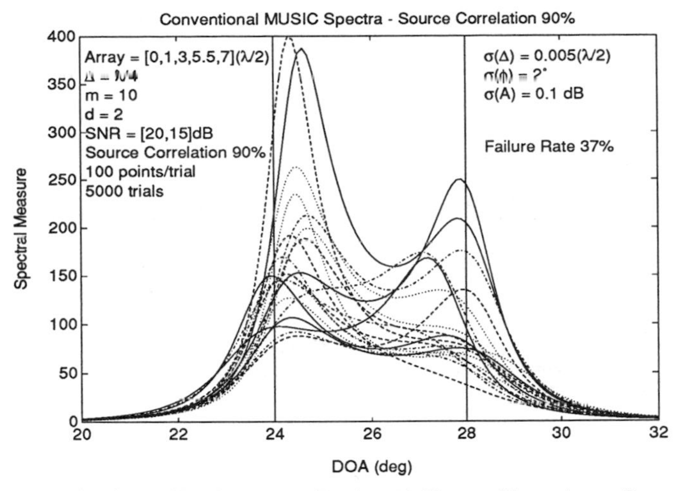

Figure 7: Conventional **MUSIC** Spectra — Random 10 Element Linear Array, Source Correlation 90%, Small Array Aperture.

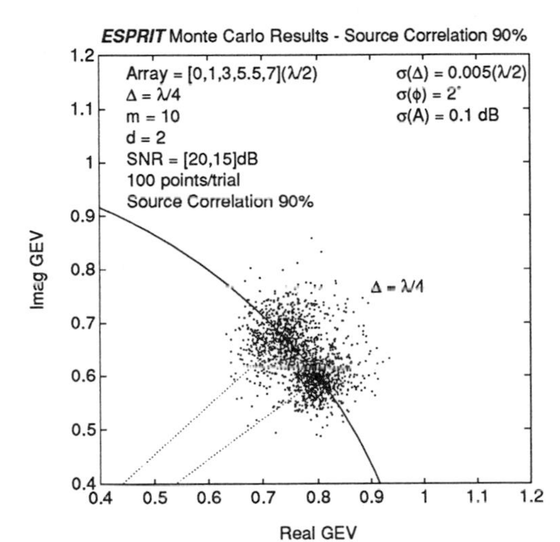

Figure 8: ***ESPRIT*** GEVs — Random 10 Element Linear Array, Source Correlation 90%, Small Array Aperture ($\Delta = \lambda/4$).

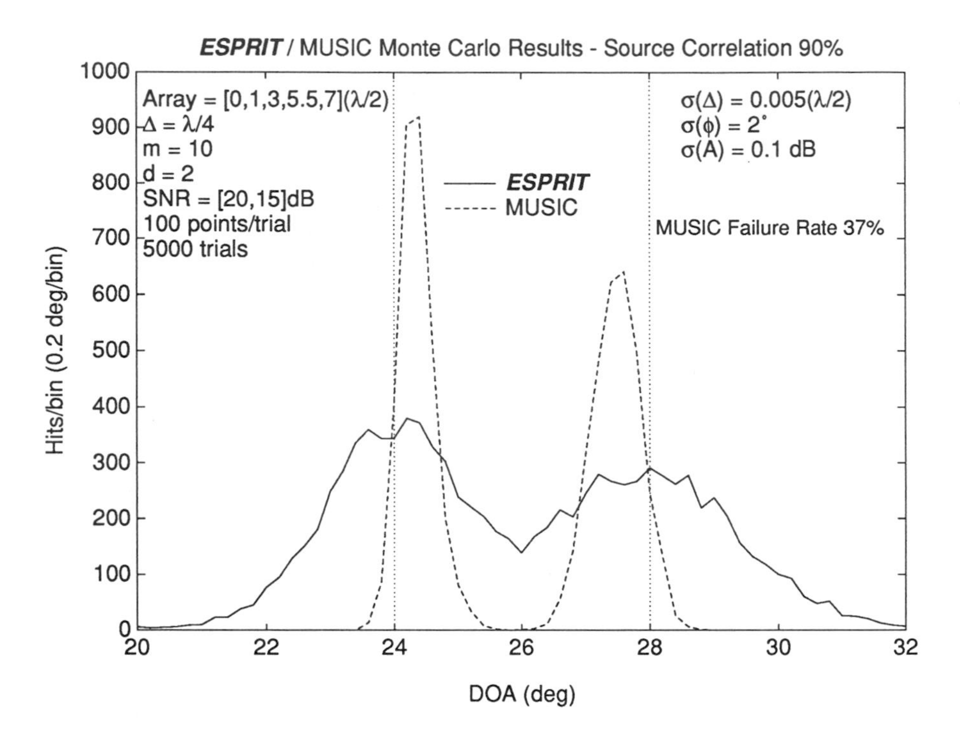

Figure 9: Histogram of **MUSIC** and ***ESPRIT*** Results — Random 10 Element Linear Array, Source Correlation 90%, Small Array Aperture ($\Delta = \lambda/4$).

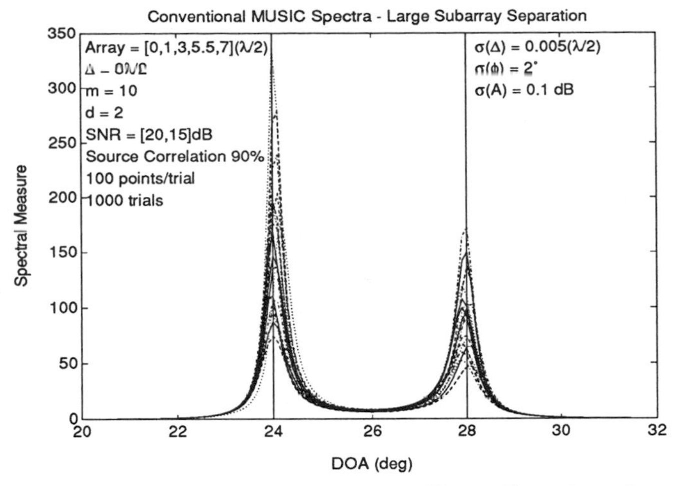

Figure 10: Conventional **MUSIC** Spectra — Random 10 Element Linear Array, Source Correlation 90%, Large Array Aperture.

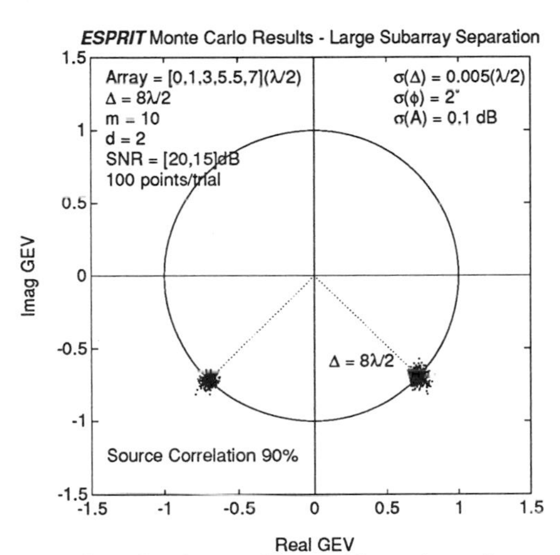

Figure 11: **ESPRIT** GEVs — Random 10 Element Linear Array, Source Correlation 90%, Large Array Aperture ($\Delta = 4\lambda$).

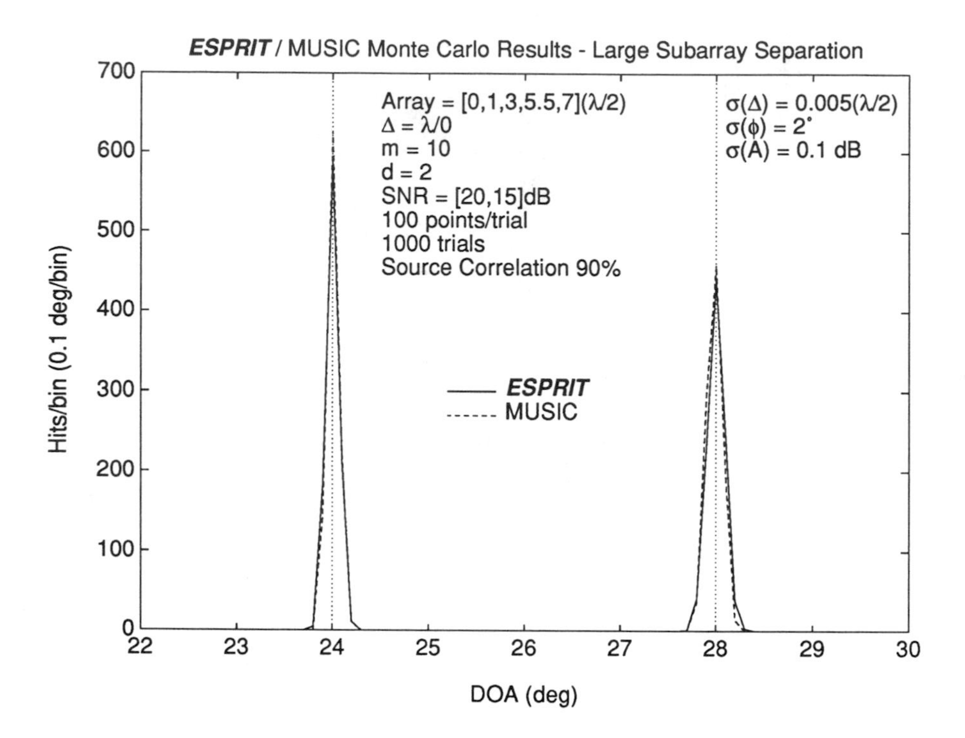

Figure 12: Histogram of **MUSIC** and ***ESPRIT*** Results — Random 10 Element Linear Array, Source Correlation 90%, Large Array Aperture ($\Delta = 4\lambda$).

Estimator	*Source DOA Estimates*	
	$\hat{\theta}_1$ $(\theta_1 = 24^\circ)$	$\hat{\theta}_2$ $(\theta_2 = 28^\circ)$
MUSIC	$24.011^\circ \pm 0.056^\circ$	$27.986^\circ \pm 0.078^\circ$
ESPRIT	$24.003^\circ \pm 0.062^\circ$	$28.002^\circ \pm 0.089^\circ$

Table 3: ***ESPRIT*** and **MUSIC** DOA Sample Means and Sigmas — Random 10 Element Linear Array, Source Powers [20,15] dB, Source Correlation 90%, $\Delta = 4\lambda$.

Numerical Concerns

The earlier versions of ***ESPRIT*** described in [11, 12] sought generalized eigenvalues of a pair of matrices that ideally shared a common $(2m - d)$-dimensional nullspace, a fact that leads to some concern over potential numerical difficulties in solving the generalized eigenproblem. A preliminary solution was to find a least-squares estimate of an $m \times m$ operator whose action was restricted to a d-dimensional subspace. Imposing the subspace restriction (*e.g.*, by subspace *deflation*) as a constraint prior to solving the generalized eigenproblem mitigates these numerical concerns; however, the least-squares property of the estimate is retained [1, pp. 133-145]. In many cases, the difference between the LS and TLS parameter estimates is insignificant, the difference becoming notable at very low SNRs. The LS estimates, as predicted, are biased, while the TLS estimates are not. The major advantage of the TLS approach at higher SNRs is clearly numerical. By properly formulating the estimation problem in a subspace of the appropriate dimension (determined by the dimension of the signal subspace where the observed signal *power* is maximum), numerical problems associated with subspace deflation are mitigated altogether.

ESPRIT and Null-steering

Many of the previous high-resolution parameter estimation techniques are based on *steering beams* toward signal directions and, in some cases, simultaneously attempting to otherwise minimize the power in some weighted combination of sensor outputs. Parameter estimates are associated with DOAs at which peaks in the power output occur. Though intuitively appealing at first glance, there is a much more powerful alternative philosophy, that of *null-steering*. It is well-known that deep, sharp notches in directivity patterns and filter gain functions are much easier to achieve than sharp peaks. Interferometers exploit this fact to obtain accurate estimates of source parameters by finding the relative phase required to cancel signal components in two channels. In this context, ***ESPRIT*** can be interpreted as a *multi-dimensional null-steering* parameter estimation algorithm. Calculation of the eigenvalues of the (rotation) operator $\boldsymbol{\Psi}$, which are the roots of its characteristic polynomial, can be interpreted as multi-dimensional null-steering. Instead of steering broad beams, ***ESPRIT*** steers sharp nulls at all sources simultaneously and does so without relying on knowledge of the array manifold!

Computational Advantages of *ESPRIT*

The primary computational advantage of ***ESPRIT*** is that it eliminates the search procedure inherent in all previous methods (ML, ME, **MUSIC**). ***ESPRIT*** produces signal parameter estimates directly in terms of (generalized) eigenvalues. As noted previously, this involves computations of the order d^3. On the other hand, **MUSIC** and the other high resolution techniques require a search over $\mathcal{A}$, and it is this search that is computationally expensive.[27]

A simple example serves to demonstrate the notable advantage of ***ESPRIT***. Consider a DOA estimation application with a sensor array composed of twenty (20) elements. Assume that the parameter space is one dimensional, *i.e.,* that azimuth estimates are to be obtained, and that two signals are present. Further assume the aperture covers an arc of 2 radians and that a one-half milliradian resolution in azimuth is required. Neglecting comparison operations in searching for the four largest peaks in the **MUSIC** *spectrum*, a search of the entire array manifold requires on the order of $20 \times 4000 \approx 10^5$ multiplications and additions (operations). The resulting ***ESPRIT*** eigenproblems (SVDs) require on the order of $10 \times 2^3 \approx 10^2$ operations. The resulting computational advantage of ***ESPRIT*** is on the order of 10^3 over **MUSIC**.[28] Note that an exhaustive search of $\mathcal{A}$ is assumed for the search required. **MUSIC** would certainly benefit from a more sophisticated search technique, but the advantage of ***ESPRIT*** is still overwhelming.

VII. Directions of Current and Future Research

Multidimensional *ESPRIT*

The planar array of sensors in Figure 3 has, as far as ***ESPRIT*** is concerned, DOA sensitivity in one dimension only, *i.e.,* azimuth sensitivity. If the sources are actually parameterized by an azimuth and an elevation angle, *i.e.,* a DOA estimation problem in three-dimensional space is being addressed, another sensor array of doublets with a displacement vector independent of the first is required. If these two arrays have no elements in common and are sampled independently, then independent processing of the array outputs is *nearly optimal.* The first sensor array yields *azimuth* DOA estimates which locate the sources on *cones of ambiguity* in three-space. Assuming that the displacement vector of the second array is orthogonal to the first (though it need not be), the *elevation* DOA estimates from the second array reduce the ambiguity to that of a pair of DOA {azimuth,elevation} tuple estimates that are symmetric with respect to the plane defined by $\boldsymbol{\Delta}_{az} \wedge \boldsymbol{\Delta}_{el}$. Practically, this ambiguity can be resolved in many situations by physical reasoning. The important point to be made is that for each *dimension in parameter space* in which an estimate is desired, a *displacement*

[27] The significant computational advantage of ***ESPRIT*** becomes even more pronounced in multi-dimensional parameter estimation where the computational load grows linearly with dimension in ***ESPRIT***, while that of **MUSIC** grows exponentially. If r_l is the resolution (*i.e.,* number of vectors) required in the calibration of $\mathcal{A}$ for the lth dimension in Θ, the computation required to search over L dimensions for d parameter vectors is proportional to $\prod_{l=1}^{L} r_l$. For $r_l = r$, the computational load is r^L.

[28] For a two-dimensional manifold, *e.g.,* azimuth and elevation, the advantage improves to a remarkable factor of 10^7.

invariance is required in that dimension. This invariance manifests itself in the data model as a complex scaling of the elements of $\mathcal{A}$, a fact that is the key to ***ESPRIT***.

There is an important issue to be addressed when employing the decoupled procedure described above in the presence of multiple sources. The estimates from the two subarrays must be *paired* to obtain a tuple associated with a particular source. One possible approach is to perform signal copy in both azimuth and elevation separately, and correlate signal estimates. This suboptimal strategy can be employed only if the arrays are sampled such that the signal components are highly correlated (recall the assumption of temporally independent snapshots). Such a correlation results if the arrays are *nearly* simultaneously sampled in the sense that the distance travelled by the signals between azimuth and elevation array sampling times is small compared to the extent of the entire array.

Note that practically, azimuth and elevation estimates can be obtained from an array of *triplets* of sensors, a generalization of the commonly used rectangular array of sensors (*cf.* phased-arrays) for source location in two dimensions. If frequency estimates are desired as well, tapped delay lines with pairs of taps separated by a fixed *time increment behind each sensor* can be used to obtain the necessary information. Practically this might be implemented by a sampler that at regular or irregular intervals obtains a pair of samples separated by a fixed time increment. In such practical implementations, decoupled estimation is clearly suboptimal.

For the purposes of illustration, consider a regular rectangular array of identical sensors (commonly referred to as a phased-array). The array clearly possesses many *invariances* which can be exploited. Consider for the moment however, only the three subarrays depicted in Figure 13. Though it is certainly possible to choose subarrays that do not share any elements in common, the subarrays shown are maximally *overlapping*, a condition that has some interesting associated mathematical consequences as discussed later on. By combining subarrays 0 and 1 and subarrays 0 and 2, two arrays with orthogonal displacement vectors ($\mathbf{\Delta}_1 \perp \mathbf{\Delta}_2$) result. Furthermore, the two arrays share a common subarray, subarray 0, by construction, so that by simultaneously sampling the output of each sensor in the entire array, the following *data model/signal subspace relationships* are easily derived.

$$\begin{bmatrix} \mathbf{E}_0 \\ \mathbf{E}_1 \\ \mathbf{E}_2 \end{bmatrix} = \begin{bmatrix} \mathbf{B} \\ \mathbf{B\Psi}_1 \\ \mathbf{B\Psi}_2 \end{bmatrix} = \begin{bmatrix} \mathbf{AT} \\ \mathbf{ATT}^{-1}\mathbf{\Phi}_1\mathbf{T} \\ \mathbf{ATT}^{-1}\mathbf{\Phi}_2\mathbf{T} \end{bmatrix}. \tag{39}$$

$\mathbf{E}_i$ is an $m \times d$ matrix that spans the signal subspace observed by subarray i.

In the presence of noisy data, equation (39) does not hold since $\mathbf{E}$ must be estimated. As in the one-dimensional case, a TLS approach is adopted and the objective is to find

$$\min_{\mathbf{B},\mathbf{\Psi}_1,\mathbf{\Psi}_2} \left\| \begin{bmatrix} \mathbf{E}_0 \\ \mathbf{E}_1 \\ \mathbf{E}_2 \end{bmatrix} - \begin{bmatrix} \mathbf{B} \\ \mathbf{B\Psi}_1 \\ \mathbf{B\Psi}_2 \end{bmatrix} \right\|_F^2, \tag{40}$$

subject to the constraint that

$$[\mathbf{\Psi}_1, \mathbf{\Psi}_2] \stackrel{\text{def}}{=} \mathbf{\Psi}_1\mathbf{\Psi}_2 - \mathbf{\Psi}_2\mathbf{\Psi}_1 = \mathbf{0}.$$

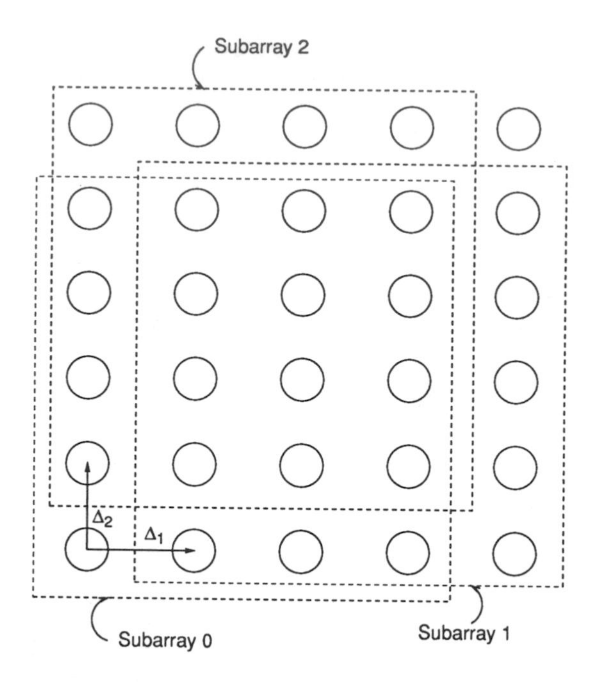

Figure 13: Possible Subarray Structure of a Regular Rectangular Array

that is, $\mathbf{\Psi}_1$ commutes with $\mathbf{\Psi}_2$, a constraint that is essentially equivalent to the constraint that they share a common set of eigenvectors, $\mathbf{T}$. A computationally efficient solution to this optimization problem is an area of current investigation. Several iterative algorithms have been derived, but the unavoidable issues of local minima, numerical stability, and convergence rates make these approaches unattractive (unsuitable) for real-time autonomous implementation. The objective is derive an algorithm in the spirit of ***ESPRIT***, involving SVDs and eigendecompositions as the fundamental operations.

Alternatives to the SVD and Eigendecompositions

In estimating the signal subspace, the indicated technique in all current high resolution signal subspace algorithms is to perform an SVD of the data matrix or an eigendecomposition of the sample covariance matrix. Once the measurements are reduced to yield $\mathbf{E}_z$, unless d is exceedingly large, the majority of the computational effort required to obtain parameter estimates is complete. The bulk of the remaining effort as far as ***ESPRIT*** is concerned involves an eigendecomposition of a matrix in $\mathbb{C}^{2d \times 2d}$, where d is the estimated number of signals present (*i.e.,* the number of columns of $\mathbf{E}_Z$).

As should be clear from the developments and discussions to this point, eigenvectors or singular vectors are *not required* in the implementation of ***ESPRIT***, or any of the other signal subspace algorithms for that matter, in obtaining estimates of the signal subsapce. They can be replaced by *any* set of vectors (not necessarily orthonormal) that span the signal subspace, *i.e.,* any set of vectors $\bar{\mathbf{E}}_Z$ that satisfy

$$\bar{\mathbf{E}}_Z = \mathbf{E}_Z \bar{\mathbf{T}}\,, \tag{41}$$

for *any* nonsingular $\bar{\mathbf{T}}$ is informationally to $\mathbf{E}_Z$ as far as ***ESPRIT*** is concerned.[29] Thus, any efficient algorithm for obtaining any set of linearly independent vectors spanning the (estimated) signal subspace can be employed. An area for further research involves deriving a computationally efficient (*i.e.,* more efficient than a full eigendecomposition) algorithm for obtaining such a set of vectors. It might also prove fruitful to investigate applications of multilinear algebra to this problem, since an object that quite succinctly characterizes the signal subspace is the *wedge* or *Grassmann* product of any set of vectors that span the subspace!

APPENDICES

A. EQUIVALENT TLS FORMULATIONS

In this appendix, the equivalence between two formulations of the TLS ***ESPRIT*** solution is demonstrated. Though certainly of pedagogical interest, the equivalence also yields a method for obtaining TLS estimates of the array manifold vectors, as well as a method for extending ***ESPRIT*** to arrays with multiple invariances.

The most common statement of the TLS linear parameter estimation problem as stated, for example, in [2] is as follows. Given $\mathbf{A}, \mathbf{B} \in \mathbb{C}^{m \times d}$, $m > d$, find $\mathbf{R_A}$ and $\mathbf{R_B}$ of minimum Frobenius[30] norm, and $\mathbf{X}$ such that

$$[\mathbf{A} + \mathbf{R_A}]\,\mathbf{X} = \mathbf{B} + \mathbf{R_B}\,. \tag{42}$$

Note that in general $\mathbf{B}$ need not have the same number of columns as $\mathbf{A}$ though for the applications described herein, $\mathbf{A}$ and $\mathbf{B}$ have the same dimensions and $\mathbf{X}$ is square. With reference to ***ESPRIT***, $\mathbf{R}_{\{\mathbf{A},\mathbf{B}\}}$ represent errors in the subspace estimates $\mathbf{E}_X = \mathbf{A}$ and $\mathbf{E}_Y = \mathbf{B}$ respectively, and $\mathbf{X}$ is the operator $\mathbf{\Psi}$ whose eigenvalues are parameters of interest. Thus, the TLS ***ESPRIT*** estimation problem can be stated as follows:

> *Given subspace estimates* $\mathbf{E}_X$ *and* $\mathbf{E}_Y$*, find an operator* $\mathbf{\Psi}$ *and residual matrices* $\mathbf{R}_{E_X}, \mathbf{R}_{E_Y}$ *to minimize*
>
> $$J = \|[\mathbf{R}_{E_X} \,\vdots\, \mathbf{R}_{E_Y}]\|_F^2 \tag{43}$$
>
> *subject to*
>
> $$[\mathbf{E}_X + \mathbf{R}_{E_X}]\,\mathbf{\Psi} = \mathbf{E}_Y + \mathbf{R}_{E_Y}\,. \tag{44}$$

---

[29] The structure of $\mathcal{A}$ being exploited by ***ESPRIT*** is the linear structure manifest in the *row space*, and is independent of *any linear* operations in the *column space*, *i.e.,* linear operations on the *right.*

[30] If knowledge of the covariance of the errors is available (in general in the form of a 4-tensor), a weighted Frobenius norm can be employed. Since for the problems considered herein it can be argued that the errors are independent and identically distributed (at least asymptotically), the *weighting matrix* or metric is proportional to the identity.

The objective is to show that the resulting estimate, $\mathbf{\Psi}$, is equivalent to that obtained from the constrained minimization problem posed in Section IV, specifically:

Given subspace estimates $\mathbf{E}_X$ *and* $\mathbf{E}_Y$*, find a matrix* $\mathbf{F} \in \mathbb{C}^{2d \times d}$ *to minimize*

$$J = \|[\mathbf{E}_X \,\vdots\, \mathbf{E}_Y]\,\mathbf{F}\|_F^2 \tag{45}$$

subject to

$$\mathbf{F}^*\mathbf{F} = \mathbf{I}\,, \tag{46}$$

where

$$\mathbf{F} \stackrel{\text{def}}{=} \begin{bmatrix} \mathbf{F}_X \\ \mathbf{F}_Y \end{bmatrix}, \tag{47}$$

$$\mathbf{\Psi} = -\mathbf{F}_X[\mathbf{F}_Y]^{-1}\,. \tag{48}$$

Starting with the TLS formulation in equations (43) and (44), define $\mathbf{B} \stackrel{\text{def}}{=} \mathbf{E}_X + \mathbf{R}_{E_X}$. Substituting into equation (44) gives $\mathbf{E}_Y + \mathbf{R}_{E_Y} = \mathbf{B}\mathbf{\Psi}$. Rearranging these equations gives

$$\mathbf{R}_{E_X} = \mathbf{E}_X - \mathbf{B}\,, \tag{49}$$

$$\mathbf{R}_{E_Y} = \mathbf{E}_Y - \mathbf{B}\mathbf{\Psi}\,, \tag{50}$$

so the minimization becomes

$$\min_{\{\mathbf{B},\mathbf{\Psi}\}} J = \min_{\{\mathbf{B},\mathbf{\Psi}\}} \|[\mathbf{E}_X - \mathbf{B} \,\vdots\, \mathbf{E}_Y - \mathbf{B}\mathbf{\Psi}]\|_F^2\,, \tag{51}$$

which using standard properties of the trace operator (Tr) can be written as follows:

$$\min_{\{\mathbf{B},\mathbf{\Psi}\}} J = \min_{\{\mathbf{B},\mathbf{\Psi}\}} Tr\left\{\left[\mathbf{E}_{XY} - \mathbf{B}\bar{\mathbf{\Psi}}\right]^* \left[\mathbf{E}_{XY} - \mathbf{B}\bar{\mathbf{\Psi}}\right]\right\}\,, \tag{52}$$

where

$$\mathbf{E}_{XY} \stackrel{\text{def}}{=} [\mathbf{E}_X \,\vdots\, \mathbf{E}_Y]\,, \tag{53}$$

$$\bar{\mathbf{\Psi}} \stackrel{\text{def}}{=} [\mathbf{I} \,\vdots\, \mathbf{\Psi}]\,. \tag{54}$$

The minimization problem can be solved by setting[31] $\partial_{\mathbf{B}} J = \mathbf{0}$ and solving for $\mathbf{B}$ in terms of $\bar{\mathbf{\Psi}}$ which yields[32]

$$\mathbf{B} = \mathbf{E}_{XY}\bar{\mathbf{\Psi}}^* \left[\bar{\mathbf{\Psi}}\bar{\mathbf{\Psi}}^*\right]^{-1}\,. \tag{55}$$

---

[31] Since complex conjugation is not analytic, care must be taken in the calculation of the derivative. It turns out, however, that treating $\mathbf{B}^*$ as an independent variable yields the correct equations, though there is certainly nothing mathematically rigorous in this approach.

[32] Note that for any $\delta\mathbf{B}$, $Tr\{\delta\mathbf{B}\bar{\mathbf{\Psi}}^*\bar{\mathbf{\Psi}}\delta\mathbf{B}\} > 0$, so the stationary point is actually a minimum.

Substituting this expression for $\mathbf{B}$ back into equation (52), and employing the cyclic property of the trace operator gives

$$\min_{\bar{\Psi}} J = \min_{\bar{\Psi}} Tr\left\{\mathbf{E}_{XY}^{*}\mathbf{E}_{XY}\left(\mathbf{I} - \bar{\Psi}^{*}\left[\bar{\Psi}\bar{\Psi}^{*}\right]^{-1}\bar{\Psi}\right)\right\}. \tag{56}$$

A similar sequence of manipulations on equation (45) yields

$$\min_{\mathbf{F}} J = \min_{\mathbf{F}} Tr\left\{\mathbf{E}_{XY}^{*}\mathbf{E}_{XY}\left(\mathbf{F}\left[\mathbf{F}^{*}\mathbf{F}\right]^{-1}\mathbf{F}^{*}\right)\right\}, \tag{57}$$

where an identity derived from the constraint equation (46) has been inserted to demonstrate that an equivalent constraint is $\mathbf{F}$ be full-rank d so that $\mathbf{F}^{*}\mathbf{F}$ is invertible. Since $\bar{\Psi}$ satisfies this condition by construction, $\mathbf{I} - \bar{\Psi}^{*}\left[\bar{\Psi}\bar{\Psi}^{*}\right]^{-1}\bar{\Psi} \in \mathbb{C}^{2d\times 2d}$ is rank d as well, a fact easily proven using eigendecompositions. In general, for $\mathbf{A} \in \mathbb{C}^{m\times n}$ and $m > n$,

$$P_{\mathbf{A}} \stackrel{\text{def}}{=} \mathbf{A}\left[\mathbf{A}^{*}\mathbf{A}\right]^{-1}\mathbf{A}^{*} \tag{58}$$

defines a *projection operator* $P_{\mathbf{A}}$ on $\mathbb{C}^{m}$, mapping elements of $\mathbb{C}^{m}$ to the subspace of $\mathbb{C}^{m}$ spanned by the columns of $\mathbf{A}$, and $\mathbf{I} - P_{\mathbf{A}}$ defines a projection operator $P_{\mathbf{A}}^{\perp}$ into the orthogonal complement of the subspace spanned by the columns of $\mathbf{A}$. Employing this notation convention, the minimization problems are manifestly equivalent,[33] since both are minimizations of the same functional form over rank d projection matrices in $\mathbb{C}^{2d\times 2d}$. Furthermore, the solution is easily visualized geometrically (and proven algebraically in Appendix B) to be the sum of the d smallest eigenvalues[34] of $\mathbf{E}_{XY}^{*}\mathbf{E}_{XY}$ (squares of the d smallest singular values of $\mathbf{E}_{XY}$), obtained by choosing to project into the subspace spanned by the d corresponding right eigenvectors (right singular-vectors of $\mathbf{E}_{XY}$). Letting

$$\mathbf{E}_{XY}^{*}\mathbf{E}_{XY} = [\mathbf{E}_1 \,\vdots\, \mathbf{E}_2]\,\mathbf{\Lambda}\,[\mathbf{E}_1 \,\vdots\, \mathbf{E}_2]^{*}, \tag{59}$$

where $\mathbf{\Lambda} = \text{diag}\{\lambda_1, \ldots, \lambda_{2d}\}$ and $\lambda_1 \geq \lambda_2 \geq \ldots \geq \lambda_{2d}$, $P_{\mathbf{F}} = P_{\bar{\Psi}^{*}}^{\perp} = P_{\mathbf{E}_2}$.

Having solved the minimization problem, equivalence of the estimates of Ψ remains to be shown. Defining

$$[\mathbf{E}_1 \,\vdots\, \mathbf{E}_2] \stackrel{\text{def}}{=} \begin{bmatrix} \mathbf{E}_{11} & \mathbf{E}_{12} \\ \mathbf{E}_{21} & \mathbf{E}_{22} \end{bmatrix}, \tag{60}$$

the TLS ***ESPRIT*** estimate of Ψ is easily seen to be $-\mathbf{E}_{12}\mathbf{E}_{22}^{-1}$. By orthogonality of the eigenvectors of $\mathbf{E}_{XY}^{*}\mathbf{E}_{XY}$, $P_{\bar{\Psi}^{*}}^{\perp} = P_{\mathbf{E}_2}$ implies $P_{\bar{\Psi}^{*}} = P_{\mathbf{E}_1}$, and therefore there exists a non-singular matrix $\mathbf{T}$ such that

$$\begin{bmatrix} \mathbf{I} \\ \Psi^{*} \end{bmatrix} = \begin{bmatrix} \mathbf{E}_{11} \\ \mathbf{E}_{21} \end{bmatrix} \mathbf{T}. \tag{61}$$

---

[33] There is a subtle difference in the constraints however. Mathematically speaking, the set of matrices of the form $\bar{\Psi}^{*}$ is a subset of all full-rank matrices in $\mathbb{C}^{2d\times d}$ since full-rank matrices whose upper $d \times d$ block is not invertible can not be converted into the form of $\bar{\Psi}^{*}$. For the problems considered herein, this set has zero probability of occurring and need not be considered further.

[34] This sum is clearly an achievable lower bound since replacing any of the d smallest eigenvalues with any of the remaining eigenvalues results in an increase in J, unless the dth smallest eigenvalue has multiplicity greater than one. Though mathematically possible, by the problem construction this also will occur with probability zero, and therefore need not be considered.

Thus, the estimate of $\mathbf{\Psi}$ is given by $\mathbf{E}_{11}^{-*}\mathbf{E}_{21}^{*}$. The orthogonality relation

$$\mathbf{E}_{11}^{*}\mathbf{E}_{12} + \mathbf{E}_{21}^{*}\mathbf{E}_{22} = \mathbf{0} \tag{62}$$

now easily establishes the equivalence of the two estimates of $\mathbf{\Psi}$ as was to be shown. Note that substituting the equation for $\mathbf{\Psi}$ into the equation for $\mathbf{B}$ yields

$$\mathbf{B} = \mathbf{E}_{XY}\mathbf{E}_{1}\mathbf{E}_{11}^{*} \tag{63}$$

using the easily proven fact that $\bar{\mathbf{\Psi}}\bar{\mathbf{\Psi}}^{*} = \mathbf{E}_{11}^{-*}\mathbf{E}_{11}^{-1}$. Also note that if the SVD of $\mathbf{E}_{XY}$ is used to obtain $[\mathbf{E}_1 \,\vdots\, \mathbf{E}_2]$ as the matrix of right singular-vectors, defining $[\mathbf{U}_1 \,\vdots\, \mathbf{U}_2]$ as the matrix of left-singular vectors, and $\mathbf{\Sigma}_{XY}$ as the diagonal matrix of ordered singular values,

$$\mathbf{E}_{XY} = [\mathbf{U}_1 \,\vdots\, \mathbf{U}_2]\mathbf{\Sigma}_{XY}[\mathbf{E}_1 \,\vdots\, \mathbf{E}_2]^{*} \,. \tag{64}$$

Substituting this expression into equation (63) yields

$$\mathbf{B} = \mathbf{U}_1\mathbf{\Sigma}_1\mathbf{E}_{11}^{*} \,, \tag{65}$$

where $\mathbf{\Sigma}_1 = \text{diag}\{\sigma_1, \ldots, \sigma_d\}$, the d *largest* singular values of $\mathbf{E}_{XY}$. Thus, once the SVD of $\mathbf{E}_{XY}$ has been computed, $\mathbf{B}$ can be computed without computing $\mathbf{\Psi}$.

B. TLS Estimation of $\mathbf{\Psi}$

In this appendix, a short derivation of the TLS estimate of the (rotation) operator $\mathbf{\Psi}$ is presented. The objective is to find a matrix $\bar{\mathbf{\Psi}} \in \mathcal{C}^{2d \times d}$ satisfying

$$\bar{\mathbf{\Psi}}^{*}\bar{\mathbf{\Psi}} = \mathbf{I},$$

that minimizes the Frobenius norm of $\mathbf{E}_{XY}\bar{\mathbf{\Psi}}$. Mathematically,

$$\begin{aligned} \bar{\mathbf{\Psi}} &= \arg \min_{\bar{\mathbf{\Psi}}^{*}\bar{\mathbf{\Psi}}=\mathbf{I}} \|\mathbf{E}_{XY}\bar{\mathbf{\Psi}}\|_F, \\ &= \arg \min_{\bar{\mathbf{\Psi}}^{*}\bar{\mathbf{\Psi}}=\mathbf{I}} Tr\{\bar{\mathbf{\Psi}}^{*}\mathbf{E}_{XY}^{*}\mathbf{E}_{XY}\bar{\mathbf{\Psi}}\}. \end{aligned}$$

Defining the cost function J

$$J \stackrel{\text{def}}{=} Tr\{\bar{\mathbf{\Psi}}^{*}\mathbf{E}_{XY}^{*}\mathbf{E}_{XY}\bar{\mathbf{\Psi}}\},$$

and appending the constraint (inside the trace) with a matrix of Lagrange multipliers $\mathbf{\Lambda}$,

$$J = Tr\{\bar{\mathbf{\Psi}}^{*}\mathbf{E}_{XY}^{*}\mathbf{E}_{XY}\bar{\mathbf{\Psi}} - \mathbf{\Lambda}(\bar{\mathbf{\Psi}}^{*}\bar{\mathbf{\Psi}} - \mathbf{I})\}.$$

At a minimum of J, the gradient of J with respect $\bar{\mathbf{\Psi}}$ is zero, *i.e.*,

$$\frac{\partial J}{\partial \bar{\mathbf{\Psi}}} = \mathbf{E}_{XY}^{*}\mathbf{E}_{XY}\bar{\mathbf{\Psi}} - \bar{\mathbf{\Psi}}\mathbf{\Lambda} = \mathbf{0}. \tag{66}$$

This implies that $\bar{\Psi}$ spans an *invariant subspace* of $\mathbf{E}_{XY}^*\mathbf{E}_{XY}$. Multiplying equation (66) on the left by $\bar{\Psi}^*$ and using the constraint $\bar{\Psi}^*\bar{\Psi} = \mathbf{I}$ gives

$$\Lambda = \bar{\Psi}^*\mathbf{E}_{XY}^*\mathbf{E}_{XY}\bar{\Psi}. \tag{67}$$

Clearly, Λ is a positive semi-definite Hermitian matrix and therefore diagonalizable. Thus, there exists a non-singular matrix $\mathbf{T}$, whose columns are the right eigenvectors of Λ, such that

$$\mathbf{T}^{-1}\Lambda\mathbf{T} = \operatorname{diag}\{\lambda_1, \ldots, \lambda_n\} \stackrel{\text{def}}{=} \Lambda_d. \tag{68}$$

Multiplying equation (66) on the right by $\mathbf{T}$ leads to the following *partial* eigenproblem:

$$\mathbf{E}_{XY}^*\mathbf{E}_{XY}\bar{\Psi}\mathbf{T} = \bar{\Psi}\mathbf{T}\mathbf{T}^{-1}\Lambda\mathbf{T} = \bar{\Psi}\mathbf{T}\Lambda_d. \tag{69}$$

Thus, the columns of $\bar{\Psi}\mathbf{T}$ are eigenvectors of $\mathbf{E}_{XY}^*\mathbf{E}_{XY}$ and the associated eigenvalues are the diagonal elements of Λ_d. Since the objective is to minimize

$$J = Tr\{\bar{\Psi}^*\mathbf{E}_{XY}^*\mathbf{E}_{XY}\bar{\Psi}\} = Tr\,\Lambda_d,$$

Λ_d must contain the d *smallest* eigenvalues of $\mathbf{E}_{XY}^*\mathbf{E}_{XY}$. $\bar{\Psi}$ therefore spans the d-dimensional *invariant subspace* with the *smallest volume*, *i.e.*, the total *least*-squared error. The estimate of the operator Ψ_{XY} is obtained as described in Appendix A.

C. The SVD TLS *ESPRIT* Algorithm

As mentioned earlier, it is often preferable to avoid *squaring* data to obtain covariance matrices, and instead to operate directly on the *data* itself. Benefits accrue from the resulting reduction in matrix condition numbers. With infinite precision computation, the algorithm is theoretically the same whether eigendecompositions or singular value decompositions (SVD) are used. The resulting parameter estimates are identical. Practically, finite precision arithmetic must be employed, and numerical issues can become important. For large arrays composed of hundreds of sensors, and for large amounts of data, round-off and overflow are potential problems to be aware of when forming covariance matrices. Often such problems can be overcome by operating directly on the data. Applying ***ESPRIT*** to data matrices directly requires the use of the SVD. The SVD version of ***ESPRIT*** is obtained by simply replacing all the eigendecompositions in the covariance version with SVDs[35] except for the final eigendecomposition[36] of Ψ. The following is a summary of the algorithm.

Summary of the TLS *ESPRIT* SVD Algorithm

1. Form the matrix $\mathbf{Z}$ from the available measurements.

[35] See [2] for a discussion of the SVD.

[36] Note that SVD's must sooner or later give way to eigendecompositions since SVD's do not yield phase information in the singular values. The extra degrees of freedom in the SVD allow the positive real singular value constraint without loss of generality.

2. Compute the SVD of $\mathbf{Z}$:
$$\bar{\mathbf{E}}^*\mathbf{Z}\bar{\mathbf{V}} = \mathbf{\Sigma},$$
where $\mathbf{\Sigma} = \text{diag}\{\sigma_1, \ldots, \sigma_{2m}\}$, $\sigma_1 \geq \cdots \geq \sigma_{2m}$, and $\bar{\mathbf{E}} = [\mathbf{e}_1 \,\vdots\, \cdots \,\vdots\, \mathbf{e}_{2m}]$.

3. If necessary, estimate the number of sources $\hat{d}$ (*cf.* Appendix E.

4. Obtain the signal subspace estimate $\widehat{\mathcal{S}}_z = \mathcal{R}\{\mathbf{E}_z\}$, where
$$\mathbf{E}_z = [\,\mathbf{e}_1 \,\vdots\, \cdots \,\vdots\, \mathbf{e}_{\hat{d}}\,] = \begin{bmatrix} \mathbf{E}_X \\ \mathbf{E}_Y \end{bmatrix}.$$

5. Compute the SVD of $\{[\mathbf{E}_X \,\vdots\, \mathbf{E}_Y]\} = \mathbf{U}\mathbf{\Sigma}\mathbf{V}^*$,
$$\mathbf{U} = [\,\mathbf{U}_X \,\vdots\, \mathbf{U}_Y\,];\ \mathbf{\Sigma} = \text{diag}\{\sigma_1, \ldots, \sigma_{2d}\};\ \mathbf{V} = \begin{bmatrix} \mathbf{V}_{11} & \mathbf{V}_{12} \\ \mathbf{V}_{21} & \mathbf{V}_{22} \end{bmatrix}.$$

6. Calculate the eigenvalues of $-\mathbf{V}_{11}\mathbf{V}_{22}^{-1}$;
$$\widehat{\phi}_k = \lambda_k(-\mathbf{V}_{12}\mathbf{V}_{22}^{-1}).$$

7. Estimate the signal parameters $\hat{\theta}_k = f^{-1}(\widehat{\phi}_k)$.

Note that if the noise covariance is $\sigma^2\mathbf{\Sigma}_n$, a Mahalanobis transformation on the measurements, *i.e.*, $\mathbf{Z} \Longleftarrow \mathbf{\Sigma}_n^{-\frac{1}{2}}\mathbf{Z}$, to *whiten the noise* can be used, or the SVD performed to obtain $\mathbf{E}_z$ can be replaced by a generalized SVD of the matrix pair $\{\mathbf{Z}^*, \mathbf{\Sigma}_n^{\frac{*}{2}}\}$. Further details on this variant of ***ESPRIT*** can be found in [1]. A thorough discussion of the GSVD is given in [2].

D. The Maximum Likelihood Estimator

For the class of problems considered herein, the maximum likelihood estimator is simple to derive analytically, though in most practical real-time applications, computationally prohibitive. For deterministic (non-random) signals in Gaussian noise with covariance $\mathbf{\Sigma}_n$, $\mathbf{z}(t) = \mathbf{A}(\boldsymbol{\theta})\mathbf{s}(t) + \mathbf{n}(t)$, the likelihood function is easily written [13, 1]:

$$\mathcal{L}[\mathbf{z}(t)] = -\ln\left[P\{\mathbf{z}(t)|\mathbf{z}(t) = \mathbf{A}(\boldsymbol{\theta})\mathbf{s}(t) + \mathbf{n}\}\right] \tag{70}$$

$$\propto -[\mathbf{z}(t) - \mathbf{A}(\boldsymbol{\theta})\mathbf{s}(t)]^*\mathbf{\Sigma}_n^{-1}[\mathbf{z}(t) - \mathbf{A}(\boldsymbol{\theta})\mathbf{s}(t)]. \tag{71}$$

The maximization of $\mathcal{L}$ is over $\{\mathbf{s}(t); t \in [0, N]\} \times \{\boldsymbol{\theta} \in \Theta\}$, and is therefore a nonlinear optimization problem. It belongs, however, to the class of *separable* nonlinear optimization problems. Golub and Pereyra [14] prove that the optimization can be carried out in two steps. A solution for the optimal $\mathbf{s}(t)$ is sought as a function of $\boldsymbol{\theta}$, then the maximization over $\boldsymbol{\theta}$ is performed. Employing this procedure gives $\hat{\mathbf{s}}(t) = \mathbf{w}^*(\boldsymbol{\theta})\mathbf{z}(t)$.

where $\mathbf{w}(\boldsymbol{\theta}) = \boldsymbol{\Sigma}_n^{-1}\mathbf{A}(\boldsymbol{\theta})[\mathbf{A}^*(\boldsymbol{\theta})\boldsymbol{\Sigma}_n^{-1}\mathbf{A}(\boldsymbol{\theta})]^{-1}$. Substituting the expression for $\mathbf{s}(t)$ back into (71) and using standard properties of the trace operator,

$$\mathcal{L}(\boldsymbol{\theta}) \propto -\; Tr\left\{P_{\mathbf{A}^{\perp}(\boldsymbol{\theta})}\mathbf{R}_{zz}\boldsymbol{\Sigma}_n^{-1}\right\}, \qquad (72)$$

where $P_{\mathbf{A}^{\perp}(\boldsymbol{\theta})}$ is the *oblique* projection operator onto the complement of the space spanned by $\mathbf{A}(\boldsymbol{\theta})$ (in the metric $\boldsymbol{\Sigma}_n$). Maximization of this criterion is equivalent to finding

$$\max_{\boldsymbol{\theta}} Tr\left\{P_{\mathbf{A}(\boldsymbol{\theta})}\mathbf{R}_{zz}\boldsymbol{\Sigma}_n^{-1}\right\}, \qquad (73)$$

as can be easily verified.[37] Though easy to describe analytically, the computational burden of actually carrying out the multi-dimensional projection and maximization over $\boldsymbol{\theta}$ is generally prohibitive, resulting in the need for *reasonable* approximate solutions such as **MUSIC** and ***ESPRIT***.

E. Estimating Subspace Dimensions

The estimation of the dimension of the signal subspace $\mathcal{S}_Z$ is not required as a preliminary step in ***ESPRIT***. In the absence of *a priori* knowledge, the maximum number of signals (equal to the number of doublets m) can be assumed to be present and that many parameters estimated. However, an accurate estimate of the true signal subspace dimension improves parameter estimates in the sense of reducing the root-mean-square (RMS) error of the parameter estimates.

If the signal subspace dimension is to be estimated, the likelihood ratio techniques described in detail in [1] can be applied directly. These techniques rely on the fact that when the covariances of the measurements and noise are known, the minimum repeated generalized eigenvalue is the noise power. When the covariances must be estimated from a finite number of measurements, a sequence of hypotheses can be tested against a common null hypothesis to obtain an estimate of d. If the noise covariance is completely known, likelihood threshold tests can be devised for a given level of significance.

If the noise power is not known, more sophisticated statistical criteria such as An Information Criterion (AIC) and the Minimum Description Length (MDL) criteria can be applied as well. These criteria are basically generalized likelihood ratios (GLRs) for testing the equality of the k smallest generalized eigenvalues augmented by terms that attempt to account for the increasing *complexity of the model* with increasing number of signals. Since the GLRs are generally monotonically decreasing functions of the hypothesized number of signals, augmenting them with terms that are monotonically increasing functions of the number of signals generally yields a convex function with a minimum in $(1, d_{\max})$ though the minimum can be attained at one of the boundary points in some cases.

[37] This development assumes that the number of signals is known. When the number of signals is not known *a priori* the maximum likelihood estimator must be redefined [13. 1].

For reference, the MDL criterion (under the stochastic signal assumption[38]) is given by

$$\text{(74)} \qquad \mathrm{MDL}(k) = -\log \left\{ \frac{\prod_{i=k+1}^{m} \hat{\lambda}_i^{\frac{1}{m-k}}}{\frac{1}{m-k} \sum_{i=k+1}^{m} \hat{\lambda}_i} \right\}^{(m-k)N} + \frac{k}{2}(2m-k)\log N\,;$$

where $\hat{\lambda}_i$ are the generalized eigenvalues of $\{\mathbf{R}_{zz}, \mathbf{\Sigma}_n\}$. The use of this criterion was first proposed by Wax and Kailath (1985) [15]. The estimate $(\hat{d})$ of the number of sources d is given by the value of k for which the MDL function is minimized.

References

[1] R. H. Roy, ***ESPRIT*** – *Estimation of Signal Parameters via Rotational Invariance Techniques*, PhD thesis, Stanford University, Stanford, CA., 1987.

[2] G. H. Golub and C. F. Van Loan, *Matrix Comptutations*, Johns Hopkins University Press, Baltimore, MD., 1984.

[3] H. L. Van Trees, *Detection, Estimation, and Modulation Theory*, John Wiley and Sons, New York, 1971.

[4] R. O. Schmidt, *A Signal Subspace Approach to Multiple Emitter Location and Spectral Estimation*, PhD thesis, Stanford University, Stanford, CA., 1981.

[5] J.J. Dongarra, J.R. Bunch, C.B. Moler, and G.W. Stewart, *LINPACK Users' Guide*, SIAM, Philadelphia, PA, 1979.

[6] A. J. Barabell, J. Capon, D. F. Delong, J. R. Johnson, and K. Senne, Performance comparison of superresolution array processing algorithms, Technical Report TST-72, Lincoln Laboratory, M.I.T., 1984.

[7] T. W. Anderson, Asymptotic theory for principal component analysis, *Ann. Math. Statist.*, **34**:122–148, 1963.

[8] T. W. Anderson, *An Introduction to Multivariate Statistics Analysis*, John Wiley and Sons, Inc., New York, 2nd edition, 1985.

[9] A. Paulraj, R. Roy, and T. Kailath, Patent Application: *Methods and Means for Signal Reception and Parameter Estimation*, Stanford University, Stanford, CA, 1985.

[38] The likelihood ratios obtained when the signals are assumed to be parameters to be estimated are different from those obtained when the signals are modeled as stochastic processes whose covariance is to be estimated (or is known). Since little is currently known about even the asymptotic properties of the MDL or AIC criteria under the deterministic signal assumption, the stochastic formulation is presented herein.

[10] R. Roy, B. Ottersten, L. Swindlehurst, and T. Kailath, Multiple invariance ***ESPRIT***, In *Proc.* 22nd *Asilomar Conference of Signals, Systems, and Computers*, Asilomar, CA., November 1988

[11] R. Roy, A. Paulraj, and T. Kailath, ***ESPRIT*** – A subspace rotation approach to estimation of parameters of cisoids in noise, *IEEE Trans. on ASSP*, **34**(4):1340–1342, October 1986.

[12] A. Paulraj, R. Roy, and T. Kailath, A subspace rotation approach to signal parameter estimation, *Proceedings of the IEEE*, pages 1044–1045, July 1986.

[13] M. Wax, *Detection and Estimation of Superimposed Signals*, PhD thesis, Stanford University, Stanford, CA., 1985.

[14] G. H. Golub and A. Pereyra, The differentiation of pseudo-inverses and nonlinear least squares problems whose variables separate, *Siam J. Numer. Anal.*, **10**:413–432, 1973.

[15] M. Wax and T. Kailath, Detection of signals by information theoretic criteria, *IEEE Trans. on ASSP*, **33**(2):387–392, April 1985.